Bioactive Molecules: Structures, Functions and Potential Uses for Cancer Prevention and Targeted Therapies (2nd Edition)

Bioactive Molecules: Structures, Functions and Potential Uses for Cancer Prevention and Targeted Therapies (2nd Edition)

Guest Editor

Qingping Dou

MDPI

Basel • Beijing • Wuhan • Barcelona • Belgrade • Novi Sad • Cluj • Manchester

Guest Editor
Qingping Dou
Department of Oncology
Wayne State University
Detroit
USA

Editorial Office
MDPI AG
Grosspeteranlage 5
4052 Basel, Switzerland

This is a reprint of the Special Issue, published open access by the journal *Biomolecules* (ISSN 2218-273X), freely accessible at: https://www.mdpi.com/journal/biomolecules/special_issues/C0039B708M.

For citation purposes, cite each article independently as indicated on the article page online and as indicated below:

Lastname, A.A.; Lastname, B.B. Article Title. *Journal Name* **Year**, *Volume Number*, Page Range.

ISBN 978-3-7258-5965-8 (Hbk)
ISBN 978-3-7258-5966-5 (PDF)
https://doi.org/10.3390/books978-3-7258-5966-5

Contents

About the Editor

Qingping Dou

Qingping Dou is Professor of Oncology and Pharmacology, Barbara Ann Karmanos Cancer Institute, Wayne State University School of Medicine, Detroit, MI. Dr. Dou obtained his B.S. degree in Chemistry from Shandong University in 1981, his Ph.D. degree in Chemistry from Rutgers University in 1988, and postdoctoral training at the Dana-Farber Cancer Institute and Harvard Medical School from 1988 to 1993 (Mentor: Arthur B. Pardee). Dr. Dou has extensive experience in the fields of drug discovery, chemoprevention, natural products, proteasome inhibitors, cell cycle, and apoptosis. He has published ~276 peer-reviewed research and review articles, many of which are in journals of the highest quality, and also holds numerous patents. He has extensive experience in professional service, including various study sections, advisory and editorial boards, and committees. He has mentored numerous graduate students, undergraduates, post-doctoral fellows, and visiting scholars.

biomolecules

MDPI

Article

Dissecting the tRNA Fragment tRF3E–Nucleolin Interaction: Implications in Breast Cancer

Maurizio Falconi [1,*,†], Junbiao Wang [1,†], Andrea Costamagna [2], Mara Giangrossi [1], Sunday Segun Alimi [1], Emilia Turco [2], Massimo Bramucci [3], Luana Quassinti [3], Rossana Petrilli [1], Michela Buccioni [3], Gabriella Marucci [3], Augusto Amici [1], Paola Defilippi [2], Roberta Galeazzi [4,‡] and Cristina Marchini [1,*,‡]

[1] School of Biosciences and Veterinary Medicine, University of Camerino, 62032 Camerino, Italy; junbiao.wang@unicam.it (J.W.); mara.giangrossi@unicam.it (M.G.); sundaysegun.alimi@studenti.unicam.it (S.S.A.); rossana.petrilli@unicam.it (R.P.); augusto.amici@unicam.it (A.A.)

[2] Department of Molecular Biotechnology and Health Sciences, University of Turin, 10126 Turin, Italy; a.costamagna@unito.it (A.C.); emilia.turco@unito.it (E.T.); paola.defilippi@unito.it (P.D.)

[3] School of Pharmacy, University of Camerino, 62032 Camerino, Italy; massimo.bramucci@unicam.it (M.B.); luana.quassinti@unicam.it (L.Q.); michela.buccioni@unicam.it (M.B.); gabriella.marucci@unicam.it (G.M.)

[4] Department of Life and Environmental Sciences, Marche Polytechnic University, 60131 Ancona, Italy; r.galeazzi@staff.univpm.it

* Correspondence: maurizio.falconi@unicam.it (M.F.); cristina.marchini@unicam.it (C.M.); Tel.: +39-0737-403274 (M.F.); +39-0737-403275 (C.M.)

† These authors contributed equally to this work.

‡ These authors contributed equally as last authors.

Abstract

Nucleolin (NCL), an RNA-binding protein which regulates critical cellular processes, is frequently dysregulated in human cancers, including breast cancer, making it an attractive therapeutic target. However, molecular details of the RNA-NCL interaction have not been investigated yet. A tRNA fragment named tRF3E, displaying tumor suppressor roles in breast cancer, was found to bind NCL with high affinity displacing NCL-controlled transcripts. Here, we investigated the determinants and cooperativity of tRF3E-NCL interaction by Electrophoretic Mobility Shift Assays and in silico docking analysis, using wild-type or mutated tRF3E. We found that NCL, through its RNA-binding domains (RBD1–2 and RBD3–4), binds simultaneously two tRF3E molecules, giving rise to an energetically favored complex. Instead, a mutant form of tRF3E (M19–24), in which the NCL recognition element in position 19–24 has been disrupted, contacts NCL exclusively at RBD3–4, causing the loss of cooperativity among RBDs. Importantly, when expressed in MCF7 breast cancer cells, tRF3E significantly reduced cell proliferation and colony formation, confirming its role as tumor suppressor, but tRF3E functional properties were lost when the 19–24 motif was mutated, suggesting that cooperativity among multiple domains is required for the NCL-mediated tRF3E antitumor function. This study sheds light on the dynamic of RNA-NCL interaction and lays the foundations for using tRF3E as a promising NCL-targeted biodrug candidate.

Keywords: nucleolin; tRNA fragments; RNA–protein interaction; breast cancer

1. Introduction

Nucleolin (NCL) is an evolutionarily conserved RNA-binding protein, highly expressed in the nucleolus, where it plays essential roles in ribosome biogenesis; it is also found in the nucleoplasm, cytoplasm and cell membrane [1]. NCL is a multifunctional protein able to modulate crucial molecular processes such as cell proliferation and survival [2]. The multiple functions of NCL reflect its complex structure. NCL is composed of 710 amino acid residues, and it can be divided into three main structural regions endowed with specific activities. The N- and C-terminal regions are involved in the interaction with other proteins, including components of the pre-rRNA processing complex and ribosomal proteins. The central region of NCL contains four tandem RNA-binding domains (RBS1–2 and RBS3–4) that mediate its interaction with target RNAs, such as ribosomal RNA precursor (pre rRNA), and permit the post-transcriptional regulation (processing, stability and translation) of recognized messenger RNAs [3]. The NCL recognition element (NRE) is characterized by a G-rich signature motif (U/G)CCCG(A/G) within a stem-loop structure [4,5]. NCL is also involved in pathological processes, particularly cancer and viral infections [6]. Indeed, it is highly expressed and associated with poor prognosis in several different human cancers, including gastric cancer [7], breast cancer [8], non-small-cell lung cancer [9], pancreatic cancer [10,11] and endometrial cancer [12]. The p53 and BCL-2 mRNAs emerge among the NCL-regulated transcripts having a role in cancer. The p53 mRNA translation is inhibited by NCL binding [13,14], whereas BCL-2 mRNA stability is enhanced by the interaction with NCL [15,16]. Given its implication in oncogenic processes, NCL has attracted attention as a druggable target. Various therapeutic molecules have been developed to control NCL activity, such as NCL-directed anticancer aptamers. Particularly, the aptamer AS1411, a 26-mer unmodified guanosine (G)-rich oligonucleotide, is among the synthetic compounds most extensively studied [17] and was tested as an anticancer agent in Phase II clinical trials, showing good therapeutic efficacy and tolerability [18,19].

tRNA-derived RNA fragments (tRFs) represent an important new category of small non-coding RNAs with regulatory roles. A dysregulation of tRFs has been reported in tumor cells, where they may have oncogenic or tumor-suppressor functions. tRFs can operate through different mechanisms [20,21]. They can inhibit gene expression through miRNA-like silencing or modulate translation by displacing eukaryotic initiation factors from mRNAs [22] or by competing with mRNA for ribosome binding [23]. Moreover, tRFs can exert their function by interacting with proteins, in particular RNA-binding proteins. A Cysteine tRNA fragment (5′-tRF^{Cys}) was recently found to increase during breast cancer metastatic progression. 5′-tRF^{Cys} promotes the oligomerization of NCL into a transcript-stabilizing ribonucleoprotein complex, protecting these messenger RNAs from degradation [24]. Conversely, a specific set of tRFs induced by hypoxia, with tumor-suppressive and metastasis-suppressive activity, was identified in breast cancer cells by Goodarzi et al. [25]. These tRFs, including a tRF^{Glu}, were found to specifically interact with the RNA-binding protein Y-box Binding Protein 1 (YBX1), leading to a competitive displacement of endogenous oncogenic transcripts from YBX1, thereby promoting their destabilization.

In a recent study, we identified a 32 nt tRF (named tRF3E), derived from mature tRNA^{Glu}, which is downregulated in breast cancer. tRF3E acts as a tumor-suppressor and operates through a mechanism that is dependent on the physical interaction with NCL [26]. Indeed, tRF3E, thanks to its binding properties, is potentially able to displace specific transcripts controlled by NCL. In particular, we provided evidence that tRF3E can compete with p53 mRNA for NCL binding. As a consequence, p53 mRNA is released by NCL and translated, leading to a reduction in cell proliferation [26]. According to the determinants required for the formation of stable RNA-NCL complexes, tRF3E bears two nucleotide stretches matching the identified NRE. The high binding affinity of tRF3E for NCL is demonstrated by the low dissociation constant ($K_D \cong 120$ nM) of the NCL-tRF3E complex, as calculated by Electrophoretic Mobility Shift Assays (EMSAs) [26]. Such a K_D value is even lower than that estimated for the binding sites B1 and B2 of pre-rRNA, a well-known natural target of NCL [4,5]. Moreover, NCL protein can form a complex containing two molecules of tRF3E that simultaneously occupy RBS1–2 and RBS3–4 [26]. Thus, tRF3E exerts tumor-suppressor functions by competitive displacement of NCL-regulated transcripts.

Here, we analyzed in depth the critical determinants of the tRF3E-NCL complex and the functional role of the binding cooperativity, aiming to validate tRF3E as an NCL natural modulator capable of interfering with cell oncogenic behavior. tRF3E might represent a promising NCL-targeted biodrug against breast cancer.

2. Materials and Methods

2.1. General Procedures

DNA and RNA samples were quantified with NanoDrop (Thermo Fisher Scientific, Waltham, MA, USA). RNA oligonucleotide wild-type (wt) tRF3E and its mutants (shown in Figure 1A) were [^{32}P]-labeled using the T4 polynucleotide kinase. Radioactive bands on solid supports were detected and quantified by Molecular Imager (model FX; Bio-Rad, Hercules, CA, USA). Purified human NCL was provided by MyBioSource (San Diego, CA, USA). This protein is missing the N-terminal region, and its molecular mass is 55 versus 70 kDa of wild-type protein. NCL is expressed with a $-6\times$ His tag at N-terminus.

2.2. Electroblotting

MCF-7-tRF3E and MCF-7-M19–24 cells were left untreated (control) or induced with 1 µg/mL doxycycline for different times (24 h, 48 h, 72 h and 168 h); then, RNA was extracted from each cell line, using EuroGOLD Trifast^{TM} Kit (Euroclone, Pero (MI), Italy) following the manufacturer's instructions, and quantified. The RNA samples were resuspended in loading buffer (95% formamide, EDTA 10 mM, bromophenol-blu and xylene cyanol 0.01%), denaturated at 65 °C for 5 min, loaded on 8% PAGE-urea gel in Tris-borate EDTA buffer 0.5 X and transferred to a neutral nylon Hybond-N membrane (Amersham, Little Chalfont, UK) by semidry electroblotting (Trans-Blot Turbo Transfer System; Bio-Rad, Hercules, CA, USA) at 0.3 A for 30 min. The membrane was hybridized with a 20 mer [^{32}P]-labeled tRF3E probe (5′-CACCGGGAGTCGAACCCGGGCCGCC-3′) at 56 °C, essentially as described by Sambrook and Russell [27].

2.3. Electrophoretic Mobility Shift Assay (EMSA)

EMSA was made in 15 µL of Gel Retardation buffer (20 mM Tris-HCl, pH 7.5, 50 mM KCl, 10% glycerol, 0.3 mg/mL bovine serum albumin, 0.02% Nonidet P-40) incubating 2 pmoles (unless otherwise stated) of [^{32}P]–labeled synthetic RNAs with the indicated amounts of NCL at 20 °C for 30 min. Competitive EMSAs were performed under the same experimental conditions, preincubating the [^{32}P]-labeled tRF3E with a fixed concentration of NCL (0.5 µM) for 30 min before adding increasing amounts of unlabeled RNAs (wt tRF3E and M19–24) as indicated. Then, the incubation was prolonged for an additional 30 min. Samples were loaded on native 7% polyacrylamide gel run at 4 °C. To efficiently transfer the NCL protein from the EMSA gel to the polyvinylidene difluoride (PVDF) membrane (Immobilion P, Millipore, Burlington, MA, USA) for subsequent immunodetection, the polyacrylamide gel was incubated for 2 h in 40 mL of a solution containing 300 mM Tris-HCl, pH 8.0 and 1% SDS, to make the protein negatively charged before the electrotransfer. The immunodetection of NCL was achieved with an anti-His tag antibody (His-probe (H-3): sc-8036) from Santa Cruz Biotechnology (Dallas, TX, USA) that was visualized with a peroxidase-conjugated second antibody, as described below in the Western blot paragraph.

2.4. Cell Cultures and Generation of MCF-7-tRF3E and MCF-7-M19–24 Cell Lines

SK-BR-3 (estrogen receptor-negative, progesteron receptor-negative and human epidermal growth factor receptor (HER2)-positive) human breast cancer cells were cultured in Dulbecco's Modified Essential Medium (DMEM, CORNING, Corning, NY, USA) supplemented with 10% fetal bovine serum (FBS, Gibco, Thermo Fisher Scientific, Waltham, MA, USA) and 1% penicillin–streptomycin (Gibco, Thermo Fisher Scientific, Waltham, MA, USA). MCF-7 (estrogen receptor-positive, progesteron receptor-positive and HER2-negative) human breast cancer cells were cultured in RPMI (Roswell Park Memorial Institute; CORNING, Corning, NY, USA) supplemented with 10% FBS (Gibco, Thermo Fisher Scientific, Waltham, MA, USA), 1X L-Glutamine and 1% penicillin–streptomycin (Gibco, Thermo Fisher Scientific, Waltham, MA, USA). MCF-7 cells were used as transduction target cells. In brief, MCF-7-tRF3E, MCF-7-M19–24 and MCF-7 control cells were generated by lentiviral transduction of recombinant pLKO-Tet-On vectors, encoding for wt tRF3E or the mutant form of tRF3E (M19–24) or empty vector, respectively. First, the sequences encoding for wt tRF3E or M19–24 were cloned into Tet-pLKO-puro vector (Lenti Tet-pLKO-puro Addgene Plasmid#21915) [28], using AgeI and EcoRI restriction enzymes, in order to be expressed under the control type III RNA Pol III promoter (H1) regulated by doxycycline in stably transduced cells. For this purpose, the forward and reverse oligonucleotides indicated below were annealed. The row of T at the end of the forward oligonucleotides are recognized as stop signals by the RNA polymerase III.

TRF3E Forward 5′-CCGGAGGCGGCCCGGGTTCGACTCCCGGTGTGGGAATTTTT-3′
TRF3E Reverse 5′-AATTAAAAATTCCCACACCGGGAGTCGAACCCGGGCCGCCT-3′
M19–24 Forward 5′-CCGGAGGCGGCCCGGGTTCGACTAAATGTGTGGGAATTTTT-3′
M19–24 Reverse 5′-AATTAAAAATTCCCACACATTTAGTCGAACCCGGGCCGCCT-3′

Lentiviruses were produced according to the manufacturer's instructions and used to transduce MCF-7 cells. To select transduced cells, puromycin (Sigma Aldrich/Merck, Darmstadt, Germany) (0.5 μg/mL) was added 24 h after infection with recombinant lentiviral vectors as already described [29]. MCF-7-tRF3E, MCF-7-M19–24 and MCF-7 control cells were maintained in RPMI supplemented with 10% FBS, 1X L-Glutamine, 1% penicillin–streptomycin and 0.5 μg/mL puromycin, in a humidified atmosphere with 5% CO_2 at 37 °C. Doxycycline hyclate (Sigma Aldrich/Merck, Darmstadt, Germany) (1 μg/mL) was added to complete medium to induce tRF3E or M19–24 expression. Induction medium containing doxycycline was changed every 2 days. MCF-7-tRF3E, MCF-7-M19–24 and MCF-7 (empty) control cells were generated and kindly provided by the laboratory of Prof. P. Defilippi (Department of Molecular Biotechnology and Health Sciences, University of Turin, Turin, Italy). Cells were tested for mycoplasma contamination with negative results.

2.5. Cell Viability Assay

Cell viability was evaluated by seeding 1×10^4 cells/well in 96-well plates in complete medium. The day after, fresh medium containing 2% FBS and 1 μg/mL doxycycline hyclate (Sigma Aldrich/Merck, Darmstadt, Germany) was added. Cell viability was determined after 72 h using an MTT [3-(4,5-dimethylthiazol-2-yl)-2,5-diphenyl-2H-tetrazolium bromide Sigma Aldrich, St. Louis, MO, USA] assay, which is based on the conversion of MTT to formazan by mitochondrial enzymes. The formazan deposits were dissolved in dimethyl sulfoxide (DMSO) and the absorbance of each well was measured at 540 nm in a FLUOstar Omega (BMG Labtech, Ortenberg, Germany) Plate Reader (version: 1.30.101.8). Each experimental condition was evaluated with 24 replicates and the experiments were repeated three times. Obtained data were analyzed using GraphPad Prism 10 software.

2.6. Cell Cycle Analysis

MCF-7-tRF3E, MCF-7-M19–24 and MCF-7 control cells were seeded onto 6-well tissue culture plates, 5×10^5 cells per well, in complete medium. The day after, cells were starved for 24 h and then treated with 1 µg/mL doxycycline hyclate for an additional 48 h. After 48 h incubation, the cells were harvested and fixed with ice-cold 70% ethanol for 1 h at 4 °C. RNA was digested by 1 mg/mL bovine RNase (Sigma Aldrich/Merck, Darmstadt, Germany) 30 min at 37 °C with shaking. Cells were then labeled with 15 mg/mL propidium iodide (PI) 30 min at 37 °C in the dark. Samples were analyzed by fluorescence-activated cell sorting (FACS) (BD FACScalibur™, BD Biosciences, Becton Dickinson and company, Franklin Lakes, NJ, USA) and data were elaborated via FlowJo software (v 8.7).

2.7. Growth Curve Analysis

MCF-7-tRF3E, MCF-7-M19–24 and MCF-7 control cells were plated in 6-well plates in duplicate (4×10^4 cells/well) in a complete medium and treated or not with 1 µg/mL doxycycline hyclate. The cell number was counted on day 2, day 5 and day 7, when cells were detached with Trypsin–EDTA and counted using Trypan blue to exclude dead cells.

2.8. Colony Assay

MCF-7-tRF3E, MCF-7-M19–24 and MCF-7 control cells were seeded at the density 600 cells/well in a 6-well plate and incubated in complete medium, with or without 1 µg/mL doxycycline hyclate. After two weeks, plates were stained with crystal violet and colonies ($\geq$50 cells) were counted, as described elsewhere [30]. The surviving fraction was calculated as the ratio of the number of colonies in the treated sample to the number of colonies in the untreated sample.

2.9. Western Blot

Cells were homogenized in RIPA buffer (0.1% SDS, 1% NP40, 0.5% CHAPS) supplemented with protease inhibitors (Sigma Aldrich/Merck, Darmstadt, Germany). For Western blotting analysis, equal amounts of protein lysates were separated onto Criterion™ TGX™ precast gels (Bio-Rad, Hercules, CA, USA) and transferred to polyvinylidene difluoride (PVDF) membranes (Immobilion P, Millipore, Burlington, MA, USA) using Criterion™ Blotter (Bio-Rad, Hercules, CA, USA). Membranes were blocked with EveryBlot Blocking Buffer (Bio-Rad, Hercules, CA, USA) and then incubated overnight with primary antibodies at 4 °C. Primary antibodies to NCL (C23 (MS-3); sc-8031, lot#E2023) and β-actin (C4; sc-47778, lot#K1617) were from Santa Cruz Biotechnology (Dallas, TX, USA); the primary antibody to p27 was from Cell Signaling Technology (Danvers, MA, USA) (Cat#3686S). Secondary antibodies conjugated with peroxidase were from Sigma-Aldrich (Darmstadt, Germany). Secondary antibody binding was performed at room temperature for 1 h. After TBS-T washing, membranes were incubated with PierceTM ECL Western Blotting Substrate (Thermo Fisher scientific, Waltham, MA, USA) and the immunoreactive proteins were detected with the ChemiDoc™ XRS-System (Bio-Rad, Hercules, CA, USA). Densitometry analysis was performed using ImageJ software (2.1.0/1.53c).

2.10. Immunofluorescence Analysis

MCF-7-tRF3E, MCF-7-M19–24 and MCF-7 control cells were seeded in a 24-well plate (2×10^5 cells/well). After a 48 h incubation with 1 µg/mL doxycycline hyclate, cells were fixed for 10 min with 4% paraformaldehyde (Sigma Aldrich/Merck, Darmstadt, Germany) in phosphate-buffered saline (PBS). After incubation in blocking buffer (PBS, 10% bovine serum albumin (BSA; Sigma Aldrich/Merck, Darmstadt, Germany)) for 20 min, cells were incubated for 1 h at 37 °C with the primary antibody against NCL (mouse monoclonal antibody C23 (MS-3); sc-8031, lot#E2023; Santa Cruz Biotechnology (Dallas, TX, USA); diluted 1:100). After washing, cells were incubated with anti-mouse IgG Alex Fluor 488^{TM} secondary antibody (Invitrogen Molecular Probes, Eugene, OR, USA) diluted at 1:200 for 1 h at 37 °C. Cells were then examined under a fluorescence microscope (Carl Zeiss GmbH, Oberkochen, Germany).

2.11. Computational Modeling

The complete structure of NCL was obtained using the model reconstructed according to the protocol already published [26]. Starting from their nucleotide sequences, the 3D RNA structures were modelled using SimRNA v.2.0 (https://genesilico.pl/SimRNAweb) (https://doi.org/10.1093/nar/gkae356 accessed on 28 March 2022). These computational tools rely on the Monte Carlo method for sampling the conformational space and employ a statistical potential to approximate the energy, thereby identifying conformations that correspond to biologically relevant structures. Moreover, the 3D RNA server [31] was used as a second control, confirming the obtained results. The 3D structures of the RNA-NCL complexes were predicted using HADDOCK 2.4 (https://wenmr.science.uu.nl/ accessed on 18 April 2022) (High Ambiguity Driven DOCKing) [32], which has demonstrated reliability as well as flexibility [32–35]. HADDOCK distinguishes itself from ab initio docking methods by incorporating information from known or predicted protein interfaces into ambiguous interaction restraints (AIRs) to guide the docking process. RNA/NCL complexes were put in a simulation box of 15 × 15 × 15 nm, adding TIP3P water molecules and NaCl, to reach the physiological conditions, taking into account the net charge of NCL protein and RNA molecules. CHARMM-GUI was used for setting the simulation conditions [36]. The systems were minimized with 10,000 cycles steepest descent (SD) followed by 5000 steps conjugate gradient (CG), obtaining a convergence of maximum force to the energy threshold of 1000 kJ/mol nm^2. Then, 6 equilibration steps let NCL gradually accommodate within the aqueous environment; the Verlet cutoff scheme for neighbor searching, combined with PME for electrostatics, was used. The cutoff for the Van der Waals forces calculation was set to 1.2 nm with force smoothly switched to zero (between 1.0 and 1.2 nm) generating the velocities at 310 K in the NVT ensemble using a Maxwell distribution function with random seed (Berendsen thermostat) (2 simulation runs, 25 ps). Then, we shifted to the NPT ensemble, maintaining the weak coupling also for pressure control (Berendsen barostat, isotropic conditions, 1 bar, time coupling 5 ps), maintained for 4 equilibration runs (50 ps). In the 100 ns production phase, we shifted to Nosé–Hoover and Parrinello–Rahman algorithms for temperature control and pressure coupling, respectively; a leapfrog algorithm and a time step of 0.002 ps were used. On the obtained trajectories, we calculated the MM-PBSA energies of all NCL-RNA systems using the gmx_MMPBSA gromacs tool [37]. The Charmm36m force field was used within the GROMACS 2024.4 software package (https://doi.org/10.5281/zenodo.10589697 accessed on 19 May 2023) [38].

2.12. Statistical Analysis

The significance of differences was evaluated with an unpaired Student *t* test when two groups were compared, while a 2-way ANOVA test followed by Šídák's multiple comparisons test or Tukey's multiple comparisons test was used to compare three or more groups. Statistical analysis was carried out with GraphPad Prism 10. Differences were considered significant at $p < 0.05$.

3. Results and Discussion

3.1. Identification of the Molecular Determinants of tRF3E-NCL Interaction

The ability of RNA molecules to bind NCL relies on the presence of specific NCL recognition elements (NREs). NREs, displaying a high binding affinity, have been identified by a selection-amplification protocol (SELEX) and are constituted by an 18-nucleotide (18-nt) exposing the single-stranded 6-nucleotide (6-nt) motif UCCCGA [4]. The tRF3E sequence contains two 6-nt stretches, located at positions 6–11 and 19–24, which exactly match NRE and might be relevant determinants that mediate the RNA-NCL interaction. Thus, to validate this hypothesis, first we confirmed the interaction of NREs with NCL using molecular docking, then we designed two RNA oligonucleotides carrying four-base substitutions on these signature motifs, at positions 7–10 or 20–23 on the tRF3E sequence, to further evaluate the contribution of each consensus motif to NCL binding. These mutated forms of tRF3E were named M6–11 and M19–24, respectively. In detail, we replaced the core CCCG motif with AAAU to disrupt the potential NCL recognition. In addition, we constructed a double mutant, called D2M, bearing the same mutations on both NCL target consensus sequences (6–11 and 19–24) (Figure 1A). Electrophoretic Mobility Shift Assays (EMSAs) were performed to estimate the binding affinity of NCL for the wild-type (wt) tRF3E and its mutants (M6–11, M19–24 and D2M). Unexpectedly, the 19–24 mutation seemed not to affect the interaction with NCL and both tRF3E and M19–24 exhibited a similar dissociation constant (K_D) of ≅280 nM (Figure 1B).

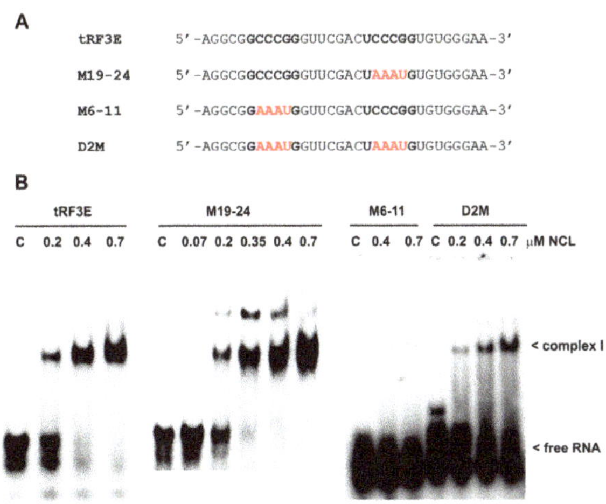

Figure 1. Comparative interaction of NCL with wild-type tRF3E and its mutants by EMSA. (**A**). The sequences of wt tRF3E and its mutants in the (U/G)CCCG(A/G) consensus motif recognized by NCL are reported. Nucleotides on tRF3E matching the NCL binding site are in black (bold) whereas mutated nucleotides are in red. (**B**). EMSA was carried out as described in Section 2, incubating 2 pmol of [^{32}P]-labeled RNA oligonucleotides with the indicated concentrations of purified NCL. "C" is the control sample in absence of protein and the electrophoretic migration of free RNA and of RNA-NCL complex I are indicated. K_D values, reported in Section 3, were estimated by Molecular Imager quantification of radioactivity associated with bound tRF3E and expressed as the minimal concentration of protein required to retard 50% of the total RNA probe.

This K_D value is higher than that previously determined by [26] because of the different experimental conditions (incubation temperature and RNA concentration) used in the EMSAs of the present study. However, a binding affinity of ≅280 nM implies a very stable RNA-NCL interaction by virtue of the fact that the K_D for natural RNA targets varies from 500 to 1000 nM. Notably, lower K_D values (≤50 nM) have been obtained only with in vitro selected sequences [4]. Unlike tRF3E and M19–24, NCL lost completely its ability to retard the M6–11 mutant, whereas it was still able to interact with the double mutant D2M, although with a very low affinity ($K_D \geq 1500$ nM). All together, these outcomes outline the importance of the 6–11 motif and clearly indicate that mutations in this region and/or modifications of RNA structure, associated with sequence changes, negatively impact the tRF3E-NCL interaction.

To correlate the secondary and tertiary RNA structure with the NCL binding properties, the RNA folding of tRF3E and its mutants was predicted in silico by RNAfold (Figure 2) and SimRNA v.2.0 (Supplementary Figure S1). These computational analyses revealed that, diversely from previous studies [4,5], which found the NCL conserved recognition sequence (U/G)CCCG(A/G) within a hairpin loop, the two tRF3E motifs 6–11 and 19–24 are instead located in base-paired regions (stems), proposed to be an adverse condition for NCL binding. Surprisingly, tRF3E and M19–24, in addition to exhibiting a similar three-dimensional (3D) conformation (Supplementary Figure S1), expose the conserved 3-nt CGA (positions 15–17) within a loop that may function as minimal NRE (Figure 2A,B). The relevant role of the 3-nt CGA is supported by NCL gel shift assays carried out with NRE variants carrying mutations in the conserved stem–loop motif. In fact, Johansson et al. [5] showed that in vitro selected NREs containing the G in the first position combined to the G → A substitution in the fifth position (resulting mutated motif: GCCCAA) and almost completely abolished the RNA-NCL interaction ($K_D \geq 15$ µM). Thus, structural homology and proper arrangement of the 3-nt CGA in tRF3E and M19–24 may explain the comparable binding affinity of NCL to these RNAs. In this respect, the D2M and M6–11 mutants show a distinct arrangement of the CGA motif; in D2M, it is located within an RNA bubble, whereas in M6–11, it lies at the junction between a double-stranded and an unstructured region (Figure 2C,D). Consistently, the residual binding specificity ($K_D \geq 1500$ nM) of D2M, which indicates a very weak interaction with NCL, and the complete loss of the ability of M6–11 to bind NCL, could reflect profound modifications in the secondary and tertiary structures of these mutants compared to tRF3E. Thus, an unfavorable conformation and the lack of exposure of the 3-nt sequence CGA within a loop may be responsible for abolishing the recognition of M6–11 and D2M by NCL. Docking predictions, carried out with the M6–11 and D2M mutants, confirmed that they were not able to fully interact with RBD1–2 or RBD3–4; both mutants marginally contacted only the RBD2 domain, leaving free RBD3 and RBD4 (Supplementary Figure S2). For these reasons, M6–11 and D2M were not further investigated in this study.

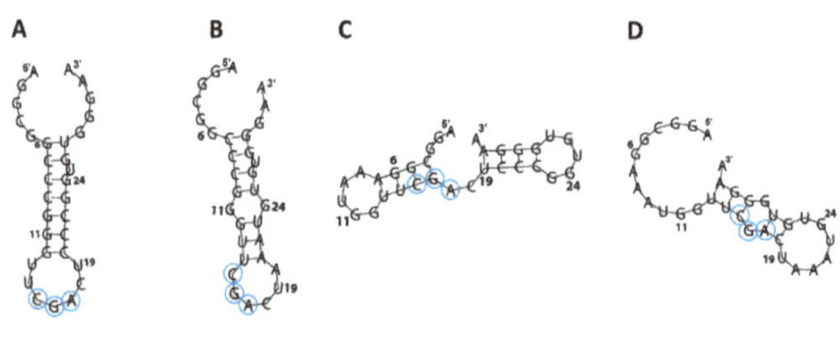

Figure 2. RNA folding of wt and mutated tRF3E. The secondary structures of tRF3E (**A**), M19–24 (**B**), M6–11 (**C**) and D2M (**D**), as predicted by RNAfold (http://rna.tbi.univie.ac.at/cgi-bin/RNAWebSuite/RNAfold.cgi accessed on 7 June 2023), are shown. Nucleotides corresponding to the minimal nucleolin recognition element (CGA), localized in a stem–loop context, are circled.

3.2. NCL Cooperatively Binds tRF3E

Recently, it has been reported that cooperativity among multiple domains of RNA-binding proteins can markedly enhance their intrinsic affinity for RNA targets with respect to that determined for individual domains [39]. We have previously shown that the four RNA-binding domains (RBD1–2 and RBD3–4) can function in pairs, providing evidence that they can be simultaneously occupied by two tRF3E molecules [26]. However, operating with an excess of NCL, the experimental conditions used in the EMSA of Figure 1B, only one of the two RNA binding sites of the protein is bound by tRF3E or M19–24. Indeed, at protein concentrations ranging from 0.4 to 0.7 μM (6–10 pmol), NCL was able to completely capture 2 pmol of tRF3E or M19–24, causing the progressive disappearance of bands corresponding to the free radiolabeled (hot) RNA. Under these experimental conditions and according to previous computational analysis [26], we assumed that only the higher affinity NCL RBD1–2 participated in tRF3E binding giving rise to the RNA-NCL complex I. A retarded band having a similar electrophoretic mobility of complex I was formed also with M19–24 (Figure 1B). Importantly, NCL-docked models predicted for the M19–24 showed that this mutant, unlike the wt tRF3E, was able to sit exclusively in the cleft between RBD3 and RBD4, leaving the RBD1–2 totally free (Figure 3). These findings suggest that the NCL sites 1–2 and 3–4 were not functionally equivalent and the formation of the two distinct complexes, tRF3E-RBD1–2 and M19–24-RBD3–4, is strongly influenced by the presence of native versus mutated NCL target motifs.

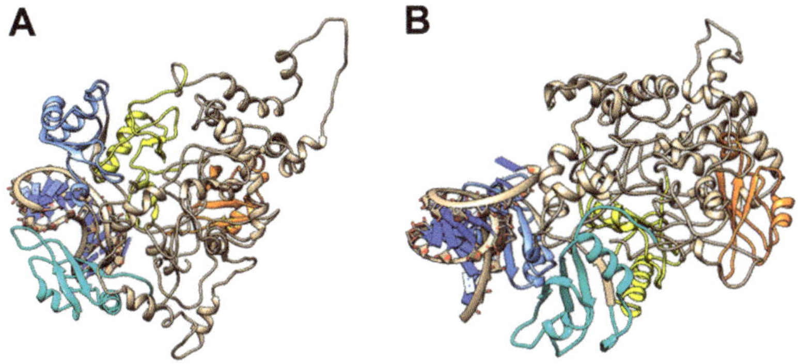

Figure 3. Molecular modelling of NCL with M19–24. The lowest energy models for M19–24 NCL complexes (**A**,**B**) as predicted by HADDOCK 2.4 (see Section 2) are shown. The RBD1 is in yellow, the RBD2 is in orange, the RBD3 is in cyan, the RBD4 is in cornflower blue and RNAs are in dark blue.

To gain deeper insight into the RNA-NCL interaction, we carried out EMSAs in which the pre-formed tRF3E-NCL complex was challenged by increasing concentrations of either wt tRF3E or M19–24. As expected, with an excess of NCL (7 pmol of protein versus 2 pmol of hot tRF3E) and in the absence of any competitor RNA, only the complex I was formed (Figure 4A, lane 0). Upon the addition of unlabeled (cold) tRF3E, complex I drastically dissociated to generate complex II, which represented 60% of total radioactivity at 20 pmol of cold wt tRF3E as competitor (Figure 4A,C). In detail, according to the proposed model, increased amounts of cold tRF3E, instead of displacing the hot tRF3E, already bound to RBD1–2, saturated both RBD1–2 and RBD3–4, originating the NCL complex II. Indeed, upon the addition of 20 pmol of cold tRF3E, corresponding to a 10-fold higher amount of hot tRF3E (2 pmol), only 25% of RNA was recovered in its free form. Differently, when the tRF3E-NCL complex I was challenged with an excess (10 and 20 pmol) of cold M19–24 as competitor, it quickly disappeared, giving rise to a low quantity of complex II (24–28%) and consequently most of the hot probe (58–68%) was released as free RNA (Figure 4A,C). These results suggest that the mixed complex II, in which tRF3E and M19–24 are respectively bound to RBD1–2 and RBD3–4, is transient and extremely less stable than the native complex II, containing two wt tRF3E molecules. Accordingly, molecular dynamics study (Section 3.3) revealed that the interaction of M19–24 with RBD3–4 leads to a consistent change in the 3D structure of the overall RNA-NCL complex, which in turn negatively impacts the binding property also of RBD1–2 for its target RNA. As a result, wt tRF3E is kicked out from RBD1–2 and the spurious complex II dissociates. In other words, the M19–24-NCL interaction causes the loss of cooperativity between the two RNA-binding domains (RBD1–2 and RBD3–4).

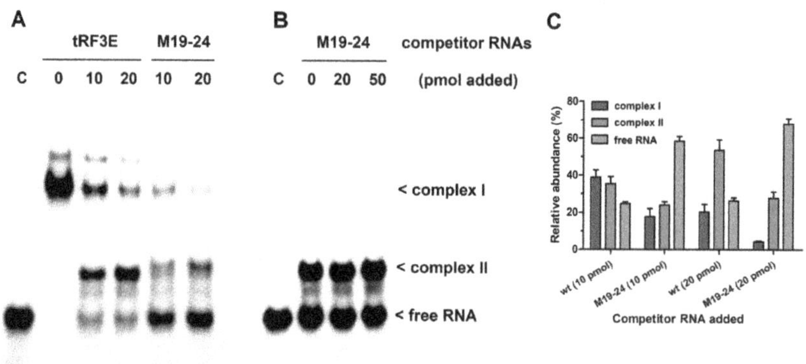

Figure 4. Competitive EMSA of wt tRF3E and M19–24 with NCL. Competitive EMSAs were carried out essentially as described in Figure 1. After a preincubation of 2 pmol (**A**) and 20 pmol (**B**) of [^{32}P]-labeled wt tRF3E with 7 pmol (0.5 μM) of NCL, the indicated amounts of not-labeled (cold) tRF3E and M19–24 were added and incubation prolonged. The sample without competition is marked with "0" whereas "C" represents the control in absence of protein. The electrophoretic migration of free RNA and RNA-NCL complexes I and II are indicated. Radioactive signals associated to complex I (blue bars), complex II (red bars) and free RNA (green bars) were quantified by imager and expressed as a percentage of total radioactivity per each lane (**C**). Values are means ± SEM obtained from panel (**A**) and EMSA reported in Supplementary Figure S3.

Undeniably, RNA molecules within cells are in excess and NCL utilizes at one time both RBD1–2 and RBD3–4 to interact with NREs on its natural targets, including messenger RNAs. To more closely mimic this condition, we carried out an EMSA using an overload of wt tRF3E (20 pmol of RNA versus 7 pmol of protein) that drastically triggers the formation of the only NCL complex II, rather than both complexes. As seen in Figure 4B, the complex II, bearing two wt tRF3E molecules bound respectively to NCL sites 1–2 and 3–4 (tRF3E-RBD1–2·RBD3–4-tRF3E), did not dissociate upon addition of increasing amounts of cold M19–24 (20 and 50 pmol). Indeed, the intensity of radioactive signals associated with the retardation band of complex II remained constant as well as the free RNA. Taken together, these findings indicate that the native tRF3E-NCL complex II is somehow "locked" and the mutant M19–24, unlike what was observed in complex I, is unable to displace wt tRF3E even from RBD3–4. Notably, the RNA-NCL complex II exhibited a higher electrophoretic mobility than complex I. It has been well established that migration of DNA- and RNA-protein complexes, on native gels under constant-field conditions, is primarily dependent on the negative charges of nucleic acids. Consequently, complex II in which NCL is bound to two tRF3E molecules moves faster than complex I, containing only one RNA molecule. To definitely prove that both complex I and II arose from the tRF3E-NCL interaction, the competitive EMSA was subjected to Western blotting and NCL was immunologically detected. As clearly shown, there is an almost perfect correspondence between the retarded bands obtained in the EMSA (Figure 5A) and the protein localization revealed by anti-NCL antibody (Figure 5B).

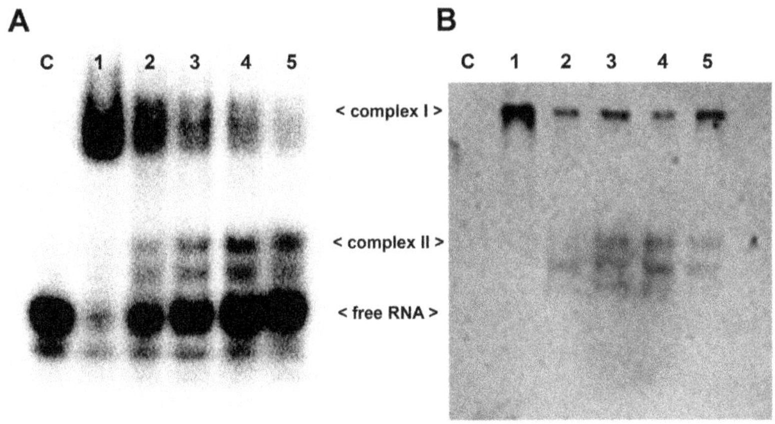

Figure 5. Identification of tRF3E-NCL complexes by EMSA-Western blotting analysis. (**A**). EMSA was carried out as described in Figure 4A with 2 pmol of [^{32}P]-labeled wt tRF3E and 7 pmol (0.5 μM) of NCL. "C" represents the control in absence of protein whereas samples from 1 to 5 contain 0, 5, 10, 20 and 30 pmol of not-labeled wt tRF3E as competitor, respectively. (**B**). After running, the polyacrylamide gel (panel A) was incubated in a 1% SDS denaturing solution for 2 h and blotted on a PVDF membrane as described in Section 2. The NCL protein in complex I and complex II was visualized thanks to the peroxidase-conjugated secondary antibody directed against the anti-His tag primary antibody, recognizing NCL.

3.3. *Molecular Dynamics of tRF3E and M19–24 NCL Complexes*

Findings obtained with tRF3E allowed us to have an insight into the key determinants governing the dynamic of RNA-NCL interaction and possibly open the way for a deeper comprehension of how NCL recognizes its target RNAs and the regulatory implications of this process. The formulated model predicts that the NCL RBD1–2 and RBD3–4 do not function in the same manner. RBDs show a high degree of cooperativity only with wt tRF3E and respond differently to native or mutated NREs present on the tested RNAs. A previous in silico study revealed that wt tRF3E strongly interacts with NCL RBD1–2, whereas its binding to RBD3–4 is less favorite. Thus, when two tRF3E molecules simultaneously filled up both binding sites, the tRF3E-NCL-tRF3E complex II underwent a drastic stabilization, as validated by a 2.6-fold reduction in the complex binding energy compared to a system containing only one tRF3E molecule bound to RBD1–2 [26].

Here, in order to better rationalize the experimental results, we compared by molecular dynamics simulations the stability of a native complex II (tRF3E-RBD1–2·RBD3–4-tRF3E) with that of a mixed one, containing tRF3E in RBD1–2 and M19–24 in RBD3–4 (tRF3E-RBD1–2·RBD3–4-M19–24) in their predicted bound state. As evident in Figure 6A,B, the overall 3D structures of the two RNA-NCL complexes differ significantly, involving modifications of the spatial organization of all four NCL RBDs. These changes affected the stability of the complex, which was greater for tRF3E-RBD1–

2·RBD3–4-tRF3E ($\Delta G_{complex} \cong -22{,}000$ kcal/mol) than for tRF3E-RBD1–2·RBD3–4-M19–24 ($\Delta G_{complex} \cong -20{,}000$ kcal/mol) (Table 1). In addition, the spurious complex exhibited a remarkably positive value of the binding free energy ($+361 \leq \Delta G_{binding\ (kcal/mol)} \leq +371$) (Figure 6D and Table 1). It is noteworthy that $\Delta G_{binding}$ expresses the difference in free energy between the bound and completely unbound states and positive values indicate an unfavorable interaction. Thus, the tRF3E-RBD1–2·RBD3–4-M19–24 complex is destabilized and prone to suddenly dissociate, releasing both tRF3E and M19–24 from the respective binding sites as evidenced by the appearance of a very intense band (≅70% of total radioactivity) corresponding to free RNA in the EMSA (Figure 4A,C). Conversely, $\Delta G_{binding}$ for the tRF3E-RBD1–2·RBD3–4-tRF3E complex was negative (from −32.88 to −41.21 kcal/mol), suggesting a strong RNA-NCL interaction (Figure 6C and Table 1).

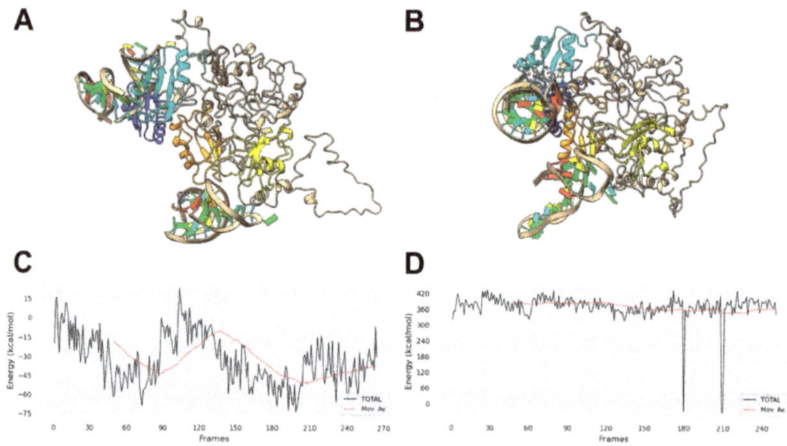

Figure 6. Molecular modelling of NCL with tRF3E and M19–24. The lowest energy models for the tRF3E-RBD1–2·RBD3–4-tRF3E complex (**A**) and for the mixed tRF3E-RBD1–2·RBD3–4-M19–24 complex (**B**) as predicted by HADDOCK 2.4 (see Section 2) are shown. The RBD1 is in orange, the RBD2 is in yellow, the RBD3 is in cyan, the RBD4 is in cornflower blue and RNA nucleotides are represented in Tube/Slab representation using red for A, cyan for U, green for G and yellow for C. The binding free energy was calculated using the gmx_MMPBSA gromacs tool (see Section 2) for the tRF3E-RBD1–2-NCL-RBD3–4-tRF3E complex (**C**) and for the mixed complex tRF3E-RBD1–2·RBD3–4-M19–24 (**D**). The simulation time was 100 ns.

Table 1. Computational ΔG estimation of RNA-NCL complexes.

	tRF3E-NCL-tRF3E	tRF3E-NCL-M19–24
$\Delta G_{complex}$ (complex stability, 100 ns)	−22,165 ± 135 kcal/mol	−20,373 ± 128 kcal/mol
$\Delta G_{complex}$ (complex stability, last 50 ns)	−22,224 ± 105 kcal/mol	−20,397 ± 130 kcal/mol
$\Delta G_{binding}$ (100 ns)	−32.88 ± 12 kcal/mol	371 ± 50 kcal/mol
$\Delta G_{binding}$ (last 50 ns)	−41.21 ± 14 kcal/mol	361 ± 66 kcal/mol

Molecular dynamics simulation fully supported the data from band shift assays, providing precious information on the kinetics of occupancy of the four NCL RBDs, RNA–protein 3D structures and the binding energies of resulting complexes.

In addition to its identified substrate (NRE) [4,5], NCL has been shown to bind also with a very high affinity to (G)-rich sequences. These DNA or RNA regions are capable of folding into highly stable four-stranded secondary structures known as G-quadruplexes

(G4s), which display specific regulatory roles in different cellular processes [17,40,41]. Given that tRF3E contains five sequential runs of Gs, a canonical feature of G4, its potential to adopt a G-quadruplex conformation was assessed using the G4RNA Screener tool (http://scottgroup.med.usherbrooke.ca/G4RNA_screener/ accessed on 8 July 2024). This in silico analysis revealed that the tRF3E molecule has a moderate propensity of rearranging into a G4 secondary structure as reflected by consistently low predictive scores (consecutive G over consecutive C ratio (cGcC), 1.4783; G4Hunter (G4H), 0.3438; G4 Neural Network (G4NN), 0.0644). Notably, when the score was calculated for a well-known DNA sequence (36-nt oligonucleotide from human telomeres) able to form G-quadruplex [42], the values were extremely higher than those obtained for tRF3E (cGcC = 1.50, G4H = 600 and G4NN = 0.9984). Therefore, we cannot exclude that the interaction between NCL and tRF3E might rely on alternative higher-order RNA configurations, but it does not seem to be mediated by the canonical G4 formation.

3.4. The Occupancy of Both NCL RBD1–2 and RBD3–4 Is Required for tRF3E Antitumor Function in Breast Cancer Cells

To clarify the functional role of tRF3E in breast cancer, SK-BR-3 and MCF-7 breast cancer cell lines were transiently transfected with tRF3E and its ability to reduce cancer cell viability was assessed by MTT assay (Supplementary Figure S4). To corroborate tRF3E's tumor suppressive function, MCF-7 cells were stably transduced with a pLKO-Tet-On vector properly engineered in order to allow the expression of tRF3E upon doxycycline induction; they were called MCF-7-tRF3E cells. In parallel, MCF-7 cells were stably transduced with the empty vector and used as a control (MCF-7 control cells). Moreover, a third MCF-7 transduced cell line was generated using a pLKO-Tet-On vector encoding the M19–24 mutant (MCF-7-M19–24 cells) that, according to in silico analysis and EMSA results, binds NCL on RBD3–4 but not on RBD1–2. Because of that, MCF-7-M19–24 cells permitted evaluation of the functional consequences of the loss of a cooperative interaction between tRF3E and NCL. MCF-7-tRF3E, MCF-7-M19–24 and MCF-7 control cells expressed similar levels of NCL, localized mainly in nuclei, as expected (Figure 7A,B). A Northern blotting analysis was performed to verify the expression of wt tRF3E and its mutated form M19–24 in MCF-7 cells upon doxycycline induction, and thus to validate this new in vitro model. As shown in Figure 7C, the expression of wt tRF3E and M19–24 starts to be clearly detectable after 48 h of doxycycline induction.

Then, the three MCF-7 transduced cell lines were characterized evaluating their proliferation rate, viability and clonogenicity upon doxycycline treatment. In particular, to compare the proliferation rates of these cell lines, a growth curve was performed by plating equal numbers of cells, at low density, in a 6-well plate in the presence of doxycycline. Cells were then collected and counted after 2, 5 and 7 days. The obtained growth curves indicate that cells expressing tRF3E exhibit a statistically significant lower proliferation rate compared to the other two cell lines (Figure 8A). These results are consistent with previous findings demonstrating the inhibitory effect of tRF3E on cell proliferation [26]. Moreover, the induction of tRF3E expression by a 72 h treatment with doxycycline caused a 20% reduction in MCF-7-tRF3E cell viability, as assessed by an MTT assay. In contrast, the viability of both control and doxycycline-induced MCF-7-M19–24 cells remained unaffected, suggesting that M19–24 lacks the functional properties of wt tRF3E (Figure 8B). Next, we examined the impact of tRF3E and M19–24 expression on cell cycle distribution by flow cytometry analysis. As expected, doxycycline treatment did not alter the percentage of MCF-7 control cells in G0/G1 phase, as well as in S or in M/G2 phases. On the contrary, induction of tRF3E resulted in a statistically significant increase in the percentage of MCF-7-tRF3E cells in the G0/G1 phase (79.2 ± 6%) with respect to uninduced cells (67.6 ± 2.5%), while the proportion of cells in S phase and M/G2 phase remained largely unchanged. The expression of the mutant M19–24 did not significantly shift the percentage of MCF-7-M19–24 cells in G0/G1 phase; however, it seemed to reduce the cell number in S phase and slightly increase the proportion of cells in G2/M phase (Figure 8C).

Finally, to further evaluate the functional consequence of tRF3E expression, a colony formation assay was performed by plating cells at low density (600 cells per well in a 6-well plate). This test permits us to assess the ability of residual tumor cells to form recurrences. As shown in Figure 8D,E, a statistically significant reduction in colony number was found in induced MCF-7-tRF3E cells as compared with the control ones. In contrast, no reduction in clonogenicity was observed in MCF-7 cells expressing the M19–24 mutant. Thus, the expression of tRF3E impaired MCF-7 cell survival and their ability to form colonies. In summary, tRF3E is capable of controlling the malignant behavior of breast cancer cells. Notably, tRF3E tumor-suppressor properties are lost upon disruption of the NCL binding motif in position 19–24, suggesting that a cooperative binding involving both RBD1–2 and RBD3–4 domains is required for NCL-mediated tRF3E antitumor functions. Interestingly, we found that the functional ability of tRF3E to control cell proliferation is associated with a significant increase in the level of p27 in MCF-7-tRF3E cells induced for 48 h with doxycycline (Figure 9). p27, encoded by the CDKN1B gene, is a cyclin-dependent kinase (CDK) inhibitor of the kinase-inhibitory protein (Kip) family, which mediates cell-cycle inhibition. In normal cells, p27 levels are tightly regulated across the cell cycle and an increase in p27 can efficiently inhibit G1–S-phase cyclin-CDKs [43,44].

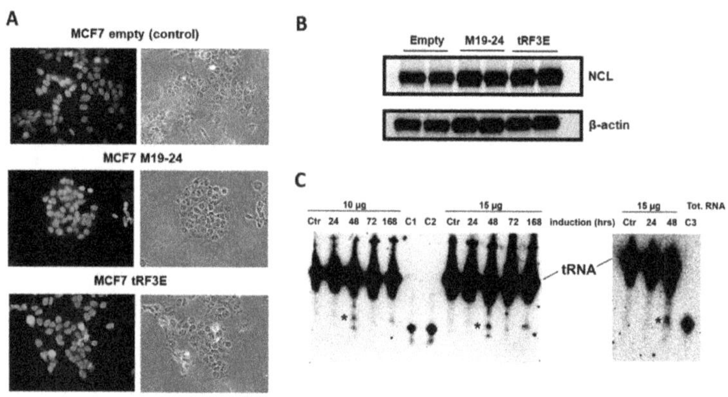

Figure 7. NCL expression and tRF3E or M19–24 induction in MCF-7 cell lines. (**A**). Immunolocalization of NCL in MCF-7-tRF3E, MCF-7-M19–24 and MCF-7 control cells. Cells were processed for immunofluorescence with anti-NCL antibody and then observed with a fluorescence microscope (Carl Zeiss GmbH, Germany). (**B**). Representative Western blot showing the expression of NCL and β-actin (loading control) in MCF-7-tRF3E, MCF-7-M19–24 and MCF-7 control cells. Samples in duplicate containing 20 µg of proteins/well were loaded. (**C**). Northern blotting analysis of wt tRF3E (left gel) and M19–24 (right gel) was performed on total RNA (10 µg and 15 µg), extracted from MCF-7-tRF3E cells left untreated (Ctr) or treated with 1 µg/mL doxycycline for the indicated times to induce tRF3E and M19–24 (marked with asterisks). Total RNA was run on a denaturing 8% PAGE-urea 7M gel and transferred by electroblotting to nylon membranes that were hybridized with a [^{32}P]-labeled tRF3E probe. As standards, known quantities of synthetic RNAs were loaded as follows: C1 and C2 contained 0.3 ng and 0.6 ng of tRF3E whereas C3 contained 0.6 ng of M19–24.

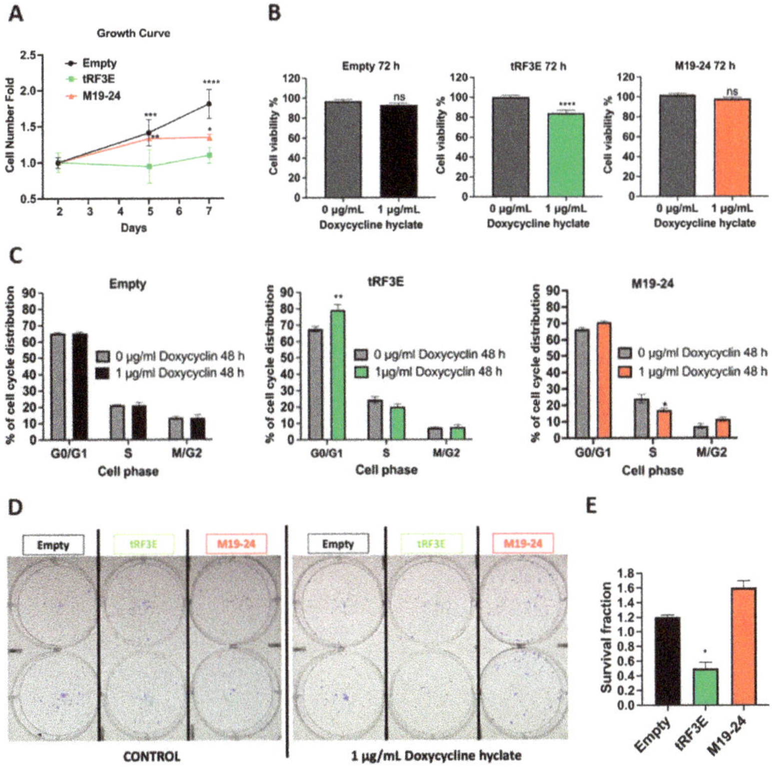

Figure 8. Functional characterization of tRF3E-expressing MCF-7 cells. (**A**). Growth curve. MCF-7 control cells (empty) and MCF-7-tRF3E cells and MCF-7-M19–24 cells were seeded in 6-well plates (40,000 cells/well) and treated with 1 µg/mL doxycycline. The cell number was counted on day 2, day 5 and day 7. The graph shows cell number fold change with respect to day 2 ($n = 4$). Values are means ± SD. Two-way ANOVA followed by Tukey's multiple comparisons test. * $p < 0.05$, ** $p < 0.01$, *** $p = 0.0001$, **** $p < 0.0001$ (MCF-7-tRF3E versus (vs.) MCF-7 control cells or vs. MCF-7-M19–24 cells). (**B**). Effect of tRF3E expression on MCF-7 cell viability determined by MTT assay. MCF-7 control cells (empty) or MCF-7-tRF3E cells or MCF-7-M19–24 cells were left untreated (grey) or induced with 1 µg/mL doxycycline for 72 h, to express tRF3E (green), or M19–24 (red) or any tRF (black). The results are expressed as percentage of viable cells with respect to control. Values are means ± SEM ($n = 20$). Unpaired t test. **** $p < 0.0001$. (**C**). Cell cycle analysis by FACS. Histograms show the percentage of MCF-7 control cells (empty) or MCF-7-tRF3E cells or MCF-7-M19–24 cells in G0/G1, S and G2/M phases, when they are not induced (grey) or induced with 1 µg/mL doxycycline for 48 h. Data are presented as the mean ± SEM of three repeats. Two-way ANOVA test followed by Šídák's multiple comparisons test. * $p < 0.05$; ** $p < 0.005$. (**D**). Colony assay of MCF-7 control cells (empty; black), MCF-7-tRF3E cells (green) and MCF-7-M19–24 cells (red) plated at low density (600 cell/well) and left untreated (control) or induced with 1 µg/mL doxycycline for 2 weeks. Colonies were stained with crystal violet and counted. Representative images from two independent experiments, performed in duplicate. (**E**). Data from panel D are expressed as survival fraction and the mean ± SEM is shown. Statistical significance was calculated using the unpaired t test. * $p < 0.05$ (tRF3E vs. empty); M19–24 vs. Empty: not statistically significant.

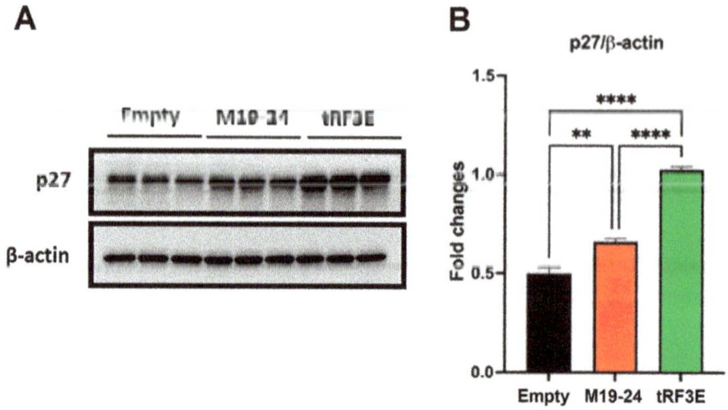

Figure 9. p27 expression increases in tRF3E-expressing MCF-7 cells. (**A**). Representative Western blot showing the expression of p27 and β-actin (loading control) in MCF-7 control cells (empty), MCF-7-M19–24 cells and MCF-7-tRF3E cells treated with 1 µg/mL doxycycline for 48 h. A total of 20 µg of proteins/well were loaded. (**B**). Densitometric quantifications of p27 expression, normalized on β-actin, are shown; data are presented as the mean ± SEM of three repeats. ** $p < 0.01$, **** $p < 0.0001$. One-way ANOVA, followed by Tukey's multiple comparisons test.

4. Conclusions

Breast cancer is the leading cause of cancer-related deaths in women worldwide. A deeper knowledge of the relevant molecular mechanisms behind the onset and progression of BC is needed in order to identify new targets and develop more effective therapies. Nucleolin (NCL) is an RNA-binding protein considered a relevant target in cancer [45]. Indeed, NCL can bind cancer-related mRNAs, controlling their stability and translation, as well as non-coding RNAs. Among NCL-interacting partners, tRNA fragments (tRFs) emerge as crucial NCL modulators, displaying functional roles in post-transcriptional gene regulation [21]. In this study, we analyzed the molecular determinants of RNA-NCL interaction, focusing on the tRF3E-NCL complex. tRF3E is a tumor-suppressor tRF derived from mature $tRNA^{Glu}$, that operates through a mechanism dependent on its physical interaction with NCL. We found that cooperativity among multiple NCL RNA-binding domains (RBD1–2 and RBD3–4) is required for NCL-mediated tRF3E antitumor functions: two tRF3E molecules can simultaneously occupy RBD1–2 and RBD3–4, leading to a drastic stabilization of the tRF3E-NCL complex. Reported results shed light on the dynamic of NCL interaction with its target RNAs and provide crucial information for the development of an RNA-based drug targeting NCL.

Supplementary Materials: The following supporting information can be downloaded at https://www.mdpi.com/article/10.3390/biom15071054/s1, Figure S1: RNA folding of wild-type and mutated tRF3E. Figure S2: 3D docked complex between NCL and mutated tRF3E M6–11 and D2M. Figure S3: Competitive EMSA of wild-type tRF3E and M19–24 with NCL. Figure S4: tRF3E affects SK-BR-3 and MCF-7 breast cancer cell viability.

Author Contributions: Conceptualization, M.F., J.W., R.G. and C.M.; formal analysis, M.F., J.W., R.G. and C.M.; investigation, J.W., M.F., M.G., R.G., A.A., R.P., S.S.A., M.B. (Michela Buccioni), G.M., M.B. (Massimo Bramucci), L.Q., A.C., P.D. and E.T.; resources, A.A. and M.F.; writing—original draft preparation, M.F. and C.M.; writing—review and editing, M.F., J.W., C.M., R.G. and P.D.; supervision, C.M. and M.F.; project administration, C.M. and M.F.; funding acquisition, M.F. and C.M. All authors have read and agreed to the published version of the manuscript.

Funding: This work was supported by the F.A.R. of the University of Camerino to M. Falconi (BVI000002) and to C. Marchini (BVI002002: "finanziato dall'Unione Europea-Next Generation UE" MUR-Fondo Promozione e Sviluppo—University Research Projects FAR 2022 PNR, under the funding of Ministerial Decree n. 737/2021 (tRNAcatching; The tRNA-fragment tRF3E: tumor suppressor role in breast cancer and diagnostic detection methods)). This work was also supported by a Fondazione Umberto Veronesi fellowship to Junbiao Wang.

Institutional Review Board Statement: Not applicable.

Informed Consent Statement: Not applicable.

Data Availability Statement: The data presented in this study are available in this article and in the Supplementary Materials.

Acknowledgments: J. Wang was supported by Fondazione Umberto Veronesi.

Conflicts of Interest: The authors declare no conflicts of interest.

Abbreviations

The following abbreviations are used in this manuscript:

NCL	Nucleolin
RBD	RNA-Binding Domain
EMSA	Electrophoretic Mobility Shift Assay
tRF	tRNA-derived Fragment

References

1. Mongelard, F.; Bouvet, P. Nucleolin: A multifaceted protein. *Trends Cell Biol.* **2007**, *17*, 80–86. [CrossRef] [PubMed]
2. Jia, W.; Yao, Z.; Zhao, J.; Guan, Q.; Gao, L. New perspectives of physiological and pathological functions of nucleolin (NCL). *Life Sci.* **2017**, *186*, 1–10. [CrossRef] [PubMed]
3. Abdelmohsen, K.; Tominaga, K.; Lee, E.K.; Srikantan, S.; Kang, M.J.; Kim, M.M.; Selimyan, R.; Martindale, J.L.; Yang, X.; Carrier, F.; et al. Enhanced translation by nucleolin via G-rich elements in coding and non-coding regions of target mRNAs. *Nucleic Acids Res.* **2011**, *39*, 8513–8530. [CrossRef] [PubMed]
4. Ghisolfi-Nieto, L.; Joseph, G.; Puvion-Dutilleul, F.; Amalric, F.; Bouvet, P. Nucleolin is a sequence-specific RNA-binding protein: Characterization of targets on pre-ribosomal RNA. *J. Mol. Biol.* **1996**, *260*, 34–53. [CrossRef] [PubMed]
5. Johansson, C.; Finger, L.D.; Trantirek, L.; Mueller, T.D.; Kim, S.; Laird-Offringa, I.A.; Feigon, J. Solution structure of the complex formed by the two N-terminal RNA-binding domains of nucleolin and a pre-rRNA target. *J. Mol. Biol.* **2004**, *337*, 799–816. [CrossRef] [PubMed]
6. Tonello, F.; Massimino, M.L.; Peggion, C. Nucleolin: A cell portal for viruses, bacteria, and toxins. *Cell. Mol. Life Sci.* **2022**, *79*, 271. [CrossRef] [PubMed]
7. Qiu, W.; Zhou, F.; Zhang, Q.; Sun, X.; Shi, X.; Liang, Y.; Wang, X.; Yue, L. Overexpression of nucleolin and different expression sites both related to the prognosis of gastric cancer. *APMIS* **2013**, *121*, 919–925. [CrossRef] [PubMed]
8. Pichiorri, F.; Palmieri, D.; De Luca, L.; Consiglio, J.; You, J.; Rocci, A.; Talabere, T.; Piovan, C.; Lagana, A.; Cascione, L.; et al. In vivo NCL targeting affects breast cancer aggressiveness through miRNA regulation. *J. Exp. Med.* **2013**, *210*, 951–968, Erratum in *J. Exp. Med.* **2017**, *214*, 1557. [CrossRef] [PubMed]
9. Xu, J.Y.; Lu, S.; Xu, X.Y.; Hu, S.L.; Li, B.; Li, W.X.; Chang, J.Y. Prognostic significance of nuclear or cytoplasmic nucleolin expression in human non-small cell lung cancer and its relationship with DNA-PKcs. *Tumour Biol.* **2016**, *37*, 10349–10356. [CrossRef] [PubMed]
10. Azman, M.S.; Alard, E.L.; Dodel, M.; Capraro, F.; Faraway, R.; Dermit, M.; Fan, W.; Chakraborty, A.; Ule, J.; Mardakheh, F.K. An ERK1/2-driven RNA-binding switch in nucleolin drives ribosome biogenesis and pancreatic tumorigenesis downstream of RAS oncogene. *EMBO J.* **2023**, *42*, e110902. [CrossRef] [PubMed]
11. Gilles, M.E.; Maione, F.; Cossutta, M.; Carpentier, G.; Caruana, L.; Di Maria, S.; Houppe, C.; Destouches, D.; Shchors, K.; Prochasson, C.; et al. Nucleolin Targeting Impairs the Progression of Pancreatic Cancer and Promotes the Normalization of Tumor Vasculature. *Cancer Res.* **2016**, *76*, 7181–7193. [CrossRef] [PubMed]

12. Lin, Q.; Ma, X.; Hu, S.; Li, R.; Wei, X.; Han, B.; Ma, Y.; Liu, P.; Pang, Y. Overexpression of nucleolin is a potential prognostic marker in endometrial carcinoma. *Cancer Manag. Res.* **2021**, *13*, 1955–1965. [CrossRef] [PubMed]
13. Takagi, M.; Absalon, M.J.; McLure, K.G.; Kastan, M.B. Regulation of p53 translation and induction after DNA damage by ribosomal protein L26 and nucleolin. *Cell* **2005**, *123*, 49–63. [CrossRef] [PubMed]
14. Chen, J.; Guo, K.; Kastan, M.B. Interactions of nucleolin and ribosomal proteinL26 (RPL26) in translational control of human p53 mRNA. *J. Biol. Chem.* **2012**, *287*, 16467–16476. [CrossRef] [PubMed]
15. Zhang, J.; Tsaprailis, G.; Bowden, G.T. Nucleolin stabilizes Bcl-X L messenger RNA in response to UVA irradiation. *Cancer Res.* **2008**, *68*, 1046–1054. [CrossRef] [PubMed]
16. Otake, Y.; Soundararajan, S.; Sengupta, T.K.; Kio, E.A.; Smith, J.C.; Pineda-Roman, M.; Stuart, R.K.; Spicer, E.K.; Fernandes, D.J. Overexpression of nucleolin in chronic lymphocytic leukemia cells induces stabilization of bcl2 mRNA. *Blood* **2007**, *109*, 3069–3075. [CrossRef] [PubMed]
17. Wu, J.; Song, C.; Jiang, C.; Shen, X.; Qiao, Q.; Hu, Y. Nucleolin targeting AS1411 modified protein nanoparticle for antitumor drugs delivery. *Mol. Pharm.* **2013**, *10*, 3555–3563. [CrossRef]
18. Doherty, C.; Wilbanks, B.; Khatua, S.; Maher, L.J. 3rd Aptamers in neuro-oncology: An emerging therapeutic modality. *Neuro Oncol.* **2024**, *26*, 38–54. [CrossRef] [PubMed]
19. Tong, X.; Ga, L.; Ai, J.; Wang, Y. Progress in cancer drug delivery based on AS1411 oriented nanomaterials. *J. Nanobiotechnol.* **2022**, *20*, 57. [CrossRef] [PubMed]
20. Gong, M.; Deng, Y.; Xiang, Y.; Ye, D. The role and mechanism of action of tRNA-derived fragments in the diagnosis and treatment of malignant tumors. *Cell Commun. Signal* **2023**, *21*, 62. [CrossRef] [PubMed]
21. Kuhle, B.; Chen, Q.; Schimmel, P. tRNA renovatio: Rebirth through fragmentation. *Mol. Cell* **2023**, *83*, 3953–3971. [CrossRef] [PubMed]
22. Ivanov, P.; Emara, M.M.; Villen, J.; Gygi, S.P.; Anderson, P. Angiogenin-induced tRNA fragments inhibit translation initiation. *Mol. Cell* **2011**, *43*, 613–623. [CrossRef] [PubMed]
23. Gebetsberger, J.; Wyss, L.; Mleczko, A.M.; Reuther, J.; Polacek, N. A tRNA-derived fragment competes with mRNA for ribosome binding and regulates translation during stress. *RNA Biol.* **2017**, *14*, 1364–1373. [CrossRef] [PubMed]
24. Liu, X.; Mei, W.; Padmanaban, V.; Alwaseem, H.; Molina, H.; Passarelli, M.C.; Tavora, B.; Tavazoie, S.F. A pro-metastatic tRNA fragment drives Nucleolin oligomerization and stabilization of its bound metabolic mRNAs. *Mol. Cell* **2022**, *82*, 2604–2617.e8. [CrossRef] [PubMed]
25. Goodarzi, H.; Liu, X.; Nguyen, H.C.; Zhang, S.; Fish, L.; Tavazoie, S.F. Endogenous tRNA-derived fragments suppress breast cancer progression via YBX1 displacement. *Cell* **2015**, *161*, 790–802. [CrossRef] [PubMed]
26. Falconi, M.; Giangrossi, M.; Zabaleta, M.E.; Wang, J.; Gambini, V.; Tilio, M.; Bencardino, D.; Occhipinti, S.; Belletti, B.; Laudadio, E.; et al. A novel 3′-tRNAGlu-derived fragment acts as a tumor suppressor in breast cancer by targeting nucleolin. *FASEB J.* **2019**, *33*, 13228–13240. [CrossRef] [PubMed]
27. Sambrook, J.; Russell, D.W. *Molecular Cloning. A Laboratory Manual*, 3rd ed.; CSHL Press, Cold Spring Harbor: Long Island, NY, USA, 2001.
28. Wiederschain, D.; Wee, S.; Chen, L.; Loo, A.; Yang, G.; Huang, A.; Chen, Y.; Caponigro, G.; Yao, Y.M.; Lengauer, C.; et al. Single-vector inducible lentiviral RNAi system for oncology target validation. *Cell Cycle* **2009**, *8*, 498–504. [CrossRef] [PubMed]
29. Bisaro, B.; Montani, M.; Konstantinidou, G.; Marchini, C.; Pietrella, L.; Iezzi, M.; Galié, M.; Orso, F.; Camporeale, A.; Colombo, S.M.; et al. p130Cas/Cyclooxygenase-2 axis in the control of mesenchymal plasticity of breast cancer cells. *Breast Cancer Res.* **2012**, *14*, R137. [CrossRef] [PubMed]
30. Segatto, I.; Zompit, M.M.; Citron, F.; D'Andrea, S.; Vinciguerra, G.L.R.; Perin, T.; Berton, S.; Mungo, G.; Schiappacassi, M.; Marchini, C.; et al. Stathmin Is Required for Normal Mouse Mammary Gland Development and Δ16HER2-Driven Tumorigenesis. *Cancer Res.* **2019**, *79*, 397–409. [CrossRef] [PubMed]
31. Wang, J.; Huang, Y.; Xiao, Y. 3dRNA v2.0: An Updated Web Server for RNA 3D Structure Prediction. *Int. J. Mol. Sci.* **2019**, *20*, 4116. [CrossRef] [PubMed]
32. Dominguez, C.; Boelens, R.; Bonvin, A.M.J.J. HADDOCK: A protein-protein docking approach based on biochemical or biophysical information. *J. Am. Chem. Soc.* **2003**, *125*, 1731–1737. [CrossRef] [PubMed]
33. Tuszynska, I.; Matelska, D.; Magnus, M.; Chojnowski, G.; Kasprzak, J.M.; Kozlowski, L.P.; Dunin-Horkawicz, S.; Bujnicki, J.M. Computational modeling of protein–RNA complex structures. *Methods* **2014**, *65*, 310–319. [CrossRef] [PubMed]
34. Liu, X.; Duan, Y.; Hong, X.; Xie, J.; Liu, S. Challenges in structural modeling of RNA-protein interactions. *Curr. Opin. Struct. Biol.* **2023**, *81*, 102623. [CrossRef] [PubMed]

35. van Zundert, G.C.P.; Rodrigues, J.P.G.L.M.; Trellet, M.; Schmitz, C.; Kastritis, P.L.; Karaca, E.; Melquiond, A.S.J.; van Dijk, M.; de Vries, S.J.; Bonvin, A.M.J.J. The HADDOCK2.2 webserver: User-friendly integrative modeling of biomolecular complexes. *J. Mol. Biol.* **2016**, *428*, 720–725. [CrossRef] [PubMed]
36. Guterres, H.; Wonpil, I. CHARMM-GUI-Based Induced Fit Docking Workflow to Generate Reliable Protein–Ligand Binding Modes. *J. Chem. Inf. Model* **2023**, *63*, 4772–4779. [CrossRef] [PubMed]
37. Valdés-Tresanco, M.S.; Valdés-Tresanco, M.E.; Valiente, P.A.; Moreno, E. gmx_MMPBSA: A New Tool to Perform End-State Free Energy Calculations with GROMACS. *J. Chem. Theory Comput.* **2021**, *17*, 6281–6291. [CrossRef] [PubMed]
38. Abraham, M.J.; Murtola, T.; Schulz, R.; Páll, S.; Smith, J.C.; Hess, B.; Lindahl, E. GROMACS: High performance molecular simulations through multi-level parallelism from laptops to supercomputers. *SoftwareX* **2015**, *1–2*, 19–25. [CrossRef]
39. Stitzinger, S.H.; Sohrabi-Jahromi, S.; Söding, J. Cooperativity boosts affinity and specificity of proteins with multiple RNA-binding domains. *NAR Genom. Bioinform.* **2023**, *5*, lqad057. [CrossRef] [PubMed]
40. Chen, L.; Dickerhoff, J.; Zheng, K.W.; Erramilli, S.; Feng, H.; Wu, G.; Onel, B.; Chen, Y.; Wang, K.B.; Carver, M.; et al. Structural basis for nucleolin recognition of MYC promoter G-quadruplex. *Science* **2025**, *388*, 6744. [CrossRef] [PubMed]
41. Varshney, D.; Spiegel, J.; Zyner, K.; Tannahill, D.; Balasubramanian, S. The regulation and functions of DNA and RNA G-quadruplexes. *Nat. Rev. Mol. Cell Biol.* **2020**, *21*, 459–474. [CrossRef] [PubMed]
42. Lago, S.; Tosoni, E.; Nadai, M.; Palumbo, M.; Richter, S.N. The cellular protein nucleolin preferentially binds long-looped G-quadruplex nucleic acids. *Biochim. Biophys. Acta Gen. Subj.* **2017**, *1861*, 1371–1381. [CrossRef] [PubMed]
43. Razavipour, S.F.; Harikumar, K.B.; Slingerland, J.M. p27 as a Transcriptional Regulator: New Roles in Development and Cancer. *Cancer Res.* **2020**, *80*, 3451–3458. [CrossRef] [PubMed]
44. Reynisdottir, I.; Polyak, K.; Iavarone, A.; Massague, J. Kip/Cip and Ink4 Cdk inhibitors cooperate to induce cell cycle arrest in response to TGF-beta. *Genes Dev.* **1995**, *9*, 1831–1845. [CrossRef] [PubMed]
45. Thongchot, S.; Aksonnam, K.; Thuwajit, P.; Yenchitsomanus, P.T.; Thuwajit, C. Nucleolin-based targeting strategies in cancer treatment: Focus on cancer immunotherapy. *Int. J. Mol. Med.* **2023**, *52*, 81. [CrossRef] [PubMed]

biomolecules

MDPI

Article

Optimization of Intra-Arterial Administration of Chemotherapeutic Agents for Glioblastoma in the F98-Fischer Glioma-Bearing Rat Model

Juliette Latulippe [1], Laurent-Olivier Roy [2], Fernand Gobeil [1] and David Fortin [2,*]

[1] Department of Pharmacology and Physiology, Faculty of Medicine and Health Sciences, Université de Sherbrooke, Sherbrooke, QC J1H 5N4, Canada; juliette.latulippe@usherbrooke.ca (J.L.); fernand.gobeil@usherbrooke.ca (F.G.)

[2] Division of Neurosurgery, Department of Surgery, Centre Hospitalier de l'Université de Sherbrooke, Sherbrooke, QC J1H 5N4, Canada; laurent-olivier.roy@usherbrooke.ca

* Correspondence: david.fortin@usherbrooke.ca

Abstract: Glioblastoma (GBM) is a difficult disease to treat for different reasons, with the blood–brain barrier (BBB) preventing therapeutic drugs from reaching the tumor being one major hurdle. The median overall survival is only 14.6 months after the standard first line of treatment. At relapse, there is no recognized standard second-line treatment. Our team uses intra-arterial (IA) chemotherapy as a means to bypass the BBB, hence achieving an overall median survival of 25 months. However, most patients eventually fail the treatment and progress. This is why we wish to expand our portfolio of options in terms of chemotherapy agents available for IA administration. In this study, we tested topotecan, cytarabine, and new formulations of carboplatin and paclitaxel by IA administration in the F98-Fischer glioma-bearing rat model as a screening tool for identifying potential candidate drugs. The topotecan IA group showed increased survival compared to the intravenous (IV) group (29.0 vs. 23.5), whereas the IV cytarabine group survived longer than the IA group (26.5 vs. 22.5). The new formulation of carboplatin showed a significant increase in survival compared to two previous studies with the conventional form (37.5 vs. 26.0 and 30.0). As for paclitaxel, it was too neurotoxic for IA administration. Topotecan and the new formulation of carboplatin demonstrated significant results, warranting their transition for consideration in clinical trials.

Keywords: chemotherapeutic agents; topotecan; cytarabine; carboplatin; paclitaxel; glioblastoma; blood–brain barrier

1. Introduction

Glioblastoma (GBM) is the most aggressive primary malignant brain tumor in adults. The overall survival is only 14.6 months, and the progression-free survival is 6.9 months after the first line of treatment, which consists of maximal surgical resection followed by concomitant radiotherapy and chemotherapy with temozolomide [1]. There is no consensus on the optimal therapeutic method to use at relapse. Hence, every clinical team offers treatment at relapse based on their own experience, biases, and resources. GBM remains challenging to treat due to the limited penetration of therapeutic drugs in the brain parenchyma, primarily restricted by the blood–brain barrier (BBB). Indeed, the BBB, a selectively impermeable barrier formed by the microvasculature of the central nervous system (CNS), safeguards the healthy brain by blocking most blood-borne substances but also hinders effective treatments of CNS diseases [2,3].

Consequently, most chemotherapeutic agents (CTAs) administered intravenously (IV) or orally fail to cross the BBB, while others achieve only partial penetration, preventing the accumulation of therapeutically effective concentrations in tumor cells [4]. To overcome this limitation, various strategies, such as regional or local therapies, have been proposed to bypass the BBB and target the bulk of the tumor more effectively [5]. At the Centre Hospitalier de l'Université de Sherbrooke (CHUS), our team elected to deliver the CTAs intra-arterially (IA) to increase the accumulation of therapeutic drugs in the diseased brain. IA delivery offers benefits that are not found in traditional IV or oral administrations, such as selective infusion, an increase in drug concentration via the first pass effect, and also offering the option of manipulating the BBB permeability via different means [6,7]. Clinical studies conducted by our team on GBM patients (n = 319) have demonstrated that IA administration has the potential to extend the overall median survival to 25 months [8], which is quite significant compared to the current standard of care first-line treatment (overall median survival: 14.6 months) [1]. Moreover, IA drug delivery is a safe procedure and decreases systemic adverse effects as the entire administration is delivered to the brain tumor area, comparatively to IV infusions [9–11]. This decreases the systemic recirculation of the drug by up to 20% [12]. Currently, only five CTAs are used in the clinic by our team, as each agent needs to be carefully selected and tested for efficacy as well as safety in IA administration. We intend to improve our drug portfolio for IA administration to increase options for our patients [8]. This will eventually allow us to use drug sensitivity testing to construct personalized drug protocol solutions for each patient in the hope of decreasing drug resistance and improving outcomes.

This study compares four different CTAs (topotecan, cytarabine, carboplatin, and paclitaxel) in two different administration routes (IA and IV). Topotecan is a topoisomerase I inhibitor that has demonstrated a beneficial effect on proliferating glioma cells [13]. It can also partially penetrate the intact BBB when administered IV [14]. Cytarabine is an S-phase-specific antimetabolite, classified as a pyrimidine analog, that is commonly used to treat CNS disorders, such as primary CNS lymphoma [15]. Carboplatin is a platinum alkylating agent currently used to treat CNS tumors in IA administration by our team, with or without BBB opening [16,17]. Considering that this agent has already shown a response in IA administration, we included a new formulation of carboplatin in this study. The carboplatin used was solubilized in a new confidential vehicle specifically developed and produced by the company Ingenew BioPharma. Finally, paclitaxel is a microtubule-stabilizing CTA that represents a great advance in oncology, with clinical responses in a variety of primary cancers. Unfortunately, this drug presents a poor penetration across the BBB when infused IV, but localized therapy with nanoparticles has shown promising results [18–20]. Paclitaxel has previously been tested by IA with Cremophor as a solubilization vehicle, but the animals did not survive the procedure as the vehicle itself seemed to produce fatal side effects (data not published). Hence, the use of the new confidential formulation by Ingenew BioPharma, incorporating an alternative solubilizing agent, was evaluated to determine whether IA administration would be feasible and beneficial.

The principal aims of this study were to (1) determine the maximal tolerable dose (MTD) and the innocuity of the previously mentioned CTAs administered IA in a healthy rat model and (2) evaluate their therapeutic effects in a glioma-bearing rat model.

2. Materials and Methods

2.1. Chemicals

Topotecan (4 mg/mL) and cytarabine (100 mg/mL) were purchased from the oncology pharmacy of the CHUS (Sherbrooke, Québec, Canada). Paclitaxel (25 mg/mL) and carboplatin (20 mg/mL) were provided by the pharmaceutical industry Ingenew BioPharma (Laval, Québec, Canada).

2.2. Cell Line and Culture Conditions

The cell line chosen for this project was the F98 glioma cells (American type culture collection—Manassas, VA, USA; ATCC CRL-2397), as this cell line is syngeneic with the Fischer rat. Thereby, we avoid variations that could arise from immune responses produced in non-immunosuppressed animals by other cell lines. The F98-Fischer model was also chosen since it recreates GBM's behavior of humans well. Maintenance and preparation for our experiments were as previously described [21].

2.3. Animals

For the evaluation of the MTD for each chemotherapeutic agent, male Wistar rats were used. These animals were not implanted with glioma cells, as the goal was to assess toxicity at different dose levels. For the survival efficacy study, male Fischer rats were used. All rats were purchased at Charles River Laboratories (Saint-Constant, Québec, Canada). For every procedure (implantation, chemotherapy infusion, euthanasia), animals were anesthetized with isoflurane (2–2.5%, O_2 at 2 L/min). Every experimental protocol conformed to the regulations set by the Canadian Council on Animal Care and was approved by the institutional ethical committee of the Université de Sherbrooke (protocols ID: 2022-2856 and 2023-3922).

2.4. Drugs Administration

The IA procedure involved drug delivery through a catheter inserted retrogradely into the right external carotid artery using PE-10 intramedic tubing, with the catheter tip positioned right above the bifurcation. This setup allowed for retrograde infusion into the external carotid artery, transitioning to orthograde flow into the internal carotid artery, ultimately directing the drug into the right hemisphere of the brain. After the infusion, the external carotid was condemned, and the animal was closed with sutures (Figure 1). This procedure was performed as previously described by Fortin et al. [22,23]. The IV infusions of CTAs were administered via the caudal tail vein. The intra-tumoral (IT) injections of CTAs were performed over the three-minute duration using the same coordinates as those employed for the implantation procedure (see below). Since the MTD study was conducted on tumor-free animals, the IA infusion was the first procedure performed. In contrast, for the survival study, animals received the treatment 10 days following the F98 glioma cell implantation.

2.5. Treatment Groups

For the MTD study, there were four independent groups with three Wistar rats each for every CTA tested. Each group corresponded to a different dose (Figure 2). As for the efficacy study, a total of eight groups were planned, with four animals in each group. However, some animals were added to different groups to increase the sample size either to compensate for the loss of some animals because of events unrelated to the tumor or to validate drug efficacy (topotecan and carboplatin) or lack thereof (cytarabine). The final number of animals per group was as follows: (1) Topotecan IA (n = 5), (2) topotecan IV (n = 4), (3) cytarabine IA (n = 6), (4) cytarabine IV (n = 4), (5) control IA (n = 4), (6) carboplatin IA (n = 8), (7) paclitaxel IT (n = 4), and (8) control IT (n = 4). Note that six animals were used for IA administration of paclitaxel at different doses, while two received only the vehicle; however, the route of administration had to be changed due to toxicity.

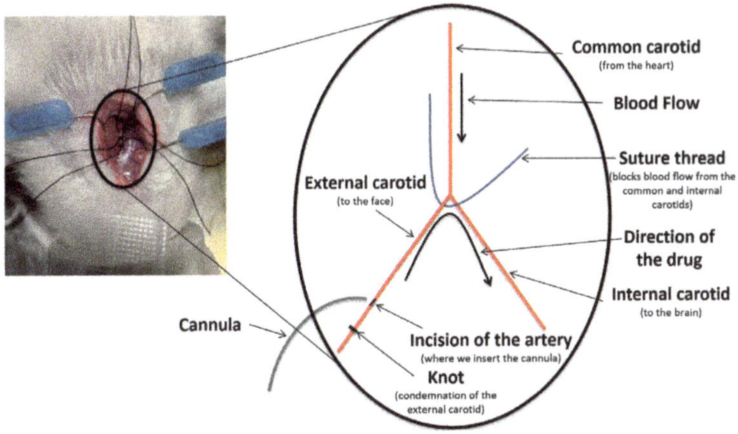

Figure 1. Schematic representation of the IA surgery.

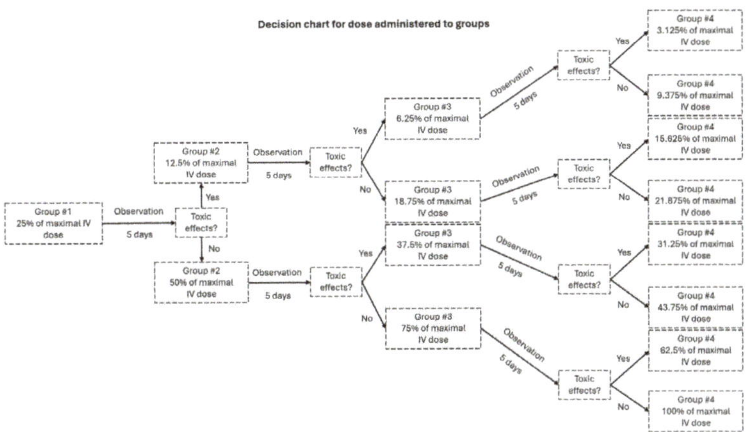

Figure 2. Graphic illustrating the decision chart representing the administered dose for each group.

2.6. MTD Evaluation

The starting dose for the first group was 25% of the maximal IV dose found in the literature [24,25]. After five days of observation, the animal was euthanized, and the brain, the heart, a kidney, parts of the spleen, the lungs, and the liver were harvested for histology. A decision scheme was used to select the subsequent doses (Figure 2). Briefly, if there was any sign of toxicity, behaviorally or in the organs, the second group received half the dose of the previous one (12.5% in this case). If no sign of toxicity was detected, the dose was doubled (50%). This paradigm was followed until the fourth group was reached.

2.7. Implantation

This procedure was only performed for the efficacy study using the F98-Fischer rat model. Confluent F98 cells were harvested and resuspended in non-supplemented warm Dulbecco's Modified Eagle medium (DMEM) at a concentration of 2000 cells/μL. This procedure was performed as described by Blanchard et al. [21]. Briefly, 10,000 cells (5 μL) were implanted in the right frontal lobe, 3 mm laterally and 1 mm anteriorly from the bregma at a depth of 5 mm.

2.8. Evaluation of Mean Survival Time

Animals were monitored daily. Loss of self-grooming (periocular secretion accumulation), landing ability, coordination, stability, mobility, and weight measurement were performed. The presence of convulsions was also monitored. The animals were euthanized if they reached a score of 3/10 or less in 3 categories or if they lost more than 20% of their initial weight. Euthanasia was performed under anesthesia with isoflurane (2–2.5%, O_2 at 2 L/min) via a 60 mL intracardiac perfusion of 4% paraformaldehyde (PFA). The brain was harvested and conserved in 4% PFA before being transferred in 70% ethanol 24 h later and maintained at 4 °C until preparation for histology.

2.9. Histology

Haemotoxylin and eosin (H&E) staining was performed by the Université de Sherbrooke's histology platform on all harvested specimens. For histological examination, tissue samples were preserved in 10% buffered formalin, embedded in paraffin, and sectioned into 4 µm-thick slices, which were then stained with hematoxylin and eosin (H&E). High-resolution images of the H&E-stained sections were obtained using a Nanozoomer 2-slide scanner (Olympus) and analyzed with the NDP view 2 imagingsoftware (version 2.6.13) [26]. The H&E staining was used to determine if there was any sign of toxicity in the MTD study and to identify the necrotic and tumoral areas in the survival study.

2.10. Statistical Analysis

Data for the survival study were analyzed by the Kaplan–Meier survival curves using a Log-Rank test. *P*-values under 0.05 were considered statistically significant.

3. Results

3.1. MTD Study

Since carboplatin's IA dosage had already been determined by our group, only topotecan, cytarabine, and paclitaxel were tested in this study [27].

Topotecan was obtained from a 1 mg/mL solution. The IV dose of 4 mg/kg was selected based on the literature [24]. As per the dose decision scheme (Figure 2), the first IA group received 1 mg/kg (Table 1). There was no treatment-related complication nor sign of toxicity in the organs in the first group. The other three groups showed a similar outcome. No abnormal signs in the animals' behavior were observed. Likewise, no toxicity was observed in the brain, heart, kidney, spleen, lung, or liver based on the H&E sections (Figure 3). Therefore, following the dose escalation scheme described above, the selected treatment dose for IA infusion was 4 mg/kg, representing 100% of the IV dose.

Table 1. First and final dosage used in the MTD and survival study.

CTA	1st Dosage	4th Dosage	% of the IV Dosage	Dosage Used for Survival Studies
Topotecan	1 mg/kg	4 mg/kg	100%	4 mg/kg
Cytarabine	10 mg/kg	8.75 mg/kg	21.88%	8.75 mg/kg
Paclitaxel	3.3 mg/kg	0.825 mg/kg	-	Too toxic

Cytarabine's first tested dose was 10 mg/kg from a 100 mg/mL solution, as the IV dose was 40 mg/kg (Table 1) [25]. This dosage was considered too toxic as the animals did not survive the entire procedure. Indeed, they presented heavy tachycardia and died immediately or soon after the end of the IA infusion. The subsequent dose was, thus, reduced by half. The second, third, and fourth groups tolerated their dosages well, and there were no apparent signs of toxicity (Figure 3). The IA cytarabine MTD was, therefore, 8.75 mg/kg, which represented almost 22% of the maximal IV dose.

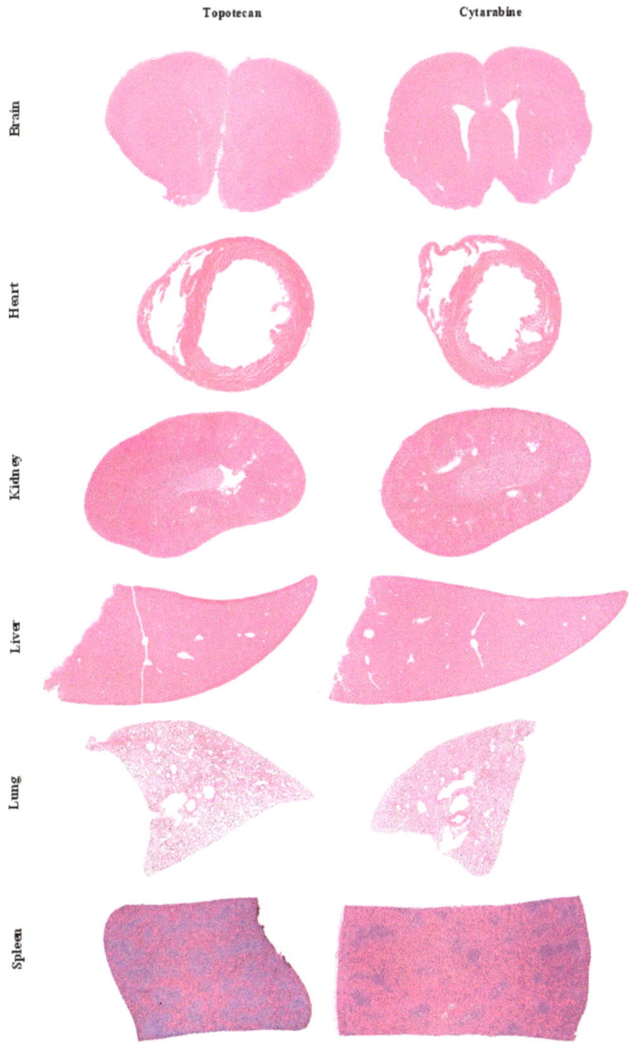

Figure 3. H&E images of vital organs at the end of treatments showing the non-toxicity of topotecan and cytarabine when administered IA at 4 mg/kg and 8.75 mg/kg, respectively, which were their highest doses tested. Magnification: 20×.

As for paclitaxel, no animals survived the IA infusions. During the first administration (3.3 mg/kg), animals displayed abnormally irregular and jerky breathing (Table 1). The second and third doses (1.65 and 0.825 mg/kg) also resulted in mortality of the animals during the infusions. We repeated this procedure using only the vehicle at a volume matching the last tested dosage to verify whether this effect was caused by the CTA or not. Unfortunately, the animals also died during the administration. From that point on, IA administration was substituted for IT infusions for survival studies using paclitaxel.

3.2. Efficacy/Survival Study

To evaluate the efficacy of topotecan and cytarabine using IA infusion at selected doses (Table 1), we conducted survival studies comparing these groups' survival with IV and control groups. Topotecan showed a significant increase in survival for the IA group, with a median of 29.0 days compared to 23.5 days for the IV group ($p = 0.015$). However, there was no significant difference between the IA group and the control group (Figure 4, Table 2). The control group's median survival (25.5 days) is consistent with our previously reported standardization experiments, which described a median survival of 26 ± 2 days for untreated animals [28].

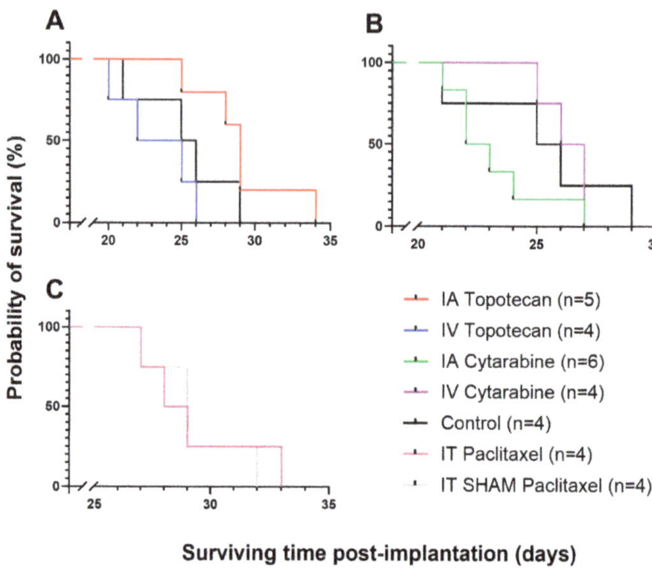

Figure 4. Kaplan–Meier survival analysis of F98 glioma-bearing rats following treatment. (**A**) Topotecan by IA and IV and its control. (**B**) Cytarabine by IA and IV and its control. (**C**) IT paclitaxel and SHAM.

Table 2. Median survival times of the different groups.

Chemotherapy Agent and Administration Route	Median Survival (Days)
Topotecan—IA	29.0
Topotecan—IV	23.5
Cytarabine—IA	22.5
Cytarabine—IV	26.5
Control	25.5
Carboplatin (Ingenew's formulation)—IA	37.5
Paclitaxel (Ingenew's formulation)—IT	28.5
Paclitaxel (Ingenew's formulation) SHAM—IT	29.0

The group receiving IA cytarabine showed a lower survival rate with a median of 22.5 days compared to the IV and control groups at 26.5 and 25.5 days, respectively (Figure 4, Table 2). These results demonstrate that IA administration appears more toxic than IV delivery, suggesting that this approach is not advisable for clinical use.

Carboplatin was used at a dose previously described by our group (20 mg/kg) from 20 mg/mL [27,29]. This new solution showed interesting results, as the median survival of the IA group was 37.5 days (n = 8). This CTA was not compared with an IV group since the goal of this study was to determine if this new formulation was more efficient than the carboplatin diluted in saline (median survival of 26.0 days, n = 8) or 5% dextrose (median survival of 30.0 days, n = 9). Table 3 illustrates the results of this study alongside two others conducted by our group using the same animal model and the other two vehicles [27,29]. The newly tested carboplatin formulation demonstrated improved efficacy in enhancing the survival of the animals.

Table 3. Median survival times of animals treated with carboplatin in three different studies.

Study	Côté et al. (2013) [29]	Charest et al. (2013) [27]	Present Study
Drug	Carboplatin (saline)	Carboplatin (5% dextrose)	Carboplatin (Ingenew's formulation)
Median survival (days)	26.0 ***	30.0 **	37.5

** $p < 0.01$, *** $p < 0.001$ vs. new carboplatin formulation from Ingenew BioPharma, based on Log-Rank test.

Finally, since paclitaxel could not be administered IA, we tested its effect when administered via the IT route (375 µg/kg from a 25 mg/mL solution [30]). A SHAM group, in which animals received an IT infusion of the vehicle alone, was included. The group treated with paclitaxel achieved a median survival of 28.5 days, while the SHAM group survived a median of 29.0 days, showing no statistically significant difference in survival outcomes (Figure 3, Table 2).

4. Discussion

GBM patients still present a short median survival despite decades of research. One of the main reasons for this dreadful outcome is the selective impermeability of the BBB, preventing a significant buildup of therapeutic concentration in the CNS. Even though IA infusions of CTAs increase survival, most patients who respond eventually recur and progress while developing chemoresistance [8]. Since IA chemotherapy has proven effective as a treatment at first relapse, we aim to expand the range of available CTAs that can be administered via this route.

Hence, to improve our therapeutic portfolio of available CTAs, four distinct drugs were selected and tested for IA perfusion in the F98-Fischer preclinical syngeneic model of GBM. These agents were chosen as they either were already used to treat brain tumors IV, cross partly the BBB, show good potential to treat aggressive tumors, or were presented under a new formulation that could possibly lead to better results than those already used in clinical care [8,14,15,18].

In the treatment of GBMs, the administration routes are classically the oral or IV route. However, these delivery strategies administer the CTAs systemically, thus decreasing their accumulation in the brain and the tumor. Indeed, the survival results of our IV studies using topotecan and cytarabine are consistent with those obtained in the control group, suggesting that they can barely, if at all, cross the BBB (Figure 4). On the other hand, Charest et al. showed that platinum compounds administered IA would accumulate significantly more in the tumor compared to IV infusions [27]. IA drug delivery has also been shown to be safe and efficacious in brain tumors in the clinical setting when the CTAs were adequately selected [10,11,31,32]. It appears to be as safe as diagnostic cerebral angiography, as the complications arise mostly from the catheter placement and are extremely unusual (0.75%) [11]. This route has also been utilized in children and was well tolerated [33]. Additionally, CTAs are not the only compounds that can be administered IA. Many clinical trials have used monoclonal antibodies for different types of brain diseases [31,33]. Furthermore, BBB disruption (BBBD) prior to IA infusions is a well-established procedure that can increase up to 300 times the local concentration of CTAs in the brain compared to IV infusion [7]. Hence, BBBD serves as a valuable adjunct to IA infusion. Although BBBD was not utilized in this study, it is a strategy worth considering for future investigations.

Considering that topotecan, cytarabine, and paclitaxel have never been tested IA, the first step was to determine their maximal tolerable dose with no observable adverse effects. Our results showed that neither topotecan nor cytarabine produced neurotoxicity when tested at 4 mg/kg and 8.75 mg/kg, their respective highest tolerated dose. The same could be said of systemic innocuity, as the heart, the kidneys, the liver, the lungs, and the spleen were also surveyed to confirm the absence of toxic effects of these drugs. Our results suggest that these drugs can be safely administered IA at these doses without causing any adverse effects. However, none of the animals survived the IA administration of paclitaxel. Preliminary experiments revealed that the vehicle was responsible for the observed neurotoxicity when infused via the IA route, prompting a switch to IT infusions for our survival studies.

The IA topotecan group showed an increase in survival compared to the IV group. Indeed, while the median survival of the IV group was not significantly different from the controls, the IA group presented a median survival significantly higher with 29 days (p = 0.015). Hence, considering that this drug seems to be well tolerated by the animals and produced an increase in survival, it would be interesting to do a Phase 0–1 clinical trial (escalating dose and innocuity assessment) to determine its usefulness in our population of relapsing GBM patients. This is not the first attempt to use this drug with a regional delivery strategy. In fact, other research groups have used convention-enhanced delivery (CED) of topotecan in both small and large animals and reported great results with significantly higher survival using this method [34]. Our results align with theirs while using a single IA infusion of the therapeutic agent. Moreover, Spinazzi et al. complemented these previous findings in a clinical study where they used the same agent and method. While further studies would be needed to prove favorable outcomes, their small group showed that CED is a potentially safe and effective treatment for recurrent GBM patients [13]. Hence, topotecan could be a good option for regional therapy via IA administration in relapsing GBM patients.

In contrast, the cytarabine groups revealed that the IV-treated rats survived longer than those treated via IA administration, although the difference was not statistically significant ($p = 0.0523$). Additionally, the IV group did not exhibit any significant differences in survival compared to the control group. This suggests that cytarabine is ineffective despite the treatment being well tolerated by the animals. It is also important to note that the control group used with topotecan and cytarabine displayed irregularities. More specifically, four animals were used in this group and had survival times of 21, 25, 26, and 29 days, respectively, resulting in a median survival of 25.5 days. While this is technically within the range reported in the literature (26 $\pm$ 2 days), such variation has never been observed previously in our lab [28].

The new IA carboplatin formulation (Ingenew BioPharma) showed improved survival compared to the standard formulation (Novopharm) used in our laboratory in preclinical and clinical studies, with a median survival of 37.5 days. In the past, certain diverging results obtained by our laboratory hinted at the fact that the solubilization vehicle might influence results. Hence, using the same F98-Fischer glioma-bearing rat model, Côté et al.

described a median survival of 26 days with carboplatin in saline, while Charest et al. reported a median survival of 30 days when 5% dextrose was used as a vehicle [27,29]. The new carboplatin formulation, submitted for patent by Ingenew BioPharma, appears to surpass these results and could potentially replace the formulation currently used in our patients in the clinic, pending confirmation in a clinical comparative study.

The standard vehicle for paclitaxel is Cremophor, as the drug is not soluble in saline. However, Cremophor itself is a toxic compound when delivered via the IA route. Indeed, our lab has previously attempted to use the IA route for the delivery of paclitaxel with its usual vehicle, but severe instances of neurotoxicity occurred and resulted in fatal outcomes. In fact, Cremophore caused severe convulsions during the infusion (data not published). Hence, the IA infusion of placlitaxel with this compound was promptly interrupted. In this study, paclitaxel was dissolved using a new formulation developed by Ingenew BioPharma. Since Cremophor was identified as the cause of the toxic effect observed previously, testing this new formulation for IA delivery was of interest. Unfortunately, the new vehicle was also poorly tolerated by the animals, as none survived the treatment. IT administration of paclitaxel was, thus, instead implemented to see if better results could be obtained, but it did not show a significantly higher survival compared to the SHAM group (28.5 vs. 29 days). The results from the SHAM group were unexpected, as these animals received only the vehicle and were, therefore, expected to exhibit the same median survival as the control group. Interestingly, all animals that received either the drug or the vehicle via the IT route showed brain hemorrhage. This could, of course, be attributed to the needle used for the procedure (23G); however, the same needle size is routinely used for the implantation procedure without any adverse events. The administration of paclitaxel was performed on day 10 post-tumor cell implantation. Given that tumors are more highly vascularized compared to healthy brain parenchyma, it is possible that the needle might have damaged blood vessels in the tumor, leading to hemorrhages.

In conclusion, two of the four drugs tested in this study showed an increase in median survival rates, while the other two were either too toxic or unsuitable for IA infusion in the F98-Fischer glioma-bearing rat model. BBBD could be considered in future studies to further enhance these efficacy results. Phase 0–1 clinical trials will be planned to evaluate the new Ingenew BioPharma carboplatin formulation and topotecan.

Author Contributions: Conceptualization, L.-O.R. and D.F.; methodology, J.L., L.-O.R. and D.F.; formal analysis, J.L., L.-O.R. and D.F.; investigation, J.L.; resources, D.F.; data curation, J.L.; writing—original draft preparation, J.L.; writing—review and editing, J.L., L.-O.R., F.G. and D.F.; visualization, J.L.; supervision, F.G. and D.F.; project administration, L.-O.R., F.G. and D.F.; funding acquisition, D.F. The authors thank the Histology Research Core of the Faculté de Médecine et des Sciences de la Santé at the Université de Sherbrooke for their histology and phenotyping services. All authors have read and agreed to the published version of the manuscript.

Funding: This work was supported by the National Bank research chair for the treatment of brain tumors, the Fondation Coeur en Tête, the Fondation du CHUS, and the Fondation de l'Université de Sherbrooke.

Institutional Review Board Statement: This animal study protocol was approved by the institutional ethical committee of the Université de Sherbrooke (protocol codes 2022-2856 (2023-02-27) and 2023-3922 (2023-11-27)) and followed guidelines of the Canadian Council on Animal Care.

Data Availability Statement: The data used and/or analyzed in the current study are available from the corresponding author upon reasonable request.

Acknowledgments: This work was supported by Canadian Institutes of Health Research and the company Ingenew BioPharma. The histological platform of Medicine and Health Sciences of the Université Sherbrooke also supported this work. David Fortin, Fernand Gobeil, and Laurent-Olivier

Roy are members of the Centre de recherche du Centre Hospitalier de l'Université de Sherbrooke supported by the Fonds de la Recherche en Santé au Québec.

Conflicts of Interest: Ingenew BioPharma had no role in this study's design, analysis, or interpretation of the results. Ingenew BioPharma was given the opportunity to review this manuscript for medical and scientific accuracy regarding Ingenew BioPharma's compounds and to address any intellectual property concerns.

Abbreviations

The following abbreviations are used in this manuscript:

BBB	Blood–brain barrier
CED	Convection-enhanced delivery
CHUS	Centre Hospitalier de l'Université de Sherbrooke
CNS	Central nervous system
CTA	Chemotherapy agent
DMEM	Dulbecco's modified eagle medium
GBM	Glioblastoma multiform
H&E	Hematoxylin and eosin
IA	Intra-arterial/intra-arterially
IT	Intratumoral
IV	Intravenous
MTD	Maximal tolerable dose
BBBD	Blood–brain barrier disruption
PFA	Paraformaldehyde

References

1. Stupp, R.; Mason, W.P.; van den Bent, M.J.; Weller, M.; Fisher, B.; Taphoorn, M.J.B.; Belanger, K.; Brandes, A.A.; Marosi, C.; Bogdahn, U.; et al. Radiotherapy plus Concomitant and Adjuvant Temozolomide for Glioblastoma. *N. Engl. J. Med.* **2005**, *352*, 987–996. [CrossRef]
2. Blanchette, M.; Tremblay, L.; Lepage, M.; Fortin, D. Impact of Drug Size on Brain Tumor and Brain Parenchyma Delivery after a Blood–Brain Barrier Disruption. *J. Cereb. Blood Flow Metab* **2014**, *34*, 820–826. [CrossRef] [PubMed]
3. Wu, D.; Chen, Q.; Chen, X.; Han, F.; Chen, Z.; Wang, Y. The Blood-Brain Barrier: Structure, Regulation, and Drug Delivery. *Signal Transduct. Target. Ther.* **2023**, *8*, 217. [CrossRef] [PubMed]
4. Alifieris, C.; Trafalis, D.T. Glioblastoma multiforme: Pathogenesis and treatment. *Pharmacol. Ther.* **2015**, *152*, 63–82. [CrossRef] [PubMed]
5. van Solinge, T.S.; Nieland, L.; Chiocca, E.A.; Broekman, M.L.D. Advances in local therapy for glioblastoma—Taking the fight to the tumour. *Nat. Rev. Neurol.* **2022**, *18*, 221–236. [CrossRef]
6. D'amico, R.S.; Khatri, D.; Reichman, N.; Patel, N.V.; Wong, T.; Fralin, S.R.; Li, M.; Ellis, J.A.; Ortiz, R.; Langer, D.J.; et al. Super selective intra-arterial cerebral infusion of modern chemotherapeutics after blood–brain barrier disruption: Where are we now, and where we are going. *J. Neuro-Oncology* **2020**, *147*, 261–278. [CrossRef]
7. Pinkiewicz, M.; Pinkiewicz, M.; Walecki, J.; Zawadzki, M. A systematic review on intra-arterial cerebral infusions of chemotherapeutics in the treatment of glioblastoma multiforme: The state-of-the-art. *Front. Oncol.* **2022**, *12*, 950167. [CrossRef]
8. Fortin, D. Drug Delivery Technology to the CNS in the Treatment of Brain Tumors: The Sherbrooke Experience. *Pharmaceutics* **2019**, *11*, 248. [CrossRef]
9. Boockvar, J.A.; Tsiouris, A.J.; Hofstetter, C.P.; Kovanlikaya, I.; Fralin, S.; Kesavabhotla, K.; Seedial, S.M.; Pannullo, S.C.; Schwartz, T.H.; Stieg, P.; et al. Safety and maximum tolerated dose of superselective intraarterial cerebral infusion of bevacizumab after osmotic blood-brain barrier disruption for recurrent malignant glioma. *J. Neurosurg.* **2011**, *114*, 624–632. [CrossRef]
10. Ruan, J.; Shi, Y.; Luo, P.; Li, L.; Huang, J.; Chen, J.; Yang, H. Safety and feasibility of intra-arterial delivery of teniposide to high grade gliomas after blood–brain barrier disruption: A case series. *J. NeuroInterventional Surg.* **2023**, *16*, 1152–1156. [CrossRef]
11. Uluc, K.; Ambady, P.; McIntyre, M.K.; Tabb, J.P.; Kersch, C.N.; Nerison, C.S.; Huddleston, A.; Liu, J.J.; Dogan, A.; Priest, R.; et al. Safety of intra-arterial chemotherapy with or without osmotic blood–brain barrier disruption for the treatment of patients with brain tumors. *Neuro-Oncology Adv.* **2022**, *4*, vdac104. [CrossRef] [PubMed]

12. Drapeau, A.; Fortin, D. Chemotherapy Delivery Strategies to the Central Nervous System: Neither Optional nor Superfluous. *Curr. Cancer Drug Targets* **2015**, *15*, 752–768. [CrossRef] [PubMed]
13. Spinazzi, E.F.; Argenziano, M.G.; Upadhyayula, P.S.; Banu, M.A.; Neira, J.O.; Higgins, D.M.; Wu, P.B.; Pereira, B.; Mahajan, A.; Humala, N.; et al. Chronic convection-enhanced delivery of topotecan for patients with recurrent glioblastoma: A first-in-patient, single-centre, single-arm, phase 1b trial. *Lancet Oncol.* **2022**, *23*, 1409–1418. [CrossRef]
14. Kollmannsberger, C.; Mross, K.; Jakob, A.; Kanz, L.; Bokemeyer, C. Topotecan—A Novel Topoisomerase I Inhibitor: Pharmacology and Clinical Experience. *Oncology* **1999**, *56*, 1–12. [CrossRef]
15. Baker, W.J.; Royer, G.L., Jr.; Weiss, R.B. Cytarabine and neurologic toxicity. *J. Clin. Oncol.* **1991**, *9*, 679–693. [CrossRef] [PubMed]
16. Dréan, A.; Lemaire, N.; Bouchoux, G.; Goldwirt, L.; Canney, M.; Goli, L.; Bouzidi, A.; Schmitt, C.; Guehennec, J.; Verreault, M.; et al. Temporary blood–brain barrier disruption by low intensity pulsed ultrasound increases carboplatin delivery and efficacy in preclinical models of glioblastoma. *J. Neuro-Oncology* **2019**, *144*, 33–41. [CrossRef]
17. Carpentier, A.; Stupp, R.; Sonabend, A.M.; Dufour, H.; Chinot, O.; Mathon, B.; Ducray, F.; Guyotat, J.; Baize, N.; Menei, P.; et al. Repeated Blood-Brain Barrier Opening with a Nine-Emitter Implantable Ultrasound Device in Combination with Car-boplatin in Recurrent Glioblastoma: A Phase I/II Clinical Trial. *Nat. Commun.* **2024**, *15*, 1650. [CrossRef]
18. Alqahtani, F.Y.; Aleanizy, F.S.; El Tahir, E.; Alkahtani, H.M.; Alquadeib, B. Paclitaxel. In *Profiles of Drug Substances, Excipients and Related Methodology*; Elsevier BV: Amsterdam, The Netherlands, 2019; Volume 44, pp. 205–238.
19. Ranganath, S.H.; Kee, I.; Krantz, W.B.; Chow, P.K.-H.; Wang, C.-H. Hydrogel Matrix Entrapping PLGA-Paclitaxel Microspheres: Drug Delivery with Near Zero-Order Release and Implantability Advantages for Malignant Brain Tumour Chemotherapy. *Pharm. Res.* **2009**, *26*, 2101–2114. [CrossRef]
20. Li, K.W.; Dang, W.; Tyler, B.M.; Troiano, G.; Tihan, T.; Brem, H.; Walter, K.A. Polilactofate Microspheres for Paclitaxel De-livery to Central Nervous System Malignancies. *Clin. Cancer Res.* **2003**, *9*, 3441–3447.
21. Blanchard, J.; Mathieu, D.; Patenaude, Y.; Fortin, D. MR-Pathological Comparison in F98-Fischer Glioma Model Using a Human Gantry. *Can. J. Neurol. Sci. / J. Can. des Sci. Neurol.* **2006**, *33*, 86–91. [CrossRef]
22. Charest, G.; Sanche, L.; Fortin, D.; Mathieu, D.; Paquette, B. Glioblastoma Treatment: Bypassing the Toxicity of Platinum Compounds by Using Liposomal Formulation and Increasing Treatment Efficiency With Concomitant Radiotherapy. *Int. J. Radiat. Oncol.* **2012**, *84*, 244–249. [CrossRef] [PubMed]
23. Fortin, D.; Adams, R.; Gallez, A. A blood-brain barrier disruption model eliminating the hemodynamic effect of ketamine. *Can. J. Neurol. Sci. / J. Can. des Sci. Neurol.* **2004**, *31*, 248–253. [CrossRef]
24. Jeong, S.-H.; Jang, J.-H.; Lee, Y.-B. Pharmacokinetic Comparison of Three Different Administration Routes for Topotecan Hydrochloride in Rats. *Pharmaceuticals* **2020**, *13*, 231. [CrossRef] [PubMed]
25. Groothuis, D.R.; Benalcazar, H.; Allen, C.V.; Wise, R.M.; Dills, C.; Dobrescu, C.; Rothholtz, V.; Levy, R.M. Comparison of cytosine arabinoside delivery to rat brain by intravenous, intrathecal, intraventricular and intraparenchymal routes of administration. *Brain Res.* **2000**, *856*, 281–290. [CrossRef] [PubMed]
26. Valdes, J.; Gagné-Sansfaçon, J.; Reyes, V.; Armas, A.; Marrero, G.; Moyo-Muamba, M.; Ramanathan, S.; Perreault, N.; Ilangumaran, S.; Rivard, N.; et al. Defects in the expression of colonic host defense factors associate with barrier dysfunction induced by a high-fat/high-cholesterol diet. *Anat. Rec.* **2022**, *306*, 1165–1183. [CrossRef]
27. Charest, G.; Sanche, L.; Fortin, D.; Mathieu, D.; Paquette, B. Optimization of the route of platinum drugs administration to optimize the concomitant treatment with radiotherapy for glioblastoma implanted in the Fischer rat brain. *J. Neuro-Oncology* **2013**, *115*, 365–373. [CrossRef]
28. Mathieu, D.; Lecomte, R.; Tsanaclis, A.M.; Larouche, A.; Fortin, D. Standardization and Detailed Characterization of the Syngeneic Fischer/F98 Glioma Model. *Can. J. Neurol. Sci. / J. Can. des Sci. Neurol.* **2007**, *34*, 296–306. [CrossRef]
29. Côté, J.; Savard, M.; Neugebauer, W.; Fortin, D.; Lepage, M.; Gobeil, F. Dual kinin B1 and B2 receptor activation provides enhanced blood–brain barrier permeability and anticancer drug delivery into brain tumors. *Cancer Biol. Ther.* **2013**, *14*, 806–811. [CrossRef]
30. Vinchon-Petit, S.; Jarnet, D.; Paillard, A.; Benoit, J.-P.; Garcion, E.; Menei, P. In vivo evaluation of intracellular drug-nanocarriers infused into intracranial tumours by convection-enhanced delivery: Distribution and radiosensitisation efficacy. *J. Neuro-Oncology* **2009**, *97*, 195–205. [CrossRef]
31. Dashti, S.R.; Kadner, R.J.; Folley, B.S.; Sheehan, J.P.; Han, D.Y.; Kryscio, R.J.; Carter, M.B.; Shields, L.B.E.; Plato, B.M.; La Rocca, R.V.; et al. Single low-dose targeted bevacizumab infusion in adult patients with steroid-refractory radiation necrosis of the brain: A phase II open-label prospective clinical trial. *J. Neurosurg.* **2022**, *137*, 1676–1686. [CrossRef]
32. Gahide, G.; Vendrell, J.-F.; Massicotte-Tisluck, K.; Caux, S.; Deschamps, S.; Noël-Lamy, M.; Belzile, F.; Roy, L.-O.; Fortin, D. Safety of Cerebral Intra-Arterial Chemotherapy for the Treatment of Malignant Brain Tumours. *J. Clin. Med.* **2025**, *14*, 524. [CrossRef] [PubMed]

33. McCrea, H.J.; Ivanidze, J.; O'connor, A.; Hersh, E.H.; Boockvar, J.A.; Gobin, Y.P.; Knopman, J.; Greenfield, J.P. Intraarterial delivery of bevacizumab and cetuximab utilizing blood-brain barrier disruption in children with high-grade glioma and diffuse intrinsic pontine glioma: Results of a phase I trial. *J. Neurosurgery Pediatr.* **2021**, *28*, 371–379. [CrossRef] [PubMed]
34. Upadhyayula, P.S.; Spinazzi, E.F.; Argenziano, M.G.; Canoll, P.; Bruce, J.N. Convection Enhanced Delivery of Topotecan for Gliomas: A Single-Center Experience. *Pharmaceutics* **2020**, *13*, 39. [CrossRef] [PubMed]

biomolecules

MDPI

Article

PLGA-Nano-Encapsulated Disulfiram Inhibits Cancer Stem Cells and Targets Non-Small Cell Lung Cancer In Vitro and In Vivo

Kate Butcher [1,2], Zhipeng Wang [1], Sathishkumar Kurusamy [1,3], Zaixing Zhang [1,4], Mark R. Morris [1], Mohammad Najlah [5], Christopher McConville [6], Vinodh Kannappan [1,2,*] and Weiguang Wang [1,2,*]

1 Faculty of Science and Engineering, University of Wolverhampton, Wolverhampton WV1 1LY, UK
2 Disulfican Ltd., Wolverhampton WV9 5HD, UK
3 School of Biosciences, Division of Natural Sciences, University of Kent, Canterbury CT2 7NZ, UK
4 Department of Otorhinolaryngology Head and Neck Surgery, Affiliated Hospital of North China University of Science and Technology, Tangshan 063000, China
5 Faculty of Health, Medicine and Social Care, Anglia Ruskin University, Cambridge CB1 1PT, UK; mohammad.najlah@aru.ac.uk
6 School of Biomedical Sciences, Ulster University, Coleraine BT52 1SA, UK
* Correspondence: v.kannappans@wlv.ac.uk (V.K.); w.wang2@wlv.ac.uk (W.W.)

Abstract: Cancer stem cells (CSCs) play a key role in non-small cell lung cancer (NSCLC) chemoresistance and metastasis. In this study, we used two NSCLC cell lines to investigate the regulating effect of hypoxia in the induction and maintenance of CSC traits. Our study demonstrated hypoxia-induced stemness and chemoresistance at levels comparable to those in typical CSC sphere culture. Activation of the NF-κB pathway (by transfection of NF-κB-p65) plays a key role in NSCLC CSCs and chemoresistance. Disulfiram (DS), an anti-alcoholism drug, showed a strong *in vitro* anti-CSC effect. It blocked cancer cell sphere reformation and clonogenicity, synergistically enhanced the cytotoxicity of four anti-NSCLC drugs (doxorubicin, gemcitabine, oxaliplatin and paclitaxel) and reversed hypoxia-induced resistance. The effect of DS on CSCs is copper-dependent. A very short half-life in the bloodstream is the major limitation for the translation of DS into a cancer treatment. Our team previously developed a poly lactic-co-glycolic acid (PLGA) nanoparticle encapsulated DS (DS-PLGA) with a long half-life in the bloodstream. Intra venous injection of DS-PLGA in combination with the oral application of copper gluconate has strong anticancer efficacy in a metastatic NSCLC mouse model. Further study may be able to translate DS-PLGA into cancer applications.

Keywords: non-small cell lung cancer; cancer stem cells; chemoresistance; disulfiram; copper gluconate; PLGA nano-delivery

1. Introduction

Lung cancer is the second most common malignancy and the leading cause of cancer-related deaths worldwide, with an estimated 2.2 million new cases and 1.8 million deaths. Non-small cell lung cancer (NSCLC) accounts for about 80–85% of all lung cancer cases [1,2]. Despite advances in lung cancer screening, more than 75% of NSCLC patients are diagnosed at a locally advanced or metastatic stage [3]. Although the therapeutic outcomes of NSCLC have been significantly improved in the last two decades, with the development of treatments against driver mutations and immune check point inhibitors, the 5-year survival rate of NSCLC patients remains at 26% in all stages and only 8% in distant metastatic patients [2]. Chemoresistance remains one of the major causes of therapeutic failure in NSCLC. Therefore, elucidation of underlying resistant mechanisms to currently available drugs and the development of novel anti-NSCLC drugs to reverse chemoresistance are of clinical importance and urgency.

Increasing evidence suggests that a small population of cells known as cancer stem cells (CSCs) plays a pivotal role in de novo and acquired resistance to chemo and targeted therapies [4]. NSCLC is a highly heterogeneous tumor with cells expressing putative stem cell markers such as aldehyde dehydrogenase (ALDH), ABCG2, CD44, CD133 and embryonic stem cell (ESC) markers, e.g., Sox2, Oct4 and Nanog [5]. These cells are resistant to all currently available anticancer drugs and are responsible for cancer metastasis, due to their self-renewal ability, uncontrolled proliferation and genomic instability [4]. Therefore, targeting CSCs may improve the therapeutic outcomes of NSCLC patients. Recent studies demonstrate that CSCs are located in hypoxic niches where hypoxic signaling plays a crucial role in the induction and maintenance of their stemness [6]. CSCs exist in a transient and reversible CSC-like state rather than a permanent entity and can originate from the dedifferentiation of cancer cells driven by various factors such as a hypoxic tumor microenvironment [7]. Hypoxia induces epigenetic alterations which change the phenotype of the cancer cells and make them acquire CSC characteristics and chemoresistance [8,9]. Elucidation of the molecular links between hypoxia and CSCs may identify potential targetable CSC-manipulating pathways to overcome chemoresistance in NSCLC. The cellular response to a hypoxic stimulus activates nuclear factor-κB (NF-κB), a principal transcription factor [10]. The NF-κB pathway is activated by numerous external stimuli, oncogenic mutations and various crosstalk signals [11,12] and is constitutively activated in CSCs to promote various survival mechanisms [13–15]. Suppression of the NF-κB pathway decreases CSC characteristics, sensitizes cancer cells to chemotherapy, inhibits tumor growth and blocks metastasis [8,16–18].

Therefore, the development of drugs that target the CSC population is key to improving the therapeutic outcomes for NSCLC patients. New drug discovery and development is a time-consuming and expensive procedure. Anticancer drug development has a persistent attrition rate of 95% due to safety issues [19]. Drug repurposing is an ideal process to identify new uses for FDA-approved drugs outside of the original intended medical use [20]. Disulfiram (DS) is an anti-alcoholism drug used in the clinic for more than 70 years with excellent safety records [21]. DS has been demonstrated significant anticancer activity in various cancers *in vitro* and *in vivo* [18,22–25]. Previous studies revealed that the cytotoxicity of DS is entirely copper (Cu) dependent [26,27]. The chelation of DS with Cu generates reactive oxygen species (ROS) which are highly toxic to cancer cells. Both DS plus Cu and copper-diethyldithiocarbamate (Cu-DDC), the final product of the Cu and DS chelation, inhibit NF-κB activity and target CSCs [16,17,22,23]. The sulfhydryl group of DS is essential for its reaction with Cu and anticancer activity [26]. Only an oral version of DS is currently available in the clinic [21]. After oral administration, DS is enriched and metabolized in the liver where the sulfhydryl group of DS is promptly methylated, glucuronidated and degraded and loses its anticancer efficacy in cancer patients [26]. This explains the discrepancy between the anticancer activities of DS in the laboratory and in cancer clinical trials. Therefore, the very short half-life of DS in the bloodstream became the bottleneck for the translation of DS into cancer treatment. In order to apply DS to cancer treatment, we first suggested using the nano-drug-delivery system to protect the sulfhydryl group and deliver the intact DS into cancer tissues [28]. We have previously shown that encapsulation of DS into liposome or poly lactic-co-glycolic acid (PLGA) significantly extends the half-life of DS and improves its *in vivo* anticancer efficacy in mouse breast, liver cancer and glioblastoma models [8,18,23].

In this study, we demonstrate that hypoxia-induced CSCs are responsible for chemoresistance in NSCLC cell lines. NF-κB plays pivotal roles in hypoxia-induced cancer cell stemness and chemoresistance. DS/Cu strongly targets CSCs and demonstrated high anti-NSCLC activity *in vitro*. In combination with copper, PLGA encapsulated DS (DS-PLGA) showed very strong *in vivo* anti-NSCLC efficacy in a mouse metastatic model. Our data indicated that further study may translate DS-PLGA into NSCLC clinical treatment.

2. Materials and Methods

2.1. Cell Lines and Reagents

The NSCLC cell lines A549 and NCI-H23 were purchased from ATCC (Teddington, UK). DS, copper (II) chloride ($CuCl_2$), copper gluconate (CuGlu), doxorubicin (DOX), gemcitabine (dFdC), oxaliplatin (OXA), paclitaxel (PTX), poly-2-hydroxyethyl methacrylate (poly-HEMA), propidium iodide (PI) were purchased from Sigma (Dorset, UK). DS-PLGA was developed and characterized in our lab [23].

2.2. 1-(4,5-. Dimethylthiazol-2-yl)-3,5-Diphenylformazan (MTT) Cytotoxicity Assay and CI-Isobologram Analysis

All cell lines were cultured in DMEM (Lonza, Wokingham, UK) supplemented with FBS (10%), L-glutamine (2 mM), penicillin/streptomycin/amphotericin (1%, *v*/*v*) (Lonza, Slough, UK) and maintained in normoxic conditions at 37 °C and 5% CO_2. Hypoxic-cultured cells were incubated at 37 °C, 5% CO_2 and 1% O_2 for 4 days. For cytotoxicity assay, cells (2.5×10^4 cells/mL) were cultured in 96-well flat-bottomed microtiter plates overnight and then exposed to drugs for 72 h in both normoxic and hypoxic conditions before subjected to a standard MTT assay [29]. CI isobologram was used to determine the synergistic effect between anticancer drugs and DS/Cu and calculation of Combination Index (CI). Cells were exposed to various concentrations of anticancer drugs, DS/Cu or in a combination of anticancer drug and DS/Cu [consistent concentration (10 μM) of CuGlu] and then subjected to MTT assay. A constant ratio of anticancer drug:DS of 1:1 (DOX, dFdC and PTX) or 100:1 (OXA) was used. The combined cytotoxicity was determined by CI isobologram using CalcuSyn software (Biosoft, Cambridge, UK) [30]. Mutually exclusive equations were used to determine the CI.

All cell lines were cultured in DMEM (Lonza, Wokingham, UK) supplemented with FBS (10%), L-glutamine (2 mM), penicillin/streptomycin/amphotericin (1%, *v*/*v*) (Lonza, Slough, UK) and maintained in normoxic conditions at 37 °C and 5% CO_2. Hypoxic-cultured cells were incubated at 37 °C, 5% CO_2 and 1% O_2 for 4 days. For cytotoxicity assay, cells (2.5×10^4 cells/mL) were cultured in 96-well flat-bottomed microtiter plates overnight and then exposed to drugs for 72 h in both normoxic and hypoxic conditions before subjected to a standard MTT assay [29]. CI isobologram was used to determine the synergistic effect between anticancer drugs and DS/Cu and calculation of Combination Index (CI). Cells were exposed to various concentrations of anticancer drugs, DS/Cu or in a combination of anticancer drug and DS/Cu [consistent concentration (10 μM) of CuGlu] and then subjected to MTT assay. A constant ratio of anticancer drug:DS of 1:1 (DOX, dFdC and PTX) or 100:1 (OXA) was used. The combined cytotoxicity was determined by CI isobologram using CalcuSyn software (Biosoft, Cambridge, UK) [30]. Mutually exclusive equations were used to determine the CI.

2.3. Spheroid Culture and Sphere Reformation Assay

The NSCLC cells were cultured as spheres in poly-HEMA-coated ultra-low adherence flasks containing stem cell medium (SCM) which contains DMEM-F12 supplemented with 2 mM L-glutamine, 1% (*v*/*v*) penicillin/streptomycin/amphotericin mix (Lonza), B-27 (Invitrogen, Paisley, UK), 10 ng/mL basic fibroblast growth factor (b-FGF) (R&D system, Abingdon, UK), 10 ng/mL epidermal growth factor (EGF, Sigma), 20 μg/mL insulin, 20 μg/mL heparin, and 4.5 g/L D-glucose (Sigma). For the sphere reformation assay, the 6-day-cultured spheroid cells were trypsinised and the single cell suspension was exposed to different drugs for 6 h at 37 °C, 5% CO_2. The treated cells were cultured in drug-free medium at a low cell density (2000 cells/well) in poly-HEMA-coated 24 well plates for 7 days. The total numbers of the reformed spheres were counted and images were taken at 4× magnification.

2.4. Clonogenic Assay

Normoxic and hypoxic-cultured cells were exposed to IC_{50} concentrations of DOX, dFdC, OXA, PTX, Cu, DS and DS/Cu for 48 h. The drug-treated cells (5×10^3/well) were re-seeded in 6-well plates and cultured in drug-free medium under normoxic and hypoxic conditions, respectively, for 10 days. The cells were fixed and stained with crystal violet. The colonies containing at least 50 cells were counted.

2.5. Detection of CSC Markers

The activity and expression levels of ALDH, CD133 and ABCG2 were determined in spheroid, normoxia and hypoxia-cultured A549 and NCI-H23 cells. The CSC markers in spheroid and hypoxia-cultured cells were also examined after 6 h DS/Cu treatment. The ALDH-positive population was detected by the ALDEFLUOR kit (StemCell Tech, Cambridge, UK) following the supplier's instruction. The specificity of the kit was verified with diethylaminobenzaldehyde (DEAB) treatment. The CD133 and ABCG2-positive populations were detected by FITC conjugated anti-CD133 (Macs Miltenyi Biotec, Surrey, UK) and APC conjugated anti-ABCG2 (BD Biosciences, Berkshire, UK) antibodies.

2.6. Measurement of the Hypoxic Cell Population

The Hypoxyprobe™-1 Plus Kit (Hypoxyprobe, Inc, Massachusetts, USA) was used to detect and quantify the hypoxic population in A549 and NCI-H23 cell lines. The cells were incubated overnight in a 100 µM HPP-containing medium. The cells were fixed with ice-cold ethanol and stained with anti-HPP-FITC antibody (1:1000) in flow buffer in the dark and then analyzed by flow cytometry. For immunocytochemistry assay, the cells were fixed using an Image-iT® Fix Perm Kit (Invitrogen, Waltham, MA, USA) and stained with anti-HPP-FITC antibody (1:100) and counterstained with PI.

2.7. Western Blot Analysis

The expression levels of NF-κB-p65 in whole and nuclear protein were determined using primary NF-κB-p65 antibody (Abcam, Cambridge, UK) and HRP conjugated secondary antibody (Sigma). β-actin (Sigma) (1:5000) and nucleolin (Santa Cruz, TX, USA) (1:1000) were used to detect housekeeping proteins in whole and nuclear proteins, respectively. The Western blotting signal was detected using the EZ-ECL enhanced chemiluminescence kit (Biological Industries, Cromwell, CT, USA).

2.8. Quantitative Real Time RT-PCR (qRT-PCR)

qRT-PCR was performed using Taqman™ assays according to the manufacturer's instructions. RNA was extracted from cultured NSCLC cells using an RNA purification kit (Norgen Biotek, Thorold, ON, Canada)) following the manufacturer's instructions, as a template for cDNA synthesis using Multiscribe Reverse Transcriptase (Applied Biosystems, Woburn, MA, USA). The following primers/probes were used: Taqman™ gene expression assay ID P65 Hs01042014_m1, HIF1A Hs00936371_m1, HIF2A Hs01026149_m1, SOX2 Hs01053049_s1, OCT4 Hs00999632_g1, NANOG Hs02387400_g1, HPRT1 Hs99999909_m1. Gene expression was normalized for control gene expression (HPRT1) and calculated according to the comparative $2^{-\Delta\Delta CT}$ method.

2.9. Luciferase Reporter Gene Assay

A dual luciferase assay kit was used for luciferase reporter gene assay in 96-well plates following the instruction from the supplier (Promega, Southampton, UK). The transcriptional activity of NF-κB and hypoxia response element (HRE) was determined using pNFκB-Tal-Luc (BD Biosciences) and pGL4.4-LUC2p/HRE (Promega) luciferase vectors, respectively. pGL3-Basic (BD Biosciences) vector was used as the background control. All of the luciferase vectors were co-transfected with pSV40-Renilla (BD Biosciences) an internal control for normalization of the transcriptional activity. The luciferase activity in each well was normalized to pSV40-Renilla using Ln = L/R (Ln: normalized luciferase unit;

L: luciferase activity reading; R: Renilla activity reading). The transcriptional specificity was monitored by the transcriptional activity of the pGL3-Basic. All transfections were performed in triplicate with at least duplicate independent experiments.

2.10. NSCLC Xenograft and In Vivo DS-PLGA Treatment

Five-week-old female BALB/c Nu/Nu athymic nude mice (Biotechnology and Cell Biology Shanghai, Shanghai, China) were housed under pathogen-free conditions according to the animal care guidelines of Fourth Military Medical University (FMMU) China. All experiments were reviewed and approved by the FMMU Ethical Committee. The A549 NSCLC metastatic xenograft model was used in this study. A549 cells (5×10^6 in 100 μL of PBS) were injected into the tail vein. Three days after tumor cells injection, the tumor-bearing mice were randomly subdivided into 2 groups (7 mice/group) and subjected to the following treatments. 1. Control: no treatment; 2. CuGlu 5 mg/kg p.o and DS-PLGA 5 mg/kg i.v (4 h after CuGlu) 2 times per week for 3 weeks. The mouse body weight was monitored three times per week and all the mice were sacrificed on day 21. The lungs were dissected, weighted and then formalin-fixed for paraffin embedment, sectioning and H&E staining.

2.11. Statistical Analysis

The software GraphPad Prism was used for statistical analysis. The statistical significance of treatment outcomes was assessed using the Student's *t*-test and one-way ANOVA; *p*-value scores of <0.05 (*) and <0.01 (**) were considered statistically significant. Data are displayed as mean values ± standard deviation.

3. Results

3.1. Spheroid-Cultured NSCLC Cells Are Hypoxic with CSC Characters and Resistant to Anti-NSCLC Drugs

In this study, we used spheroid cultures to induce the CSC population in two NSCLC cell lines. After 6 days of culture in SCM at low attachment condition, A549 and NCI-H23 cells formed a distinct spheroid morphology (Figure 1a). In comparison with the attached cells, a significantly larger population of the sphere-cultured cells expresses $ALDH^+$, $CD133^+$ and $ABCG2^+$ (Figure 1b–g). The stemness of the sphere-cultured cells was further proved by the detection of other ESC markers, e.g., SOX2, OCT4 and NANOG, using RT-PCR (Figure 1h). Furthermore, the sensitivity of the spheroid-cultured cells to the four most commonly used NSCLC chemotherapy drugs was examined using the sphere reformation assay (Figure 1i,j). DOX and dFdC did not inhibit the sphere reformation in both cell lines and no inhibition by PTX was observed in the A549 cell line. Although the sphere reformation was inhibited by OXA in both cell lines and PTX in the NCI-H23 cell line, significant numbers of the reformed spheres were still detected. These results demonstrate that sphere-cultured CSCs are resistant to anti-NSCLC drugs. The hypoxic status in the sphere-cultured cells was examined using Hypoxyprobe immunocytochemistry staining and flow cytometry analysis. In comparison to the attached cells, a significantly higher proportion of hypoxic cells were detected in the sphere-cultured cells (Figure 1k–m).

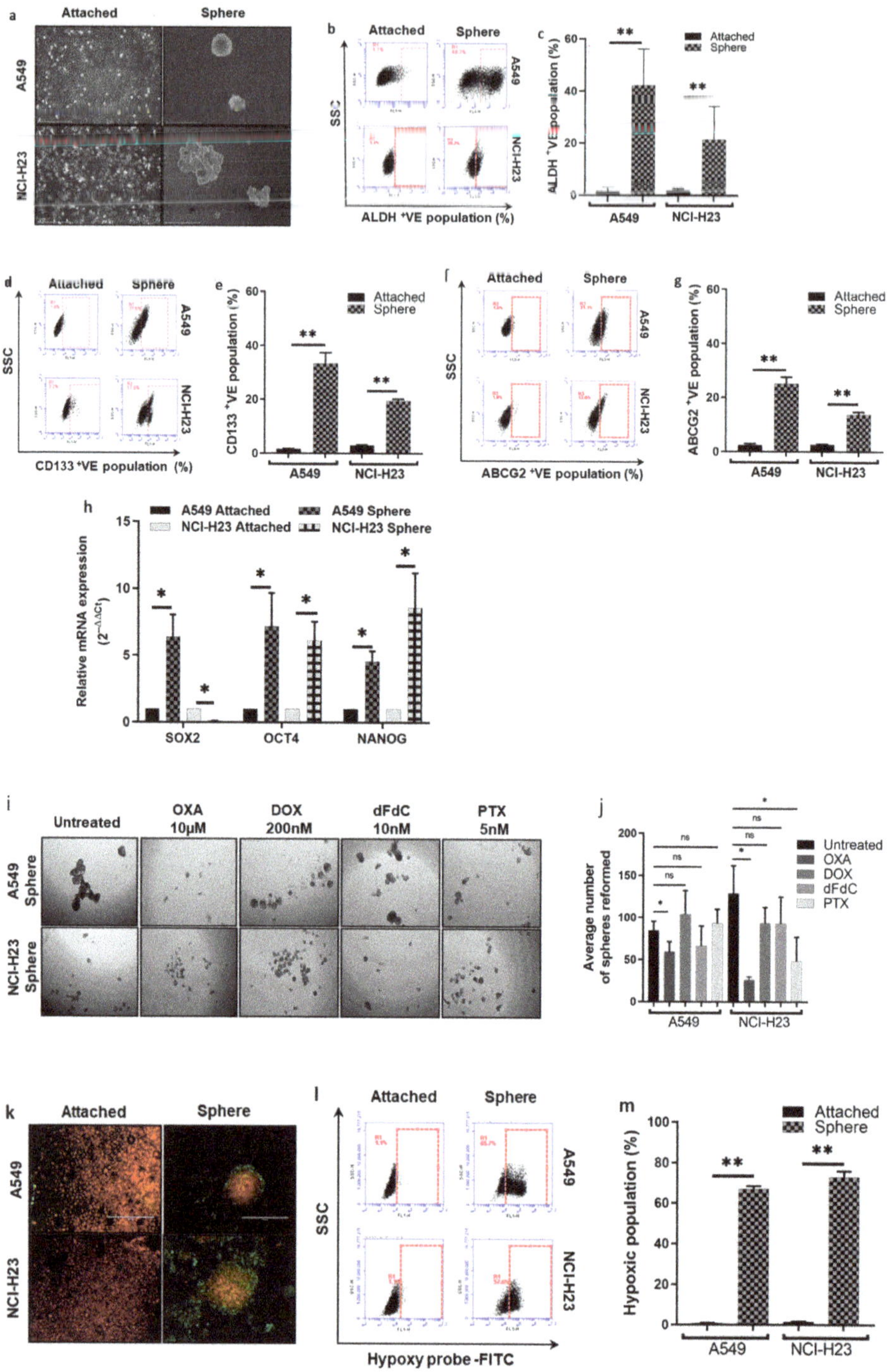

Figure 1. Sphere-cultured NSCLC cells demonstrate CSC and ESC markers, chemoresistance and contain high proportion of hypoxic cell population. (**a**) Morphology of the attached and sphere-cultured

cells (×10 magnification). (**b–g**). Flow cytometric analysis of the expression of CD133, ALDH and ABCG2 CSC markers in attached and sphere-cultured NSCLC cells (mean ± SD; n = 3). (**h**) qRT-PCR analysis of ESC markers in attached and sphere-cultured cells (mean ± SD; n = 3). (**i,j**) Effect of four chemotherapy drugs on the sphere reformation ability (×10 magnification, mean ± SD; n = 3). (**k**) Morphology of hypoxic cells in the spheres (×10 magnification). Hypoxic cells were detected by Hypoxyprobe and FITC-conjugated anti-Hypoxyprobe MAb staining at Exc 492 nm and Emi 520 nm (green, cytoplasm). The nuclei were counterstained by PI (red) Exc 530 nm and Emi 618 nm. (**l,m**) Flow cytometry comparison of the proportions of Hypoxyprobe stained hypoxic population in monolayer and sphere-cultured cells (mean ± SD; n = 3). *ns = non-significant*; * $p < 0.05$ and ** $p < 0.01$.

3.2. Hypoxia-Cultured NSCLC Cells Show CSC Traits and Are Resistant to Anticancer Drugs

Hypoxia and CSC traits were co-detected in sphere-cultured cells (Figure 1). To determine the relationship between hypoxia and CSC traits, we grew the monolayer-cultured cancer cell lines in hypoxic conditions. The hypoxia was confirmed using the Hypoxyprobe kit. Figure 2a shows the immunocytochemistry image of the hypoxia-cultured monolayer cells. Flow cytometric analysis demonstrated a significantly higher percentage of hypoxic cells after being cultured in hypoxia conditions (Figure 2b,c). In comparison with normoxia-cultured cells, a significantly higher percentage of CSC markers (ALDH, CD133 and ABCG2) positive cells was detected in hypoxia-cultured cells (Figure 2d–i). Significantly higher levels of mRNA for ESC markers (SOX2, OCT4 and NANOG) were detected in hypoxia-cultured cells (Figure 2j). The cytotoxicity of DOX, dFdC, OXA and PTX after 72 h treatment was analysed using the MTT assay (Figure 2k). The IC_{50} values of the anti-NSCLC drugs in normoxia and hypoxia-cultured cancer cell lines are shown in Table 1. Hypoxia-cultured cells are highly resistant to all of these clinically used drugs. These results suggest that hypoxia-cultured NSCLC cells developed more chemoresistant CSCs.

Table 1. IC_{50} values in A549 and NCI-H23 cells treated with drugs.

	A549		H23	
Drug	**Normoxia**	**Hypoxia**	**Normoxia**	**Hypoxia**
OXA (μM)	1.4 ± 0.3	>25	9.0 ± 5.2	>25
DOX (nM)	56.6 ± 9.3	>1000	292.5 ± 8.2	>1000
dFdC (nM)	12.7 ± 2.7	>100	9.6 ± 3.2	>100
PTX (nM)	4.2 ± 3.4	>100	8.7 ± 6.1	>100
DS/Cu nM	1.0 ± 0.3	5.9 ± 3.0	13.8 ± 2.3	8.7 ± 1.0

The normoxia and hypoxia-cultured cells were exposed to different drugs for 72 h and subjected to MTT assay.

3.3. NF-κB Pathway Plays a Pivotal Role in CSC Traits and Chemoresistance

In comparison with normoxia-cultured cells, both hypoxia and sphere-cultured A549 and NCI-H23 cell lines express significantly higher levels of NF-κB-p65 mRNA, the key player of the canonical NF-κB activation pathway involving in chemoresistance (Figure 3a). The luciferase reporter gene assay demonstrated significantly higher NF-κB transcriptional activity in hypoxia-cultured cell lines (Figure 3b). These results are in line with our previous findings in human glioblastoma cell lines [8] and indicate the relationship between NF-κB activation and hypoxia-induced NSCLC CSCs. To investigate the causal relationship between the NF-κB pathway, CSC traits and chemoresistance, we stably transfected pcDNA3.1/NF-κB-p65 plasmid into the A549 cell line to overexpress NF-κB-p65. The pcDNA3.1 empty vector-transfected cells were used as mock control. Compared with a mock clone, two positive clones (C7 and C13) expressed significantly higher NF-κB-p65 mRNA and protein (whole and nuclear) (Figure 3c,d). Higher NF-κB activity was also detected by luciferase reporter gene assay (Figure 3e). In comparison with the mock clone, the

ALDH activity and CD133 expression were significantly induced by NF-κB-p65 transfection in C7 and C13 clones (Figure 3f–i). There was no significant difference in the expression of ABCG2 (Figure 3j,k). The RT-PCR results also show an increase in the mRNA levels of the ESC transcription factors SOX2, OCT4 and NANOG in the NF-κB-p65 transfected clones (Figure 3l). These results strongly indicate the positive causal relationship between NF-κB activity and CSC traits in NSCLC cells. Chemoresistance is one of the major characteristics of hypoxia-induced CSCs. To establish the role of NF-κB in chemoresistance, the cytotoxicity of anti-NSCLC drugs in NF-κB-p65 transfected A549 cell lines was compared with that in the mock clone. The NF-κB-p65 transfected clones demonstrate resistance to dFdC, OXA, PTX and DOX, although it does not reach statistical significance in DOX-treated cells (Figure 3m and Table 2).

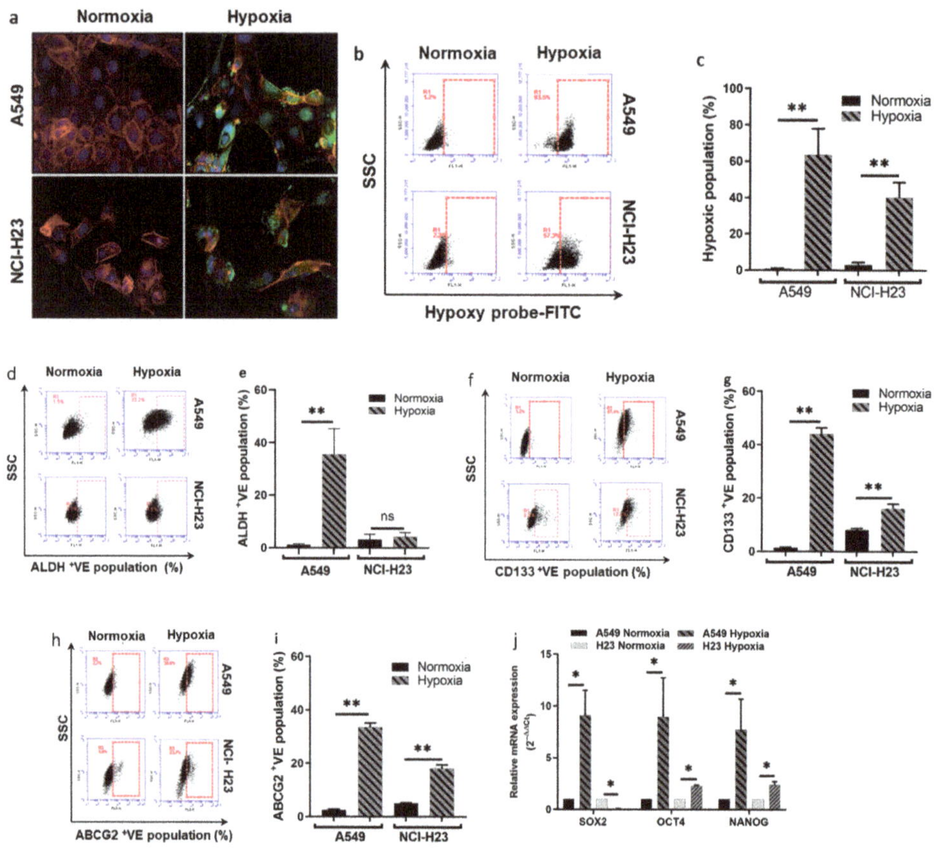

Figure 2. *Cont.*

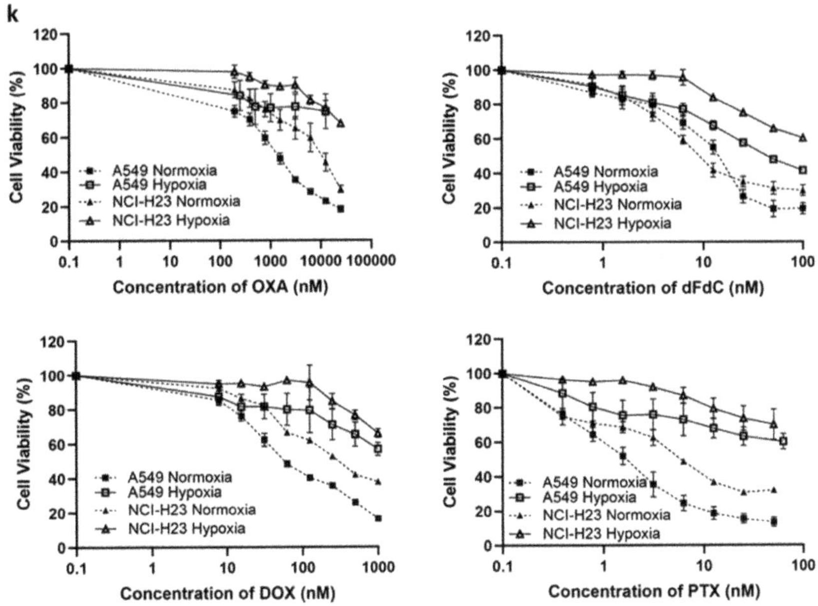

Figure 2. Hypoxic culture induces expression of CSC and ESC markers and chemoresistance. (**a**) Images of Hypoxyprobe-stained monolayer cells cultured in normoxic and hypoxic conditions (×40 magnification). (**a**) Hypoxic cells were detected by Hypoxyprobe and FITC-conjugated anti-Hypoxyprobe MAb staining (Exc 492 nm and Emi 520 nm; green, cytoplasm). The nuclei were counterstained by DAPI (Exc 345 nm and Emi 455 nm; Blue). The cytoplasm was stained with ActinRed 555 (Exc 540 nm and Emi 565 nm; red, cytoplasm) (**b**,**c**) Flow cytometric analysis of Hypoxyprobe-stained cells (mean ± SD; n = 3). (**d**–**i**) Flow cytometric analysis of the expression of CD133, ALDH and ABCG2 CSC markers in attached and sphere-cultured cells (mean ± SD; n = 3). (**j**) qRT-PCR analysis of ESC markers in attached and sphere-cultured cells (mean ± SD; n = 3). (**k**) Comparison of cytotoxicity of four chemotherapy drugs in normoxia and hypoxia-cultured cells (mean ± SD; n = 3). *ns = non-significant;* * $p < 0.05$, ** $p < 0.01$.

Table 2. The IC_{50} anti-NSCLC drugs in the transfected clones.

Drug	Mock	NFκB-p65	
		C7	C13
OXA (μM)	6.7 ± 1.5	>25	>25
DOX (nM)	453.6 ± 240.3	585.2 ± 241.9	706.7 ± 287.9
dFdC (nM)	24.9 ± 20.4	>100	>100
PTX (nM)	2.4 ± 0.5	23.5 ± 23.3	37.2 ± 22.2

3.4. DS/Cu Targets CSCs and Reverses Chemoresistance in NSCLC Cell Lines

The above data strongly suggest that NF-κB plays a critical role in maintaining CSC status and chemoresistance. Our previous findings indicate that DS/Cu inhibits NF-κB activity, targets CSCs and reverses chemoresistance in other types of cancer [8,18,27]. Here, we examine the cytotoxic effect of DS/Cu on CSCs in A549 and NCI-H23 cell lines. Most of the sphere- and hypoxia-induced CSC markers in A549 cell lines are significantly inhibited by DS/Cu. DS/Cu inhibits the expression of ALDH, CD133 and ABCG2 in the sphere and hypoxia-cultured A549 cells (Figure 4a–f). The expression of ALDH and CD133 in sphere-cultured NCI-H23 cells is inhibited by DS/Cu (Figure 4a–d). The expression of CD133 and ABCG2 in hypoxia-cultured NCI-H23 cells is also inhibited by DS/Cu (Figure 4c–f).

Unexpectedly, the expression of ALDH in NCI-H23 cells was not induced by hypoxic culture (Figure 2b) and DS/Cu did not show an inhibiting effect on ABCG2 expression in sphere-cultured NCI-H23 cells (Figure 4e,f). Although hypoxia induces resistance to four anti-NSCLC drugs, the cytotoxicity of DS/Cu is comparable in normoxia and hypoxia-cultured cells (Figure 4g). The effect of DS/Cu on sphere reforming ability was examined. DS and Cu, in combination, completely blocked the sphere reformation ability in A549 and NCI-H23 cell lines. In contrast, these effects were not observed in DS or Cu singly treated cells (Figure 4h,i). The clonogenicity is also the character of CSCs. A clonogenic assay was used to determine the clonogenicity of normoxic and hypoxic NSCLC cells treated with NSCLC chemotherapy drugs. The clinically used NSCLC chemotherapy drugs inhibited the clonogenicity in normoxia-cultured cells but the hypoxia-cultured cells were resistant to the treatment (Figure 4j,l). The combination of DS/Cu completely blocked the colony formation in both normoxia and hypoxia-cultured cells. Whereas, the inhibiting effect was not observed in DS or Cu singly treated cells (Figure 4k,l). These results indicate that DS/Cu targets NSCLC CSCs in which Cu is indispensable. Furthermore, the effect of DS/Cu on the hypoxia-induced anti-NSCLC drug resistance in A549 and NCI-H23 cell lines was examined. DS/Cu significantly enhances the cytotoxicity of four first-line anti-NSCLC drugs and completely reverses the hypoxia-induced chemoresistance in both cell lines (Figure 4m and Table 3). Isobologram analysis shows a synergistic combination effect between DS/Cu and anticancer drugs at a wide concentration range (Table 3).

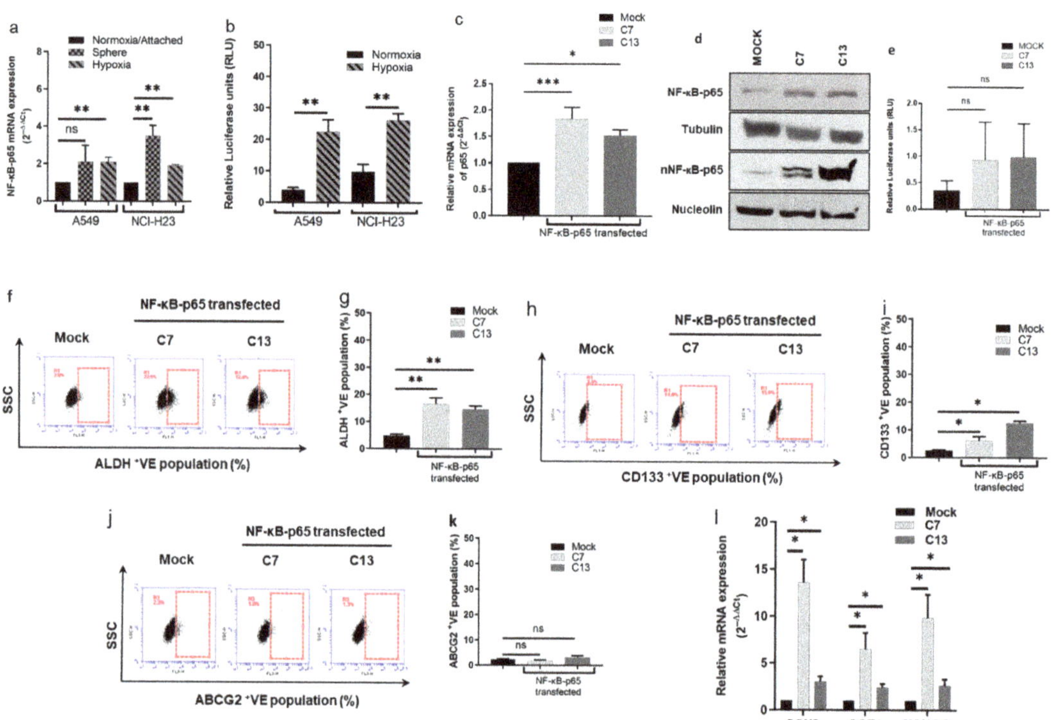

Figure 3. *Cont.*

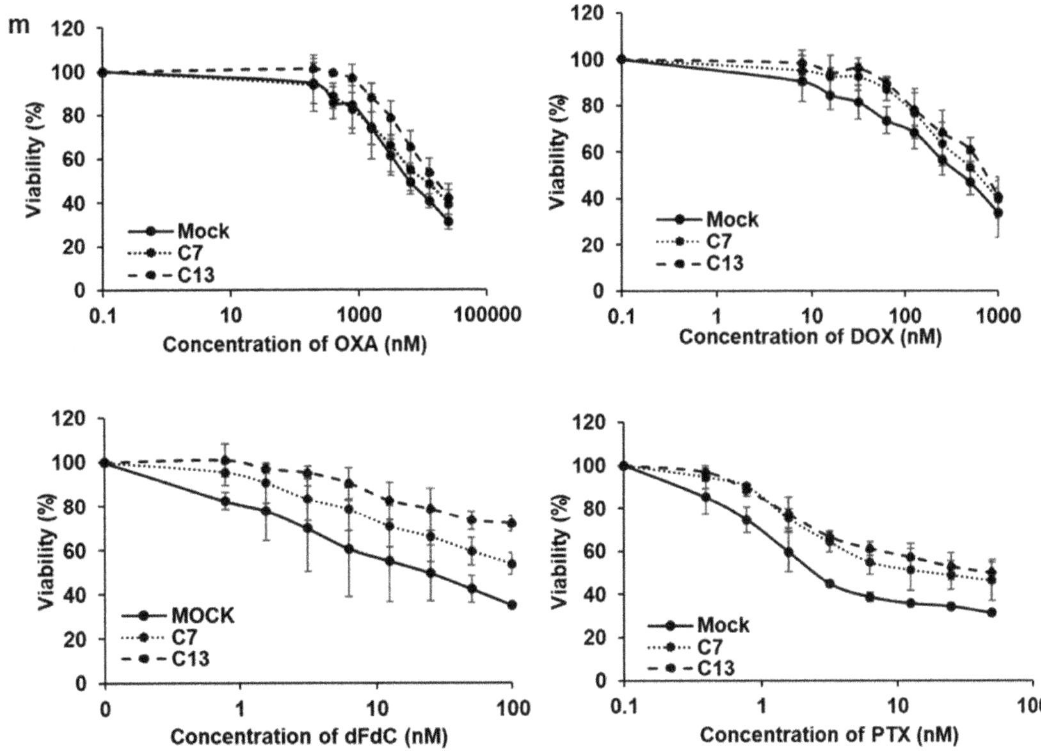

Figure 3. Stable transfection of NFκB-p65 induces the expression of CSC markers and chemoresistance in the A549 cell line. (**a**,**b**) Hypoxic culture induces NFκB-p65 mRNA expression and NFκB transcriptional activity. (**c**–**e**) qRT-PCR and Western blot show that NFκB-p65 stably transfected clones express high levels of NFκB-p65 mRNA (**c**) and protein (**d**) and demonstrate high NFκB transcriptional activity (**e**). (**f**–**k**) Flow cytometry detection of the expression status of CSC markers in the NFκB-p65 transfected clones. (**l**) qRT-PCR detection of the mRNA expression of the ESC markers in NFκB-p65 transfected clones. (**m**) MTT analysis of the cytotoxicity of four chemotherapy drugs on the NFκB-p65 transfected clones (mean ± SD; $n = 3$). *ns = non-significant*; * $p < 0.05$, ** $p < 0.01$. Western blot original images are in the supplementary materials.

Table 3. Synergistic effect of chemotherapy drugs and DS/Cu on NSCLC cell lines.

Drug	A549 Hypoxia					H23 Hypoxia				
	IC_{50}		CI Values			IC_{50}		CI Values		
	Sing	*Com*	ED50	ED75	ED90	*Sing*	*Com*	ED50	ED75	ED90
DOX (nM)	>100	*3.4 ± 2.1*	0.92	0.86	0.81	>100	*10.6 ± 5.2*	0.81	0.79	0.77
dFdC (nM)	59.5 ± *29.1*	*1.8 ± 0.5*	0.66	0.34	0.18	>100	*7.0 ± 2.1*	0.64	0.52	0.48
OXA (μM)	>10	*1.2 ± 11.1*	0.93	0.98	1.02	>10	*0.8 ± 0.3*	0.84	0.83	0.83
PTX (nM)	8.9 ± *3.0*	*4.5 ± 0.7*	0.95	0.69	0.78	>100	*6.7 ± 1.5*	0.80	0.63	0.58

Sing: Single drug; Com: Drug in combination with DS/Cu.

3.5. DS-PLGA/Cu Targets Cancer Xenografts in a Metastatic NSCLC Mouse Model

To develop an injectable DS, we invented a PLGA-encapsulated DS (DS-PLGA) [23]. Our previous studies show that DS-PLGA significantly improved the half-life of DS in serum [23]. In combination with oral CuGlu, intravenous administration of DS-PLGA demonstrated significant anticancer efficacy in liver cancer and glioblastoma mouse models [8,23]. In this study, the same DS-PLGA formulation was used to assess the anti-NSCLC efficacy in a metastatic A549 NSCLC mouse model. Two mice in the control group (n = 7) died on days 15 and 18. All the mice in the treated group (n = 7) tolerated the treatment very well and survived to the end of the experiment (day 21). No body weight loss was observed in DS-PLGA/CuGlu-treated mice. The lung macro morphologies in both groups are shown in Figure 5a. Tumor nodules were identified in the lung of the control mice, especially around the edge (arrows in Figure 5a). Due to the cancer cell infiltration, the average lung weight in the control mice is significantly heavier than that in the treated mice (Figure 5b). In comparison with the treated mice, massive cancer cell infiltration (arrows) was observed in the lung, especially around the edge, of the control mice (Figure 5c). Figure 5d shows the typical histological images of the lung in the control and treated groups. Cancer cells were observed in the lung of the control group (arrows).

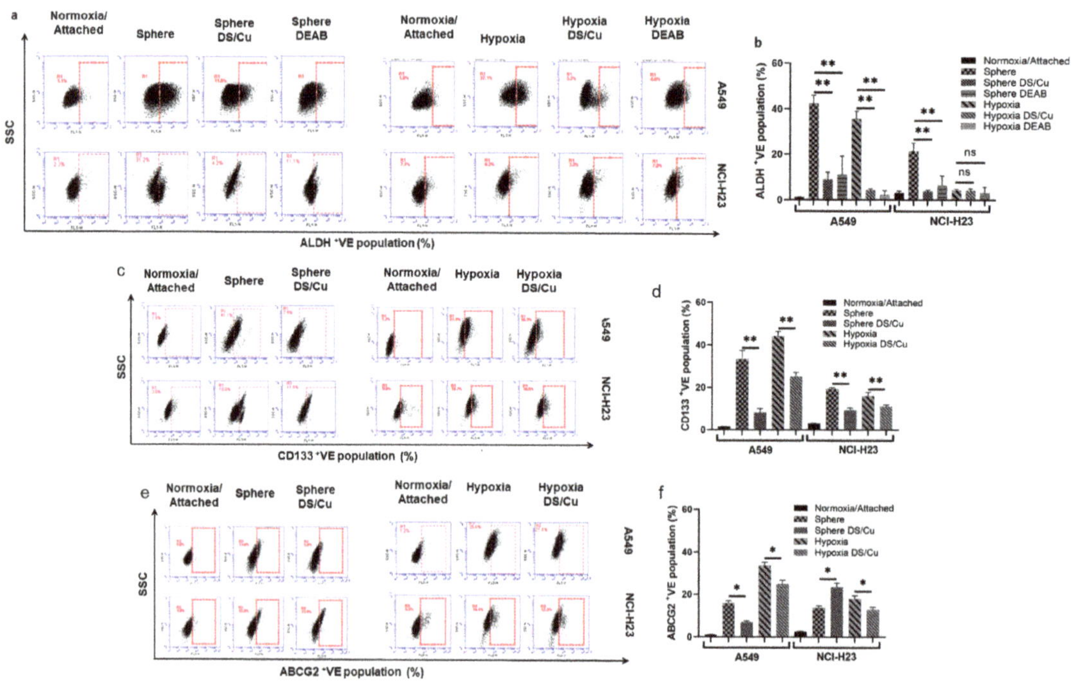

Figure 4. *Cont.*

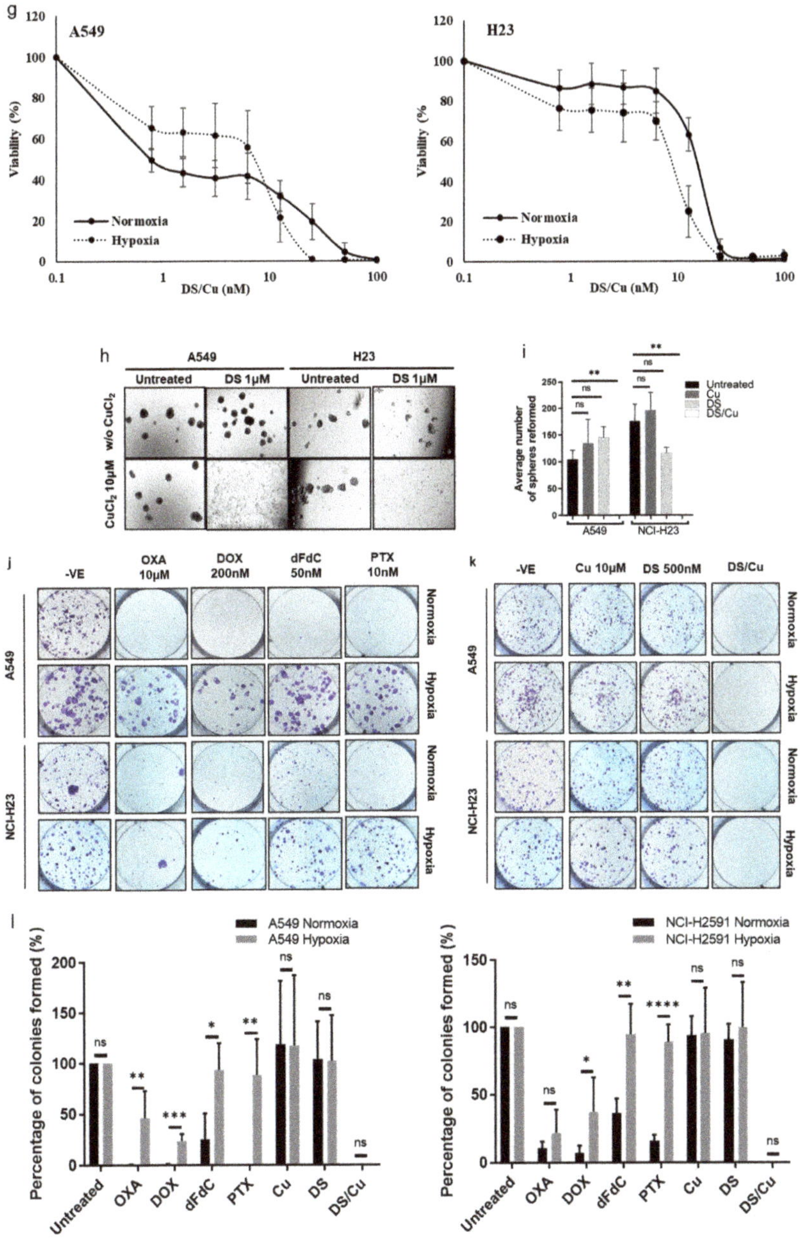

Figure 4. DS/Cu targets the CSC population and reverses chemoresistance in vitro. (**a–f**) DS/Cu inhibits the expression of CSC markers in sphere and hypoxia-cultured cells (mean ± SD; n = 3). (**g**) MTT analysis of DS/Cu on normoxia and hypoxia-cultured cell lines. (**h**,**i**) DS/Cu blocks sphere reformation ability in NSCLC cell lines (mean ± SD; n = 3). (**j–l**) Clonogenic assay: The chemotherapy drugs inhibit clonogenicity in normoxia-cultured cells but not in hypoxia-cultured cells (**j**). The clonogenicity in both cell lines was blocked by DS/Cu in both normoxic and hypoxic conditions in a copper-dependent manner (**k**). The colony numbers were counted and analyzed (mean ± SD; n = 3) (**l**). *ns = non-significant*; * $p < 0.05$; ** $p < 0.01$; *** $p < 0.001$; **** $p < 0.0001$.

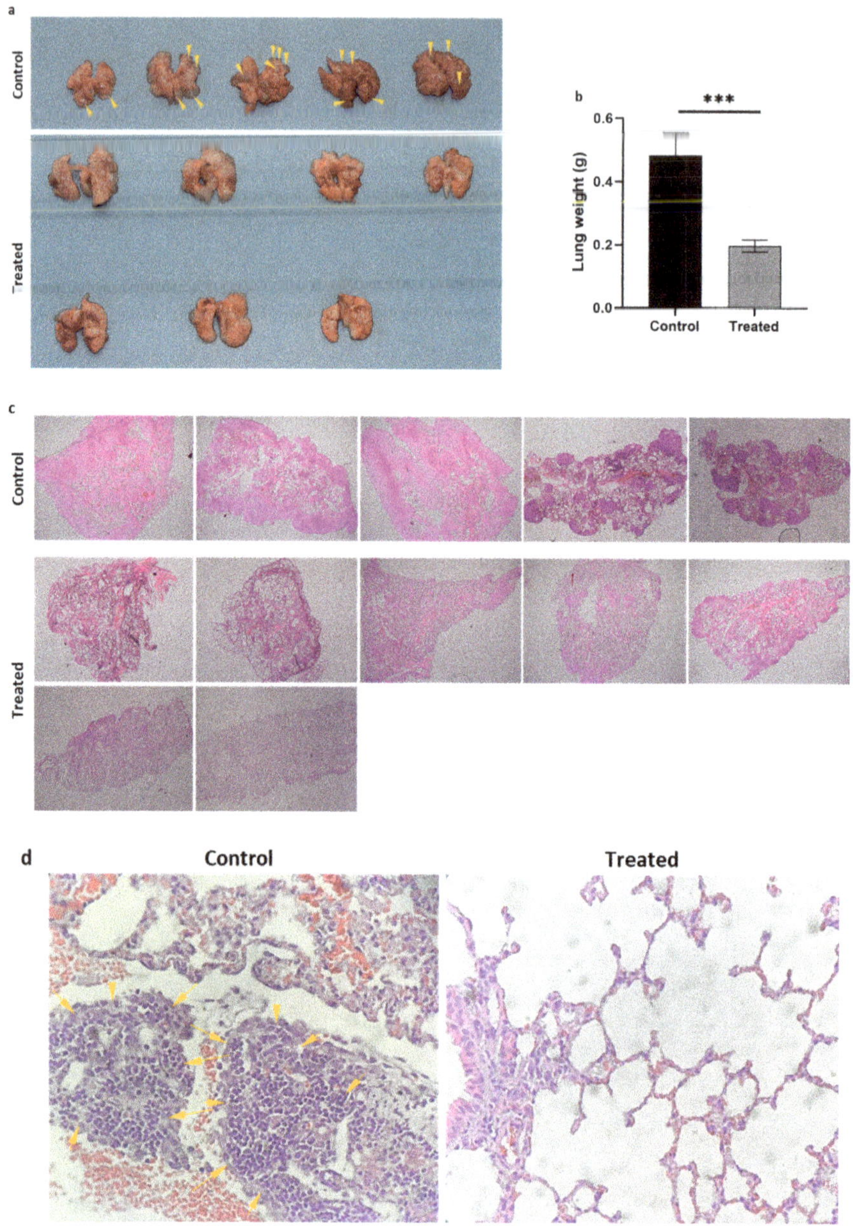

Figure 5. Anti-NSCLC efficacy of DS-PLGA/Cu in mouse NSCLC metastatic xenograft model. (**a**) The macro-morphology of the lungs from control and treated mice. Yellow arrows show tumor nodules. (**b**) The weight of the lungs from control and treated groups. (**c**) The micro-morphology shows multiple tumor nodules in the lung sections of the control mice. No or very few nodules were detected in the sections of the lungs from the treated group (×4 magnification). (**d**) Typical pathological images of the H&E stained lung sections showing cancer cell clusters (yellow arrows) in the lungs of control mice (×100 magnification). *** $p < 0.001$.

4. Discussion

In combination with immunotherapy and targeted therapy, cytotoxic chemotherapy is currently still the first-line therapeutic arm for advanced/metastatic NSCLC patients [31]. Neoadjuvant or adjuvant chemotherapy often achieves good initial therapeutic outcomes. Due to the acquired chemoresistance, most NSCLC patients will relapse [32]. NSCLC is highly heterogenous containing a very small population of CSCs with increased capacity for invasion, metastasis and chemoresistance [8,17,18,33–35]. The first identification of human lung CSCs was reported 40 years ago [36]. Due to the genotypic and histological varieties of lung cancer, lung CSCs have been studied much less than in other types of cancer [34]. In this study, we investigated the chemosensitivity of NSCLC CSCs to the four most used chemotherapeutic agents (PTX, DOX, dFdC and OXA). A classical *in vitro* spheroid suspension CSC culture from two NSCLC cell lines was examined (Figure 1a). In comparison with the attached culture counterparts, the sphere-cultured cells express high levels of CSC markers (CD133, ALDH and ABCG2) (Figure 1b–g). Significantly higher mRNA expression of the embryonic stem cell markers (SOX2, OCT4 and NANOG) was also detected in the sphere-cultured cells (Figure 1h). These results are in line with previous reports [37–40] indicating that NSCLC CSCs were enriched by spheroid culture. Furthermore, the anti-CSC activity of the four anti-NSCLC chemotherapy drugs was examined using a sphere reformation assay (Figure 1i,j). No inhibition of sphere reformation was observed in both cell lines after treatment with DOX and dFdC as well as the PTX-treated A549 cell line. Although the sphere number was significantly reduced after OXA treatment in both cell lines and PTX in the NCI-H23 cell line, a considerable number of spheres was still reformed in all treated cells. This result is consistent with our previous data derived from other types of cancers [8,16–18,22,23] indicating that the chemoresistant NSCLC CSC population is the source of cancer recurrence.

It is widely accepted that both CSCs and somatic stem cells are harboured in a specific microenvironment named stem cell niches [41]. Unlike somatic stem cells of which the differentiation is irreversible, CSC is a dynamically reversible state rather than an entity [8,42]. The molecular and cellular characters of CSCs are modified by microenvironmental factors and the lack of blood vasculature in rapidly expanded solid cancer mass results in the formation of a hypoxic CSC niche [43] which is identified as a hallmark of the tumor microenvironment (TME) [44]. The hypoxic niche plays a pivotal role in the induction and maintenance of the status of CSCs [44]. In line with our previous studies on other types of cancer [8,18], sphere-cultured NSCLC cells showed a hypoxic core region with a high population of hypoxic cells detected (Figure 1k–m). This finding indicates that hypoxia may also play a key role in the maintenance of stemness in NSCLC CSCs. Hypoxia-monolayer-cultured cells expressed high levels of CSC markers and embryonic stem cell markers (Figure 2d–j). In contrast, ALDH and SOX2 were not induced by hypoxic culture. This further indicates that no single marker is completely reliable for NSCLC CSC detection [37]. Consistent with sphere reformation assay results, the hypoxia-cultured cells demonstrated strong resistance to all four chemotherapy drugs (Table 1 and Figure 2k) and showed significantly higher invasive activity (Figure 4i,k).

The activation of the NF-κB pathway plays a central role in the development, survival and proliferation of CSCs in many types of cancers [8,16,18,22,23]. Figure 3a and b show that both sphere and hypoxia-cultured NSCLC cell lines expressed high levels of NF-κB-p65 mRNA. To examine the effect of the NF-κB pathway on NSCLC CSC traits, A549 cells were stably transfected with NF-κB-p65. The transfected clones showed high expression of NF-κB-p65 mRNA, protein expression and NF-κB activity (Figure 3c–e). These clones expressed high levels of ALDH, CD133 and embryonic stem cell markers (Figure 3f–l). The transfected clones are also significantly resistant to all four tested chemotherapy drugs (Table 3 and Figure 3m). These results indicate that the NF-κB pathway plays a key role in the maintenance of stemness in NSCLC CSCs.

DS is a medicine that has been used in alcoholism treatment for over 70 years [45], and it shows very strong anticancer potential [46,47]. Our previous studies indicate that it is a strong NF-κB pathway inhibitor and anti-CSC agent in multiple types of cancer [8,16–18,22,23,27,48,49]. In this study, we demonstrated that DS reverses hypoxia-induced stemness in NSCLC cell lines (Figure 4a–f). DS completely blocked the sphere reformation ability in both cell lines (Figure 4g,h). Hypoxia induced significantly higher clonogenic activity in both cell lines which could not be reversed by the clinically used chemotherapeutic drugs (Figure 4i,k). In contrast, the sphere reformation and hypoxia-induced clonogenicity were completely blocked by DS/Cu (Figure 4g–k). In line with our previous reports [26,27], the effect of DS on NSCLC sphere reformation and clonogenicity is copper-dependent. In combination with copper, DS reverses hypoxia-induced chemoresistance to four tested drugs (Figure 4l). Isobologram indicates a synergistic effect between DS/Cu and the anticancer drugs (Table 3). These results are highly consistent with our previous studies in other types of cancer [8,16–18,22,23,27,48,49].

Although DS shows strong cytotoxicity *in vitro*, the clinical application of DS in cancer therapeutics has been limited by its very short half-life in the bloodstream [23,50]. To overcome this bottleneck, our team developed a PLGA-nanoparticle encapsulated DS to protect DS from degradation in the bloodstream which shows very optimistic efficacy in mouse GBM and liver cancer models [8,23]. In this study, we examined the efficacy of DS-PLGA in a metastatic NSCLC mouse model. After intravenous injection of A549 cells, multiple cancer nodules were observed in the lungs of the control mice, especially on the edge of the lung (Figure 5a). The micrographic images show solid infiltration of cancer cells in the lungs of the control mice (Figure 5c,d). This observation was further supported by the heavier lung weight of the control mice (Figure 5b). In comparison with the control group, very few cancer nodules were observed in the lungs of the DS-PLGA/Cu treated mice (Figure 5a,c,d). In line with our previous studies [8,23], all the mice in the treated group tolerated the treatment very well and no body weight loss was observed.

5. Conclusions

Our findings revealed that hypoxia is the key regulator of the stemness in NSCLC cells. The hypoxia-induced NSCLC CSCs are highly resistant to the clinically used anti-NSCLC chemotherapy drugs. Disulfiram, an anti-alcoholism medicine, demonstrated strong anti-CSC activity and reversed hypoxia-induced chemoresistance. The promising *in vivo* anticancer efficacy indicates that further investigation might translate DS-PLGA into NSCLC clinical application.

Supplementary Materials: The following are available online at https://www.mdpi.com/article/10.3390/biom14121651/s1.

Author Contributions: Conceptualization, W.W.; Funding Acquisition, W.W. and C.M.; Project organisation, W.W., V.K.; Consultancy, C.M.; Methodology, V.K., K.B., Z.W., S.K., Z.Z.; Investigation, V.K., K.B., Z.W., M.N. and M.R.M.; Writing: Original Draft Preparation, W.W., V.K., K.B.; Writing Review and Editing, M.R.M., K.B.; Supervision, W.W., M.R.M. and V.K. All authors have read and agreed to the published version of the manuscript.

Funding: The grant (RG14-8) was awarded to WW and CM by British Lung Foundation. VK was the Postdoctoral Fellow supported by the grant.

Institutional Review Board Statement: The animal study protocol was approved and registered by the Animal Study Advisory Group, Ethical Committee of the Fourth Military Medical University, China (No. 010611-2013FMMU) and performed under the project license of corresponding author WW (Approval code: SP2001-206; Approval date: 2013.12.9)..

Informed Consent Statement: Not applicable.

Data Availability Statement: Data are available upon request.

Conflicts of Interest: WW is a director and academic shareholder of Disulfican Ltd. VK is an employee and academic shareholder of Disulfican Ltd. KB is an employee of Disulfican Ltd. Disulfican Ltd is a university of Wolverhampton spin out company for commercial translation of PLGA Disulfiram for mesothelioma. The authors did not receive any funding support from the company for this work on

NSCLC as this work was completed before the company was established. All other authors declare no conflicts of interest.

References

1. World Cancer Research Fund International. Lung Cancer Statistics. Available online: https://www.wcrf.org/cancer-trends/lung-cancer-statistics/ (accessed on 16 December 2024).
2. Kratzer, T.B.; Bandi, P.; Freedman, N.D.; Smith, R.A.; Travis, W.D.; Jemal, A.; Siegel, R.L. Lung cancer statistics, 2023. *Cancer* **2024**, *130*, 1330–1348. [CrossRef] [PubMed]
3. Provencio, M.; Carcereny, E.; Rodriguez-Abreu, D.; Lopez-Castro, R.; Guirado, M.; Camps, C.; Bosch-Barrera, J.; Garcia-Campelo, R.; Ortega-Granados, A.L.; Gonzalez-Larriba, J.L.; et al. Lung cancer in Spain: Information from the Thoracic Tumors Registry (TTR study). *Transl. Lung Cancer Res.* **2019**, *8*, 461–475. [CrossRef] [PubMed]
4. Makena, M.R.; Ranjan, A.; Thirumala, V.; Reddy, A.P. Cancer stem cells: Road to therapeutic resistance and strategies to overcome resistance. *Biochim. Biophys. Acta Mol. Basis Dis.* **2020**, *1866*, 165339. [CrossRef]
5. Raniszewska, A.; Kwiecien, I.; Rutkowska, E.; Rzepecki, P.; Domagala-Kulawik, J. Lung Cancer Stem Cells-Origin, Diagnostic Techniques and Perspective for Therapies. *Cancers* **2021**, *13*, 2996. [CrossRef]
6. Keith, B.; Johnson, R.S.; Simon, M.C. HIF1alpha and HIF2alpha: Sibling rivalry in hypoxic tumour growth and progression. *Nat. Rev. Cancer* **2011**, *12*, 9–22. [CrossRef]
7. Zipori, D. The nature of stem cells: State rather than entity. *Nat. Rev. Genet.* **2004**, *5*, 873–878. [CrossRef]
8. Kannappan, V.; Liu, Y.; Wang, Z.; Azar, K.; Kurusamy, S.; Kilari, R.S.; Armesilla, A.L.; Morris, M.R.; Najlah, M.; Liu, P.; et al. PLGA-Nano-Encapsulated Disulfiram Inhibits Hypoxia-Induced NF-kappaB, Cancer Stem Cells, and Targets Glioblastoma In Vitro and In Vivo. *Mol. Cancer Ther.* **2022**, *21*, 1273–1284. [CrossRef] [PubMed]
9. Das, B.; Tsuchida, R.; Malkin, D.; Koren, G.; Baruchel, S.; Yeger, H. Hypoxia enhances tumor stemness by increasing the invasive and tumorigenic side population fraction. *Stem Cells* **2008**, *26*, 1818–1830. [CrossRef]
10. Hayden, M.S.; Ghosh, S. Shared principles in NF-kappaB signaling. *Cell* **2008**, *132*, 344–362. [CrossRef] [PubMed]
11. Oeckinghaus, A.; Hayden, M.S.; Ghosh, S. Crosstalk in NF-kappaB signaling pathways. *Nat. Immunol.* **2011**, *12*, 695–708. [CrossRef] [PubMed]
12. Taylor, C.T.; Cummins, E.P. The role of NF-kappaB in hypoxia-induced gene expression. *Ann. N. Y. Acad. Sci.* **2009**, *1177*, 178–184. [CrossRef] [PubMed]
13. Vazquez-Santillan, K.; Melendez-Zajgla, J.; Jimenez-Hernandez, L.; Martinez-Ruiz, G.; Maldonado, V. NF-kappaB signaling in cancer stem cells: A promising therapeutic target? *Cell. Oncol.* **2015**, *38*, 327–339. [CrossRef] [PubMed]
14. Prasad, S.; Ravindran, J.; Aggarwal, B.B. NF-kappaB and cancer: How intimate is this relationship. *Mol. Cell Biochem.* **2010**, *336*, 25–37. [CrossRef] [PubMed]
15. Wang, W.; Cassidy, J.; O'Brien, V.; Ryan, K.M.; Collie-Duguid, E. Mechanistic and predictive profiling of 5-Fluorouracil resistance in human cancer cells. *Cancer Res.* **2004**, *64*, 8167–8176. [CrossRef] [PubMed]
16. Liu, P.; Brown, S.; Goktug, T.; Channathodiyil, P.; Kannappan, V.; Hugnot, J.P.; Guichet, P.O.; Bian, X.; Armesilla, A.L.; Darling, J.L.; et al. Cytotoxic effect of disulfiram/copper on human glioblastoma cell lines and ALDH-positive cancer-stem-like cells. *Br. J. Cancer* **2012**, *107*, 1488–1497. [CrossRef] [PubMed]
17. Liu, P.; Kumar, I.S.; Brown, S.; Kannappan, V.; Tawari, P.E.; Tang, J.Z.; Jiang, W.; Armesilla, A.L.; Darling, J.L.; Wang, W. Disulfiram targets cancer stem-like cells and reverses resistance and cross-resistance in acquired paclitaxel-resistant triple-negative breast cancer cells. *Br. J. Cancer* **2013**, *109*, 1876–1885. [CrossRef] [PubMed]
18. Liu, P.; Wang, Z.; Brown, S.; Kannappan, V.; Tawari, P.E.; Jiang, J.; Irache, J.M.; Tang, J.Z.; Armesilla, A.L.; Darling, J.L.; et al. Liposome encapsulated Disulfiram inhibits NFκB pathway and targets breast cancer stem cells in vitro and in vivo. *Oncotarget* **2014**, *5*, 7471–7485. [CrossRef]
19. Jentzsch, V.; Osipenko, L.; Scannell, J.W.; Hickman, J.A. Costs and Causes of Oncology Drug Attrition With the Example of Insulin-Like Growth Factor-1 Receptor Inhibitors. *JAMA Netw. Open* **2023**, *6*, e2324977. [CrossRef]
20. Pushpakom, S.; Iorio, F.; Eyers, P.A.; Escott, K.J.; Hopper, S.; Wells, A.; Doig, A.; Guilliams, T.; Latimer, J.; McNamee, C.; et al. Drug repurposing: Progress, challenges and recommendations. *Nat. Rev. Drug Discov.* **2019**, *18*, 41–58. [CrossRef]
21. Suh, J.J.; Pettinati, H.M.; Kampman, K.M.; O'Brien, C.P. The status of disulfiram: A half of a century later. *J. Clin. Psychopharmacol.* **2006**, *26*, 290–302. [CrossRef]
22. Yip, N.C.; Fombon, I.S.; Liu, P.; Brown, S.; Kannappan, V.; Armesilla, A.L.; Xu, B.; Cassidy, J.; Darling, J.L.; Wang, W. Disulfiram modulated ROS-MAPK and NFkB pathways and targeted breast cancer cells with cancer stem cell like properties. *Br. J. Cancer* **2011**, *104*, 1564–1574. [CrossRef] [PubMed]
23. Wang, Z.; Tan, J.; McConville, C.; Kannappan, V.; Tawari, P.E.; Brown, J.; Ding, J.; Armesilla, A.L.; Irache, J.M.; Mei, Q.B.; et al. Poly lactic-co-glycolic acid controlled delivery of disulfiram to target liver cancer stem-like cells. *Nanomedicine* **2017**, *13*, 641–657. [CrossRef] [PubMed]
24. Chen, D.; Cui, Q.C.; Yang, H.; Dou, Q.P. Disulfiram, a clinically used anti-alcoholism drug and copper-binding agent, induces apoptotic cell death in breast cancer cultures and xenografts via inhibition of the proteasome activity. *Cancer Res.* **2006**, *66*, 10425–10433. [CrossRef]

25. Skrott, Z.; Mistrik, M.; Andersen, K.K.; Friis, S.; Majera, D.; Gursky, J.; Ozdian, T.; Bartkova, J.; Turi, Z.; Moudry, P.; et al. Alcohol-abuse drug disulfiram targets cancer via p97 segregase adaptor NPL4. *Nature* **2017**, *552*, 194–199. [CrossRef]
26. Tawari, P.E.; Wang, Z.; Najlah, M.; Tsang, C.W.; Kannappan, V.; Liu, P.; McConville, C.; He, B.; Armesilla, A.L.; Wang, W. The cytotoxic mechanisms of disulfiram and copper(II) in cancer cells. *Toxicol. Res.* **2015**, *4*, 1439–1442. [CrossRef]
27. Butcher, K.; Kannappan, V.; Kilari, R.S.; Morris, M.R.; McConville, C.; Armesilla, A.L.; Wang, W. Investigation of the key chemical structures involved in the anticancer activity of disulfiram in A549 non-small cell lung cancer cell line. *BMC Cancer* **2018**, *18*, 753. [CrossRef] [PubMed]
28. Wang, W. Disulfiram Formulation and Uses Thereof. WO 2010/076897 A1, 2010.
29. Plumb, J.A.; Milroy, R.; Kaye, S.B. Effects of the pH dependence of 3-(4,5-dimethylthiazol-2-yl)-2,5-diphenyl-tetrazolium bromide-formazan absorption on chemosensitivity determined by a novel tetrazolium-based assay. *Cancer Res.* **1989**, *49*, 4435–4440.
30. Chou, T.C.; Talalay, P. Quantitative analysis of dose-effect relationships: The combined effects of multiple drugs or enzyme inhibitors. *Adv. Enzym. Regul.* **1984**, *22*, 27–55. [CrossRef] [PubMed]
31. Sridhar, A.; Khan, H.; Yohannan, B.; Chan, K.H.; Kataria, N.; Jafri, S.H. A Review of the Current Approach and Treatment Landscape for Stage III Non-Small Cell Lung Cancer. *J. Clin. Med.* **2024**, *13*, 2633. [CrossRef]
32. Kim, E.S. Chemotherapy Resistance in Lung Cancer. *Adv. Exp. Med. Biol.* **2016**, *893*, 189–209. [CrossRef]
33. Reya, T.; Morrison, S.J.; Clarke, M.F.; Weissman, I.L. Stem cells, cancer, and cancer stem cells. *Nature* **2001**, *414*, 105–111. [CrossRef]
34. Sullivan, J.P.; Minna, J.D.; Shay, J.W. Evidence for self-renewing lung cancer stem cells and their implications in tumor initiation, progression, and targeted therapy. *Cancer Metastasis Rev.* **2010**, *29*, 61–72. [CrossRef]
35. Bajaj, J.; Diaz, E.; Reya, T. Stem cells in cancer initiation and progression. *J. Cell Biol.* **2020**, *219*, e201911053. [CrossRef] [PubMed]
36. Carney, D.N.; Gazdar, A.F.; Bunn, P.A., Jr.; Guccion, J.G. Demonstration of the stem cell nature of clonogenic tumor cells from lung cancer patients. *Stem Cells* **1982**, *1*, 149–164. [PubMed]
37. Leung, E.L.; Fiscus, R.R.; Tung, J.W.; Tin, V.P.; Cheng, L.C.; Sihoe, A.D.; Fink, L.M.; Ma, Y.; Wong, M.P. Non-small cell lung cancer cells expressing CD44 are enriched for stem cell-like properties. *PLoS ONE* **2010**, *5*, e14062. [CrossRef]
38. Ho, M.M.; Ng, A.V.; Lam, S.; Hung, J.Y. Side population in human lung cancer cell lines and tumors is enriched with stem-like cancer cells. *Cancer Res.* **2007**, *67*, 4827–4833. [CrossRef] [PubMed]
39. Eramo, A.; Lotti, F.; Sette, G.; Pilozzi, E.; Biffoni, M.; Di Virgilio, A.; Conticello, C.; Ruco, L.; Peschle, C.; De Maria, R. Identification and expansion of the tumorigenic lung cancer stem cell population. *Cell Death Differ.* **2008**, *15*, 504–514. [CrossRef]
40. Liu, Y.P.; Yang, C.J.; Huang, M.S.; Yeh, C.T.; Wu, A.T.; Lee, Y.C.; Lai, T.C.; Lee, C.H.; Hsiao, Y.W.; Lu, J.; et al. Cisplatin selects for multidrug-resistant CD133+ cells in lung adenocarcinoma by activating Notch signaling. *Cancer Res.* **2013**, *73*, 406–416. [CrossRef]
41. Liu, Q.; Guo, Z.; Li, G.; Zhang, Y.; Liu, X.; Li, B.; Wang, J.; Li, X. Cancer stem cells and their niche in cancer progression and therapy. *Cancer Cell Int.* **2023**, *23*, 305. [CrossRef]
42. Rossi, F.; Noren, H.; Jove, R.; Beljanski, V.; Grinnemo, K.H. Differences and similarities between cancer and somatic stem cells: Therapeutic implications. *Stem Cell Res. Ther.* **2020**, *11*, 489. [CrossRef] [PubMed]
43. Maciaczyk, D.; Picard, D.; Zhao, L.; Koch, K.; Herrera-Rios, D.; Li, G.; Marquardt, V.; Pauck, D.; Hoerbelt, T.; Zhang, W.; et al. CBF1 is clinically prognostic and serves as a target to block cellular invasion and chemoresistance of EMT-like glioblastoma cells. *Br. J. Cancer* **2017**, *117*, 102–112. [CrossRef]
44. Kim, H.; Lin, Q.; Glazer, P.M.; Yun, Z. The hypoxic tumor microenvironment in vivo selects the cancer stem cell fate of breast cancer cells. *Breast Cancer Res.* **2018**, *20*, 16. [CrossRef] [PubMed]
45. Johansson, B. A review of the pharmacokinetics and pharmacodynamics of disulfiram and its metabolites. *Acta Psychiatr. Scand. Suppl.* **1992**, *369*, 15–26. [CrossRef]
46. Kannappan, V.; Ali, M.; Small, B.; Rajendran, G.; Elzhenni, S.; Taj, H.; Wang, W.; Dou, Q.P. Recent Advances in Repurposing Disulfiram and Disulfiram Derivatives as Copper-Dependent Anticancer Agents. *Front. Mol. Biosci.* **2021**, *8*, 741316. [CrossRef] [PubMed]
47. Lu, C.; Li, X.; Ren, Y.; Zhang, X. Disulfiram: A novel repurposed drug for cancer therapy. *Cancer Chemother. Pharmacol.* **2021**, *87*, 159–172. [CrossRef]
48. McConville, C.; Tawari, P.; Wang, W. Hot melt extruded and injection moulded disulfiram-loaded PLGA millirods for the treatment of glioblastoma multiforme via stereotactic injection. *Int. J. Pharm.* **2015**, *494*, 73–82. [CrossRef]
49. Wang, W.; McLeod, H.L.; Cassidy, J. Disulfiram-mediated inhibition of NF-kappaB activity enhances cytotoxicity of 5-fluorouracil in human colorectal cancer cell lines. *Int. J. Cancer* **2003**, *104*, 504–511. [CrossRef]
50. Eneanya, D.I.; Bianchine, J.R.; Duran, D.O.; Andresen, B.D. The actions of metabolic fate of disulfiram. *Annu. Rev. Pharmacol. Toxicol.* **1981**, *21*, 575–596. [CrossRef] [PubMed]

biomolecules

MDPI

Article

Chaga Mushroom Triterpenoids Inhibit Dihydrofolate Reductase and Act Synergistically with Conventional Therapies in Breast Cancer

Junbiao Wang [1,*,†], Daniela Beghelli [1,†], Augusto Amici [1], Stefania Sut [2], Stefano Dall'Acqua [3], Giulio Lupidi [4], Diego Dal Ben [5], Onelia Bistoni [6], Daniele Tomassoni [1], Barbara Belletti [7], Sanaa Musa [8,9], Jamal Mahajna [8,10], Stefania Pucciarelli [1,*,†] and Cristina Marchini [1,†]

1 School of Biosciences and Veterinary Medicine, Via Gentile III da Varano, University of Camerino, 62032 Camerino, Italy; daniela.beghelli@unicam.it (D.B.); augusto.amici@unicam.it (A.A.); daniele.tomassoni@unicam.it (D.T.); cristina.marchini@unicam.it (C.M.)

2 DAFNAE Dipartimento di Agronomia, Animali, Alimenti, Risorse Naturali e Ambiente, University of Padova, 35020 Legnaro, Italy; stefania.sut@unipd.it

3 DSF Department of Pharmaceutical and Pharmacological Sciences, University of Padova, 35121 Padova, Italy; stefano.dallacqua@unipd.it

4 School of Pharmacy, University of Camerino, 62032 Camerino, Italy; giulio.lupidi@unicam.it

5 School of Pharmacy-Chemistry Interdisciplinary Project (CHIP), University of Camerino, 62032 Camerino, Italy; diego.dalben@unicam.it

6 Rheumatology Unit, Department of Medicine and Surgery, University of Perugia, 06123 Perugia, Italy; onelia.bistoni@ospedale.perugia.it

7 Molecular Oncology Unit, Centro di Riferimento Oncologico di Aviano (CRO), IRCCS, National Cancer Institute, 33081 Aviano, Italy; bbelletti@cro.it

8 Natural Compounds and Organic Synthesis, Migal-Galilee Research Institute, Kiryat Shmona 11016, Israel; sanaa@migal.org.il (S.M.); jamalm@migal.org.il (J.M.)

9 Department of Biotechnology, Tel Hai College, Kiryat Shmona 1220800, Israel

10 Cancer Drug Discovery Program, Migal, Galilee Research Institute, P.O. Box 831, Kiryat Shmona 11016, Israel

* Correspondence: junbiao.wang@unicam.it (J.W.); stefania.pucciarelli@unicam.it (S.P.)

† These authors contributed equally to this work.

Abstract: *Inonotus obliquus* (Chaga) is a medicinal mushroom with several pharmacological properties that is used as a tea in traditional Chinese medicine. In this study, Chaga water extract was digested in vitro to mimic the natural processing and absorption of its biocomponents when it is consumed as functional beverage, and its anticancer activities were evaluated in breast cancer (BC) cell lines, representing HER2-positive and triple-negative subtypes. After chemical characterization by liquid chromatography/mass spectrometry (HR-QTOF) analysis, the effect of Chaga biocomponents on cell viability and cell cycle progression was assessed by MTT assay, FACS analysis, and Western blot. Dihydrofolate reductase (DHFR) activity was measured by an enzymatic assay. Four highly bioactive triterpenoids (inotodiol, trametenolic acid, 3-hydroxy-lanosta-8,24-dien-21-al, and betulin) were identified as the main components, able to decrease BC cell viability and block the cell cycle in G0/G1 by inducing the downregulation of cyclin D1, CDK4, cyclin E, and phosphorylated retinoblastoma protein. DHFR was identified as their crucial target. Moreover, bioactive Chaga components exerted a synergistic action with cisplatin and with trastuzumab in SK-BR-3 cells by inhibiting both HER2 and HER1 activation and displayed an immunomodulatory effect. Thus, *Inonotus obliquus* represents a source of triterpenoids that are effective against aggressive BC subtypes and display properties of targeted drugs.

Keywords: Chaga mushroom (*Inonotus obliquus*); triterpenoids; breast cancer; DHFR; cell cycle regulation

1. Introduction

Breast cancer (BC) can be classified into four main molecular subtypes based on the expression of estrogen receptor (ER), progesterone receptor (PR), and human epidermal growth factor receptor 2 (HER2) [1,2]. Among them, HER2-positive and triple-negative (ER-negative, PR-negative, and HER2-negative) BCs are associated with the poorest patient survival [3]. Triple-negative BC has the worst prognosis and cannot benefit from targeted therapies, while HER2-positive BC has aggressive behavior, but it can be treated with HER2-targeted therapies such as trastuzumab. However, HER2-targeted therapies have limitations such as intrinsic or acquired drug resistance [4]. Traditional Chinese medicine (TCM) may represent a source of new anticancer compounds, potentially more effective and less toxic than conventional cancer drugs [5,6]. Chinese people have been using TCM for more than 5000 years, and a wealth of information on natural compounds used for the treatment of a variety of diseases including cancer was recorded in ancient medical books. Fungi (sporophores or fruiting bodies) are traditionally used in Chinese medicine as antitumor remedies [7], but the underlying pharmacological mechanisms are not yet completely understood. Recently, many studies have revealed promising anticancer activities of some mushrooms and their active compounds, but there are still no fungal products approved as anticancer therapeutics [8]. Thus, to better exploit their antitumor potential, it is crucial to understand the mechanisms of action of fungal extracts on cancer cells and identify the molecular targets of specific therapeutic components. *Inonotus obliquus*, commonly known as Chaga due to its irregularly formed sterile conk with a burnt charcoal-like appearance, is an edible mushroom belonging to the Hymenochaetaceae family of Basidiomycetes. The presence of Chaga mushrooms is restricted to cold habitats at latitudes of 45° N–50° N, including North America, Central and Northern Europe, Russia (West Siberia), northeast China, and Japan [9]. Chaga mushroom is a parasitic fungus that grows on the bark of various boreal deciduous angiosperms such as birch (*Betula* spp.) and beech (*Fagus* spp.), and it has been used as a folk remedy to treat various diseases such as cancer, cardiovascular diseases, diabetes, and gastrointestinal disorders since the 12th century [10]. Many bioactive constituents, including triterpenoids, polysaccharides, and polyphenols, have been identified in Chaga extracts, and their antitumor, anti-inflammatory, antioxidant, hypoglycemic, and immunomodulatory properties have been described [9,11,12]. Chaga has been reported to exert cytotoxic effects against various types of cancer, including sarcoma [13], lung adenocarcinoma [14], colon cancer [15], melanoma [16], and hepatocellular carcinoma [17]. In this study, we investigated the anti-neoplastic action and the underlying molecular mechanisms of *Inonotus obliquus* extracts against HER2-positive and triple-negative BC. The traditional preparation to consume this fungus is a functional beverage (tea) obtained by the infusion with water of pulverized fungal material to obtain a Chaga water extract. In our work, Chaga water extracts were first treated to simulate human digestion. Bioactive triterpenoids were identified as the main components. Digested Chaga extract was able to block the cell cycle in G0/G1 and exert a synergistic action when administered in combination with trastuzumab in SK-BR-3 (ER-/PR-/HER2+) cells and with cisplatin in both SK-BR-3 and MDA-MB-231 (ER-/PR-/HER2-) cells.

Dihydrofolate reductase (DHFR), a key enzyme in the amino acids, purines and nucleotide biosynthesis, was identified as a crucial target of *Inonotus obliquus* extract components. Betulinic acid (3β, hydroxyl-lup-20(29)-en-28-oic acid, BA), a pentacyclic triterpene, which is formed by the oxidation of the triterpenoid betulin, one of the major components of digested Chaga extract, was found to affect DHFR activity in both the SK-BR-3 and MDA-MB-231 cell lines.

2. Materials and Methods

2.1. Inonotus Obliquus Extract

The dried Chaga mushroom (*Inonotus obliquus*) used in this study was purchased in Changchun, Jilin province, China. To obtain an aqueous extract, 10 g of Chaga mushroom was powdered and heated in water at 65 °C for 2 and a half hours with continuous stirring, imitating the traditional preparation system. Subsequently, the solution was filtered, frozen at −20 °C, and eventually lyophilized. The next step involved the in vitro digestion of the lyophilized powder. In vitro digestion helps in understanding and predicting the behavior of food components in the gastrointestinal tract, mimicking what happens in our bodies every time we ingest food or liquids. The simulated digestion method used in this study was described by Minekus et al. and includes the oral, gastric, and small intestinal phases [18]. This method tries to mimic physiological conditions in vivo, considering the presence of digestive enzymes and their concentrations, pH, digestion time, and salt concentrations. The digested solution was then dialyzed using a membrane with a 3500 Da cut-off and placed against water for 12 h to simulate small intestinal absorption. At the end of the incubation process, two solutions were obtained: the one flowing outside dialysis tubing represented the serum-available portion, or in other words, the one absorbed (IN), while the other represented the colon-available solution, which is the non-absorbable sample (OUT). Both solutions, IN and OUT, corresponding to MW < 3500 Da and a MW > 3500 Da extracts of Chaga mushroom, respectively, were collected and lyophilized for further analysis.

2.2. Chaga Chemical Characterization

The analysis of Chaga samples was performed by combining the data from High-Resolution Quadrupole Time of Flight (HR-QTOF) analysis for qualitative approaches and two different liquid-chromatography–mass-spectrometry (LC-MS) methods: one using Electrospray Ion Source (ESI), for the quantification of hydrophilic constituents as sugar derivatives and phenolics, and one using the liquid chromatography atmospheric pressure chemical ionization (LC-APCI) method for the analysis of triterpene and inotodiol related compounds. To perform a qualitative analysis of phytocomplex by LC-QTOF and by LC-ESI-Ion trap, lyophilized samples (100 mg) were suspended in methanol and sonicated for 5 min (15 mL) and then centrifuged. Liquids were used for the analysis. LC-MS analysis was performed in different conditions, and several compounds were identified combining the QTOF-MS and ion trap multiple-stage mass spectrometry (MSn) approaches. Furthermore, ESI and APCI sources have been used to obtain the maximum opportunity to detect and identify constituents. To perform an analysis of terpenoids by LC-APCI-MS, 100 mg of powder was weighed in a flask, and 5 mL of ethyl acetate was added. Flask was sonicated for 10 min and centrifuged for 5 min. Supernatant was removed and collected separately, and the procedure was repeated three times. At the end, the supernatant was pooled, and the ethyl acetate fraction was dried under vacuum with a rotary evaporator. The residual material from the extraction was extracted again with methanol (5 mL three times), and the supernatant was dried under vacuum. Then, the residual material was transferred to a round-bottom flask and dried under vacuum to obtain a dried powder. Five milligrams of powder was weighed and diluted with 1.5 mL of water and sonicated and prepared for the analysis of polysaccharides. Yields for each step were calculated (Supplementary Table S1). Ethyl acetate extracts and methanol extracts were dissolved in methanol (1 mL) and analyzed by LC-APCI-MS. LC was performed using an Agilent 1260 chromatograph equipped with a 1260 diode array (DAD) and an Agilent/Varian MS-500 ion trap (Santa Clara, CA, USA) as detectors. An SB-Aq C18 4.6 × 50 mm 1.8 µm (Agilent, Santa Clara, CA, USA) column was used as a stationary phase and acetonitrile (A), methanol (B), and 0.1% formic acid (FA) in water (C) were used as mobile phases. The elution gradient was set as follows: from 65/1/34% A/B/C to 70/30/0% A/B/C, 0–4 min; isocratic elution for 4–13 min; re-equilibration with the initial solvent ratio for 3 min. The flow rate was 0.6 mL/min, and the injection volume was 10 µL. At the end of the column, a T connector split the flow rate to the DAD and MS detector. MS spectra were recorded in negative ion mode in 50–2000 Da range using an APCI ion source. The turbo data-dependent scanning (TDDS) function allowed us to obtain the fragmentation of the main ionic species. The identification of compounds was based on the fragmentation spectra, as well as the comparison of the fragmentation pattern with the literature and the injection of reference compounds, when available. The DAD chromatograms were monitored at λ = 350, 330, 280, and 254 nm. MS parameters: corona current 5 Amp, neb gas pressure 45 psi, drying gas pressure 15 psi, vaporizing gas pressure 20 psi, drying gas temperature from 320 to 285 °C in 10 min, vaporizing temperature 350, RF loading 81%, capillary 95, and positive ion mode. To establish the presence and the distribution of molecular weight of polysaccharides, we used LC Size-Exclusion Chromatography (SEC) analysis. Water extracts were filtered, and for the analysis, an Agilent 1100 equipped with an Evaporative Light-Scattering Detector (Sedex LX60) was used. As the stationary phase, a Tosohas G3000 was used, and water 0.1% formic acid was used as the mobile phase. This method allows the separation of polymers, and the standards used were dextran 270 KDa, 12 KDa, and 1 KDa. The water extract of each sample was analyzed. All the samples presented a complex chemical composition with compounds with MW < 1 KDa; thus, in the extraction condition, there was poor or no extraction of large-molecular-weight carbohydrates.

2.3. Cell Cultures

Human MDA-MB-231, SK-BR-3, CCD 841 CoN (ATCC CRL-1790), and HEK-293 cells were cultured in Dulbecco's Modified Essential Medium (DMEM, CORNING, Mediatech, New York, NY, USA) supplemented with 10% fetal bovine serum (FBS, Gibco, Life Technologies, Carlsbad, CA, USA) and 1% penicillin–streptomycin (Gibco, Life Technologies). MCF10A cells were cultured in mammary epithelial cell growth medium (PromoCell, Heidelberg, Germany) and 1% penicillin–streptomycin. Cells were cultured at 37 °C under humidified atmosphere with 5% CO_2. The HEK-293 cell line comprises immortalized human embryonic kidney cells, and they were used as a non-cancerous control cell line, as well as the MCF10A cells, which are a non-tumorigenic mammary epithelial cell line. CCD 841 CoN cells were isolated from normal human colon tissue, and they can be considered as normal colon epithelial cells. Cell lines were kindly provided by the laboratory of Dr. B. Belletti (Division of Molecular Oncology, CRO of Aviano, IRCCS, National Cancer Institute, Aviano, Italy) and tested for mycoplasma contamination with negative results.

2.4. Cell Viability Assay

Cell viability was evaluated by seeding MDA-MB-231 cells (7000 cells/well) or SK-BR-3 cells (1×10^4 cells/well), or HEK-293 cells (1×10^4 cells/well) in 96-well plates using complete medium (DMEM supplemented with 10% FBS and 1% penicillin–streptomycin). The day after, fresh medium containing appropriate concentrations of Chaga extract (non-digested or digested), ranging from 0.1 to 5 mg/mL, was added; to evaluate synergistic effects, Chaga extract was administered in combination with trastuzumab (Herceptin, Genentech, San Francisco, CA, USA) or platinum drugs (cisplatin and its derivative RJY13 [19], kindly provided by Prof. Jamal Mahajna and Prof. Sanaa Musa, Galilee Research Institute, Kiryat Shmona, Israel). Betulinic acid (Sigma Aldrich, St. Louis, MO, USA) was tested in a range from 0 to 60 μM. Cell viability was determined, after 24 h, 48 h or 72 h, using an MTT [3-(4,5-dimethylthiazol-2-yl)-2,5-diphenyl-2H-tetrazolium bromide Sigma Aldrich, St. Louis, MO] assay, which is based on the conversion of MTT to formazan by mitochondrial enzymes [20]. The formazan deposits were dissolved in DMSO, and the absorbance of each well was measured at 540 nm in Multiskan Ascent 96/384 Plate Reader. Each drug concentration was evaluated with six replicates, and the experiments were repeated three times. IC_{50} values were calculated for each of the cell lines tested by fitting the concentration–effect curve data obtained in the three experiments with the sigmoid-Emax model using nonlinear regression, weighted by the reciprocal of the square of the predicted effect.

To evaluate drug interaction, the Bliss Independence model was considered using the equation $E(x, y) = Ex + Ey - (Ex * Ey)$, where E is the fractional effect (between 0 and 1), and x and y are the doses (or concentrations) of drugs in the combination. Observed effects greater than E(x, y) indicated synergistic interactions [21].

2.5. Western Blot Analysis

Cells were homogenized in RIPA buffer (0.1% SDS, 1% NP40, 0.5% CHAPS) supplemented with protease inhibitors (Sigma-Aldrich, St. Louis, MO, USA). For Western blot analysis, an equal amount of protein lysates was separated onto Criterion™ TGX™ precast gels (Bio-Rad, Hercules, CA, USA) and transferred to a polyvinylidene difluoride (PVDF) membrane (Millipore, Burlington, MA, USA) using Criterion™ Blotter (Bio-Rad). Membranes were blocked with EveryBlot Blocking Buffer (Bio-Rad, Hercules, CA, USA) and then incubated overnight with primary antibodies at 4 °C. Primary antibodies to Src (cat. #2109s, lot 4), p-Src (cat. #2101s, lot 20), and pHER1 (cat. #3777s, lot 10) were from Cell Signaling Technology (1:1000). Primary antibodies to β-actin (sc-47778, lot #K1607), HER2 (sc-284, lot #I0507), p-HER2 (sc-12352-R, lot #D2512), pRb (Ser780) (rabbit sc-12901), p53 (mouse sc-126), Cyclin E2 (mouse sc-28351), and CDK4 (mouse sc-260) were from Santa Cruz Biotechnology. Primary antibody to Cyclin D1(mouse cc12) and secondary antibodies conjugated with peroxidase were from Sigma-Aldrich (Sigma-Aldrich/Merck, Darmstadt, Germany). Secondary antibody binding was performed at room temperature for 1 h. After TBS-T washing, membranes were incubated with PierceTM ECL Western Blotting Substrate (Thermo Scientific, Boston, MA, USA), and the immunoreactive proteins were detected with ChemiDoc™ XRS-System (Bio-Rad, Hercules, CA, USA). Densitometry analysis was performed through ImageJ software (Version: 2.1.0/1.53C).

2.6. Cell Cycle Analysis

A total of 5×10^5 SK-BR-3 and MDA-MB-231 cells per well were seeded onto 6-well tissue culture plates. The day after, fresh medium containing 0.5 mg/mL or 1 mg/mL Chaga extract (MW < 3500 Da) was added. After 24 h incubation, the cells were harvested and fixed with ice-cold 70% ethanol, 1 h at 4 °C. RNA was digested by 1 mg/mL bovine RNase (Sigma) 30 min at 37 °C with shaking. Cells were then labeled with 15 mg/mL propidium iodide (PI) 30 min at 37 °C in the dark. Samples were analyzed by fluorescence activated cell sorting (FACS) (BD FACScalibur™, BD Biosciences, San Jose, CA, USA), and data were elaborated via BD CellQuest software (Becton Dickinson and company, Franklin Lakes, NJ, USA, v 8.7).

2.7. DHFR Enzymatic Assay

SK-BR-3 and MDA-MB-231 cells were plated onto 25 cm^2 flasks (2×10^6 cells/flask). The day after, cells were treated with 0.5 mg/mL or 1 mg/mL digested Chaga extract (MW < 3500 Da) in DMEM supplemented with 2% FBS (Invitrogen, Carlsbad, CA, USA) for 4–6 h. Cell lysates were obtained using Cell Culture Lysis Reagent (Promega, Madison, WI, USA). DHFR activity was assessed by both spectrophotometric assay and discontinuous HPLC assay. In the spectrophotometric assay, the activity of the DHFR was followed by recording the absorbance with a Shimadzu UV-2450 (UV–vis) spectrophotometer. The enzyme assays were carried out in a quartz cuvette by using similar experimental conditions: 100−200 µL of the cell lysate, NADPH 60−80 µM, and dihydrofolate (DHF) 50 µM. After the mixture was stored for 5 min at 37 °C, DHF was added. The decreasing absorbance at 340 nm, due to the oxidation of NADPH to NADP+, was detected after the DHF was added, and expressed in U mL^{-1} using the equation U mL^{-1} = $\Delta abs/\Delta t/11.8 \times$ dilution factor, where 11.8 is the mMolar extinction coefficient when NADPH and DHF are simultaneously present in solution. In the discontinuous HPLC enzymatic assay, performed as previously described [22], an Agilent 1100 system was used to detect and quantify $NADP^+$, as one of the products of DHFR-catalyzed reaction, in cell lysates. Enzyme activities were normalized by the protein content determined by Bradford assay [23]. Betulinic acid (Merck, Rome, Italy) was prepared as a 2 mM stock solution in 100% methanol and used in the 24 h timespan.

2.8. Measurements of Cytokines in Conditioned Medium

The pro- and anti-inflammatory cytokines IL-1α, IL-1β, IL-2, IL-4, IL -5, IL-6, IL-8, IL-10, IL-12, IL-13, IL-15, IL-17, IL-23, IFNγ, TNF-α, and TNF-β were estimated in the culture-conditioned medium of MDA-MB-231 and SK-BR-3 cells by using multiplex immunoassay (Q-Plex Human Cytokine—Screen 16-plex, Quansys Biosciences, Technogenetics Srl, Milan, Italy), Q-View Imager LS, Q-View software Version 3.11, and following the manufacturer's instructions. The culture medium was obtained by seeding 0.5×10^6 cells in 6-well plates containing complete growing cell culture medium. On the following day, the medium in each well was replaced with 2 mL serum-free medium, and the cells were incubated with or without 0.5 mg/mL Chaga extract MW < 3500 Da, each condition in triplicate for an additional 24 h. Then, the cell culture conditioned medium was collected and centrifuged to remove all dead cells and debris and stored at −80 °C until further analysis.

2.9. Statistical Analysis

Quantitative data are presented as either means ± SD or means ± SE from three independent experiments. The significance of differences was evaluated with an unpaired Student *t* test when two groups were compared, while one-way ANOVA test followed by Tukey's or Dunnett's post-test was used to compare three or more groups. Statistical analysis was carried out with GraphPad Prism 8.

2.10. Molecular Modeling

The crystal structure of the human DHFR (pdb code: 1U72; 1.90 Å resolution) [24] was imported into Molecular Operating Environment (MOE) software (version 2022.02) [25] and added to hydrogen atoms. The orientation of the hydrogen atoms was then optimized with energy minimization using the AMBER14 force field and keeping the coordinates of the heavy atoms fixed. The minimizations were performed by steepest descent steps followed by conjugate gradient minimization until the RMS gradient of the potential energy was less than 0.05 kJ mol^{-1} $Å^{-1}$. The betulinic acid's molecular structure was docked into the binding site of the DHFR using the MOE docking tool by setting the Alpha Triangle placement method and the Alpha HB scoring function with the generation of 50 docking poses. Each pose was then locally energetically minimized (by keeping fixed the target coordinates—Rigid Receptor refinement protocol) and then rescored with Alpha HB scoring function.

3. Results and Discussion

3.1. Chemical Characterization of Chaga Extract's Components

High-Performance Liquid Chromatography (HPLC) analysis provided insights into the Chaga mushroom's major components. Among them, phenolic acids, terpenes, and carbohydrates proved to be the most abundant. Some other classes of compounds such as coumarins, iridoids, and chalcones were also detected in smaller amounts. Table 1 summarizes the most abundant species in the Chaga samples, identified before in vitro digestion (non-digested), using an untargeted QTOF-HR-ESI-MS approach.

Under these analytical conditions, the peaks ascribable to inotodiol and its derivatives are detectable but suffer interference due to other compounds and the matrix; thus, a more specific targeted instrumental analysis was adopted to obtain quantitative data about the most important terpenoids in the samples. As an example of terpenoid identification, the chromatogram of the ion at m/z 425 in positive ion mode ascribable to inotodiol (retention time 6.5 min), in agreement with Kim J.H. et al. [26], is reported in Figure 1B, while the MS fragmentation pathway of betulin in positive ion mode showing the species at m/z 425, and fragment ions at m/z 407 and 191, in agreement with the work of Zhang et al. [27], is shown in Figure 1D. The most important identified terpenoids (inotodiol, trametenolic acid, 3-hydroxy-lanosta-8,24-dien-21-al, and betulin) are shown in Figure 1A. Triterpenoids can also be considered the main components of digested Chaga water extract.

Table 1. Qualitative analysis of the compounds detected in Chaga.

RT (min)	m/z	Tentative Identification	Adduct Ion	Chemical Class
4.17	177.0532	4-Methoxycinnamic acid	[M-H]$^{-}$	Phenolic acid
4.51	511.144	3-(4-Hydroxy-3-methoxyphenyl)-1,2-propanediol 2-O-(galloyl-glucoside)	[M-H]$^{-}$	Phenolic acid
3.38	179.0327	Caffeic acid	[M-H]$^{-}$	Phenolic acid
4.68	535.1082	Lyoniresinol 9′-sulfate	[M+Cl]$^{-}$	Phenolic acid
4.4	539.1753	Orientaloside	[M-H]$^{-}$	Phenolic acid
4.61	197.0428	Syringic acid	[M-H]$^{-}$	Phenolic acid
3.88	285.0595	Uralenneoside	[M-H]$^{-}$	Phenolic acid
5.14	735.213	Feruloylquinic acid	[2M-H]$^{-}$	Phenolic acid
3.91	555.1705	Cassitoroside	[M-H]$^{-}$	Phenolic acid
3.25	311.0385	Caftaric acid	[M-H]$^{-}$	Phenolic acid
4.99	649.2119	Egonol gentiobioside	[M-H]$^{-}$	Phenolic acid
4.49	423.1283	Gibberellin A32	[M+FA-H]$^{-}$	Terpene
6.42	533.3084	Ganoderic acid L	[M-H]$^{-}$	Terpene
4.99	737.229	Polyporusterone B/C	[M+FA-H]$^{-}$	Terpene
6.96	549.3419	Protobassic acid	[M+FA-H]$^{-}$	Terpene
7.71	533.3464	Ganoderiol D	[M+FA-H]$^{-}$	Terpene
4.85	509.1291	D-Galactopyranosyl-(1->3)-D-galactopyranosyl-(1->3)-L-arabinose	[M+Cl]$^{-}$	Carbohydrate
4.61	391.1012	Galactopinitol A	[M+Cl]$^{-}$	Carbohydrate
4.27	449.1071	a-L-Arabinofuranosyl-(1->3)-b-D-xylopyranosyl-(1->4)-D-xylose	[M+Cl]$^{-}$	Carbohydrate
3.91	449.1068	a-L-Arabinofuranosyl-(1->3)-[a-L-arabinofuranosyl-(1r5)]-L-arabinose	[M+Cl]$^{-}$	Carbohydrate
4.91	391.1008	Galactopinitol B	[M+Cl]$^{-}$	Carbohydrate
4.51	369.0804	5-Hydroxy-6-methoxycoumarin 7-glucoside	[M-H]$^{-}$	Coumarin
4.17	383.0959	Eleutheroside B1	[M-H]$^{-}$	Coumarin
4.76	537.1604	Lippioside I	[M-H]$^{-}$	Iridoid
4.3	553.1552	Lippioside II	[M-H]$^{-}$	Iridoid
4.91	421.112	2′,4′,3,4,alpha-Pentahydroxydihydrochalcone 3′-C-xyloside	[M-H]$^{-}$	Chalcone
3.83	273.0381	1,3,6-Trihydroxy-5-methoxyxanthone	[M-H]$^{-}$	Xanthone
7.66	505.3156	2-deoxy-20-hydroxy-5alpha-ecdysone 3-acetate	[M-H]$^{-}$	Ecdysteroid
7.34	549.3422	Desglucocoroloside	[M+FA-H]$^{-}$	Cardenolide
4.02	245.0064	Glucaric acid	[M+Cl]$^{-}$	Polyol
3.82	251.0531	Methionyl-Cysteine	[M-H]$^{-}$	Dipeptide

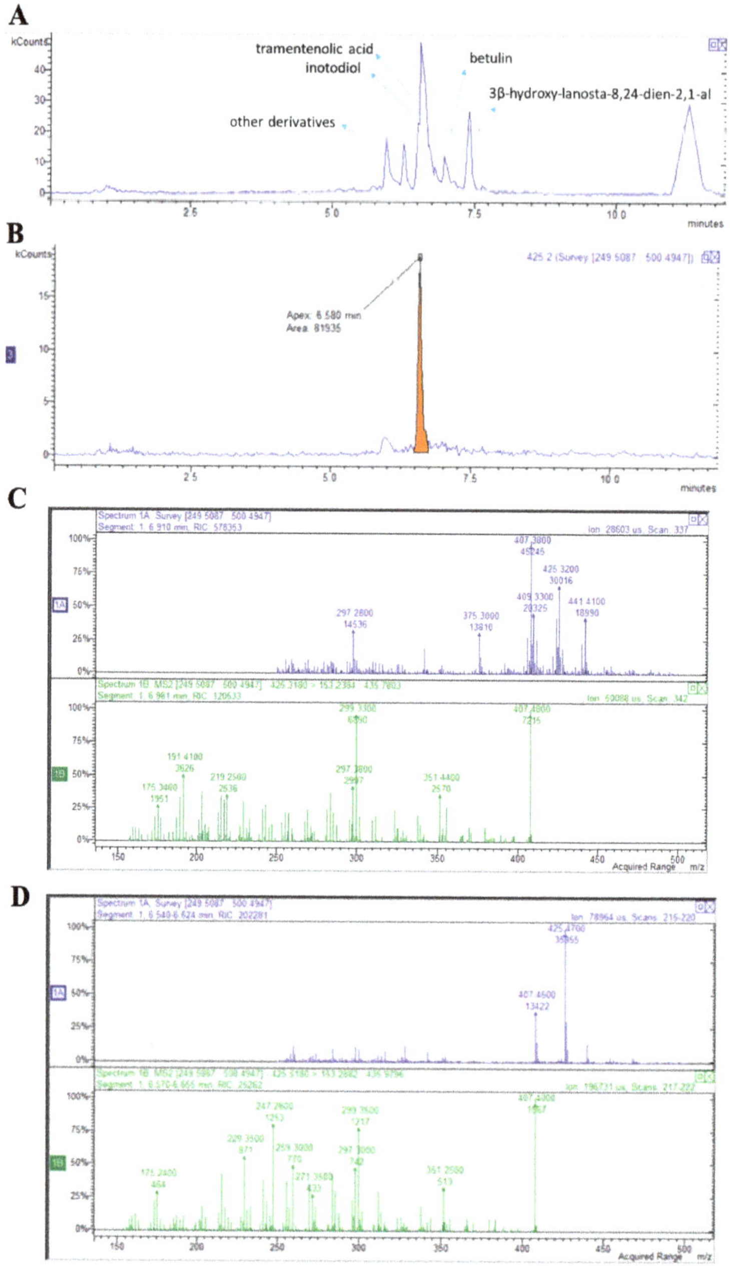

Figure 1. HPLC/MS identification of the terpenoid components of Chaga extract: (**A**) chromatogram with the identified peaks; (**B**) chromatogram recorded in positive mode in which a peak at 6.5 min, with m/z of 425, can be identified as inotodiol; (**C**) MS fragmentation pathway of betulin in positive ion mode showing the species at m/z 425 and fragment ions at m/z 407 and 191; (**D**) MS fragmentation pathway of inotodiol in positive ion mode 425 m/z and fragment ions at m/z 407 and 247.

3.2. Digested Chaga Extract Decreased the Cell Viability of Breast Cancer Cells

To evaluate the anticancer effect of Chaga against HER2-positive BC, SK-BR-3 cell viability was analyzed upon treatment with increasing concentrations of Chaga water extract (non-digested) for 24 h, 48 h, or 72 h. As shown in Supplementary Figure S1A, Chaga treatment decreased SK-BR-3 cell viability in a dose- and time-dependent manner, with IC_{50} values of 0.946 mg/mL after 48 h and 0.671 mg/mL after 72 h. Similar results were obtained by treating triple-negative BC cells with Chaga. Indeed, MDA-MB-231 cell viability was reduced in a dose- and time-dependent fashion by increasing concentrations of Chaga water extract (non-digested) administered for 24 h, 48 h, or 72 h, reaching an IC_{50} value of 0.537 mg/mL at 72 h (Supplementary Figure S1B).

Considering that *Inonotus obliquus* is used as a functional beverage, Chaga water extracts were treated to simulate human digestion in the gastro-intestinal tract, as described by Minekus et al. [18]. After dialysis, two molecular fractions, a high-molecular-weight fraction (MW > 3500 Da) and a low-molecular-weight fraction (MW < 3500 Da), were obtained, and their anticancer effect was tested on both SK-BR-3 (HER2+) and MDA-MB-231 (HER2-) cells. As shown in Figure 2A, the high-molecular-weight digested Chaga extract (MW > 3500 Da) was able to decrease SK-BR-3 cell viability only at the highest tested concentrations (IC_{50} values of 2.59 mg/mL after 48 h and 2 mg/mL after 72 h). Of note, the low-molecular-weight digested Chaga extract (MW < 3500 Da), instead, induced a strong reduction in SK-BR-3 cells' viability already after 24 h incubation, showing an IC_{50} value of 0.858 mg/mL, which was further reduced to about 0.46 mg/mL after 48 h (Figure 2B).

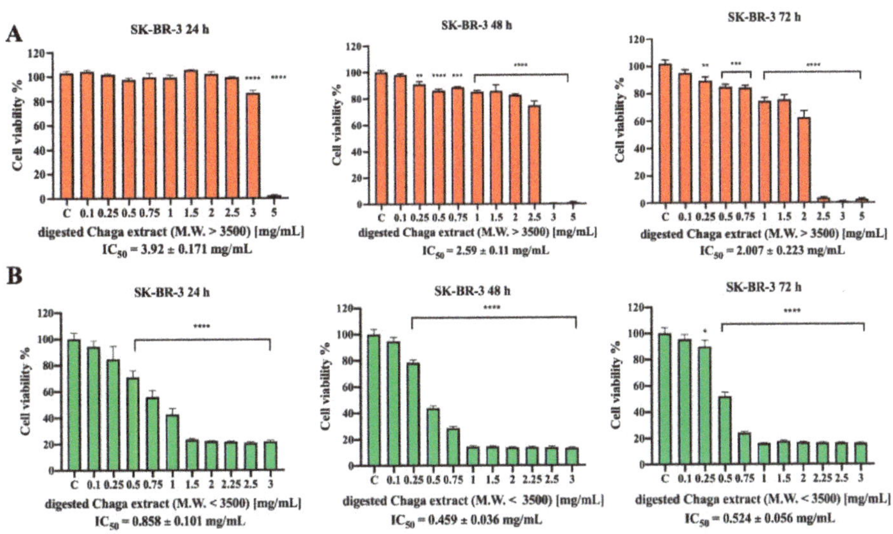

Figure 2. Effect of digested Chaga extract on SK-BR-3 cell viability. SK-BR-3 cells were left untreated (control) or incubated for 24 h, 48 h, or 72 h in the presence of increasing concentrations of the high-molecular-weight fraction (MW > 3500 Da) (**A**) or the low-molecular-weight fraction (MW < 3500 Da) (**B**) of digested Chaga water extract; cell viability was determined by MTT assay. The results are expressed as the percentage of living cells with respect to control. Columns: mean of three separate experiments wherein each treatment was repeated in 6 wells. Bars: SE. * $p < 0.05$, ** $p < 0.01$, *** $p < 0.001$, **** $p < 0.0001$. One-way ANOVA followed by Dunnett's multiple comparison test.

Analogously, the high-molecular-weight fraction of Chaga digested water extract (MW > 3500 Da) decreased MDA-MB-231 cell viability in a time-dependent and dose-dependent fashion (Figure 3A), but the most efficient fraction was shown to be the one with MW < 3500 Da. Indeed, the low-molecular-weight digested Chaga extract (MW < 3500 Da) induced a strong reduction in MDA-MB-231 cells' viability already after 24 h incubation, showing an IC_{50} value of 1.112 mg/mL, which was further reduced to about 0.626 mg/mL after 48 h and to 0.545 mg/mL after 72 h incubation (Figure 3B). Digestion fluid (without Chaga), after dialysis, was also tested on both SK-BR-3 and MDA-MB-231 cells, and the results obtained by MTT assay confirmed that it did not affect cell viability per se (Supplementary Figure S2). However, digested Chaga extract (MW < 3500 Da) also inhibited the viability of HEK-293 cells (IC_{50} = 0.85 ± 0.08 mg/mL at 24 h) and MCF-10A human breast epithelial cells (IC_{50} = 0.086 ± 0.009 mg/mL at 24 h) (Supplementary Figure S3A,B); thus, it seems not to show any apparent selective cytotoxicity to tested cancer cell lines with respect to non-cancer ones. Recently, in contrast to our findings, it has been reported that Chaga extracts were selectively cytotoxic to the MCF-7 breast cancer cell line while sparing the corresponding healthy cell lines (MCF-10A cells). However, when tested at the same concentrations on other non-cancer cell lines, they inhibited their viability [28]. Thus, further investigation is required to assess the selective cytotoxicity of digested Chaga extract to cancer cell lines, extending the study to other healthy cell lines besides HEK-293 and MCF-10A cells. Of note, CCD 841 CoN normal colon epithelial cells appeared much less sensitive to digested Chaga extract (MW < 3500 Da) than the other cells used in this study, showing an IC_{50} of 2.03 ± 0.72 mg/mL after 72 h incubation (Supplementary Figure S3C).

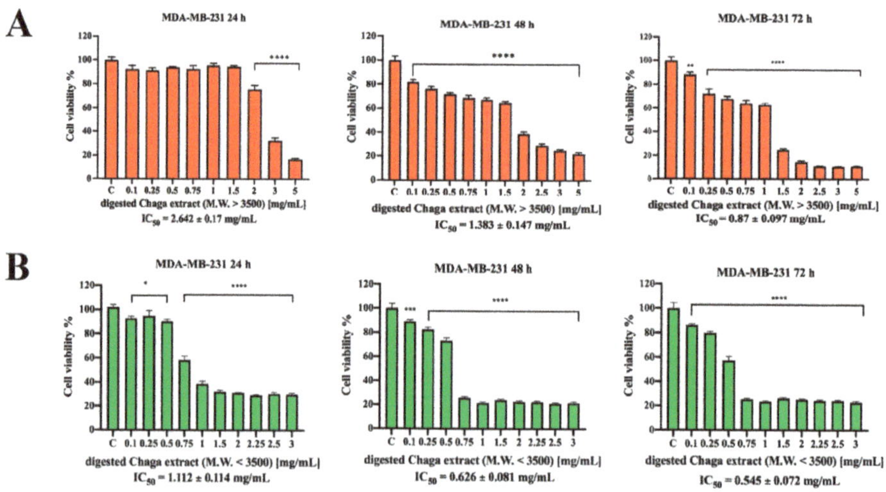

Figure 3. Effect of digested Chaga extract on MDA-MB-231 cell viability. MDA-MB-231 cells were left untreated (control) or incubated for 24 h, 48 h, or 72 h in the presence of increasing concentrations of the high-molecular-weight fraction (MW > 3500 Da) (**A**) or the low-molecular-weight fraction (MW < 3500 Da) (**B**) of digested Chaga water extract; cell viability was determined by MTT assay. The results are expressed as the percentage of living cells with respect to the control. Columns: mean of three separate experiments wherein each treatment was repeated in 6 wells. Bars: SE. * $p < 0.05$, *** $p < 0.001$, **** $p < 0.0001$. One-way ANOVA followed by Dunnett's multiple comparison test.

3.3. Digested Chaga Extract Induced G0/G1 Cell Cycle Arrest in Breast Cancer Cells

Next, we investigated the effect of digested Chaga extract (MW < 3500 Da) on SK-BR-3 cells' distribution in the cell cycle phases by flow cytometry analysis. As shown in Figure 4A, the treatment of SK-BR-3 cells with 0.5 mg/mL of digested Chaga extract (MW < 3500 Da) for 24 h resulted in a higher number of cells in the G0/G1 phase (76.1 ± 1.76%, on average) compared to the control (56.5 ± 1.53%, on average). This increase was coupled with a decreased percentage of digested Chaga-treated SK-BR-3 cells in the S phase (8.85 ± 1.29%, on average) with respect to untreated control cells (24.6 ± 1.68%, on average). Similar results were obtained by treating the SK-BR-3 cells with 1 mg/mL of digested Chaga extract (MW < 3500 Da) for 24 h (Supplementary Table S2). These results suggest that digested Chaga extract induced a G0/G1 arrest of SK-BR-3 cells. To analyze the underlying biochemical mechanisms involved in the G0/G1 cell cycle arrest, we investigated the levels of G0/G1 regulatory cyclins and cyclin-dependent kinases (CDKs) in SK-BR-3 cells treated with digested Chaga extract (MW < 3500 Da) by Western blot (Figure 4, panels B, C). Complexes containing cyclins D and E, the regulatory units, and CDK2, CDK4, or CDK6, the catalytic units, play important roles in the progression of cells through the G0/G1 phase of the cell cycle. Indeed, CDK4/cyclin D complexes phosphorylate retinoblastoma protein (pRb) in mid G1, while CDK2/cyclin E complexes phosphorylate pRb at the G1-to-S transition. The status of pRb phosphorylation is crucial for E2F activity because only the hypophosphorylated form of pRb is associated with E2F transcription factors, which are key regulators of genes required for cell cycle progression. In other words, pRb has a growth-suppressive role (it is active) only when hypophosphorylated, whereas the hyperphosphorylation of Rb protein by cyclins/CDKs results in its inactivation, causing the release of the transcription factor E2F and thus determining cell proliferation [29]. Western blot results showed a significant time-dependent decrease in the levels of cyclin D1, cyclin E2, and CDK4 in SK-BR-3 cells treated with 0.5 mg/mL digested Chaga extract (MW < 3500 Da) for 24 h and 48 h. A further reduction in the levels of the analyzed proteins was observed by increasing the Chaga concentration to 1 mg/mL. Of note, Rb protein was hyperphosphorylated (Ser 780) in untreated cells, whereas digested Chaga extract was effective at reactivating Rb function by decreasing protein phosphorylation in a dose-and-time dependent way, confirming the ability of digested Chaga extract to block the cell cycle progression (Figure 4, panels B, C).

Treatment with 0.5 mg/mL digested Chaga extract (MW < 3500 Da) for 24 h resulted also in the G0/G1 arrest of MDA-MB-231 cells. Indeed, the percentage of MDA-MB-231 cells in the G0/G1 phase upon Chaga treatment was 78.3 $\pm$ 2.12%, on average, whereas it was 62.2 $\pm$ 1.63% in the control condition. This increase was associated with the decreased percentage of treated cells in the S phase and in the G2 phase. In particular, the percentage of MDA-MB-231 cells in the S phase was almost double in the control condition (11.88 $\pm$ 0.51%) with respect to cells treated with digested Chaga (5.98 $\pm$ 0.7%), while 25.9 $\pm$ 1.14% of control cells were in the G2 phase versus 15.6 $\pm$ 1.9% of digested Chaga-extract-treated MDA-MB-231 cells (Figure 5A). Thus, the expression of two key player proteins involved in the control of cell cycle progression, p53 protein and phosphorylated pRb, was analyzed by Western blotting in MDA-MB-231 cells untreated or treated with 0.5 mg/mL of digested Chaga extract (MW < 3500 Da) for 24 h or 48 h or with 1 mg/mL for 24 h. As shown in Figure 5 (panels B, C), Rb protein was hyperphosphorylated (Ser 780) in untreated cells, whereas digested Chaga extract was effective at reactivating the Rb function by decreasing protein phosphorylation in a dose- and time-dependent way, likely reducing MDA-MB-231 cell proliferation. A similar trend was observed for p53, although a significant decrease in the level of p53 protein was obtained only upon treatment with 1 mg/mL Chaga extract for 24 h (Figure 5, panels B, C). Of note, MDA-MB-231 cells express high levels of a mutant form of p53 (R280K) (mt-p53), which loses the ability to bind responsive elements on DNA, thus becoming defective for oncosuppressor functions. Indeed, accumulating evidence underscores the role of mutant p53 in promoting transformation and metastasis [30]. Thus, the obtained results indicate that digested Chaga extract not only prevented the hyperphosphorylation of pRb in MDA-MB-231 cells, blocking cell cycle progression, but also lowered mt-p53 levels by either stimulating degradation or downregulating the expression of mt-p53. Considering that the tyrosine kinase Src is overexpressed in triple-negative BC and its activation induces tumor cell growth and metastasis [31], we investigated the ability of digested Chaga extract (MW < 3500 Da) to downregulate Src activation. As shown in Figure 5 (panels D, E), Src phosphorylation decreased in a time-dependent way upon the treatment of MDA-MB-231 cells.

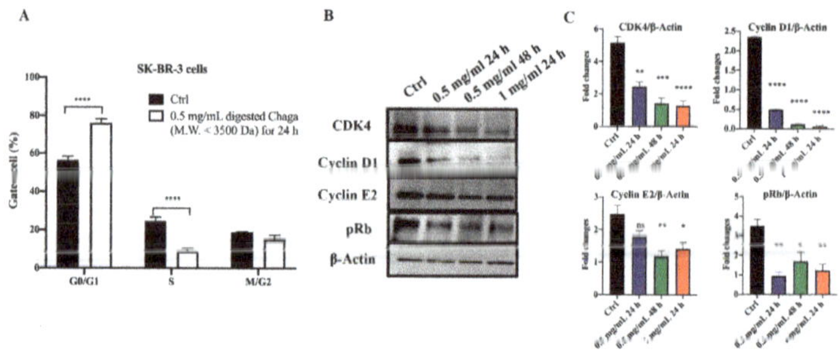

Figure 4. Digested Chaga extract (MW < 3500 Da) induced cell cycle arrest in the G0/G1 phase in SK-BR-3 cells. (**A**) Histograms showing the percentage of SK-BR-3 cells in the G0/G1, S, and G2/M phases in control condition or following 24 h treatment with 0.5 mg/mL of digested Chaga extract (MW < 3500 Da) as assessed by FACS cell cycle analysis. Data are presented as the mean ± SD of three repeats. **** $p < 0.0001$ vs. control. ANOVA followed by Sidak's multiple comparison test. (**B**) Representative Western blotting showing the expression of CDK4, cyclin D1, cyclin E2, phosphorylated Rb protein (Ser 780), and β-actin (loading control) in SK-BR-3 cells left untreated (Ctrl) or treated with 0.5 mg/mL or 1 mg/mL of digested Chaga extract (MW < 3500 Da) for 24 h or 48 h. Twenty micrograms of proteins/well were loaded. (**C**) Densitometric analysis of each assessed protein. Data are presented as the mean ± SE of three repeats. * $p < 0.05$, ** $p < 0.01$, *** $p < 0.001$, **** $p < 0.0001$. ns: not significant. One-way ANOVA followed by Dunnet's multiple comparison test.

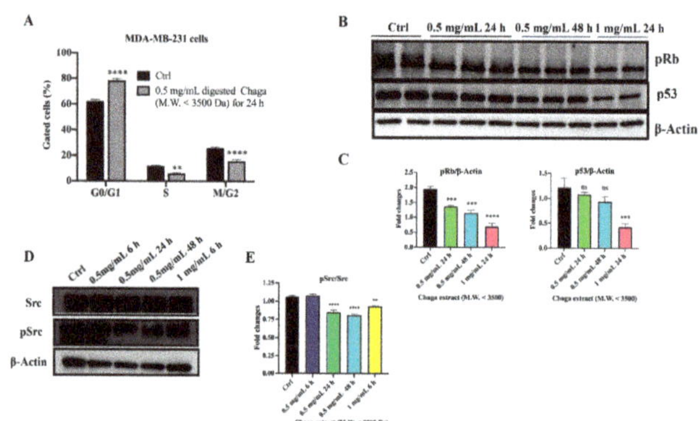

Figure 5. Digested Chaga extract (MW < 3500 Da) induced cell cycle arrest in the G0/G1 phase in MDA-MB-231 cells. (**A**) Histograms showing the percentage of MDA-MB-231 cells in the G0/G1, S, and G2/M phases, in the control condition or following 24 h treatment with 0.5 mg/mL of digested Chaga extract (MW < 3500 Da) assessed by FACS cell cycle analysis. Data are presented as the mean ± SD of three repeats. ** $p < 0.01$, **** $p < 0.0001$ vs. control. ANOVA followed by Sidak's multiple comparison test. (**B**) Representative Western blotting showing the expression of phosphorylated Rb protein (Ser 780), p53, and β-actin (loading control) in MDA-MB-231 cells, left untreated (Ctrl) or treated with 0.5 mg/mL or 1 mg/mL of digested Chaga extract (MW < 3500 Da) for 24 h or 48 h. Twenty micrograms of proteins/well were loaded. (**C**) Densitometric quantifications of pRb and p53 expression, normalized on β-actin, are shown; data are presented as the mean ± SD of three repeats. *** $p < 0.001$, **** $p < 0.0001$. ns: not significant. One-way Anova, followed by Dunnet's multiple comparison test. (**D**) Representative Western blotting showing the expression of phosphorylated Src protein (pSrc), Src, and β-actin (loading control) in MDA-MB-231 cells left untreated (Ctrl) or treated with digested Chaga extract (MW < 3500 Da) at the indicated time and concentrations. (**E**) Densitometric quantifications of pSrc/Src from three independent experiments are shown; data are presented as the mean ± SE of three repeats. ** $p < 0.01$, **** $p < 0.0001$. One-way ANOVA, followed by Dunnet's multiple comparison test.

3.4. Digested Chaga Extract Decreased DHFR Enzymatic Activity in Breast Cancer Cells

Due to the central role of the enzyme dihydrofolate reductase (DHFR) in regulating cell viability and proliferation, the effects of Chaga on DHFR activity were also investigated. The enzymatic activity of DHFR was significantly inhibited by Chaga in both SK-BR-3 and MDA-MB-231 cells. In Figure 6, it is possible to observe that the specific activity of DHFR (expressed in enzymatic units/mg of proteins) is strongly affected by low-molecular-weight components of Chaga extract; in particular, 1 mg/mL of digested Chaga in culture medium reduced the residual activity of DHFR to about 50% in both SK-BR-3 and MDA-MB-231 cells. In Figure 6C, the residual enzymatic activity of DHFR (as % of the control) is reported in the presence of Chaga extract 1 mg/mL in comparison with the residual activity obtained in the presence of the mixture of digestive fluids used to prepare the Chaga extract at 1 mg/mL, as described in Materials and Methods. The results show a non-significant effect of digestive fluids in both SK-BR-3 and MDA-MB-231 cells differently from the significant effect exerted by Chaga extract in accordance with the data reported in Figure 6A,B. This result is consistent with the observed effect of the tested Chaga extract on the cell cycle arrest in the G0/G1 phase, as the DHFR enzyme expression levels are known to increase in the G1/S boundary [32]. Furthermore, we can also hypothesize that some of the components of the extract can directly inhibit DHFR enzymatic activity, as observed by pre-incubating the BC cell lysates with 0.1 mg/mL of a low-molecular-weight fraction (MW < 3500 Da) of digested Chaga water extract directly in the reaction mixture (Supplementary Figure S6).

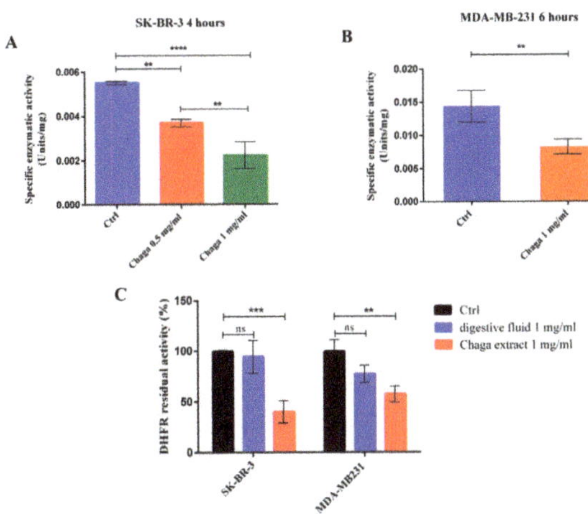

Figure 6. Digested Chaga extract inhibited DHFR enzymatic activity in BC cells. Residual enzymatic activity of DHFR was analyzed in SK-BR-3 (**A**) and MDA-MB-231 (**B**) cells after a 4–6 h treatment with 0.5 mg/mL or 1 mg/mL digested Chaga extract (MW < 3500 Da). (**C**) The residual enzymatic activity of DHFR (as % of the control) is reported in the presence of Chaga extract 1 mg/mL in comparison with the residual activity obtained in the presence of the mixture of digestive fluids used to prepare the Chaga extract at 1 mg/mL. Data are reported as the average of three replicates ± SE, ** $p \leq 0.01$; *** $p \leq 0.001$; **** $p \leq 0.0001$. ns: not significant. One-way ANOVA followed by Tukey's multiple comparison test.

3.5. Digested Chaga Displayed a Synergistic Activity When Combined with Trastuzumab and Cisplatin in Breast Cancer Cells

The molecular mechanisms underlying the Chaga anticancer effect were further investigated in SK-BR-3 cells, with a focus on the impact of digested Chaga extract (MW < 3500 Da) on HER2 expression and activation. Considering that growth factor receptors do not act as single proteins but as homo- or hetero-dimers, the level of activation of HER1 was also evaluated. Chaga was able to significantly decrease in a time-dependent manner the activation of both HER2 and HER1, as indicated by the downregulation of their phosphorylated forms (Figure 7A). This result provided the rationale for evaluating the synergistic anticancer effect of digested Chaga extract and trastuzumab, a well-established HER2-directed monoclonal antibody used as targeted therapy in HER2-positive BC patients. As shown in Figure 7B, the combination of 0.25 mg/mL digested Chaga extract (MW < 3500 Da) with trastuzumab shows a synergistic antitumor activity, significantly reducing SK-BR-3 cell viability with respect to trastuzumab treatment alone after 24 h incubation. Interestingly, Chaga was also able to synergistically enhance the action of cisplatin, a conventional chemotherapeutic agent, in both SK-BR-3 and MDA-MB-231 cells (Figure 7C,D, Supplementary Figure S3). Indeed, the cell viability of SK-BR-3 and MDA-MB-231 cells decreased significantly only upon treatment with the combination of 0.25 mg/mL Chaga and 0.5 µM or 1 µM cisplatin, whereas cisplatin alone was not effective at all when given at 0.5 µM or exerted a low inhibition at 1 µM. Similar results were obtained using the cisplatin derivative RJY13 [19], a platinum (IV)–fatty acid conjugate ((cis, cis, trans-[diamminedichloro-bisoctanoatoplatinum (IV)]) (Supplementary Figures S4 and S5).

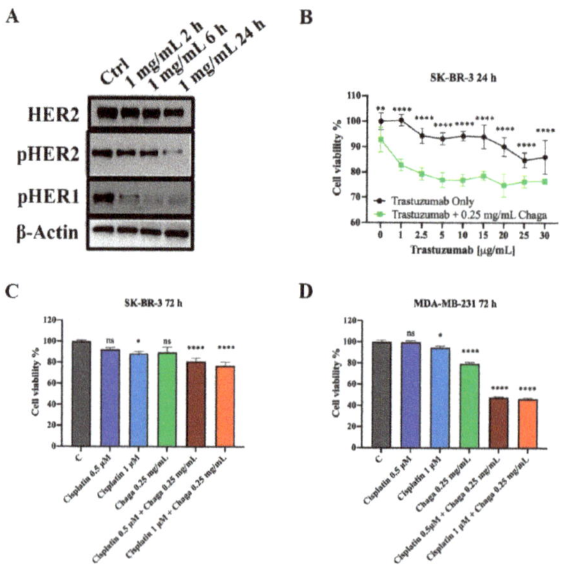

Figure 7. Digested Chaga extract impaired HER2 activation and acted synergistically with trastuzumab and cisplatin. (**A**) Representative Western blotting showing the expression of HER2, phospho-HER2, phospho-HER1, and β-actin (loading control) in SK-BR-3 cells, left untreated (Ctrl) or treated with 1 mg/mL of digested Chaga extract (MW < 3500 Da) for 2 h, 6 h, or 24 h. Twenty micrograms of proteins/well were loaded. (**B**) SK-BR-3 cells were plated onto 96-well plates, treated with increasing concentrations of trastuzumab alone or in combination with a fixed, sub-toxic concentration (0.25 mg/mL) of digested Chaga extract (MW < 3500 Da); cell viability was determined by MTT assay. (**C**) SK-BR-3 cells and (**D**) MDA-MB-231 cells were treated with the indicated concentrations of cisplatin alone or in combination with 0.25 mg/mL digested Chaga extract (MW < 3500 Da); cell viability was determined by MTT assay. Drug interaction was evaluated by the Bliss Independence model; the observed effects of the drug combination indicated synergistic interaction (calculations are reported in Table S3). Bars: SE. * $p < 0.05$, ** $p < 0.01$, **** $p < 0.0001$. ns: not significant. One-way ANOVA followed by Dunnett's multiple comparison test. Data show a representative of three independent experiments.

Drug interaction was evaluated by Bliss Independence model, which provided evidence that the observed effects of drug combinations were greater than the sum of the individual effects of each drug, indicating synergistic interactions (Table S3). The synergistic effect between cisplatin and Chaga extract observed in MDA-MB-231 cells is particularly important, considering that platinum-containing regimens are recommended in the treatment of early and advanced triple-negative BC [33]. Thus, the combination of cisplatin with Chaga extract has the potential to improve chemotherapy efficacy, to decrease the risk of cancer treatment resistance, and to reduce drug side effects thanks to the lowering of its doses [34].

In SK-BR-3 cells, the synergistic action between Chaga and platinum-based compounds was associated with a statistically significant reduction in the level of phosphorylated HER1 and phosphorylated HER2, as well as with a lower level of phosphorylated Rb (Figure 8A). In MDA-MB-231 cells, sub-toxic concentrations of cisplatin alone (1 μM) or its RJY13 derivative (0.01 μM) were not able to affect the activation of Src, the hyperphosphorylation of Rb, or the expression of mt-p53, but the expression of these key molecules governing cell proliferation and survival significantly decreased when platinum drugs were administered in combination with Chaga extract (Figure 8B). However, in this case, the combination of cisplatin and Chaga extract was no more effective than Chaga extract alone.

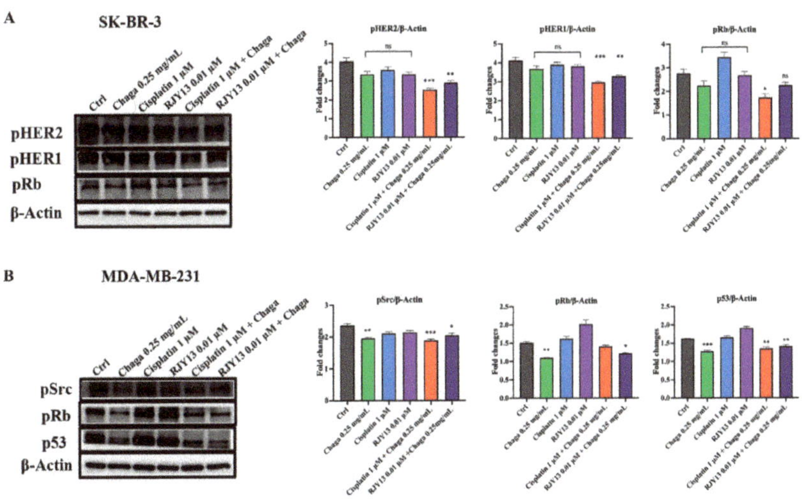

Figure 8. Oncogenic pathways in SK-BR-3 and MDA-MB-231 cells treated with Chaga extract in combination with platinum-based chemotherapeutics. (**A**) Left panel: representative Western blotting showing the expression of phosphorylated (p) HER1, pHER2, pRb protein (Ser 780), and β-actin (loading control) in SK-BR-3 cells left untreated (Ctrl) or treated with 0.25 mg/mL of digested Chaga extract (MW < 3500 Da) alone, 1μM cisplatin or its derivative 0.01 μM RJY13 alone, or their combination for 72 h. Twenty micrograms of proteins/well were loaded. Right panel: densitometric quantifications of pHER1, pHER2, and pRb expression, normalized on β-actin, are shown; data are presented as the mean ± SE of three repeats. (**B**) Left panel: representative Western blotting showing the expression of phosphorylated (p) Src, pRb protein (Ser 780), p53, and β-actin (loading control) in MDA-MB-231 cells left untreated (Ctrl) or treated with 0.25 mg/mL of digested Chaga extract (MW < 3500 Da) alone, 1 μM cisplatin or its derivative 0.01 μM RJY13 alone, or their combination for 72 h. Twenty micrograms of proteins/well were loaded. Right panel: densitometric quantifications of pSrc, pRb, and p53 expression, normalized on β-actin, are shown; data are presented as the mean ± SE of three repeats. * $p < 0.05$, ** $p < 0.01$, *** $p < 0.001$. ns: not significant. One-way ANOVA, followed by Dunnet's multiple comparison test.

3.6. Betulinic Acid Exerts Anticancer Effects on SK-BR-3 and MDA-MB-231 Cancer Cells and Impairs DHFR Enzymatic Activity

The terpenoid betulinic acid was administered to the SK-BR-3 and MDA-MB-231 cancer cell lines to evaluate the anticancer effect over a 48-h time course at concentrations ranging from 5 to 60 µM. As shown in Figure 9A, in the two cell lines, the IC_{50} was very similar and was shown to be 25.15 ± 1.21 µM in SK-BR-3 and 27.24 ± 2.58 µM in MDA-MB-231 cells. These values of IC_{50} are in good agreement with data reported elsewhere [35].

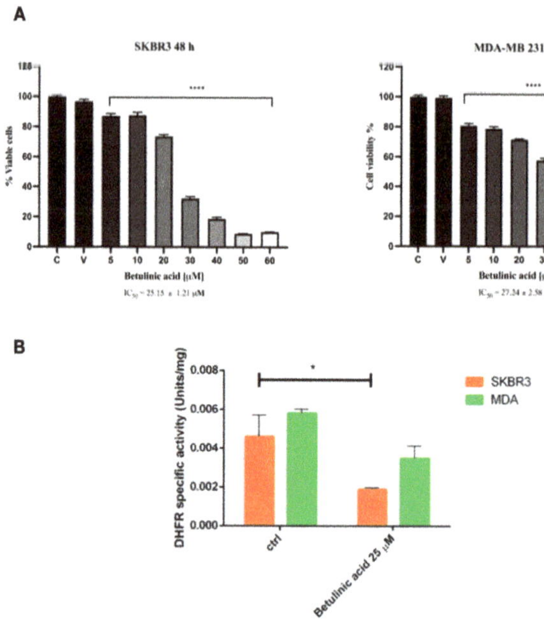

Figure 9. Sensitivity of breast cancer cells SK-BR-3 and MDA-MB-231 to betulinic acid in terms of (**A**) cell viability and (**B**) DHFR enzymatic activity. Data are reported as the average of three replicates ± SE, * $p < 0.05$, **** $p < 0.0001$. One-way ANOVA followed by Tukey's multiple comparison test.

Based on the IC_{50} values, SK-BR-3 and MDA-MB-231cancer cells were treated with 25 µM of betulinic acid, and the DHFR enzymatic activity was assayed in the cell lysates, as described in the Materials Methods, under saturating concentrations of both NADPH and dihydrofolic acid.

In Figure 9B, it is possible to observe how the SKBR3 and MDA-MB-231 cancer cells respond to the treatment with 25 µM betulinic acid by reducing the enzymatic activity of DHFR of about 40–50% in both cell lines.

A docking study was performed to simulate and analyze the binding mode of the betulinic acid molecule within the DHFR binding cavity. Hence, the crystal structure of the human enzyme in complex with methotrexate and the NADPH co-factor was downloaded from the PDB database (pdb code: 1U72; 1.90 Å resolution) [24] and added to hydrogen atoms that were energetically minimized within the Molecular Operating Environment (MOE) [25]. Docking experiments were also performed within MOE, with a Rigid Receptor refinement protocol aimed at energetically minimizing the generated docking poses.

The docking result for betulinic acid is shown in Figure 10. The top-score docking pose presents the ligand inserted in the binding side with its isopropenyl group in the depth of the cavity. The interaction with the enzyme is largely non-polar, given by contacts between the ligand scaffold and hydrophobic residues like Phe31, Phe34, and Pro61. Polar interactions are present between the carboxyl function of betulinic acid and Ser 59 and between the hydroxyl group of the same molecule and the DHFR residues Gln35 and Arg70.

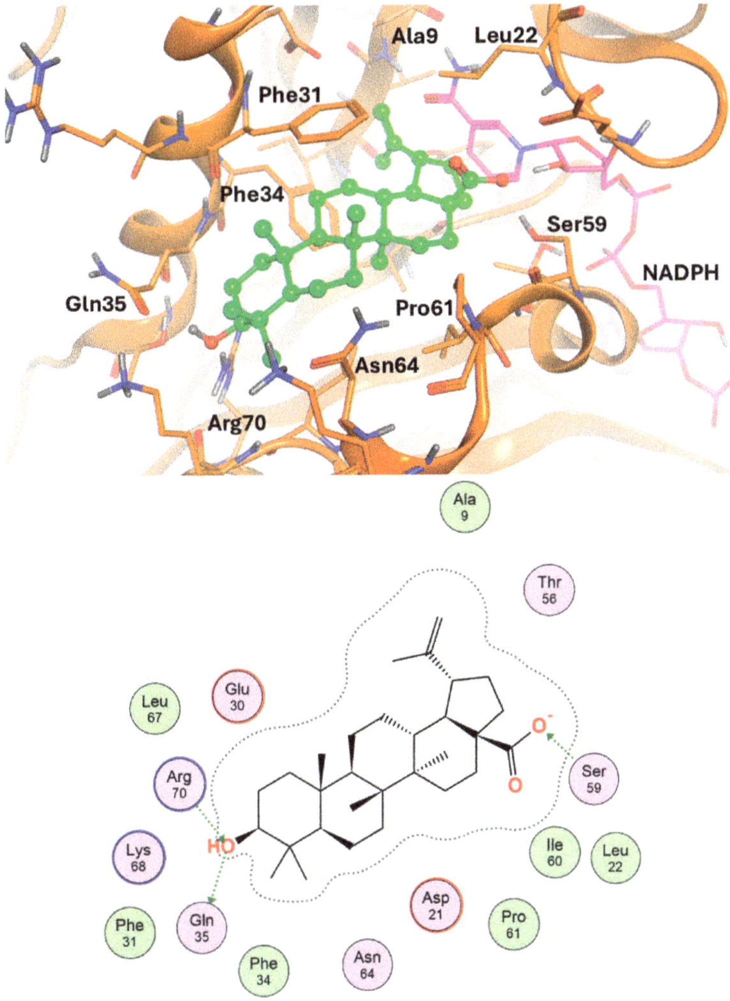

Figure 10. (**Above**): simulated binding mode of the betulinic acid (green) at the binding cavity of the DHFR; the key residues for ligand–target interaction are shown. (**Below**): schematic plot of ligand–target interaction.

3.7. Digested Chaga Displayed Immunomodulatory Properties in Breast Cancer Cells

To evaluate the ability of digested Chaga extract (MW < 3500 Da) to modulate the pro- and anti-inflammatory cytokine secretion by SK-BR-3 and MDA-MB-231 cells, the concentration of a panel of 16 cytokines (IL-1α, IL-1β, IL-2, 4, IL-5, IL-6, IL-8, IL-10, IL-12, IL-13, IL-15, IL-17, IL-23, IFNγ, TNF-α, and TNF-β) was measured in culture supernatants by a multiplex ELISA assay. IL-2 and IL-8 were the only cytokines detectable in the culture medium of 0.5 mg/mL Chaga-treated SK-BR-3 cells, whereas all the other cytokines were below the detection level. The treatment of MDA-MB-231 cells with 0.5 mg/mL Chaga induced not only an increase in the levels of IL-2 and IL-8 in the culture medium but also an elevation of IL-1α, IL-5, IL-6, IL-12, and TNFα, as well as IL-6 and IL-8 (both out of kit's scale in Chaga-treated MDA-MB-231 cells), although these last two cytokines were already detectable in control samples (Figure 11).

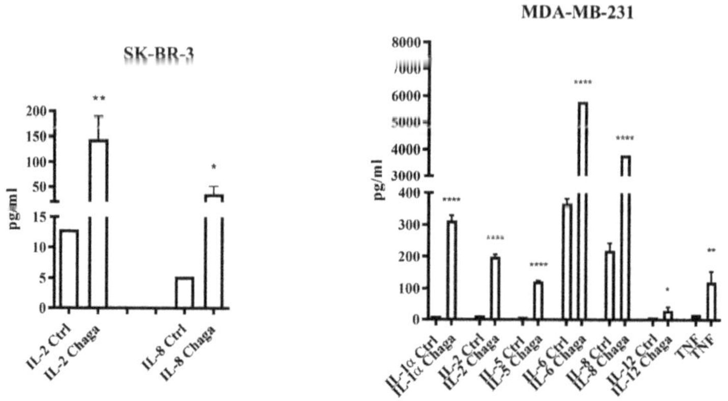

Figure 11. Digested Chaga extract (MW < 3500 Da) induced cytokine release in the culture medium of both SK-BR-3 and MDA-MB-231 cultures but differently according to the cell line. Cytokines were quantified by using a multiplex immunoassay, as described in Materials and Methods section. Data were analyzed by Student's *t*-test. * $p < 0.05$, ** $p < 0.01$, **** $p < 0.0001$, with respect to the corresponding Ctrl supernatant cytokine concentration.

The production of cytokines by different tumor cell lines, although cultivated under identical conditions, can considerably vary [36]. Our results are consistent with the data obtained from Hartman et al. [37], which indicate that the MDA-MB-231 cell line expresses high amounts of IL-6, and both pro- and anti-apoptotic functions of this cytokine have been reported [38]. Although it is known that the aqueous extract of *Inonotus obliquus* exerts anti-inflammatory effects by down-regulating the expression of pro-inflammatory mediators [39], in the present study, depending on the cell line considered, the Chaga-treated cells over-secreted these mediators (IL-1, IL-2, IL-5, IL-6, IL-8, and TNF-α). However, accumulating data have nowadays shown that cytokines play an important role in both induction and protection in BC [40]. On the one hand, the IL-2 induces the expansion of $CD4^+$ and $CD8^+$ T lymphocytes and thus could reinforce the anticancer immune responses; on the other hand, the role of IL-8 in antitumor immune responses is more controversial. Indeed, although IL-8 has been reported to favor cancer progression and metastases [41], it was also shown to be responsible for recruiting neutrophils and macrophages, which in turn can kill antibody-opsonized cancer cells by a mechanism of cytotoxicity called trogoptosis [42]. Of interest also is the increased level of IL-5 in the supernatant of MDA-MB-231 cells treated with Chaga. High levels of IL-5 were found in tumor interstitial fluid samples and were associated with a worse BC prognosis [43]. Nevertheless, an increased IL-5 production was recently found to be crucial for systemic eosinophil expansion and tumor infiltration in patients with metastatic triple-negative BC responding to immune checkpoint blockade treatment [44]. These results demonstrate that digested Chaga extract has immunomodulatory properties, in agreement with other studies reporting that medicinal mushroom components can modulate the immune system via a variety of molecular processes, including cytokine induction [45]. In particular, Chaga has been found to activate innate immunity, enhancing the phagocytosis of macrophages [46] and the maturation of dendritic cells [47].

4. Conclusions

Chaga mushrooms (*Inonotus obliquus*) are commonly used as traditional treatments in Asia due to their diverse pharmacological effects and in some cases are claimed to some anti-tumor effects. Considering that Chaga is usually consumed as a functional beverage (tea), we investigated the anticancer properties of "digested" Chaga water extracts against BC. Triterpenoids were identified as the main components of digested Chaga water extract, which was able to reduce cancer cell viability, to interfere with oncogenic signaling pathways and to induce a cell cycle G0/G1-phase arrest in both triple-negative (MDA-MB-231) and HER2-positive (SK-BR-3) BC cell lines. These data are consistent with previous investigations reporting the anti-tumor potential of triterpenoids, among Chaga phytochemical constituents [8,12]. These effects were associated with immunomodulatory actions and with the inhibition of the enzymatic activity of the enzyme dihydrofolate reductase (DHFR), which has a key role in the de novo synthesis of purines and thymidylate and thus regulates cell viability and proliferation. Moreover, digested Chaga treatment was able to act synergistically with trastuzumab and cisplatin.

In addition, we could identify in the triterpene betulinic acid that originates from betulin, as one of the putative bioactive components of Chaga extract able to impair breast cancer cell viability and inhibit DHFR activity at micromolar concentrations.

In conclusion, the present study provides evidence that digested Chaga extract is effective against aggressive BC subtypes, targeting key molecules associated with the malignant phenotype, and demonstrates that *Inonotus obliquus* can represent a good source of alternative antitumor drugs or a remedy that, in combination with conventional drugs, may increase their effectiveness or reduce their dosage.

Supplementary Materials: The following supporting information can be downloaded at https://www.mdpi.com/article/10.3390/biom14111454/s1. Table S1: Yields of Chaga extracts with ethyl acetate, methanol, and water. Figure S1: Chaga water extract decreased breast cancer cell viability. Figure S2: Digested fluid (without Chaga) did not affect SK-BR-3 and MDA-MB-231 cell viability. Table S2: Digested Chaga extract (MW < 3500 Da) induced cell cycle arrest in the G0/G1 phase in SK-BR-3 cells. Figure S3: Effect of digested Chaga extract on cell viability of non-cancerous cell lines. Figure S4: Effect of cisplatin treatment on SK-BR-3 and MDA-MB-231 cell viability. Figure S5: Synergistic effect between Chaga and RJY13 (cisplatin derivative) treatment on SK-BR-3 and MDA-MB-231 cell viability. Table S3: Drug interaction evaluated by the Bliss Independence model. Figure S6: Digested Chaga extract inhibited DHFR enzymatic activity in BC cells lysates. The original images of Western blot can be found in Supplementary Materials.

Author Contributions: Conceptualization, J.W., C.M., D.B., J.M. and S.P.; formal analysis, J.W., C.M., D.B., D.D.B. and S.P.; investigation, J.W., D.B., A.A., S.S., S.D., G.L., O.B., D.D.B., D.T., S.M. and S.P.; resources, B.B., S.S. and J.M.; writing—original draft preparation, J.W. and C.M.; writing—review and editing, J.W., C.M., D.B., S.P. and J.M.; supervision, C.M. and S.P.; project administration, C.M.; funding acquisition, J.W. and C.M. All authors have read and agreed to the published version of the manuscript.

Funding: This research received no external funding.

Institutional Review Board Statement: Not applicable.

Informed Consent Statement: Not applicable.

Data Availability Statement: The data presented in this study are available in this article and Supplementary Material.

Acknowledgments: J. Wang was supported by Fondazione Umberto Veronesi. We would like to thank Giulio Caracciolo (La Sapienza University, Roma, Italy) for kindly providing CCD 841 CoN (ATCC CRL-1790) normal colon epithelial cells.

Conflicts of Interest: The authors declare no conflicts of interest.

References

1. Perou, C.M.; Sørlie, T.; Eisen, M.B.; van de Rijn, M.; Jeffrey, S.S.; Rees, C.A.; Pollack, J.R.; Ross, D.T.; Johnsen, H.; Akslen, L.A.; et al. Molecular portraits of human breast tumours. *Nature* **2000**, *406*, 747–752. [CrossRef] [PubMed]
2. Sørlie, T.; Perou, C.M.; Tibshirani, R.; Aas, T.; Geisler, S.; Johnsen, H.; Eisen, M.B.; Hastie, T.; van de Rijn, M.; Jeffrey, S.S.; et al. Gene expression patterns of breast carcinomas distinguish tumor subclasses with clinical implications. *Proc. Natl. Acad.Sci. USA* **2001**, *98*, 10869–10874. [CrossRef] [PubMed]

3. Millar, E.K.; Graham, P.H.; O'Toole, S.A.; McNeil, C.M.; Browne, L.; Morey, A.L.; Eggleton, S.; Beretov, J.; Theocharous, C.; Capp, A.; et al. Prediction of local recurrence, distant metastases, and death after breast-conserving therapy in early-stage invasive breast cancer using a five-biomarker panel. *J. Clin. Oncol.* **2009**, *27*, 4701–4708. [CrossRef] [PubMed]
4. Rexer, B.N.; Arteaga, C.L. Intrinsic and acquired resistance to HER2-targeted therapies in HER2 gene-amplified breast cancer: Mechanisms and clinical implications. *Crit. Rev. Oncog.* **2012**, *17*, 1–16. [CrossRef] [PubMed]
5. Liu, S.H.; Chen, P.S.; Huang, C.C.; Hung, Y.T.; Lee, M.Y.; Lin, W.H.; Lin, Y.C.; Lee, A.Y. Unlocking the Mystery of the Therapeutic Effects of Chinese Medicine on Cancer. *Front. Pharmacol.* **2021**, *11*, 601785. [CrossRef]
6. Wang, J.; Iannarelli, R.; Pucciarelli, S.; Laudadio, E.; Galeazzi, R.; Giangrossi, M.; Falconi, M.; Cui, L.; Navia, A.M.; Buccioni, M.; et al. Acetylshikonin isolated from *Lithospermum erythrorhizon* roots inhibits dihydrofolate reductase and hampers autochthonous mammary carcinogenesis in Δ16HER2 transgenic mice. *Pharmacol. Res.* **2020**, *161*, 105123. [CrossRef]
7. Xu, J.; Shen, R.; Jiao, Z.; Chen, W.; Peng, D.; Wang, L.; Yu, N.; Peng, C.; Cai, B.; Song, H.; et al. Current Advancements in Antitumor Properties and Mechanisms of Medicinal Components in Edible Mushrooms. *Nutrients* **2022**, *14*, 2622. [CrossRef]
8. Yuan, S.; Gopal, J.V.; Ren, S.; Chen, L.; Liu, L.; Gao, Z. Anticancer fungal natural products: Mechanisms of action and biosynthesis. *Eur. J. Med. Chem.* **2020**, *202*, 112502. [CrossRef]
9. Fordjour, E.; Manful, C.F.; Javed, R.; Galagedara, L.W.; Cuss, C.W.; Cheema, M.; Thomas, R. Chaga mushroom: A super-fungus with countless facets and untapped potential. *Front. Pharmacol.* **2023**, *14*, 1273786. [CrossRef]
10. Szychowski, K.A.; Skóra, B.; Pomianek, T.; Gmiński, J. Inonotus obliquus—From folk medicine to clinical use. *J. Tradit. Complement. Med.* **2020**, *11*, 293–302. [CrossRef] [PubMed]
11. Lu, Y.; Jia, Y.; Xue, Z.; Li, N.; Liu, J.; Chen, H. Recent Developments in *Inonotus obliquus* (Chaga mushroom) Polysaccharides: Isolation, Structural Characteristics, Biological Activities and Application. *Polymers* **2021**, *13*, 1441. [CrossRef] [PubMed]
12. Plehn, S.; Wagle, S.; Rupasinghe, H.P.V. Chaga mushroom triterpenoids as adjuncts to minimally invasive cancer therapies: A review. *Curr. Res. Toxicol.* **2023**, *5*, 100137. [CrossRef] [PubMed]
13. Chung, M.J.; Chung, C.K.; Jeong, Y.; Ham, S.S. Anticancer activity of subfractions containing pure compounds of Chaga mushroom (*Inonotus obliquus*) extract in human cancer cells and in Balbc/c mice bearing Sarcoma-180 cells. *Nutr. Res. Pract.* **2010**, *4*, 177–182. [CrossRef] [PubMed]
14. Baek, J.; Roh, H.S.; Baek, K.H.; Lee, S.; Lee, S.; Song, S.S.; Kim, K.H. Bioactivity-based analysis and chemical characterization of cytotoxic constituents from Chaga mushroom (*Inonotus obliquus*) that induce apoptosis in human lung adenocarcinoma cells. *J. Ethnopharmacol.* **2018**, *224*, 63–75. [CrossRef]
15. Lee, S.H.; Hwang, H.S.; Yun, J.W. Antitumor activity of water extract of a mushroom, *Inonotus obliquus*, against HT-29 human colon cancer cells. *Phytother. Res.* **2009**, *23*, 1784–1789. [CrossRef]
16. Youn, M.J.; Kim, J.K.; Park, S.Y.; Kim, Y.; Park, C.; Kim, E.S.; Park, K.I.; So, H.S.; Park, R. Potential anticancer properties of the water extract of Inonotus [corrected] obliquus by induction of apoptosis in melanoma B16-F10 cells. *J. Ethnopharmacol.* **2009**, *121*, 221–228. [CrossRef]
17. Youn, M.J.; Kim, J.K.; Park, S.Y.; Kim, Y.; Kim, S.J.; Lee, J.S.; Chai, K.Y.; Kim, H.J.; Cui, M.X.; So, H.S.; et al. Chaga mushroom (*Inonotus obliquus*) induces G0/G1 arrest and apoptosis in human hepatoma HepG2 cells. *World J. Gastroenterol.* **2008**, *14*, 511–517. [CrossRef]
18. Minekus, M.; Alminger, M.; Alvito, P.; Ballance, S.; Bohn, T.; Bourlieu, C.; Carrière, F.; Boutrou, R.; Corredig, M.; Dupont, D. A standardised static in vitro digestion method suitable for food—An international consensus. *Food Funct.* **2014**, *5*, 1113–1124. [CrossRef]
19. Ratzon, E.; Najajreh, Y.; Salem, R.; Khamaisie, H.; Ruthardt, M.; Mahajna, J. Platinum (IV)-fatty acid conjugates overcome inherently and acquired Cisplatin resistant cancer cell lines: An in-vitro study. *BMC Cancer* **2016**, *16*, 140. [CrossRef]
20. Vistica, V.T.; Skehan, P.; Scudiero, D.; Monks, A.; Pittman, A.; Boyd, M.R. Tetrazolium-based assays for cellular viability: A critical examination of selected parameters affecting formazan production. *Cancer Res.* **1991**, *51*, 2515–2520.
21. Calzetta, L.; Matera, M.G.; Cazzola, M. Pharmacological mechanisms leading to synergy in fixed-dose dual bronchodilator therapy. *Curr. Opin. Pharmacol.* **2018**, *40*, 95–103. [CrossRef] [PubMed]
22. Kalogris, C.; Garulli, C.; Pietrella, L.; Gambini, V.; Pucciarelli, S.; Lucci, C.; Tilio, M.; Zabaleta, M.E.; Bartolacci, C.; Andreani, C.; et al. Sanguinarine suppresses basal-like breast cancer growth through dihydrofolate reductase inhibition. *Biochem. Pharmacol.* **2014**, *90*, 226–234. [CrossRef] [PubMed]
23. Kielkopf, C.L.; Bauer, W.; Urbatsch, I.L. Bradford Assay for Determining Protein Concentration. *Cold Spring Harb. Protoc.* **2020**, *2020*, 102269. [CrossRef] [PubMed]
24. Cody, V.; Luft, J.R.; Pangborn, W. Understanding the role of Leu22 variants in methotrexate resistance: Comparison of wild-type and Leu22Arg variant mouse and human dihydrofolate reductase ternary crystal complexes with methotrexate and NADPH. *Acta Crystallogr. D Biol. Crystallogr.* **2005**, *61*, 147–155. [CrossRef]
25. *Molecular Operating Environment (MOE), 2022.02*; Chemical Computing Group ULC: Montreal, QC, Canada, 2024.
26. Kim, J.H.; Gao, D.; Cho, C.W.; Hwang, I.; Kim, H.M.; Kang, J.S. A Novel Bioanalytical Method for Determination of Inotodiol Isolated from *Inonotus obliquus* and Its Application to Pharmacokinetic Study. *Plants* **2021**, *10*, 1631. [CrossRef]
27. Sun, Y.; Feng, F.; Nie, B.; Cao, J.; Zhang, F. High throughput identification of pentacyclic triterpenes in Hippophae rhamnoides using multiple neutral loss markers scanning combined with substructure recognition (MNLSR). *Talanta* **2019**, *205*, 120011. [CrossRef]

28. Wagle, S.; Lee, J.A.; Rupasinghe, H.P.V. Synergistic Cytotoxicity of Extracts of Chaga Mushroom and Microalgae against Mammalian Cancer Cells In Vitro. *Oxid. Med. Cell Longev.* **2024**, *2024*, 7944378. [CrossRef]
29. VanArsdale, T.; Boshoff, C.; Arndt, K.T.; Abraham, R.T. Molecular Pathways: Targeting the Cyclin D-CDK4/6 Axis for Cancer Treatment. *Clin. Cancer Res.* **2015**, *21*, 2905–2910. [CrossRef]
30. Girardini, J.E.; Napoli, M.; Piazza, S.; Rustighi, A.; Marotta, C.; Radaelli, E.; Capaci, V.; Jordan, L.; Quinlan, P.; Thompson, A. A Pin1/mutant p53 axis promotes aggressiveness in breast cancer. *Cancer Cell.* **2011**, *20*, 79–91. [CrossRef]
31. Luo, J.; Zou, H.; Guo, Y.; Tong, T.; Ye, L.; Zhu, C.; Deng, L.; Wang, B.; Pan, Y.; Li, P. SRC kinase-mediated signaling pathways and targeted therapies in breast cancer. *Breast Cancer Res.* **2022**, *24*, 99. [CrossRef]
32. Feder, J.N.; Assaraf, Y.G.; Seamer, L.C.; Schimke, R.T. The pattern of dihydrofolate reductase expression through the cell cycle in rodent and human cultured cells. *J. Biol. Chem.* **1989**, *264*, 20583–20590. [CrossRef] [PubMed]
33. Bian, L.; Yu, P.; Wen, J.; Li, N.; Huang, W.; Xie, X.; Ye, F. Survival benefit of platinum-based regimen in early stage triple negative breast cancer: A meta-analysis of randomized controlled trials. *NPJ Breast Cancer* **2021**, *7*, 157. [CrossRef] [PubMed]
34. Lin, S.R.; Chang, C.H.; Hsu, C.F.; Tsai, M.J.; Cheng, H.; Leong, M.K.; Sung, P.J.; Chen, J.C.; Weng, C.F. Natural compounds as potential adjuvants to cancer therapy: Preclinical evidence. *Br. J. Pharmacol.* **2020**, *177*, 1409–1423. [CrossRef] [PubMed]
35. Lou, H.; Li, H.; Zhang, S.; Lu, H.; Chen, Q. A Review on Preparation of Betulinic Acid and Its Biological Activities. *Molecules* **2021**, *26*, 5583. [CrossRef] [PubMed]
36. Taghavi Pourianazar, N.; Gunduz, U. Changes in apoptosis-related gene expression and cytokine release in breast cancer cells treated with CpG-loaded magnetic PAMAM nanoparticles. *Int. J. Pharm.* **2016**, *515*, 11–19. [CrossRef]
37. Hartman, Z.C.; Poage, G.M.; den Hollander, P.; Tsimelzon, A.; Hill, J.; Panupinthu, N.; Zhang, Y.; Mazumdar, A.; Hilsenbeck, S.G.; Mills, G.B.; et al. Growth of triple-negative breast cancer cells relies upon coordinate autocrine expression of the proinflammatory cytokines IL-6 and IL-8. *Cancer Res.* **2013**, *73*, 3470–3480. [CrossRef]
38. Schafer, Z.T.; Brugge, J.S. IL-6 involvement in epithelial cancers. *J. Clin. Investig.* **2007**, *117*, 3660–3663. [CrossRef]
39. Mishra, S.K.; Kang, J.H.; Kim, D.K.; Oh, S.H.; Kim, M.K. Orally administered aqueous extract of Inonotus obliquus ameliorates acute inflammation in dextran sulfate sodium (DSS)-induced colitis in mice. *J. Ethnopharmacol.* **2012**, *143*, 524–532. [CrossRef]
40. Esquivel-Velázquez, M.; Ostoa-Saloma, P.; Palacios-Arreola, M.I.; Nava-Castro, K.E.; Castro, J.I.; Morales-Montor, J. The role of cytokines in breast cancer development and progression. *J. Interferon Cytokine Res.* **2015**, *35*, 1–16. [CrossRef]
41. Teijeira, A.; Garasa, S.; Ochoa, M.C.; Villalba, M.; Olivera, I.; Cirella, A.; Eguren-Santamaria, I.; Berraondo, P.; Schalper, K.A.; de Andrea, C.E.; et al. IL8, Neutrophils, and NETs in a Collusion against Cancer Immunity and Immunotherapy. *Clin. Cancer Res.* **2021**, *27*, 2383–2393. [CrossRef]
42. Matlung, H.L.; Babes, L.; Zhao, X.W.; van Houdt, M.; Treffers, L.W.; van Rees, D.J.; Franke, K.; Schornagel, K.; Verkuijlen, P.; Janssen, H.; et al. Neutrophils Kill Antibody-Opsonized Cancer Cells by Trogoptosis. *Cell Rep.* **2018**, *23*, 3946–3959.e6. [CrossRef]
43. Espinoza, J.A.; Jabeen, S.; Batra, R.; Papaleo, E.; Haakensen, V.; Timmermans Wielenga, V.; Møller Talman, M.L.; Brunner, N.; Børresen-Dale, A.L.; Gromov, P.; et al. Cytokine profiling of tumor interstitial fluid of the breast and its relationship with lymphocyte infiltration and clinicopathological characteristics. *Oncoimmunology* **2016**, *5*, e1248015. [CrossRef] [PubMed]
44. Blomberg, O.S.; Spagnuolo, L.; Garner, H.; Voorwerk, L.; Isaeva, O.I.; van Dyk, E.; Bakker, N.; Chalabi, M.; Klaver, C.; Duijst, M.; et al. IL-5-producing CD4+ T cells and eosinophils cooperate to enhance response to immune checkpoint blockade in breast cancer. *Cancer Cell* **2023**, *41*, 106–123.e10. [CrossRef] [PubMed]
45. Pathak, M.P.; Pathak, K.; Saikia, R.; Gogoi, U.; Ahmad, M.Z.; Patowary, P.; Das, A. Immunomodulatory effect of mushrooms and their bioactive compounds in cancer: A comprehensive review. *Biomed. Pharmacother.* **2022**, *149*, 112901. [CrossRef] [PubMed]
46. Fang, J.; Gao, S.; Islam, R.; Teramoto, Y.; Maeda, H. Extracts of Phellinus linteus, Bamboo (Sasa senanensis) Leaf and Chaga Mushroom (*Inonotus obliquus*) Exhibit Antitumor Activity through Activating Innate Immunity. *Nutrients* **2020**, *12*, 2279. [CrossRef]
47. Maza, P.; Lee, J.H.; Kim, Y.S.; Sun, G.M.; Sung, Y.J.; Ponomarenko, L.P.; Stonik, V.A.; Ryu, M.; Kwak, J.Y. Inotodiol From Inonotus obliquus Chaga Mushroom Induces Atypical Maturation in Dendritic Cells. *Front. Immunol.* **2021**, *12*, 650841. [CrossRef]

biomolecules

MDPI

Review

Defibrotide for Protecting Against and Managing Endothelial Injury in Hematologic Malignancies and COVID-19

Edward Richardson [1], Clifton C. Mo [2], Eleonora Calabretta [3,4,5], Francesco Corrado [2,3,4,6], Mehmet H. Kocoglu [7,8], Rebecca M. Baron [9], Jean Marie Connors [10], Massimo Iacobelli [11], Lee-Jen Wei [12], Emily J. Benjamin [2], Aaron P. Rapoport [7,8], Maribel Díaz-Ricart [13,14], Antonio José Martínez-Mellado [15], Carmelo Carlo-Stella [3,4], Paul G. Richardson [2,*,†] and José M. Moraleda [16,†]

1 Department of Medicine, Warren Alpert Medical School at Brown University, Providence, RI 02903, USA
2 Department of Medical Oncology, Dana-Farber Cancer Institute, Jerome Lipper Center for Multiple Myeloma Research, Harvard Medical School, Boston, MA 02215, USA
3 Department of Biomedical Sciences, Humanitas University, 20089 Milan, Italy
4 IRCCS Humanitas Research Hospital, 20089 Milan, Italy
5 Department of Medical Oncology, Dana-Farber Cancer Institute, Boston, MA 02215, USA
6 Broad Institute of Massachusetts Institute of Technology (MIT) and Harvard, Cambridge, MA 02142, USA
7 University of Maryland Greenebaum Comprehensive Cancer Center, Department of Medicine, University of Maryland School of Medicine, Baltimore, MD 21201, USA
8 Transplant and Cellular Therapy Program, University of Maryland Greenebaum Comprehensive Cancer Center, Baltimore, MD 21201, USA
9 Division of Pulmonary and Critical Care Medicine, Brigham and Women's Hospital, Harvard Medical School, Boston, MA 02115, USA
10 Division of Hematology, Brigham and Women's Hospital, Boston, MA 02115, USA
11 Techitra S.r.l., 20123 Milan, Italy
12 Department of Biostatistics, Harvard T.H. Chan School of Public Health, Boston, MA 02115, USA
13 Hematopathology, Pathology Department, CDB, Hospital Clinic, IDIBAPS, 08036 Barcelona, Spain
14 Barcelona Endothelium Team, 08036 Barcelona, Spain
15 Department of Hematology, University Hospital Virgen de la Arrixaca, IMIB-Pascual Parrilla, University of Murcia, 30120 Murcia, Spain
16 Department of Medicine, Faculty of Medicine, Institute of Biomedical Research (IMIB-Pascual Parrilla), University of Murcia, 30120 Murcia, Spain
* Correspondence: paul_richardson@dfci.harvard.edu
† Co-senior authors.

Abstract

Defibrotide, which is approved for treating hepatic veno-occlusive disease (VOD)/sinusoidal obstruction syndrome (SOS), exhibits pleiotropic anti-inflammatory, anti-thrombotic, and fibrinolytic properties, conferring broad endothelial protective effects. Given these mechanisms, defibrotide has potential utility in various conditions involving endothelial injury or activation. In this review we outline the endothelial-protective mechanisms of defibrotide and comprehensively summarize current evidence supporting its applications in hematologic malignancies, including the prevention and treatment of hepatic VOD/SOS, graft-versus-host disease, and transplant-associated thrombotic microangiopathy. Additionally, we discuss its role in mitigating key toxicities linked to chimeric antigen receptor (CAR) T-cell therapies and bispecific antibodies, such as cytokine release syndrome (CRS) and immune effector cell-associated neurotoxicity syndrome (ICANS). We also explore emerging evidence on defibrotide's potential in SARS-CoV-2 infection-associated endotheliopathies, including acute COVID-19 and post-acute sequelae of SARS-CoV-2 infection ("long-COVID"), and the endothelial protective activity of defibrotide in these settings. Finally, we highlight potential future applications of defibrotide in hematologic malignancies and viral infections, emphasizing its multimodal mechanism of action.

 https://doi.org/10.3390/biom15071004

Keywords: CAR T-cell therapy; bispecific antibodies/bispecific T-cell engagers; COVID-19; CRS; endotheliitis; endotheliopathy; endothelial injury; inflammation; TA-TMA; VOD/SOS

1. Introduction

Defibrotide (DF), a polydisperse oligonucleotide mixture derived from controlled depolymerization of porcine gut mucosa, comprises approximately 90% single-stranded and 10% double-stranded phosphodiester oligonucleotides [1,2]. DF has a broad range of anti-inflammatory, anti-thrombotic, and fibrinolytic properties. Through these mechanisms, DF provides protective effects in multiple settings of endothelial injury or activation. DF is approved in the US and the EU for the treatment of hepatic veno-occlusive disease (VOD)/sinusoidal obstruction syndrome (SOS) [1–3]. DF's range of endothelial protective mechanisms nonetheless extends its utility to other conditions mediated by endothelial dysfunction. These include managing common toxicities associated with immune effector cell therapies—chimeric antigen receptor (CAR) T-cell therapies and bispecific antibodies—used for the treatment of multiple myeloma (MM) and other hematologic malignancies. SARS-CoV-2 infection—acute COVID-19 and post-acute sequelae of SARS-CoV-2 infection (PASC, or "long COVID")—also features endothelial disruption as a central process of pathobiology, and DF may offer benefits in potentially reversing endotheliitis in these diseases. Here, we review the effects of defibrotide on the endothelium, its protective activity, and potential roles in managing and preventing endothelial damage in patients with hematologic malignancies, COVID-19, and/or PASC.

2. Mechanisms of Endothelial Protection with Defibrotide

2.1. Functions of the Endothelium and Impacts of Injury/Activation

The endothelium and endothelial cells have a variety of functions that may differ according to tissue location [4–7]; in the context of this review, the key roles addressed are as follows. The endothelium is critical for the regulation of host defense and inflammation through the expression of adhesion molecules including intercellular cell adhesion molecule 1 (ICAM-1), vascular cell adhesion molecule 1 (VCAM-1), E-selectin, and P-selectin, as well as through cytokine expression, including tumor necrosis factor alpha (TNFα) and interleukin (IL) 1β. Endothelial cells produce and respond to vascular endothelial growth factor (VEGF), basic fibroblast growth factor (bFGF)/FGF2, and angiopoietin-2 (Ang-2) to undergo angiogenesis, the formation of novel vessels from existing ones [8]. The endothelium also contributes to vascular homeostasis through the expression of angiopoietin-1 and 2 (Ang-1/Ang-2) and the receptor tyrosine kinase Tie-2. These mechanisms impact vascular permeability through tight junction modulation, claudin 14, and junctional adhesion molecule (JAM) expression [4,9]. The integrity of the vascular endothelium maintains a selective barrier between the bloodstream and tissues, including the blood–brain barrier protecting the central nervous system (CNS). An additional key role of the endothelium is in hemostasis, as endothelial cells directly regulate coagulation, thrombogenesis, and fibrinolysis; specifically, endothelial cells maintain the intricate balance between the expression/release of procoagulant factors, such as tissue factor (TF), von Willebrand factor (vWF), platelet-activating mediators (e.g., ADP, thromboxane A2), thrombin, and factors VII, VIII, and X, and anticoagulant factors, such as prostacyclin, nitric oxide (NO), protein C receptor, TF pathway inhibitor (TFPI), thrombomodulin (TM), and plasminogen activator inhibitor-1 (PAI-1) [4].

Endothelial activation can occur through various mechanisms, and endothelial dysfunction may occur in the presence of multiple stimuli or noxae [10]. Pathogens, disease states, and drugs (e.g., conditioning regimens, immunotherapeutics) can activate the complement system, which subsequently drives endothelial injury and activates inflammatory and microthrombotic pathways [10]. Repeated exposure to multiple noxae results in dysregulated immune responses and pathologic endothelial activation, owing to dysregulated expression of cellular and soluble signaling mediators [11]. Endothelial dysfunction is effectively a functional imbalance reflected by changes in a number of biomarkers, including increases in markers of inflammation (such as ICAM-1, VCAM-1, E-selectin, TNFα, IL-1, IL-6), changes in markers of vascular tone and homeostatic balance (e.g., increased Ang-2/Ang-1 ratio, VEGFα, fibroblast growth factor 2 (FGF2); reduced NO and prostacyclin), and increases in procoagulant/prothrombogenic markers (including TM, vWF, TF, neutrophil extracellular traps [NETs]). These specific changes differ according to the source of endothelial injury.

Indices for determining endothelial activation, such as the Endothelial Activation and Stress Index (EASIX) and the modified EASIX (mEASIX) [12–18], use related clinical and laboratory parameters including lactate dehydrogenase (LDH), creatinine, platelet count, and C-reactive protein (CRP). These indices and other markers of endothelial damage have been associated with various different endothelial-related toxicities and sequelae [12], including VOD, transplant-associated thrombotic microangiopathy (TA-TMA), grade 2–4 acute graft-versus-host disease (GvHD) [13,14], cardiac adverse events [15], as well as with non-relapse mortality (NRM) and overall survival (OS) following allogeneic HCT [13,16]. Notably, in an abstract presented at the American Society of Hematology (ASH) 2024 meeting, a Spanish study of 110 patients with lymphoma and MM receiving CAR T-cell therapy showed a positive correlation between both EASIX and mEASIX and biomarkers of endothelial dysfunction, Ang-2 and suppression of tumorigenicity 2 (ST2). Moreover, the study showed significantly higher mEASIX scores in patients with sepsis compared to cytokine release syndrome (CRS) due to CAR T-cell therapy, a distinction that represents a common diagnostic challenge in clinical practice [18].

2.2. *Mechanisms of Action of Defibrotide*

DF has been shown to modulate multiple markers of endothelial dysfunction and thereby offer treatment of endotheliopathies in a range of scenarios, as reviewed previously (Figure 1) [19,20]. Cellular effects include decreased inflammation through reductions in proinflammatory cytokines such as IL-6, IL-12, TNFα, and thromboxane A2, and increases in anti-inflammatory cytokines including IL-10 and TNFβ. DF reduces cell adhesion by decreasing ICAM-1, VCAM-1, and P/E-selectin expression and increasing prostaglandin I2 and prostaglandin E2; DF also restores thrombo-fibrinolytic balance by reducing vWF, TF, and PAI-1 levels and increasing levels of thrombomodulin and t-PA, as well as enhancing the activity of plasmin [21]. Furthermore, DF reduces endothelial cell activation (reduced PI3K/AKT, vascular E-cadherin [VE-cadherin], p38 MAPK; increased bFGF, VEGF) and maintains vascular tone through the induction of endothelin-1 and increased production of NOS.

DF has been shown to have a protective effect against the proinflammatory and prothrombotic effects of cyclosporine A and tacrolimus plus sirolimus on microvascular endothelial cells, attenuating the increased expression of ICAM-1 and elevated extracellular matrix reactivity [22]. It has also been shown to modulate pathway activation in lipopolysaccharide-activated endothelial cells associated with leukocyte migration and activation, vasculogenesis, and inflammatory responses [23]. Furthermore, through its effects on the PI3K/Akt signaling pathway, DF prevented the upregulation of histone deacetylase expression in human umbilical cord vein endothelial cells following exposure to sera from patients with end-stage renal disease on hemodialysis; DF also inhibited upregulation of endothelial activation markers, including ICAM-1, vWF, and ROS [24].

Related findings from recent studies support and extend these mechanisms of action of DF [25–27]. For example, DF has been shown to be effective at suppressing NET formation and venous thrombosis in a mouse model of antiphospholipid syndrome [28,29]. In an abstract presented at the ASH 2024 meeting, it was also shown to increase fibrinolytic activity, as demonstrated by elevated levels of tissue plasminogen activator (tPA) and prostacyclin (PGI_2), in patients with acute chest syndrome related to sickle cell disease, highlighting its ability to reduce hypercoagulability [30].

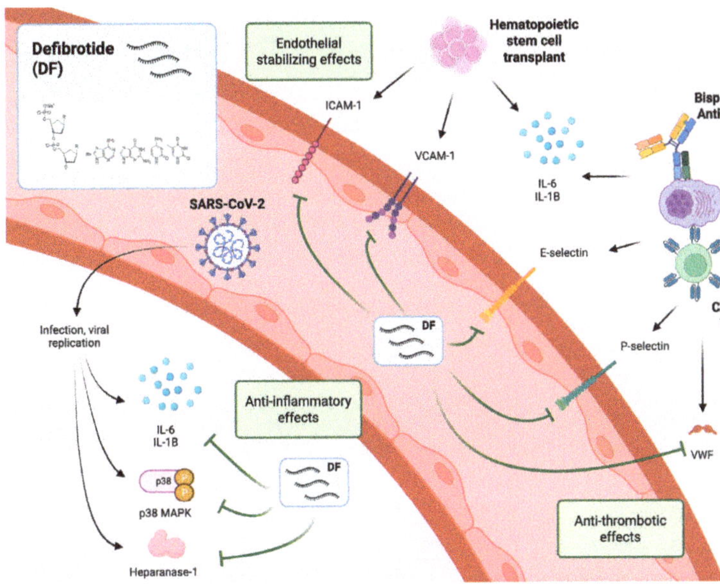

Figure 1. Defibrotide endothelial protective mechanisms in the settings of hematopoietic stem cell transplant, SARS-CoV-2 infection/PASC, and CAR T-cell therapy or bispecific antibody therapy in RRMM. Created in BioRender. Richardson, E. (2025) https://BioRender.com/hh7w5kb. Adapted and updated from Mo CC, et al. Blood Rev;2024;66:101218 [31] and Richardson PG, et al. Bone Marrow Transplant; 2021;56(12):2889–96 [19], under Creative Commons Attribution 4.0 International license (CC BY 4.0; https://creativecommons.org/licenses/by/4.0/, accessed on 19 June 2025). CAR, chimeric antigen receptor; DF, defibrotide; ICAM-1, intercellular adhesion molecule; IL, interleukin; MAPK, mitogen-activated protein kinase; PASC, post-acute sequelae of SARS-CoV-2 infection; RRMM, relapsed/refractory multiple myeloma; SARS-CoV-2, severe acute respiratory syndrome coronavirus 2; VCAM-1, vascular cell adhesion molecule-1; vWF von Willebrand factor.

3. Defibrotide for Managing and Protecting Against Endothelial Injury in Hematologic Malignancies

MM and other hematologic malignancies, such as lymphomas and leukemias, are associated with direct endothelial damage. For example, MM is characterized by elevated markers of inflammation and thrombosis/coagulopathy [32], increased cytokine signaling [31], and dysregulated signaling between MM cells and the bone marrow (BM) microenvironment [33]. Furthermore, several standard-of-care treatment modalities, including hematopoietic cell transplantation (HCT) and certain pharmacotherapies, are linked to the development of endotheliopathies [34,35], for which DF may be utilized as treatment or prophylaxis.

3.1. Hepatic VOD/SOS

VOD/SOS is a potentially life-threatening complication of HCT with an overall incidence of 2–14% and up to 60% in high-risk patients [36–40]. The pathophysiology involves primary injury to sinusoidal endothelial cells, hepatocytes, and stellate cells, giving rise to venular microthrombosis, fibrin deposition, ischemia, and fibrogenesis, with major systemic complications including portal hypertension, hepatorenal syndrome, multi-organ dysfunction (MOD), and potentially death [41,42]. The vascular endothelium is the primary target of therapeutic strategies in VOD/SOS, as toxic metabolites from high-dose chemotherapy conditioning directly affect the endothelium, resulting in increased adhesion molecule expression, cytokine signaling, and expression of procoagulant factors such as vWF [43–45].

3.1.1. Mechanism of Action of Defibrotide in Hepatic VOD/SOS

DF has demonstrated a range of effects in the setting of hepatic VOD/SOS. It provides endothelial protection through the maintenance of sinusoidal vascular integrity and the reduction in heparanase expression [45,46], and exerts anti-inflammatory effects via reductions in TNFα, VCAM-1, and p38 mitogen-activated protein kinase (MAPK) as well as Akt phosphorylation [45,47–49]. Additionally, DF helps restore the thrombotic–fibrinolytic balance through reductions in TF and PAI-1 and augmented t-PA activity [45,50–52]. DF (but not low-molecular-weight heparin [LMWH]) has also been shown to prolong survival in a rat model of monocrotaline (pyrrolizidine alkaloid)-induced SOS and to reduce levels of TNFα and PAI-1 [53].

3.1.2. Defibrotide for Managing Hepatic VOD/SOS

Multiple clinical studies and real-world analyses have demonstrated the efficacy of DF for the management of hepatic VOD/SOS (Table 1) [39,54–81]. The severity of VOD/SOS is variable; however, historical mortality in severe cases with MOD prior to DF treatment exceeded 80% [36]. As outlined below, numerous studies of DF in VOD/SOS present a marked reduction in this figure. Based on its demonstrated efficacy, DF is a cost-effective option for treating severe VOD/SOS with MOD [82].

A previous systematic review included findings from 17 prospective and retrospective studies for a total of 2598 patients with hepatic VOD/SOS treated with DF [83]. The analysis yielded a pooled day +100 survival rate of 54% overall and 56% in patients receiving standard DF dose of approximately 25 mg/kg/day. Overall, 1260 of the 2598 patients had MOD. The day +100 survival rates were 44% and 71% in patients who received the standard DF dose, with or without MOD, respectively. Respective rates in pediatric and adult patients treated with the standard DF dose were 68% and 48% [83]. A second systematic review and meta-analysis involving 3002 patients treated with DF also showed the efficacy of DF in this setting, with a complete resolution (CR) rate and a day +100 survival rate in the whole cohort of 57% and 58% and in the group with severe VOD of 39% and 44%, respectively [84].

Table 1. Clinical trials and real-world studies of DF treatment for hepatic VOD/SOS.

Study	Patients	VOD/SOS Severity	CR Rate	Survival	Safety
Clinical Trials					
Phase 3 [54]	102 (DF) vs. 32 matched controls	Severe VOD/SOS with renal and/or pulmonary failure	Day +100: 25.5% vs. 12.5%	Day +100: 38.2% vs. 25.0%	• Any hemorrhagic AEs 64% vs. 75% • Any hypotensive AEs 39% vs. 50% • Diarrhea 23.5% vs. 37.5% • Coagulopathy 2% vs. 15.6% • Possible DF-related AE leading to discontinuation 11%
Randomized dose-finding trial [70]	75 lower-dose DF/ 73 higher-dose DF	Severe VOD/SOS	Day +100: 46% (49%/43%)	Day +100: 42% (44%/39%)	• TRAEs 8% • Grade 3/4 expected AEs 54% • Grade 3–5 renal failure 31%, hypotension 29%, hypoxia 26%, other pulmonary 22%
EBMT prospective observational study [69]	104	62 with severe VOD/SOS	Day +100: 73% MOD/MOF resolution in 53%	Day +100: 73%	• SAEs 32% • Infection 24% • Bleeding 13%
Real-world studies					
Expanded access study [55]	1000 with post-HCT VOD/SOS	512 with MOD	NR	Day +100: 58.9% 49.5% with MOD 68.9% without MOD 47.1% adults 67.9% pediatric	• SAEs 53.7% • TRAEs 21.0% • Hemorrhage 29.0% • Hypotension 12.0%
International CUP [61]	710	41% with MOF 48% severe VOD/SOS	NR	Day +100: 54% 40% with MOF 65% without MOF 46% adult 65% pediatric	• AEs 53% • SAEs 51% • Withdrawal due to AE 9% • Sepsis 7% • GI hemorrhage 3%
Spanish GETH/GETMON analysis [66]	253 pediatric, 135 adult patients	Severe/very severe VOD/SOS in 173 patients, moderate VOD/SOS in 41	NR	Day +100: 62% (severe/very severe), 80% (moderate)	• Acceptable safety profile

Table 1. *Cont.*

Study	Patients	VOD/SOS Severity	CR Rate	Survival	Safety
DEFIFrance [56]	251	Severe/very severe	55% (84% pediatric, 46% adult) Day +100 rate: 74% (84% in severe cases, 63% in very severe cases)	Day +100: 61% (75% in severe disease, 49% in very severe disease)	• SAEs 29% • Infection 17% • Hemorrhage 16% • Hypotension 2%
Multicenter Australian registry study [81]	111 adult, 75 pediatric: DF use in 83/73	NR	NR	Day +100: 51.8% (adult), 90.4% (pediatric)	• NR
Italian AIEOP retrospective analysis [64]	103 pediatric, 67% received DF	Very severe or severe in all patients	NR	1-year survival: 61%	• NR
Multi-institutional study [71]	88	100% severe, 97% MOF	36%	Day +100 35%	• No worsening of clinical bleeding • No grade 3/4 AEs attributed to DF
Expanded access study [65]	82 non-transplant-associated VOD/SOS	38 VOD/SOS with MOD, 44 without MOD	NR	Day +70: 74.1% (65.8% with MOD, 81.3% without MOD)	• 66% with AEs • 25% DF-related AEs • 22% hemorrhagic AEs • 7.3% discontinued due to DF-related AEs
Korean analysis (ASH 2024 abstract) [79]	73	40 severe, 33 very severe	39.7% (52.5% severe, 24.2% very severe)	Day +100: 34.2% (40.3% severe, 26.4% very severe)	• NR
Single-center experience [60]	51 (36 adult, 15 pediatric)	Severe VOD/SOS	Day +100: 35.3%	Day +100: 56.9%	• NR
Institutional series [63]	47 RR ALL receiving inotuzumab ozogamicin pre-HCT	12 VOD/SOS: 50% very severe, 25% severe, 25% mild	67%	Day +100 mortality rate: 33% vs. 14% in patients without VOD/SOS	• NR
Retrospective multicenter study [62]	45	49% severe 51% mild or moderately severe	76% (50% in severe disease)	Day +100: 64% (36% in severe disease)	• Coagulation abnormalities 35%

Table 1. *Cont.*

Study	Patients	VOD/SOS Severity	CR Rate	Survival	Safety
Exploratory CIBMTR analysis [75]	41 (DF) vs. 55 (no DF)	Severe	Day +100: 51% vs. 29%	Day +100: 39% vs. 31%	• Day +100 acute GvHD 23% vs. 38%
DFCI/BWH experience [57]	28 post-allo HCT	11 mild-moderate-severe, 17 very severe	75%	Day +100: 64%	• Hematuria 43% • Epistaxis 18% • Hypotension 11% • Lower GI, grade III/IV pulmonary, and grade III/IV upper GI hemorrhage each 4%
UK experience [68]	27–19/8 classic/late VOD/SOS	25 very severe, 1 severe, 1 mild	NR	Day +100: 59% (58%/63% classical/late)	• NR
Retrospective series [80]	23	NR	61%	Day +100: 70%	• NR
Single-center analysis [76]	14	6 severe, 4 moderate, 4 mild	79% (50% severe, 100% moderate/mild)	Day +100: 79%	• No significant drug-related side-effects
Retrospective study of low-dose DF [58]	9/511 patients	NR—no ventilator support or dialysis required	100%	Time to resolution: 6–20 days from onset	• NR

AE, adverse event; AIEOP, Associazione Italiana di Ematologia e Oncologia Pediatrica; ALL, acute lymphocytic leukemia; allo, allogeneic; BWH, Brigham and Women's Hospital; CIBMTR, Center for International Blood and Marrow Transplant Research; CR, complete resolution; CUP, compassionate use program; DF, defibrotide; DFCI, Dana-Farber Cancer Institute; EBMT, European Group for Blood and Marrow Transplantation; GETH/GETMON, Grupo Español De Trasplante Hematopoyetico/Grupo Español De Trasplante De Medula Osea en Niños; GI, gastrointestinal; GO, gemtuzumab ozogamicin; GvHD, graft-versus-host disease; HCT, hematopoietic cell transplant; MOD, multi-organ dysfunction; MOF, multi-organ failure; NR, not reported; RR, relapsed/refractory; SAE, serious adverse event; SOS, sinusoidal obstruction syndrome; TRAE, treatment-related adverse event; VOD, veno-occlusive disease.

These findings were mirrored in a separate pooled analysis of three studies involving 651 patients with MOD post-HCT, including 233 (36%) with ventilator and/or dialysis dependence [85]. The data showed that, while the overall day +100 survival rate was 44% in this population (n = 651), survival rates were higher in those with less severe MOD: 40% in patients without ventilator or dialysis dependence (n = 418) versus 40% in those with one dependence (n = 137), 33% in patients with one or both (i.e., ventilator and/or dialysis) dependencies (n = 233), and 28% in those with both ventilator and dialysis dependence (n = 96). Nevertheless, these day +100 survival rates were higher than those in a historical control population with VOD/SOS and MOD, in which the overall rate was 25% and the rates were 28% and 14%, respectively, in patients with no dependence versus one dependence [85].

A further pooled analysis from the same three studies (incorporating 1176 patients), demonstrated the importance of continuing DF until VOD/SOS resolution [86]. Among all 390 patients who achieved CR and had data available, the median time to DF discontinuation due to CR was 22–24.5 days, and discontinuation beyond 28 days occurred in 15–40% of patients, highlighting the benefit of continuing DF treatment past the recommended minimum of 21 days. Importantly, day +100 survival was significantly higher in those who discontinued DF due to a CR compared to those who did not (92.5% vs. 37.3%), further emphasizing the value of treatment continuation to achieve CR [86].

As well as these meta-analyses and pooled analyses, the efficacy and effectiveness of DF have been shown in multiple key individual clinical studies and real-world evaluations. For example, in the pivotal phase 3 trial of DF versus historical controls in patients with post-HCT severe VOD/SOS with renal and/or pulmonary failure (MOF) [54], 102 patients were enrolled to the DF arm and compared with 32 matched contemporaneous and validated historical controls; 43% and 44% were pediatric patients, and 88% and 84% had received an allogeneic HCT. The study illustrated the challenges of undertaking a comparative trial in a rare and complex disease state, with the contemporaneous control group requiring review of 6867 patient medical charts to obtain 32 patients with an unequivocal diagnosis of VOD/SOS with MOF; with the originally planned sample size of 80 patients in the control group proving to be not feasible, statistical analysis adjustment was required. Compared to historical controls, patients receiving DF showed a significant improvement in day +100 survival rate (38.2% vs. 25.0%), the primary endpoint of the study. The estimated between-group difference (stratified by propensity score quintile via the Koch method) was 23% (95% CI 5.2–40.8, p = 0.0109). The observed rate of CR at day +100 was also significantly greater, at 25.5% vs. 12.5%. Notably, toxicity was generally manageable with DF, with lower rates of diarrhea, hemorrhagic adverse events (AEs), hypotensive AEs, and coagulopathy than in historical controls, and with only 11% of patients discontinuing due to possible DF-related AEs [54]; indeed, while the DF prescribing information includes a warning to monitor patients for bleeding and a contraindication for concomitant administration with systemic anticoagulant or fibrinolytic therapy [1], DF does not appear to significantly increase overall rates of hemorrhage in VOD/SOS [19,87], and a meta-analysis of DF studies in the non-VOD/SOS setting has demonstrated a reduced risk of bleeding events compared with controls (risk ratio 0.36) [88]. Results from an expanded-access study of DF in 1000 patients who had post-HCT VOD/SOS, including 512 with MOD, were supportive of the phase 3 trial findings [55]; overall day +100 survival rate was 58.9%, including 49.5% in patients with MOD and 68.9% in those without, and 47.1% and 67.9% in adult and pediatric patients, respectively.

Among the real-world studies demonstrating the effectiveness of DF for VOD/SOS [56,57,68], the Dana-Farber Cancer Institute (DFCI)/Brigham and Women's Hospital

(BWH) experience in 28 patients with VOD/SOS post-allogeneic HCT showed complete resolution of VOD/SOS in 75% and a day +100 survival rate of 64%, including 57% in patients with MOD [57]. Similarly, a DEFIFrance registry study in 251 patients with severe/very severe VOD/SOS demonstrated a CR rate of 55%, including 84% and 46% in pediatric and adult patients, respectively, a day +100 CR rate of 74%, 84% in severe and 63% in very severe VOD/SOS cases, and a day +100 survival rate of 61% [56].

An important element of DF treatment for VOD/SOS is prompt diagnosis and initiation of therapy [38,89,90], which has been shown to offer improved outcomes compared with delayed DF treatment [91,92]. In an analysis of data from 573 patients on the DF expanded access protocol, 31.9% received DF on the day of VOD/SOS diagnosis and 93.0% had started DF by day +7 [92]. Day +100 survival rate differences were 8.8% between patients starting DF on day 0/1 versus >1, 22.1% between patients starting DF on day ≤2 versus >2, and 20.3%, 20.2%, and 20.9% for subsequent cut-offs of day ≤3, ≤4, and ≤7, respectively. Similar findings were seen in patients with MOD. Supportive findings were seen in a multivariate analysis of a retrospective multicenter study, in which early DF intervention was the only factor associated with a CR [62] and in a single-center analysis of 111 pediatric patients who underwent HCT, in which early versus non-early DF intervention significantly reduced the peak grade of VOD/SOS [91]. Further, in a Korean retrospective real-world analysis, DF intervention within 2 days of VOD/SOS diagnosis resulted in a higher CR rate (55.6% vs. 30.4%) and better day +100 survival (37.0% vs. 26.1%) [79].

3.1.3. Defibrotide Prophylaxis Against Hepatic VOD/SOS

DF has also been evaluated as prophylaxis in HCT recipients for preventing hepatic VOD/SOS (Table 2) [66,80,81,93–107]. Of particular note, in a phase 3 trial of DF versus controls in 356 pediatric patients who had received myeloablative conditioning (MAC) for autologous or allogeneic HCT [93], the primary analysis demonstrated a rate of VOD/SOS at day +30 of 12.2% vs. 19.9%, a reduction of 7.7 percentage points (95% CI for risk difference, –15.3% to –0.1%, p = 0.049). This risk difference was –5.9% to –13.0% when evaluated separately in infants, children, and adolescents. Importantly, the safety profile showed no increase in cumulative hemorrhagic incidence and numerically lower rates of TA-TMA (3% vs. 4%) and fatal infections/infestations (3% vs. 6%) in patients receiving DF prophylaxis. Supportive findings were provided by a meta-analysis of 20 studies of DF as VOD/SOS prophylaxis [94]; in an analysis of 3005 patients, incidence of VOD/SOS with DF prophylaxis was 5% (5% in adults, 8% in pediatric patients), whereas incidence in controls (from comparative studies) was 16%, showing a risk ratio for developing VOD/SOS of 0.30 (95% CI 0.12–0.71, p = 0.006). Similarly, a network meta-analysis of primary prophylaxis options for VOD/SOS in patients receiving cell-based therapies showed an odds ratio of 0.64 for the development of VOD/SOS in those receiving DF prophylaxis, with a greater magnitude of benefit (odds ratio 0.51) in a subgroup analysis of patients receiving allogeneic HCT [107].

Table 2. Clinical trials and real-world studies of DF prophylaxis for hepatic VOD/SOS.

Study	Patients	VOD/SOS Rate	Time-to-Event Analyses	Safety
Clinical Trials				
Phase 3 HARMONY trial [95]	372 (174 aged >16 years, 198 aged ≤16 years), 190 DF vs. 182 BSC	Day +30: 4% vs. 4% Any time: 14% vs. 18%	Day +30 VOD/SOS-free survival: 67% vs. 73% Day +100: 50% vs. 57%	• Grade 3/4 stomatitis 29%/1% vs. 32%/1% • Grade 3/4 febrile neutropenia 28%/0% vs. 30%/2% • SAEs 41% vs. 35%
Phase 3 pediatric trial [93]	356, after MAC and auto/allo HCT	Day +30: 12.2% vs. 19.9% Infants 19.6% vs. 26.8% Children 11.0% vs. 16.8% Adolescents 7.0% vs. 20.0%	Median time from HCT to VOD: 17.5 vs. 14.0 days	• 22% vs. 21% cumulative hemorrhage incidence • 3% vs. 6% fatal infections/infestations • 3% vs. 4% TA-TMA through day 180 post-HCT
Meta-analysis [94]	3005 patients, 20 studies	5% (DF total) vs. 16% (controls, 8 studies)	NR	• Safety results generally consistent with known DF safety profile
Pediatric study in beta thalassemia [96]	57	1.8%	NR	• DF well tolerated
Real-world studies				
Turkish retrospective analysis [105]	1153 patients	8% vs. 66.7% in high-risk patients with vs. without DF	NR	• NR
DEFIFrance [56]	381 (178 pediatric, 203 adult)	20% (28% pediatric, 13% adult) by day +30	NR	• SAEs 25% • Hemorrhage 14% • Infection 13%
Single-center experience [100]	334 high-risk pediatric allo-HCT	5.1% (n = 17; 4 moderate, 13 mild cases)	NR	• NR
Spanish GETH/GETMON analysis [66]	253 pediatric, 135 adult patients; DF as prophylaxis in 135	NR	Day +100 survival: 89%	• Acceptable safety profile
Single-center retrospective analysis [97]	237 (DF) vs. 241 (non-DF) patients undergoing HCT	0% vs. 4.8%	1-year EFS: 38% vs. 28%	• Acute GvHD: 31% vs. 42%
Korean retrospective analysis [104]	69 DF vs. 78 historical controls	4.3% vs. 12.8% (2.9% vs. 28.6% in second HCT group)	0 vs. 3 VOD/SOS-related mortality	• NR

Table 2. *Cont.*

Study	Patients	VOD/SOS Rate	Time-to-Event Analyses	Safety
Single-center series [103]	63 high-risk adult patients	6.3% (2 cases within 21 days post-HCT, 2 late-onset cases)	2-year OS 56.5% 2-year non-relapse mortality 22.3%	• Bleeding 21.5% • DF discontinuation 6.3% • Grade II–IV acute GvHD 22.2% • TA-TMA 3.2%
Single-center analysis [98]	58	0%	Day +100 survival: 100%	• No hemorrhagic complications secondary to DF
Single-center experience [101]	56 adult allo-HCT	Day +30: 1.9%	1 death due to MOF at day +20 after very severe VOD/SOS	• NR
Korean retrospective analysis [102]	49 (34 high-risk)	2%	Day +100 transplant-related mortality: 0%	• No DF-related grade 3/4 toxicity • No worsening of clinical bleeding

Allo, allogeneic; BSC, best supportive care; DF, defibrotide; EFS, event-free survival; GETH/GETMON, Grupo Español De Trasplante Hematopoyetico/Grupo Español De Trasplante De Medula Osea en Niños; GvHD, graft-versus-host disease; HCT, hematopoietic cell transplant; MAC, myeloablative conditioning; MOF, multi-organ failure; NR, not reported; OS, overall survival; SAE, serious adverse event; SOS, sinusoidal obstruction syndrome; TA-TMA, transplant-associated thrombotic microangiopathy; VOD, veno-occlusive disease.

In contrast, the phase 3 HARMONY trial of DF versus best supportive care (BSC) in 372 adult and pediatric patients receiving SCT showed no difference in day +30 VOD/SOS-free survival (67% vs. 73%), and rates of VOD/SOS occurrence at any time of 14% versus 18% [95]. Interestingly, in an abstract presented at the ASH 2024 meeting, a retrospective analysis of prophylactic DF that stratified high-risk pediatric patients according to the HARMONY trial criteria similarly found no benefit in VOD/SOS prevention [108]. However, several potential issues were identified with regard to the HARMONY trial that could have contributed to the failure to demonstrate efficacy of DF as prophylaxis [109]. First, HARMONY included a broad patient population, with a limited representation of very high-risk pediatric patients. Second, with regard to the study design, the power calculation and sample size were insufficient based on an overestimate of VOD/SOS incidence in the enrolled population; for the primary endpoint of VOD/SOS-free survival at Day 30, only 7% of the trial population were at risk for this composite endpoint. Furthermore, the study design allowed for DF use for emergent VOD/SOS in the BSC group. Finally, there were discrepancies in VOD/SOS diagnosis between investigators and central adjudication employed in the trial. HARMONY thus provides further illustration of the challenges of designing and conducting comparative trials with appropriate statistical powering in a complex disease state such as VOD/SOS that is a relatively uncommon outcome. Nevertheless, despite findings from the HARMONY trial, collective evidence suggests that DF may have a beneficial impact when used as prophylaxis against VOD/SOS.

3.2. *Graft-Versus-Host Disease*

Acute GvHD is a complication of allogeneic HCT mediated by alloreactive T cells in the donor graft, which recognize mismatched HLA antigens on endothelial cells, leading to endothelial damage, a key component of the pathophysiology of the condition [110]. High-dose chemotherapy conditioning induces systemic inflammation and endothelial cell damage, and endothelial cells are activated by inflammatory cytokines and damage-associated molecular patterns (DAMPs) through toll-like receptor (TLR) signaling. This activation promotes the expression of adhesion molecules, facilitating the recruitment of innate and adaptive immune cells to sites of inflammation. In the lymph nodes, host dendritic cells (DCs) present allogeneic peptides, leading to the activation of CD8+ and CD4+ T cells. Cytotoxic CD8+ T cells directly damage endothelial cells, while CD4+ T cells release inflammatory cytokines, such as TNFα and interferon-γ, further activating endothelial cells. TLR signaling via MAPKs upregulates the expression of adhesion molecules (selectins, integrins), enhancing leukocyte transmigration. Simultaneously, TNFα receptor signaling on endothelial cells increases Ang-2 expression and permeability. These mechanisms collectively contribute to the progression of GvHD in target organs.

Defibrotide for Treating or Preventing GvHD

Evidence from preclinical studies supports the potential of DF for the prevention of GvHD as well as the mechanistic rationale underlying its effects [111]. Notably, in mice receiving fully MHC-mismatched allogeneic HCT, prophylactic or therapeutic administration of DF was effective in preventing T cell and neutrophil infiltration as well as acute GvHD-related tissue-specific damage in the skin, liver, colon, and tongue. Additionally, DF treatment restored the balance of inflammatory cytokines. These effects resulted in a reduction in the incidence and severity of acute GvHD, significantly improving animal survival [111]. Furthermore, mice with acute GvHD exhibited elevated levels of proinflammatory cytokines—including interferon-γ, TNFα, IL-6, and IL-12—alongside decreased concentrations of anti-inflammatory cytokines such as TGFβ and IL-10 on day

+10 post-HCT. In contrast, mice receiving prophylactic DF showed significant reductions in pro-inflammatory mediators and increased levels of anti-inflammatory cytokines compared to untreated controls [111]. Similarly, in a recently published study using a murine model of allogeneic HCT, DF treatment improved survival and reduced clinical GvHD by exerting anti-inflammatory and endothelial protective effects, as evidenced by lower levels of TNFα, IL-6, VCAM-1, ICAM-1, and Ang-2 [112]. Moreover, in vitro studies using endothelial cell lines exposed to sera from patients with acute GvHD showed that DF suppressed markers of vascular angiogenesis and endothelial activation driven by GvHD-associated patient sera [113].

Clinical studies have also demonstrated the efficacy of DF for the prevention of GvHD. In a prespecified secondary/exploratory analysis of DF for the prevention of GvHD, as part of the phase 3 pediatric VOD/SOS prophylaxis study, DF significantly reduced the rate of acute GvHD by day +30 (34% vs. 52%) and by day +100 (47% vs. 65%), with significantly lower incidence and severity (grades 1–4) compared with controls (incidence $p = 0.0057$ and severity $p = 0.0062$ at 30 days; incidence $p = 0.0046$ and severity $p = 0.0034$ at 100 days), even when grade 1 acute disease was excluded [93]. Similarly, in a phase 2 study of DF plus standard-of-care treatment versus standard-of-care treatment alone for GvHD prophylaxis in 152 patients receiving allogeneic HCT, the cumulative incidence of grade B–D acute GvHD by day +100 post-transplant was 38.4% vs. 47.1% (37.0% vs. 45.7% in a sensitivity analysis using disease relapse as a competing risk) [114].

A retrospective analysis in 38 adult patients receiving allogeneic HCT showed that DF in combination with other immunosuppressive agents (rabbit anti-T lymphocyte globulin, post-transplant cyclophosphamide, cyclosporine) may decrease the risk of GvHD—the 1-year cumulative incidence of grade III-IV acute GvHD and moderate/severe chronic GvHD were 20.6% and 5.3%, respectively [115]. These findings suggest DF might complement other prophylactic strategies, such as post-transplant abatacept. Further, a retrospective analysis of 47 vs. 44 pediatric allogeneic HCT recipients who did versus did not receive DF prophylaxis showed a significantly lower rate of acute GvHD (23% vs. 50%, including 4% vs. 39% grade II–IV); the odds ratio for developing acute GvHD with DF prophylaxis was 0.31 overall and 0.11 for moderate/severe GvHD [116]. Notably, levels of proinflammatory cytokines were significantly lower in the DF prophylaxis versus control group. Similarly, rates of acute GvHD were lowered with the use of DF in a Turkish analysis of 195 consecutive adult patients receiving allogeneic HCT [117]; in patients receiving DF prior to HCT (concurrently with conditioning), DF post HCT, or no DF, the overall rate of acute GvHD was 25.5%, 40%, and 46.5%, respectively, and the rate of grade III-IV acute GvHD was 0%, 11.2%, and 15.5%. Conversely, however, there have been other studies that have not demonstrated a benefit from DF prophylaxis on the occurrence of acute GvHD [118].

3.3. *Transplant-Associated Thrombotic Microangiopathy*

TA-TMA is associated with abnormal endothelial cell activation, complement activation, platelet-rich thrombi formation, and microvascular hemolytic anemia, ultimately leading to end-organ dysfunction [119]. It occurs following both autologous and allogeneic stem cell transplantation but is frequently observed after allogeneic transplantation. Endothelial injury caused by high-dose conditioning regimens and calcineurin/mammalian target of rapamycin (mTOR) inhibitors for GvHD prophylaxis results in elevated levels of proinflammatory cytokines (e.g., IL-2, TNFα), procoagulant factors (e.g., vWF, TF, factor VIIa), and soluble adhesion molecules, which perpetuates the activation of the complement cascade. Development of NETs following endothelial cell damage may represent a specific mechanism of complement activation in TA-TMA. Additionally, nitric oxide (NO) depletion impairs vasodilation, promoting platelet aggregation and the development of microthrombi. Of note, in the context of MM treatment, TMA has also been associated with the use of the proteasome inhibitor (PI) carfilzomib, although the precise mechanisms underlying PI-induced TMA have not been fully elucidated [120].

Defibrotide for Treating or Preventing TA-TMA

Data from a few small studies or retrospective analyses suggest that DF has activity as a treatment for or prophylaxis against TA-TMA [121,122]. In a European Society of Blood and Marrow Transplantation retrospective study of 17 adults and 22 pediatric patients with TA-TMA who received DF, TA-TMA resolved in 77% of cases, with earlier diagnosis and treatment with DF associated with higher resolution rates [123]. An Indian retrospective case series of three patients who had TA-TMA after allogeneic HCT for AML, CML-AP, or Pro-B ALL demonstrated that low-dose DF administered for 7–19 days resulted in resolution or improvement of TA-TMA in all cases [124]. A pilot study of DF prophylaxis in 25 high-risk pediatric patients, 14 of whom were receiving tandem autologous HCT for neuroblastoma and 11 of whom were undergoing allogeneic HCT, identified only one case (4%) of non-severe TA-TMA, occurring 12 days post-HCT. This incidence was significantly lower than the historical rate of TA-TMA of 18–40% in autologous/allogeneic HCT patients [125]. Additionally, a retrospective analysis of 31 patients with TA-TMA who were treated with DF $\pm$ plasmapheresis $\pm$ rituximab showed a 61% overall response rate (100% in low-risk, 25% in high-risk patients), although outcomes were poor [126].

3.4. *Defibrotide for Treating or Preventing Idiopathic Pneumonia Syndrome (IPS)*

IPS is a non-infectious acute lung injury condition occurring post-HCT [127]. Pulmonary dysfunction in IPS and acute respiratory distress syndrome (ARDS) is mediated, at least in part, by pulmonary endothelial cell injury and activation. In mouse models of IPS and lipopolysaccharide-induced ARDS, DF has been shown to substantially modulate endothelial cell injury, with reduced expression of TNFα, Ang-2, E-/P-selectin, and IL-6 [127]. Further clinical studies are warranted to investigate a potential clinical role in IPS treatment or prophylaxis.

3.5. *Immune Effector Cell Therapy-Associated Cytokine Release Syndrome and Neurotoxicity*

CAR T-cell therapies are among the standard-of-care (SOC) therapies for relapsed/refractory multiple myeloma (RRMM), as well as for leukemias and lymphomas [128]. For RRMM, the BCMA-directed therapy idecabtagene vicleucel (ide-cel) is approved for use after $\geq$2 prior lines, including a PI, an immunomodulatory drug (IMiD), and a CD38 monoclonal antibody (mAb) [129], while ciltacabtagene autoleucel (cilta-cel, also targeting BCMA) is approved for patients who have received $\geq$1 prior line, including a PI and an IMiD, and who are refractory to lenalidomide [130]. There are also several US FDA-approved CD19-directed CAR T-cell therapies for leukemia and lymphoma [131]. For large B-cell lymphoma (LBCL)/diffuse LBCL (DLBCL), axicabtagene ciloleucel (axi-cel) is approved for patients who are refractory or relapsed within 12 months of first-line chemoimmunotherapy [132], and tisagenlecleucel (tisa-cel) is approved for patients who have received $\geq$2 prior lines [133]. Axi-cel and tisa-cel are also approved for the treatment of patients with relapsed or refractory follicular lymphoma (FL) after $\geq$2 lines of therapy [134,135]. Brexucabtagene autoleucel (brexu-cel) is approved for the treatment of relapsed or refractory mantle cell lymphoma (MCL) [136], while lisocabtagene maraleucel is approved for the treatment of relapsed or refractory FL, LBCL, and MCL [137]. Additionally, tisa-cel and brexu-cel have received approval for the treatment of relapsed/refractory B-cell precursor acute lymphoblastic leukemia (ALL) in patients aged ≤25 years and adults, respectively [138,139]. Obecabtagene autoleucel is also approved for relapsed/refractory adult B-cell precursor ALL [140], and lisocabtagene maraleucel is approved for relapsed/refractory chronic lymphocytic leukemia [137]. An important element of treatment with autologous CAR T-cell therapies is that they typically require a manufacturing time of approximately 4–10 weeks. During this period patients receive bridging therapy to control or reduce their disease, followed by lymphodepleting therapy with, for example, fludarabine [141]. This may result in additional endothelial injury and predispose patients to subsequent adverse effects.

A number of bispecific antibodies/T-cell engagers are also approved for the treatment of later-relapse RRMM, including the BCMA-targeted agents teclistamab [142] and elranatamab [143] and the GPRC5D-targeted agent talquetamab [144], with several others with similar, different, or multiple targets under investigation. These dual-specific antibodies facilitate cell-to-cell interactions between MM cells expressing tumor-specific antigens and patients' T cells via engagement of CD3 (xCD3), leading to selective cell lysis [145]. Additionally, several bispecific antibodies are approved for the treatment of non-Hodgkin's lymphoma and leukemia [146,147]; these include the CD20-targeted agents epcoritamab for DLBCL and FL, glofitamab for DLBCL and LBCL arising from FL, and mosunetuzumab for FL, as well as the CD19xCD3 agent blinatumomab for B-cell precursor ALL.

Among the common AEs associated with these immune effector cell therapies are the endotheliopathy-related toxicities of CRS and immune effector cell-associated neurotoxicity syndrome (ICANS) [31,148]. The pathogenesis of CRS involves the interaction of CAR T cells with MM cells, or the engagement of MM cells and T cells by the bispecific antibody, which results in T-cell cytokine production, as well as macrophage activation and further production of proinflammatory cytokines [149]. The cytokine storm results in endothelial activation, with major inflammatory effects mediated by TNFα, IL-6, IL-1β, interferon-γ, and potentially through NET formation, as well as increases in coagulation markers such as Ang-2 and vWF. With CAR T-cell therapy, more severe CRS has been associated with fludarabine exposure prior to CAR T-cell infusion, which potentially augments endothelial injury. The pathophysiology of ICANS is similarly driven by CAR T-cell/T-cell interaction with MM cells, with endothelial activation following CAR T-cell activation and cytokine release likely increasing blood–brain barrier permeability [149]. This can lead to elevated cytokine levels in the cerebrospinal fluid and CNS, thereby driving neuroinflammation and associated neurotoxicity. Notably, a range of studies have shown that these toxicities are associated with markers of endothelial activation at baseline or during treatment [31]. The recommended management of CRS and ICANS includes anti-cytokine agents such as tocilizumab (IL-6 receptor antagonist), anakinra (IL-1 receptor antagonist), siltuximab (anti-IL-6 mAb), etanercept (TNFα inhibitor), and infliximab (anti-TNFα mAb) [150,151]. Hemophagocytic lymphohistiocytosis (HLH) is another CAR T-cell therapy-related severe adverse effect and is associated with endothelial dysfunction leading to uncontrolled endotheliitis, likely a result of a high-cytokine milieu (primarily interferon-γ). Treatment options for HLH overlap with those for CRS and ICANS and additionally may include agents such as emapalumab (anti-interferon-γ mAb) [152].

Defibrotide for Treating or Preventing Immune Effector Cell Therapy-Associated CRS and Neurotoxicity

A phase 2 study has evaluated DF for preventing CAR T-cell therapy-associated neurotoxicity in 25 patients receiving axi-cel for relapsed/refractory diffuse large B-cell lymphoma (DLBCL) [153]. Patients received DF for 3 days in tandem with lymphodepletion therapy and then from day 0 to day +7 after axi-cel. For the primary endpoint of the day +30 rate of neurotoxicity, DF demonstrated a numerically (but not statistically significantly) lower incidence of any-grade neurotoxicity of 50% (25% grade $\geq$ 3), compared with the reference rate of 64% seen in the ZUMA-1 trial of axi-cel in B-cell lymphoma. DF as a potential preventive strategy for CRS/ICANS warrants further exploration (Figure 1), including the potential benefit of combining DF with IL-6 or IL-1 blockade in more severe cases.

4. Defibrotide for Managing and Protecting Against Endotheliopathies Associated with COVID-19

4.1. Endotheliopathies Associated with SARS-CoV-2 Infection Resulting in COVID-19 and PASC

Endothelial dysfunction is a hallmark of the pathobiology of SARS-CoV-2 infection, driving COVID-19 morbidity and mortality via cytokine release, coagulopathy, and microvascular injury [154–157]. Direct infection of endothelial cells with SARS-CoV-2 occurs through ACE-2 endothelial receptors, and infection can cause endotheliitis resulting in apoptosis, with endothelial dysfunction propagated by cytokine release following infection, leading to the activation of coagulation and inflammation. Subsequent effects include increased P-selectin and vWF expression leading to platelet activation, accumulation, and production, followed by VEGF and TF release, and complement activation and increased expression of leukocyte adhesion molecules (ICAM-1, VCAM-1, E-selectin) promoting inflammation and amplifying pathologic cytokine production. Excess cytokines subsequently impair endothelial barrier functions, as IL-1β and TNFα expression promotes the loosening of inter-endothelial junctions and associated vascular leakage. Increased IL-6, IL-8, and TNFα also drive the production and release of vasoactive molecules such as thrombin, thromboxane A2, and VEGF. Following the acute phase, viral dissemination into tissue reservoirs can result in persistent residual inflammation and prolonged endothelial activation, as seen with PASC [31,158].

For patients with hematologic malignancies, there are multiple potential sources of endothelial injury in the context of endemic COVID-19 and ongoing waves of SARS-CoV-2 infections [31], including: from the malignancy itself, as in MM-associated endotheliopathy; from injury associated with commonly used treatments, such as autologous HCT in transplant-eligible newly diagnosed MM patients, CAR T-cell therapies in the early-relapse setting, and bispecific antibodies in the later-relapse setting; and from new or previous SARS-CoV-2 infections, for which patients with hematologic malignancies are at elevated risk. Indeed, one retrospective study has suggested that patients who underwent HCT following a SARS-CoV-2 infection had a significantly higher rate of TA-TMA, and a trend for higher rates of VOD/SOS and engraftment syndrome, compared with historical controls, indicating an elevated risk for endothelial-related complications post-infection [159].

4.2. Defibrotide for Endothelial Protection in the Setting of COVID-19

There are a number of proposed mechanisms for the potential endothelial protective activity of DF in acute COVID-19 [20,157,160–162] and in PASC [158] (Figure 1). DF may counteract the endothelial effects of SARS-CoV-2 infection through increased t-PA and TM expression, decreased vWF and PAI-1 expression, and platelet adhesion inhibition via increases in NO, prostaglandin I2 (PGI2), and prostaglandin E2 (PGE2). DF may also offer anti-inflammatory properties via inhibition of the p38 MAPK pathway, attenuating release of inflammatory mediators including IL-6, thromboxane A2, leukotriene B4, TNF-alpha, and ROS. DF may also inhibit expression and activity of heparanase, modulate endothelial cell injury by downregulating expression of endothelial cell adhesion molecules such as E-selectin, VCAM-1, and ICAM-1, and increase endothelial cell release of anti-inflammatory cytokines.

Preclinical findings support these proposed mechanisms and the role of DF in reversing the endotheliitis of COVID-19. As noted previously, DF results in significant decreases in proinflammatory cytokines in mice undergoing allogeneic HCT. DF has also been shown to significantly reduce the levels of adhesion molecules, including E/P-selectin, VCAM-1, and ICAM-1 [111]. Additionally, in an analysis of human dermal microvascular endothelial cells exposed to plasma from patients with acute COVID-19, DF suppressed cellular pathways associated with endothelial activation by COVID-19 plasma, including TNFα signaling, IL-17 signaling, endothelin activity, and fibrosis [9]. Further exploration of DF and other commonly used anti-inflammatory modalities for COVID-19, such as steroids and IL-6 inhibition, is warranted to determine whether their mechanisms of action may be complementary.

4.3. Clinical Findings Demonstrating the Role of Defibrotide for Endothelial Protection in the Setting of COVID-19

A safety study at the University of Michigan investigating the role of DF in the management of SARS-CoV-2-related acute respiratory distress syndrome (ARDS) suggested beneficial effects from a 7-day course of DF [163]. Among 12 patients, 10 of whom required mechanical ventilation and 6 vasopressor support at study entry, a pulmonary response at day 7 was seen in 4 patients, with D-dimer levels decreased within the first 72 h of receiving DF. For example, one patient requiring mechanical ventilation at study entry was extubated on study day 4 following DF treatment, with subsequent removal of all supplemental oxygen on day 7. Day 30 all-cause mortality was 17%, and nine patients remained alive at 64–174 days after starting DF, indicating a 75% survivorship; for context, historical 28-day mortality rates at the time for patients with SARS-CoV-2 ARDS were 26–61.5% [163]. No hemorrhagic or thrombotic complications occurred during therapy.

A phase 2 study in Italy evaluated DF in 48 patients with COVID-19 pneumonia receiving non-invasive ventilation and compared the outcomes with 153 matched case-controls [164]. All 48 patients had a WHO score of 5 on day 1. No significant hemorrhagic or bleeding episodes occurred during the study therapy. There was a trend towards longer OS and respiratory failure-free survival in the DF vs. case-control cohort on adjusted analysis and in a survival prediction model versus SOC management. DF also resulted in a significantly greater mean number of post-recovery days; i.e., in the number of COVID-19-free days out of a predefined 28-day window: 11.60 days with DF, compared with 5.29 days in the case-control observational cohort and 7.99 days predicted for SOC management.

Similarly, in an abstract presented at the 15th Congress of the European Association for Clinical Pharmacology and Therapeutics in 2022, a Spanish phase 1/2 study of DF in 150 patients with WHO grade 4–5 (72%) or 6 (28%) COVID-19 reported data consistent with the known favorable safety profile of DF in VOD/SOS and a preliminary mortality rate due to severe COVID-19 of 27%, which compares with expected mortality of >50% in historical controls [165]. An ongoing DFCI/BWH study in 39 patients (including 6 with MM) to date also confirmed the safety of DF in this setting, and preliminary analysis indicated a favorable impact of DF on cytokine markers and markers of endothelial stress (NCT04652115; Richardson PG, personal communication). In a report on two patients with pediatric inflammatory multisystem syndrome temporally associated with severe SARS-CoV-2 infection (PIMS-TA), DF was shown to be an effective treatment for the syndrome, reducing inflammation and restoring the thrombo-fibrinolytic balance [166]. Additionally, a pilot study of home-administered thromboprophylaxis in patients with COVID-19 and mild-to-moderate symptoms demonstrated that DF and LMWH were equally effective at preventing DVT and thrombotic disease and delivered similarly improved outcomes compared to a control group receiving standard management [167].

Furthermore, a recent report demonstrated the benefit of DF in two patients with RRMM and severe COVID-19 after CAR T-cell therapy [168]. Both patients had severe COVID-19 shortly after receiving CAR T-cell therapy, and experienced prolonged stays in the intensive care unit, with progressive, worsening disease despite maximal standard of care. Both patients experienced rapid improvements in their clinical condition after starting DF for 7–14 days, with intubation avoided; DF resulted in the suppression of SARS CoV-2-induced non-specific inflammatory response and related CRS, no negative impact on adaptive virus-specific antibody and/or T-cell responses, and no negative impact on persistence of CAR T cells. In both patients, their MM remains in deep and sustained remission.

5. Conclusions and Next Steps for Defibrotide

DF's pleiotropic mechanisms of action—spanning anti-inflammatory, antithrombotic, and fibrinolytic effects—support its role as a versatile endothelial protectant across multiple pathologies. DF is active in modulating and reversing the inflammatory and thrombotic/coagulation pathways that are activated following endothelial injury. DF has also shown endothelial protective effects via these mechanisms of action, through prevention of the cytokine storm that can arise following endothelial exposure to various noxae and maintenance of the thrombotic–fibrinolytic balance.

DF is approved for the management of VOD/SOS in the US and Europe and has also shown efficacy in numerous other conditions associated with endothelial injury following autologous or allogeneic HCT. Despite findings from the phase 3 HARMONY trial, DF may offer benefit as prophylaxis for VOD/SOS, particularly in high-risk cases. DF has demonstrated activity in preventing GvHD following allogeneic HCT and potentially in treating or preventing TA-TMA. Furthermore, DF has shown potential in protecting against the progressive endothelial damage that occurs in sepsis-associated organ dysfunction [169] and for reducing hypercoagulability due to loss of endothelial integrity in patients with sickle cell disease-related acute chest syndrome [30]; recently, DF has generated interest, based on its mechanisms of action, as a novel therapeutic for the key toxicities—CRS and ICANS—associated with CAR T-cell therapy and bispecific antibody therapy, and also for treating endothelial dysfunction associated with SARS-CoV-2 infection.

Based on the effects demonstrated in several small studies/case series, further evaluation of DF is warranted in multiple clinical settings. DF should be further investigated as prophylaxis against or as treatment for CRS and ICANS in high-risk patients with MM, lymphoma [148], and leukemia [12] receiving CAR T-cell or bispecific antibody therapy. Attention should be paid to the timing of DF initiation, as greater efficacy has been observed with earlier vs. later DF initiation in the treatment of VOD/SOS; additionally, combination therapies, such as with anti-IL-6 mAbs, may be beneficial in more severe cases. DF is also being studied as a treatment option for severe COVID-19, in patients with ARDS (NCT04652115), and as prophylaxis against, and treatment of, PASC, and warrants investigation in other conditions characterized by endothelial injury or dysfunction as a central component of pathobiology, as originally proposed in peripheral blood and marrow stem cell transplantation [170]. These settings might include the following:

- Other viral or infectious causes of severe acute lung injury, e.g., serious influenza;
- Inflammatory lung conditions such as IPS or other non-HCT-related lung injury;

- Prevention of microvascular ischemia and thrombosis in ischemic diseases (cardiovascular, neurological);
- Immune-mediated endothelial injury, including autoimmune diseases and antiphospholipid syndrome;
- Solid organ transplant-associated endothelial dysfunction, including ischemia-reperfusion injury and chronic allograft vasculopathy.

Treatment options that offer endothelial protection in these settings will become increasingly important for patients at risk of endothelial injury from multiple sources. This is particularly relevant given the expanding use of CAR T-cell therapy and bispecific antibodies in MM and other hematologic malignancies, the ongoing potential effects of endemic COVID-19 and possible serial waves of SARS-CoV-2 infections as well as other viruses, the growing number of patients with PASC, and the potential threats from emerging viral illnesses.

Author Contributions: Writing—review and editing: E.R., C.C.M., E.C., F.C., M.H.K., R.M.B., J.M.C., M.I., L.-J.W., E.J.B., A.P.R., M.D.-R., A.J.M.-M., C.C.-S., P.G.R., and J.M.M.; funding acquisition: P.G.R.; project administration: P.G.R.; writing—original draft preparation: P.G.R.; Conceptualization: E.R. and P.G.R.; supervision: M.I. and P.G.R. All authors have read and agreed to the published version of the manuscript.

Funding: This study received funding from the Instituto de Salud Carlos III (ISCIII), through the project COV20/00399 and the Departament de Recerca i Universitats de la Generalitat de Catalunya (2021-SGR-01118). This study also funded by Dana-Farber Cancer Institute, the RJ Corman Multiple Myeloma Research Fund, and the Paula and Rodger Riney Multiple Myeloma Research Fund.

Institutional Review Board Statement: Not applicable.

Informed Consent Statement: Not applicable.

Data Availability Statement: Not applicable.

Acknowledgments: The authors gratefully acknowledge Steve Hill, of Ashfield MedComms, an Inizio company, for medical writing and editing support.

Conflicts of Interest: E.R.: none. C.C.M.: advisory boards for AbbVie, Bristol Myers Squibb, GSK, Janssen, Karyopharm, Sanofi, and Takeda; consultancy for AbbVie, Janssen, Karyopharm, and Sanofi. E.C.: none. F.C.: none. M.H.K.: grants to institution for clinical trials from Bristol Myers Squibb/Celgene, Janssen, AbbVie, Arcellx/Kite, Roche and Poseida Therapeutics. R.M.B.: None. J.M.C.: Consulting and scientific advisory boards: Abbott, Alexion, Anthos, Bayer, Bristol Myers Squibb, Perosphere Technologies, Pfizer, Regeneron, Sanofi. M.I.: employment with Techitra s.r.l. L.-J.W.: none. E.J.B.: none. A.P.R.: None. M.D.-R.: speaker fees from Jazz Pharmaceuticals and research funding to institution from Novartis Spain, CSL Behring, and Sysmex Europe GmbH. A.J.M.-M.: None. C.C.-S.: membership of the board of directors, speakers bureau, advisory committee for ADC Therapeutics SA, Bristol Myers Squibb/Celgene, Karyopharm, Roche, AbbVie, Genmab, and SOBI; research funding from ADC Therapeutics SA, Roche, and Sanofi; honoraria from ADC Therapeutics SA, AstraZeneca, Bristol Myers Squibb, Incyte, Janssen Oncology, Roche, AbbVie, Genmab, and SOBI. P.G.R.: grants to institution for clinical trials from Bristol Myers Squibb/Celgene, Karyopharm, Jazz Pharma, and Oncopeptides; advisory committees for Bristol Myers Squibb/Celgene, GSK, Karyopharm, Oncopeptides, Adaptive Biotechnologies, and Sanofi. J.M.M.: research support from Pfizer, Gilead, Novartis, Bristol Myers Squibb, Amgen, Jazz Pharma, and Roche, and advisory committees for Rocket Pharma, Jazz Pharma, Novartis, Gilead, and Sandoz.

References

1. Jazz Pharmaceuticals Inc. DEFITELIO (Defibrotide Sodium) Injection, for Intravenous Use—Prescribing Information. Jazz Pharmaceuticals Inc., 2022. Available online: https://defitelio.com/dosing-and-administration/index (accessed on 3 May 2025).

2. Defibrotide. In *LiverTox: Clinical and Research Information on Drug-Induced Liver Injury*; National Institute of Diabetes and Digestive and Kidney Diseases: Bethesda, MD, USA, 2012.
3. Gentium Srl. Defitelio: EPAR—Product Information. Gentium Srl, 2023. Available online: https://www.ema.europa.eu/en/documents/product-information/defitelio-epar-product-information_en.pdf (accessed on 3 May 2025).
4. Augustin, H.G.; Koh, G.Y. A systems view of the vascular endothelium in health and disease. *Cell* **2024**, *187*, 4833–4858. [CrossRef] [PubMed]
5. Hildebrandt, G.C.; Chao, N. Endothelial cell function and endothelial-related disorders following haematopoietic cell transplantation. *Br. J. Haematol.* **2020**, *190*, 508–519. [CrossRef] [PubMed]
6. Gracia-Sancho, J.; Caparros, E.; Fernandez-Iglesias, A.; Frances, R. Role of liver sinusoidal endothelial cells in liver diseases. *Nat. Rev. Gastroenterol. Hepatol.* **2021**, *18*, 411–431. [CrossRef] [PubMed]
7. Poisson, J.; Lemoinne, S.; Boulanger, C.; Durand, F.; Moreau, R.; Valla, D.; Rautou, P.E. Liver sinusoidal endothelial cells: Physiology and role in liver diseases. *J. Hepatol.* **2017**, *66*, 212–227. [CrossRef]
8. Chung, A.S.; Ferrara, N. Developmental and pathological angiogenesis. *Annu. Rev. Cell Dev. Biol.* **2011**, *27*, 563–584. [CrossRef]
9. Elhadad, S.; Redmond, D.; Tan, A.; Huang, J.; Rodriguez, B.L.; Racine-Brzostek, S.E.; Subrahmanian, S.; Ahamed, J.; Laurence, J. Defibrotide mitigates endothelial cell injury induced by plasmas from patients with COVID-19 and related vasculopathies. *Thromb. Res.* **2023**, *225*, 47–56. [CrossRef]
10. Chang, J.C. Molecular Pathogenesis of Endotheliopathy and Endotheliopathic Syndromes, Leading to Inflammation and Microthrombosis, and Various Hemostatic Clinical Phenotypes Based on "Two-Activation Theory of the Endothelium" and "Two-Path Unifying Theory" of Hemostasis. *Medicina* **2022**, *58*, 1311. [CrossRef]
11. Palomo, M.; Moreno-Castano, A.B.; Salas, M.Q.; Escribano-Serrat, S.; Rovira, M.; Guillen-Olmos, E.; Fernandez, S.; Ventosa-Capell, H.; Youssef, L.; Crispi, F.; et al. Endothelial activation and damage as a common pathological substrate in different pathologies and cell therapy complications. *Front. Med.* **2023**, *10*, 1285898. [CrossRef]
12. Pennisi, M.; Sanchez-Escamilla, M.; Flynn, J.R.; Shouval, R.; Alarcon Tomas, A.; Silverberg, M.L.; Batlevi, C.; Brentjens, R.J.; Dahi, P.B.; Devlin, S.M.; et al. Modified EASIX predicts severe cytokine release syndrome and neurotoxicity after chimeric antigen receptor T cells. *Blood Adv.* **2021**, *5*, 3397–3406. [CrossRef]
13. Escribano-Serrat, S.; Rodriguez-Lobato, L.G.; Charry, P.; Martinez-Cibrian, N.; Suarez-Lledo, M.; Rivero, A.; Moreno-Castano, A.B.; Solano, M.T.; Arcarons, J.; Nomdedeu, M.; et al. Endothelial Activation and Stress Index in adults undergoing allogeneic hematopoietic cell transplantation with post-transplant cyclophosphamide-based prophylaxis. *Cytotherapy* **2024**, *26*, 73–80. [CrossRef]
14. Pedraza, A.; Salas, M.Q.; Rodriguez-Lobato, L.G.; Escribano-Serrat, S.; Suarez-Lledo, M.; Martinez-Cebrian, N.; Solano, M.T.; Arcarons, J.; Rosinol, L.; Gutierrez-Garcia, G.; et al. Easix Score Correlates with Endothelial Dysfunction Biomarkers and Predicts Risk of Acute Graft-Versus-Host Disease After Allogeneic Transplantation. *Transplant. Cell Ther.* **2024**, *30*, 187.e1–187.e12. [CrossRef] [PubMed]
15. Tolosa-Ridao, C.; Cascos, E.; Rodriguez-Lobato, L.G.; Pedraza, A.; Suarez-Lledo, M.; Charry, P.; Solano, M.T.; Martinez-Sanchez, J.; Cid, J.; Lozano, M.; et al. EASIX and cardiac adverse events after allogeneic hematopoietic cell transplantation. *Bone Marrow Transplant.* **2024**, *59*, 974–982. [CrossRef]
16. Escribano-Serrat, S.; Rodriguez-Lobato, L.G.; Suarez-Lledo, M.; Pedraza, A.; Charry, P.; Cid, J.; Lozano, M.; Esteve, J.; Rosinol, L.; Fernandez-Aviles, F.; et al. Improving the EASIX' predictive power for NRM in adults undergoing allogeneic hematopoietic cell transplantation. *Bone Marrow Transplant.* **2024**, *59*, 1022–1024. [CrossRef] [PubMed]
17. Escribano-Serrat, S.; Pedraza, A.; Suarez-Lledo, M.; Charry, P.; De Moner, B.; Martinez-Sanchez, J.; Ramos, A.; Ventosa-Capell, H.; Moreno, C.; Guardia, L.; et al. Safety and efficacy of G-CSF after allogeneic hematopoietic cell transplantation using post-transplant cyclophosphamide: Clinical and in vitro examination of endothelial activation. *Bone Marrow Transplant.* **2024**, *59*, 1466–1476. [CrossRef] [PubMed]
18. Moreno-Castaño, A.B.; Fernández, S.; Brillembourg, H.; de Moner, B.; Ventosa-Capell, H.; Martinez-Sanchez, J.; Ramos, A.; Palomo, M.; Molina, P.; Pino, M.; et al. M-Easix (Better Than EASIX) Correlates with Specific Endotheliopathy Biomarkers, Predicts Severe CAR-T Cell Toxicities and Discriminates from Sepsis. *Blood* **2024**, *144* (Suppl. S1), 3423. [CrossRef]
19. Richardson, P.G.; Palomo, M.; Kernan, N.A.; Hildebrandt, G.C.; Chao, N.; Carreras, E. The importance of endothelial protection: The emerging role of defibrotide in reversing endothelial injury and its sequelae. *Bone Marrow Transplant.* **2021**, *56*, 2889–2896. [CrossRef]
20. Richardson, E.; Garcia-Bernal, D.; Calabretta, E.; Jara, R.; Palomo, M.; Baron, R.M.; Yanik, G.; Fareed, J.; Vlodavsky, I.; Iacobelli, M.; et al. Defibrotide: Potential for treating endothelial dysfunction related to viral and post-infectious syndromes. *Expert. Opin. Ther. Targets* **2021**, *25*, 423–433. [CrossRef]

21. Echart, C.L.; Graziadio, B.; Somaini, S.; Ferro, L.I.; Richardson, P.G.; Fareed, J.; Iacobelli, M. The fibrinolytic mechanism of defibrotide: Effect of defibrotide on plasmin activity. *Blood Coagul. Fibrinolysis* **2009**, *20*, 627–634. [CrossRef]
22. Carmona, A.; Diaz-Ricart, M.; Palomo, M.; Molina, P.; Pino, M.; Rovira, M.; Escolar, G.; Carreras, E. Distinct deleterious effects of cyclosporine and tacrolimus and combined tacrolimus-sirolimus on endothelial cells: Protective effect of defibrotide. *Biol. Blood Marrow Transplant.* **2013**, *19*, 1439–1445. [CrossRef]
23. Orlando, N.; Babini, G.; Chiusolo, P.; Valentini, C.G.; De Stefano, V.; Teofili, L. Pre-Exposure to Defibrotide Prevents Endothelial Cell Activation by Lipopolysaccharide: An Ingenuity Pathway Analysis. *Front. Immunol.* **2020**, *11*, 585519. [CrossRef]
24. Palomo, M.; Vera, M.; Martin, S.; Torramade-Moix, S.; Martinez-Sanchez, J.; Belen Moreno, A.; Carreras, E.; Escolar, G.; Cases, A.; Diaz-Ricart, M. Up-regulation of HDACs, a harbinger of uraemic endothelial dysfunction, is prevented by defibrotide. *J. Cell Mol. Med.* **2020**, *24*, 1713–1723. [CrossRef]
25. Schoergenhofer, C.; Buchtele, N.; Gelbenegger, G.; Derhaschnig, U.; Firbas, C.; Kovacevic, K.D.; Schwameis, M.; Wohlfarth, P.; Rabitsch, W.; Jilma, B. Defibrotide enhances fibrinolysis in human endotoxemia—A randomized, double blind, crossover trial in healthy volunteers. *Sci. Rep.* **2019**, *9*, 11136. [CrossRef] [PubMed]
26. Shi, H.; Gandhi, A.A.; Smith, S.A.; Wang, Q.; Chiang, D.; Yalavarthi, S.; Ali, R.A.; Liu, C.; Sule, G.; Tsou, P.S.; et al. Endothelium-protective, histone-neutralizing properties of the polyanionic agent defibrotide. *JCI Insight* **2021**, *6*, e149149. [CrossRef] [PubMed]
27. Wang, X.; Pan, B.; Hashimoto, Y.; Ohkawara, H.; Xu, K.; Zeng, L.; Ikezoe, T. Defibrotide Stimulates Angiogenesis and Protects Endothelial Cells from Calcineurin Inhibitor-Induced Apoptosis via Upregulation of AKT/Bcl-xL. *Thromb. Haemost.* **2018**, *118*, 161–173. [CrossRef] [PubMed]
28. Ali, R.A.; Estes, S.K.; Gandhi, A.A.; Yalavarthi, S.; Hoy, C.K.; Shi, H.; Zuo, Y.; Erkan, D.; Knight, J.S. Defibrotide Inhibits Antiphospholipid Antibody-Mediated Neutrophil Extracellular Trap Formation and Venous Thrombosis. *Arthritis Rheumatol.* **2022**, *74*, 902–907. [CrossRef]
29. Onuora, S. Defibrotide inhibits NET-mediated thrombosis in APS models. *Nat. Rev. Rheumatol.* **2022**, *18*, 63. [CrossRef]
30. Schaefer, E.; Anderson-Crannage, M.; Ktena, Y.P.; Hochberg, J.; Kanarfogel, T.; Herrick, J.; Thatcher, E.; Shi, Q.; Hochberg, B.; Chu, Y.; et al. Defibrotide Reduces Hypercoagulable State in Patients with Sickle Cell Disease-Related Acute Chest Syndrome (IND 127812). *Blood* **2024**, *144* (Suppl. S1), 2515. [CrossRef]
31. Mo, C.C.; Richardson, E.; Calabretta, E.; Corrado, F.; Kocoglu, M.H.; Baron, R.M.; Connors, J.M.; Iacobelli, M.; Wei, L.J.; Rapoport, A.P.; et al. Endothelial injury and dysfunction with emerging immunotherapies in multiple myeloma, the impact of COVID-19, and endothelial protection with a focus on the evolving role of defibrotide. *Blood Rev.* **2024**, *66*, 101218. [CrossRef]
32. Echart, C.L.; Somaini, S.; Distaso, M.; Palumbo, A.; Richardson, P.G.; Fareed, J.; Iacobelli, M. Defibrotide blunts the prothrombotic effect of thalidomide on endothelial cells. *Clin. Appl. Thromb. Hemost.* **2012**, *18*, 79–86. [CrossRef]
33. Comerford, C.; Glavey, S.; Quinn, J.; O'Sullivan, J.M. The role of VWF/FVIII in thrombosis and cancer progression in multiple myeloma and other hematological malignancies. *J. Thromb. Haemost.* **2022**, *20*, 1766–1777. [CrossRef]
34. Leimi, L.; Jahnukainen, K.; Olkinuora, H.; Meri, S.; Vettenranta, K. Early vascular toxicity after pediatric allogeneic hematopoietic stem cell transplantation. *Bone Marrow Transplant.* **2022**, *57*, 705–711. [CrossRef]
35. Martinez-Sanchez, J.; Palomo, M.; Torramade-Moix, S.; Moreno-Castano, A.B.; Rovira, M.; Gutierrez-Garcia, G.; Fernandez-Aviles, F.; Escolar, G.; Penack, O.; Rosinol, L.; et al. The induction strategies administered in the treatment of multiple myeloma exhibit a deleterious effect on the endothelium. *Bone Marrow Transplant.* **2020**, *55*, 2270–2278. [CrossRef] [PubMed]
36. Coppell, J.A.; Richardson, P.G.; Soiffer, R.; Martin, P.L.; Kernan, N.A.; Chen, A.; Guinan, E.; Vogelsang, G.; Krishnan, A.; Giralt, S.; et al. Hepatic veno-occlusive disease following stem cell transplantation: Incidence, clinical course, and outcome. *Biol. Blood Marrow Transplant.* **2010**, *16*, 157–168. [CrossRef] [PubMed]
37. Carreras, E.; Diaz-Beya, M.; Rosinol, L.; Martinez, C.; Fernandez-Aviles, F.; Rovira, M. The incidence of veno-occlusive disease following allogeneic hematopoietic stem cell transplantation has diminished and the outcome improved over the last decade. *Biol. Blood Marrow Transplant.* **2011**, *17*, 1713–1720. [CrossRef]
38. Mohty, M.; Malard, F.; Abecasis, M.; Aerts, E.; Alaskar, A.S.; Aljurf, M.; Arat, M.; Bader, P.; Baron, F.; Basak, G.; et al. Prophylactic, preemptive, and curative treatment for sinusoidal obstruction syndrome/veno-occlusive disease in adult patients: A position statement from an international expert group. *Bone Marrow Transplant.* **2020**, *55*, 485–495. [CrossRef]
39. Ruutu, T.; Peczynski, C.; Houhou, M.; Polge, E.; Mohty, M.; Kroger, N.; Moiseev, I.; Penack, O.; Salooja, N.; Schoemans, H.; et al. Current incidence, severity, and management of veno-occlusive disease/sinusoidal obstruction syndrome in adult allogeneic HSCT recipients: An EBMT Transplant Complications Working Party study. *Bone Marrow Transplant.* **2023**, *58*, 1209–1214. [CrossRef]
40. Zaidman, I.; Barsoum, N.; Even-Or, E.; Daher, M.; Aran, A.A.; Stepensky, P.; Gefen, A. Prognostic Factors Associated With Increased Mortality in Pediatric Veno-Occlusive Disease Following Hematopoietic Cell Transplantation. *Clin. Transplant.* **2024**, *38*, e70037. [CrossRef]

41. Ho, V.T.; Linden, E.; Revta, C.; Richardson, P.G. Hepatic veno-occlusive disease after hematopoietic stem cell transplantation: Review and update on the use of defibrotide. *Semin. Thromb. Hemost.* **2007**, *33*, 373–388. [CrossRef] [PubMed]
42. Bearman, S.I. The syndrome of hepatic veno-occlusive disease after marrow transplantation. *Blood* **1995**, *85*, 3005–3020. [CrossRef]
43. Eissner, G.; Multhoff, G.; Gerbitz, A.; Kirchner, S.; Bauer, S.; Haffner, S.; Sondermann, D.; Andreesen, R.; Holler, E. Fludarabine induces apoptosis, activation, and allogenicity in human endothelial and epithelial cells: Protective effect of defibrotide. *Blood* **2002**, *100*, 334–340. [CrossRef]
44. Richardson, P.; Guinan, E. Hepatic veno-occlusive disease following hematopoietic stem cell transplantation. *Acta Haematol.* **2001**, *106*, 57–68. [CrossRef] [PubMed]
45. Richardson, P.G.; Corbacioglu, S.; Ho, V.T.; Kernan, N.A.; Lehmann, L.; Maguire, C.; Maglio, M.; Hoyle, M.; Sardella, M.; Giralt, S.; et al. Drug safety evaluation of defibrotide. *Expert. Opin. Drug Saf.* **2013**, *12*, 123–136. [CrossRef] [PubMed]
46. Mitsiades, C.S.; Rouleau, C.; Echart, C.; Menon, K.; Teicher, B.; Distaso, M.; Palumbo, A.; Boccadoro, M.; Anderson, K.C.; Iacobelli, M.; et al. Preclinical studies in support of defibrotide for the treatment of multiple myeloma and other neoplasias. *Clin. Cancer Res.* **2009**, *15*, 1210–1221. [CrossRef]
47. Palomo, M.; Diaz-Ricart, M.; Rovira, M.; Escolar, G.; Carreras, E. Defibrotide prevents the activation of macrovascular and microvascular endothelia caused by soluble factors released to blood by autologous hematopoietic stem cell transplantation. *Biol. Blood Marrow Transplant.* **2011**, *17*, 497–506. [CrossRef] [PubMed]
48. Palomo, M.; Mir, E.; Rovira, M.; Escolar, G.; Carreras, E.; Diaz-Ricart, M. What is going on between defibrotide and endothelial cells? Snapshots reveal the hot spots of their romance. *Blood* **2016**, *127*, 1719–1727. [CrossRef]
49. Inoue, Y.; Kosugi, S.; Sano, F. Improvement of High Serum Levels of Biomarkers of Endothelial Injury (Vascular Cell Adhesion Molecule-1) and Inflammation (Tumor Necrosis Factor Receptor Type I) After Allogeneic Hematopoietic Stem Cell Transplantation With Sinusoidal Obstruction Syndrome Using Defibrotide. *Am. J. Ther.* **2020**, *28*, e691–e693. [CrossRef]
50. Falanga, A.; Vignoli, A.; Marchetti, M.; Barbui, T. Defibrotide reduces procoagulant activity and increases fibrinolytic properties of endothelial cells. *Leukemia* **2003**, *17*, 1636–1642. [CrossRef]
51. Francischetti, I.M.; Oliveira, C.J.; Ostera, G.R.; Yager, S.B.; Debierre-Grockiego, F.; Carregaro, V.; Jaramillo-Gutierrez, G.; Hume, J.C.; Jiang, L.; Moretz, S.E.; et al. Defibrotide interferes with several steps of the coagulation-inflammation cycle and exhibits therapeutic potential to treat severe malaria. *Arterioscler. Thromb. Vasc. Biol.* **2012**, *32*, 786–798. [CrossRef]
52. Benimetskaya, L.; Wu, S.; Voskresenskiy, A.M.; Echart, C.; Zhou, J.F.; Shin, J.; Iacobelli, M.; Richardson, P.; Ayyanar, K.; Stein, C.A. Angiogenesis alteration by defibrotide: Implications for its mechanism of action in severe hepatic veno-occlusive disease. *Blood* **2008**, *112*, 4343–4352. [CrossRef]
53. Liu, Z.; Liang, S.; Wei, X.; Du, X.; Zhang, J. Defibrotide improved the outcome of monocrotaline induced rat hepatic sinusoidal obstruction syndrome. *BMC Gastroenterol.* **2022**, *22*, 525. [CrossRef]
54. Richardson, P.G.; Riches, M.L.; Kernan, N.A.; Brochstein, J.A.; Mineishi, S.; Termuhlen, A.M.; Arai, S.; Grupp, S.A.; Guinan, E.C.; Martin, P.L.; et al. Phase 3 trial of defibrotide for the treatment of severe veno-occlusive disease and multi-organ failure. *Blood* **2016**, *127*, 1656–1665. [CrossRef] [PubMed]
55. Kernan, N.A.; Grupp, S.; Smith, A.R.; Arai, S.; Triplett, B.; Antin, J.H.; Lehmann, L.; Shore, T.; Ho, V.T.; Bunin, N.; et al. Final results from a defibrotide treatment-IND study for patients with hepatic veno-occlusive disease/sinusoidal obstruction syndrome. *Br. J. Haematol.* **2018**, *181*, 816–827. [CrossRef]
56. Mohty, M.; Blaise, D.; Peffault de Latour, R.; Labopin, M.; Bourhis, J.H.; Bruno, B.; Ceballos, P.; Detrait, M.; Gandemer, V.; Huynh, A.; et al. Real-world use of defibrotide for veno-occlusive disease/sinusoidal obstruction syndrome: The DEFIFrance Registry Study. *Bone Marrow Transplant.* **2023**, *58*, 367–376. [CrossRef] [PubMed]
57. Nauffal, M.; Kim, H.T.; Richardson, P.G.; Soiffer, R.J.; Antin, J.H.; Cutler, C.; Nikiforow, S.; Gooptu, M.; Koreth, J.; Romee, R.; et al. Defibrotide: Real-world management of veno-occlusive disease/sinusoidal obstructive syndrome after stem cell transplant. *Blood Adv.* **2022**, *6*, 181–188. [CrossRef]
58. Bagal, B.; Chandrasekharan, A.; Chougle, A.; Khattry, N. Low, fixed dose defibrotide in management of hepatic veno-occlusive disease post stem cell transplantation. *Hematol. Oncol. Stem Cell Ther.* **2018**, *11*, 47–51. [CrossRef] [PubMed]
59. Bahoush, G.; Vafapour, M. A case report of severe veno-occlusive disease following autologous stem cell transplantation successfully treated with Defibrotide. *Eur. J. Transl. Myol.* **2020**, *30*, 9161. [CrossRef]
60. Balade Martinez, L.; Cabezuelo, M.M.; Villamanan Bueno, E.; Rodriguez Martin, E.; Herrero Ambrosio, A. Defibrotide for the treatment of severe hepatic sinusoidal obstruction syndrome: A single-centre experience. *Eur. J. Hosp. Pharm.* **2019**, *26*, 343–346. [CrossRef]
61. Corbacioglu, S.; Carreras, E.; Mohty, M.; Pagliuca, A.; Boelens, J.J.; Damaj, G.; Iacobelli, M.; Niederwieser, D.; Olavarria, E.; Suarez, F.; et al. Defibrotide for the Treatment of Hepatic Veno-Occlusive Disease: Final Results From the International Compassionate-Use Program. *Biol. Blood Marrow Transplant.* **2016**, *22*, 1874–1882. [CrossRef]

62. Corbacioglu, S.; Greil, J.; Peters, C.; Wulffraat, N.; Laws, H.J.; Dilloo, D.; Straham, B.; Gross-Wieltsch, U.; Sykora, K.W.; Ridolfi-Luthy, A.; et al. Defibrotide in the treatment of children with veno-occlusive disease (VOD): A retrospective multicentre study demonstrates therapeutic efficacy upon early intervention. *Bone Marrow Transplant.* **2004**, *33*, 189–195. [CrossRef]
63. Agrawal, V.; Pourhassan, H.; Tsai, N.C.; Ngo, D.; Koller, P.; Malki, M.M.A.; Salhotra, A.; Ali, H.; Aribi, A.; Sandhu, K.S.; et al. Post-Transplantation Sinusoidal Obstruction Syndrome in Adult Patients with B Cell Acute Lymphoblastic Leukemia Treated with Pretransplantation Inotuzumab. *Transplant. Cell Ther.* **2023**, *29*, 314–320. [CrossRef]
64. Faraci, M.; Bertaina, A.; Luksch, R.; Calore, E.; Lanino, E.; Saglio, F.; Prete, A.; Menconi, M.; De Simone, G.; Tintori, V.; et al. Sinusoidal Obstruction Syndrome/Veno-Occlusive Disease after Autologous or Allogeneic Hematopoietic Stem Cell Transplantation in Children: A retrospective study of the Italian Hematology-Oncology Association-Hematopoietic Stem Cell Transplantation Group. *Biol. Blood Marrow Transplant.* **2019**, *25*, 313–320. [CrossRef] [PubMed]
65. Kernan, N.A.; Richardson, P.G.; Smith, A.R.; Triplett, B.M.; Antin, J.H.; Lehmann, L.; Messinger, Y.; Liang, W.; Hume, R.; Tappe, W.; et al. Defibrotide for the treatment of hepatic veno-occlusive disease/sinusoidal obstruction syndrome following nontransplant-associated chemotherapy: Final results from a post hoc analysis of data from an expanded-access program. *Pediatr. Blood Cancer* **2018**, *65*, e27269. [CrossRef]
66. Gonzalez Vicent, M.; Diaz de Heredia, C.; Gonzalez de Pablo, J.; Molina, B.; Regueiro, A.; Perez Martinez, A.; Palomo, P.; Lopez Corral, L.; Garcia, E.; Fernandez, J.M.; et al. Defibrotide in hematopoietic stem cell transplantation: A multicenter survey study of the Spanish Hematopoietic Stem Cell Transplantation Group (GETH). *Eur. J. Haematol.* **2021**, *106*, 842–850. [CrossRef] [PubMed]
67. Lee, T.B.; Yang, K.; Ko, H.J.; Shim, J.R.; Choi, B.H.; Lee, J.H.; Ryu, J.H. Successful defibrotide treatment of a patient with veno-occlusive disease after living-donor liver transplantation: A case report. *Medicine* **2021**, *100*, e26463. [CrossRef] [PubMed]
68. Mehra, V.; Tetlow, S.; Choy, A.; de Lavallade, H.; Kulasekararaj, A.; Krishnamurthy, P.; Avenoso, D.; Marsh, J.; Potter, V.; Mufti, G.; et al. Early and late-onset veno-occlusive disease/sinusoidal syndrome post allogeneic stem cell transplantation—A real-world UK experience. *Am. J. Transplant.* **2021**, *21*, 864–869. [CrossRef]
69. Mohty, M.; Battista, M.L.; Blaise, D.; Calore, E.; Cesaro, S.; Maximova, N.; Perruccio, K.; Renard, C.; Wynn, R.; Zecca, M.; et al. A multicentre, multinational, prospective, observational registry study of defibrotide in patients diagnosed with veno-occlusive disease/sinusoidal obstruction syndrome after haematopoietic cell transplantation: An EBMT study. *Bone Marrow Transplant.* **2021**, *56*, 2454–2463. [CrossRef]
70. Richardson, P.G.; Soiffer, R.J.; Antin, J.H.; Uno, H.; Jin, Z.; Kurtzberg, J.; Martin, P.L.; Steinbach, G.; Murray, K.F.; Vogelsang, G.B.; et al. Defibrotide for the treatment of severe hepatic veno-occlusive disease and multiorgan failure after stem cell transplantation: A multicenter, randomized, dose-finding trial. *Biol. Blood Marrow Transplant.* **2010**, *16*, 1005–1017. [CrossRef]
71. Richardson, P.G.; Murakami, C.; Jin, Z.; Warren, D.; Momtaz, P.; Hoppensteadt, D.; Elias, A.D.; Antin, J.H.; Soiffer, R.; Spitzer, T.; et al. Multi-institutional use of defibrotide in 88 patients after stem cell transplantation with severe veno-occlusive disease and multisystem organ failure: Response without significant toxicity in a high-risk population and factors predictive of outcome. *Blood* **2002**, *100*, 4337–4343. [CrossRef]
72. Richardson, P.G.; Corbacioglu, S. Veno-occlusive disease/sinusoidal obstruction syndrome in patients with prior gemtuzumab ozogamicin: Literature analysis of survival after defibrotide treatment. *Blood Cancer J.* **2020**, *10*, 29. [CrossRef]
73. Roy Moulik, N.; Johnson, I.; Van Bruggen, L.; Petterson, T.; Mycroft, J.; Vaidya, S.J. Defibrotide treatment but not prophylaxis is useful in hepatic sinusoidal obstruction syndrome in children undergoing autologous stem cell transplant following high-dose chemotherapy: A single-center experience from the Royal Marsden Hospital, UK. *Pediatr. Blood Cancer* **2020**, *67*, e28677. [CrossRef]
74. Rudebeck, C.J.; Renard, C.; Halfon-Domenech, C.; Ouachee-Chardin, M.; Philippe, M.; Valla, F.V.; Bertrand, Y.; Penel-Page, M. Interest of the preventive and curative use of defibrotide on the occurrence and severity of sinusoidal obstruction syndrome after hematopoietic stem cell transplant in children. *EJHaem* **2022**, *3*, 885–893. [CrossRef]
75. Strouse, C.; Richardson, P.; Prentice, G.; Korman, S.; Hume, R.; Nejadnik, B.; Horowitz, M.M.; Saber, W. Defibrotide for Treatment of Severe Veno-Occlusive Disease in Pediatrics and Adults: An Exploratory Analysis Using Data from the Center for International Blood and Marrow Transplant Research. *Biol. Blood Marrow Transplant.* **2016**, *22*, 1306–1312. [CrossRef] [PubMed]
76. Sucak, G.T.; Aki, Z.S.; Yagci, M.; Yegin, Z.A.; Ozkurt, Z.N.; Haznedar, R. Treatment of sinusoidal obstruction syndrome with defibrotide: A single-center experience. *Transplant. Proc.* **2007**, *39*, 1558–1563. [CrossRef] [PubMed]
77. Yakushijin, K.; Ikezoe, T.; Ohwada, C.; Kudo, K.; Okamura, H.; Goto, H.; Yabe, H.; Yasumoto, A.; Kuwabara, H.; Fujii, S.; et al. Clinical effects of recombinant thrombomodulin and defibrotide on sinusoidal obstruction syndrome after allogeneic hematopoietic stem cell transplantation. *Bone Marrow Transplant.* **2019**, *54*, 674–680. [CrossRef]
78. Thielemans, N.; De Beule, N.; Van den Bergh, F.; Lefesvre, P.; De Becker, A. Successful Treatment of Very Severe Sinusoidal Obstruction Syndrome After Gemtuzumab Ozogamicin With Transjugular Intrahepatic Portosystemic Shunt, Defibrotide, and High-Dose Corticosteroids: A Case Report. *Cureus* **2024**, *16*, e67682. [CrossRef]

79. Yoon, J.-H.; Kwag, D.; Min, G.J.; Park, S.-S.; Park, S.; Lee, S.-E.; Cho, B.-S.; Eom, K.-S.; Kim, Y.-J.; Kim, H.J.; et al. Real-World Outcome of Defibrotide for Treatment of Adult Patients Who Developed Severe Hepatic VOD/SOS after Allogeneic Hematopoietic Cell Transplantation. *Blood* **2024**, *144* (Suppl. S1), 7326. [CrossRef]
80. Coutsouvelis, J.; Avery, S.; Dooley, M.; Kirkpatrick, C.; Spencer, A. Defibrotide for the treatment of sinusoidal obstruction syndrome: Evaluation of response to therapy and patient outcomes. *Support. Care Cancer* **2018**, *26*, 947–955. [CrossRef]
81. Coutsouvelis, J.; Kirkpatrick, C.M.; Dooley, M.; Spencer, A.; Kennedy, G.; Chau, M.; Huang, G.; Doocey, R.; Copeland, T.S.; Do, L.; et al. Incidence of Sinusoidal Obstruction Syndrome/Veno-Occlusive Disease and Treatment with Defibrotide in Allogeneic Transplantation: A Multicenter Australasian Registry Study. *Transplant. Cell Ther.* **2023**, *29*, 383.e1–383.e10. [CrossRef]
82. Carcedo Rodriguez, D.; Artola Urain, T.; Chinea Rodriguez, A.; Garcia Torres, E.; Gonzalez Vicent, M.; Gutierrez Garcia, G.; Regueiro Garcia, A.; Calvo Hidalgo, M.; Villacampa, A. Cost-effectiveness analysis of defibrotide in the treatment of patients with severe veno-occlusive disease/sinusoidal obstructive syndrome with multiorgan dysfunction following hematopoietic cell transplantation in Spain. *J. Med. Econ.* **2021**, *24*, 628–636. [CrossRef]
83. Richardson, P.; Aggarwal, S.; Topaloglu, O.; Villa, K.F.; Corbacioglu, S. Systematic review of defibrotide studies in the treatment of veno-occlusive disease/sinusoidal obstruction syndrome (VOD/SOS). *Bone Marrow Transplant.* **2019**, *54*, 1951–1962. [CrossRef]
84. Yang, L.; Qi, J.; Pan, T.; You, T.; Ruan, C.; Han, Y. Efficacy and Safety of Defibrotide for the Treatment of Hepatic Veno-Occlusive Disease after Hematopoietic Stem Cell Transplantation: A Systematic Review and Meta-Analysis. *Semin. Thromb. Hemost.* **2019**, *45*, 767–777. [CrossRef] [PubMed]
85. Richardson, P.G.; Smith, A.R.; Kernan, N.A.; Lehmann, L.; Soiffer, R.J.; Ryan, R.J.; Tappe, W.; Grupp, S. Pooled analysis of Day 100 survival for defibrotide-treated patients with hepatic veno-occlusive disease/sinusoidal obstruction syndrome and ventilator or dialysis dependence following haematopoietic cell transplantation. *Br. J. Haematol.* **2020**, *190*, 583–587. [CrossRef] [PubMed]
86. Richardson, P.G.; Smith, A.R.; Kernan, N.A.; Lehmann, L.; Ryan, R.J.; Grupp, S.A. Analysis of Time to Complete Response after Defibrotide Initiation in Patients with Hepatic Veno-Occlusive Disease/Sinusoidal Obstruction Syndrome after Hematopoietic Cell Transplantation. *Transplant. Cell Ther.* **2021**, *27*, 88.e81–88.e86. [CrossRef] [PubMed]
87. Richardson, P.G.; Grupp, S.A.; Pagliuca, A.; Krishnan, A.; Ho, V.T.; Corbacioglu, S. Defibrotide for the treatment of hepatic veno-occlusive disease/sinusoidal obstruction syndrome with multiorgan failure. *Int. J. Hematol. Oncol.* **2017**, *6*, 75–93. [CrossRef]
88. Tappe, W.; Aggarwal, S.; Topaloglu, O.; Iacobelli, M. A Meta-Analysis Evaluating the Incidence of Bleeding Events With Intravenous Defibrotide Treatment Outside the Veno-Occlusive Disease/Sinusoidal Obstruction Syndrome Setting. *Clin. Appl. Thromb. Hemost.* **2020**, *26*, 1076029620935202. [CrossRef]
89. Avenoso, D.; Kenyon, M.; Mehra, V.; Krishnamurthy, P.; Kulasekararaj, A.; Gandhi, S.; Dazzi, F.; Naresh Shah, M.; Wood, H.; Leung, Y.T.; et al. Systematic screening and focused evaluation for veno-occlusive disease/sinusoidal obstructive syndrome (VOD/SOS) following allogeneic stem cell transplant is associated with earlier diagnosis and prompt institution of defibrotide treatment. *Front. Transplant.* **2022**, *1*, 996003. [CrossRef]
90. Corbacioglu, S. Sinusoidal Obstruction Syndrome (SOS) and Defibrotide: We Are Not There Yet. *Transplant. Cell Ther.* **2023**, *29*, 287–288. [CrossRef]
91. Goto, H.; Oba, U.; Ueda, T.; Yamamoto, S.; Inoue, M.; Shimo, Y.; Yokoyama, S.; Takase, Y.; Kato, W.; Suenobu, S.; et al. Early defibrotide therapy and risk factors for post-transplant veno-occlusive disease/sinusoidal obstruction syndrome in childhood. *Pediatr. Blood Cancer* **2024**, *71*, e31331. [CrossRef]
92. Richardson, P.G.; Smith, A.R.; Triplett, B.M.; Kernan, N.A.; Grupp, S.A.; Antin, J.H.; Lehmann, L.; Miloslavsky, M.; Hume, R.; Hannah, A.L.; et al. Earlier defibrotide initiation post-diagnosis of veno-occlusive disease/sinusoidal obstruction syndrome improves Day +100 survival following haematopoietic stem cell transplantation. *Br. J. Haematol.* **2017**, *178*, 112–118. [CrossRef]
93. Corbacioglu, S.; Cesaro, S.; Faraci, M.; Valteau-Couanet, D.; Gruhn, B.; Rovelli, A.; Boelens, J.J.; Hewitt, A.; Schrum, J.; Schulz, A.S.; et al. Defibrotide for prophylaxis of hepatic veno-occlusive disease in paediatric haemopoietic stem-cell transplantation: An open-label, phase 3, randomised controlled trial. *Lancet* **2012**, *379*, 1301–1309. [CrossRef]
94. Corbacioglu, S.; Topaloglu, O.; Aggarwal, S. A Systematic Review and Meta-Analysis of Studies of Defibrotide Prophylaxis for Veno-Occlusive Disease/Sinusoidal Obstruction Syndrome. *Clin. Drug Investig.* **2022**, *42*, 465–476. [CrossRef] [PubMed]
95. Grupp, S.A.; Corbacioglu, S.; Kang, H.J.; Teshima, T.; Khaw, S.L.; Locatelli, F.; Maertens, J.; Stelljes, M.; Stepensky, P.; Lopez, P.; et al. Defibrotide plus best standard of care compared with best standard of care alone for the prevention of sinusoidal obstruction syndrome (HARMONY): A randomised, multicentre, phase 3 trial. *Lancet Haematol.* **2023**, *10*, e333–e345. [CrossRef]
96. Cappelli, B.; Chiesa, R.; Evangelio, C.; Biffi, A.; Roccia, T.; Frugnoli, I.; Biral, E.; Noe, A.; Fossati, M.; Finizio, V.; et al. Absence of VOD in paediatric thalassaemic HSCT recipients using defibrotide prophylaxis and intravenous Busulphan. *Br. J. Haematol.* **2009**, *147*, 554–560. [CrossRef]
97. Chalandon, Y.; Mamez, A.C.; Giannotti, F.; Beauverd, Y.; Dantin, C.; Mahne, E.; Mappoura, M.; Bernard, F.; de Ramon Ortiz, C.; Stephan, C.; et al. Defibrotide Shows Efficacy in the Prevention of Sinusoidal Obstruction Syndrome After Allogeneic

Hematopoietic Stem Cell Transplantation: A Retrospective Study. *Transplant. Cell Ther.* **2022**, *28*, 765.e1–765.e9. [CrossRef] [PubMed]
98. Dignan, F.; Gujral, D.; Ethell, M.; Evans, S.; Treleaven, J.; Morgan, G.; Potter, M. Prophylactic defibrotide in allogeneic stem cell transplantation: Minimal morbidity and zero mortality from veno-occlusive disease. *Bone Marrow Transplant.* **2007**, *40*, 79–82. [CrossRef]
99. Giglio, F.; Xue, E.; Greco, R.; Lazzari, L.; Clerici, D.T.; Lorentino, F.; Mastaglio, S.; Marktel, S.; Lupo-Stanghellini, M.T.; Marcatti, M.; et al. Defibrotide Prophylaxis of Sinusoidal Obstruction Syndrome in Adults Treated With Inotuzumab Ozogamicin Prior to Hematopoietic Stem Cell Transplantation. *Front. Oncol.* **2022**, *12*, 933317. [CrossRef]
100. Karagun, B.S.; Akbas, T.; Erbey, F.; Sasmaz, I.; Antmen, B. The Prophylaxis of Hepatic Veno-Occlusive Disease/Sinusoidal Obstruction Syndrome With Defibrotide After Hematopoietic Stem Cell Transplantation in Children: Single Center Experience. *J. Pediatr. Hematol. Oncol.* **2022**, *44*, e35–e39. [CrossRef]
101. Kayikci, O.; Akpinar, S.; Tekgunduz, E. Effectiveness of defibrotide in the prevention of hepatic venooclusive disease among adult patients receiving allogeneic hematopoietic cell transplantation: A retrospective single center experience. *Transfus. Apher. Sci.* **2022**, *61*, 103369. [CrossRef] [PubMed]
102. Park, M.; Park, H.J.; Eom, H.S.; Kwon, Y.J.; Park, J.A.; Lim, Y.J.; Yoon, J.H.; Kong, S.Y.; Ghim, T.T.; Lee, H.W.; et al. Safety and effects of prophylactic defibrotide for sinusoidal obstruction syndrome in hematopoietic stem cell transplantation. *Ann. Transplant.* **2013**, *18*, 36–42. [CrossRef]
103. Picod, A.; Bonnin, A.; Battipaglia, G.; Giannotti, F.; Ruggeri, A.; Brissot, E.; Malard, F.; Mediavilla, C.; Belhocine, R.; Vekhoff, A.; et al. Defibrotide for Sinusoidal Obstruction Syndrome/Veno-Occlusive Disease Prophylaxis in High-Risk Adult Patients: A Single-Center Experience Study. *Biol. Blood Marrow Transplant.* **2018**, *24*, 1471–1475. [CrossRef]
104. Roh, Y.Y.; Hahn, S.M.; Kim, H.S.; Ahn, W.K.; Han, J.H.; Kwon, S.; Lyu, C.J.; Han, J.W. Efficacy of low dose and short duration defibrotide prophylaxis for hepatic veno-occlusive disease after autologous haematopoietic stem cell transplantation. *Bone Marrow Transplant.* **2021**, *56*, 411–418. [CrossRef] [PubMed]
105. Soyer, N.; Gunduz, M.; Tekgunduz, E.; Deveci, B.; Ozdogu, H.; Sahin, H.H.; Turak, E.E.; Okay, M.; Kuku, I.; Hindilerden, I.Y.; et al. Incidence and risk factors for hepatic sinusoidal obstruction syndrome after allogeneic hematopoietic stem cell transplantation: A retrospective multicenter study of Turkish hematology research and education group (ThREG). *Transfus. Apher. Sci.* **2020**, *59*, 102827. [CrossRef] [PubMed]
106. Rahim, M.Q.; Rahrig, A.L.; Dinora, D.; Harrison, J.; Green, R.; Carter, A.; Skiles, J. The benefits of prophylactic defibrotide: Are the tides turning? *Pediatr. Blood Cancer* **2025**, *72*, e31396. [CrossRef]
107. Sousa-Pimenta, M.; Martins, A.; Estevinho, L.M.; Pinho Vaz, C.; Leite, L.; Mariz, J. Hepatic Sinusoidal Obstruction Syndrome/Veno-Occlusive Disease (SOS/VOD) Primary Prophylaxis in Patients Undergoing Hematopoietic Stem Cell Transplantation: A Network Meta-Analysis of Randomized Controlled Trials. *J. Clin. Med.* **2024**, *13*, 6917. [CrossRef] [PubMed]
108. Ramgopal, A.; Goscicki, B.; Sridar, S.; Wang, L.; Klein, D.; Kalpatthi, R.; Dalal, J. Prophylactic Defibrotide in High-Risk Pediatric HSCT: Solution or New Set of Challenges? *Blood* **2024**, *144* (Suppl. S1), 5020. [CrossRef]
109. Corbacioglu, S.; Grupp, S.A.; Richardson, P.G.; Duarte, R.; Pagliuca, A.; Ruutu, T.; Mahadeo, K.; Carreras, E. Prevention of veno-occlusive disease/sinusoidal obstruction syndrome: A never-ending story and no easy answer. *Bone Marrow Transplant.* **2023**, *58*, 839–841. [CrossRef]
110. Neidemire-Colley, L.; Robert, J.; Ackaoui, A.; Dorrance, A.M.; Guimond, M.; Ranganathan, P. Role of endothelial cells in graft-versus-host disease. *Front. Immunol.* **2022**, *13*, 1033490. [CrossRef]
111. Garcia-Bernal, D.; Palomo, M.; Martinez, C.M.; Millan-Rivero, J.E.; Garcia-Guillen, A.I.; Blanquer, M.; Diaz-Ricart, M.; Sackstein, R.; Carreras, E.; Moraleda, J.M. Defibrotide inhibits donor leucocyte-endothelial interactions and protects against acute graft-versus-host disease. *J. Cell Mol. Med.* **2020**, *24*, 8031–8044. [CrossRef]
112. Palaniyandi, S.; Kumari, R.; Strattan, E.; Huang, T.; Kohler, K.; Du, J.; Jabbour, N.; Kesler, M.; Hildebrandt, G.C. Role of Defibrotide in the Prevention of Murine Model Graft-versus-Host Disease after Allogeneic Hematopoietic Cell Transplantation. *Transplant. Cell Ther.* **2023**, *29*, 608.e1–608.e9. [CrossRef]
113. Martinez-Sanchez, J.; Hamelmann, H.; Palomo, M.; Mir, E.; Moreno-Castano, A.B.; Torramade, S.; Rovira, M.; Escolar, G.; Cordes, S.; Kalupa, M.; et al. Acute Graft-vs.-Host Disease-Associated Endothelial Activation in vitro Is Prevented by Defibrotide. *Front. Immunol.* **2019**, *10*, 2339. [CrossRef]
114. Hudspeth, M.; Mori, S.; Nachbaur, D.; Perez-Simon, J.A.; Stolzel, F.; Riches, M.; Wu, W.; Zhang, P.; Agarwal, S.; Yakoub-Agha, I. A phase II, prospective, randomized, open-label study of defibrotide added to standard-of-care prophylaxis for the prevention of acute graft-versus-host disease after allogeneic hematopoietic cell transplantation. *Haematologica* **2023**, *108*, 1026–1038. [CrossRef] [PubMed]

115. Akpinar, S.; Kayikci, O.; Tekgunduz, E. Defibrotide combined with triple therapy including posttransplant cyclophosphamide, low dose rabbit anti-t-lymphocyte globulin and cyclosporine is effective in prevention of graft versus host disease after allogeneic peripheral blood stem cell transplantation for hematologic malignancies. *Transfus. Apher. Sci.* **2022**, *61*, 103367. [CrossRef]
116. Squillaci, D.; Marcuzzi, A.; Rimondi, E.; Riccio, G.; Barbi, E.; Zanon, D.; Maximova, N. Defibrotide impact on the acute GVHD disease incidence in pediatric hematopoietic stem cell transplant recipients. *Life Sci. Alliance* **2023**, *6*, e202201786. [CrossRef]
117. Tekgunduz, E.; Kaya, A.H.; Bozdag, S.C.; Kocubaba, S.; Kayikci, O.; Namdaroglu, S.; Ugur, B.; Akpinar, S.; Batgi, H.; Bekdemir, F.; et al. Does defibrotide prophylaxis decrease the risk of acute graft versus host disease following allogeneic hematopoietic cell transplantation? *Transfus. Apher. Sci.* **2016**, *54*, 30–34. [CrossRef]
118. Tilmont, R.; Yakoub-Agha, I.; Ramdane, N.; Srour, M.; Coiteux, V.; Magro, L.; Odou, P.; Simon, N.; Beauvais, D. Impact of Defibrotide in the Prevention of Acute Graft-Versus-Host Disease Following Allogeneic Hematopoietic Cell Transplantation. *Ann. Pharmacother.* **2022**, *56*, 1007–1015. [CrossRef] [PubMed]
119. Young, J.A.; Pallas, C.R.; Knovich, M.A. Transplant-associated thrombotic microangiopathy: Theoretical considerations and a practical approach to an unrefined diagnosis. *Bone Marrow Transplant.* **2021**, *56*, 1805–1817. [CrossRef]
120. Attucci, I.; Pilerci, S.; Messeri, M.; Pengue, L.; Tomasino, G.; Caroti, L.; Vannucchi, A.M.; Antonioli, E. Carfilzomib-Induced Thrombotic Microangiopathy—Two Case Reports. *Cancer Rep.* **2024**, *7*, e2163. [CrossRef] [PubMed]
121. Martinez-Munoz, M.E.; Fores, R.; Lario, A.; Bautista, G.; Bueno, J.L.; de Miguel, C.; Navarro, B.; De Laiglesia, A.; Sanchez-Guerrero, A.; Cabrera, J.R.; et al. Use of defibrotide to treat adult patients with transplant-associated thrombotic microangiopathy. *Bone Marrow Transplant.* **2019**, *54*, 142–145. [CrossRef]
122. Wang, J.; Luo, Y.; Jia, C.; Yang, J.; Wang, B.; Zheng, J.; Jing, Y.; Chen, W.; Yang, W.; Zhu, G.; et al. Successful use of defibrotide to treat allogeneic hematopoietic stem cell transplantation associated thrombotic microangiopathy in pediatric patients: Report from Chinese single center. *Bone Marrow Transplant.* **2024**, *59*, 1483–1485. [CrossRef]
123. Yeates, L.; Slatter, M.A.; Bonanomi, S.; Lim, F.; Ong, S.Y.; Dalissier, A.; Barberi, W.; Shulz, A.; Duval, M.; Heilmann, C.; et al. Use of defibrotide to treat transplant-associated thrombotic microangiopathy: A retrospective study of the Paediatric Diseases and Inborn Errors Working Parties of the European Society of Blood and Marrow Transplantation. *Bone Marrow Transplant.* **2017**, *52*, 762–764. [CrossRef]
124. Devadas, S.K.; Toshniwal, M.; Bagal, B.; Khattry, N. Successful Treatment of Transplant Associated Thrombotic Microangiopathy (TA-TMA) with Low Dose Defibrotide. *Indian. J. Hematol. Blood Transfus.* **2018**, *34*, 469–473. [CrossRef]
125. Higham, C.S.; Shimano, K.A.; Melton, A.; Kharbanda, S.; Chu, J.; Dara, J.; Winestone, L.E.; Hermiston, M.L.; Huang, J.N.; Dvorak, C.C. A pilot trial of prophylactic defibrotide to prevent serious thrombotic microangiopathy in high-risk pediatric patients. *Pediatr. Blood Cancer* **2022**, *69*, e29641. [CrossRef]
126. Bohl, S.R.; Kuchenbauer, F.; von Harsdorf, S.; Kloevekorn, N.; Schonsteiner, S.S.; Rouhi, A.; Schwarzwalder, P.; Dohner, H.; Bunjes, D.; Bommer, M. Thrombotic Microangiopathy after Allogeneic Stem Cell Transplantation: A Comparison of Eculizumab Therapy and Conventional Therapy. *Biol. Blood Marrow Transplant.* **2017**, *23*, 2172–2177. [CrossRef]
127. Klein, O.R.; Ktena, Y.P.; Pierce, E.; Fu, H.H.; Haile, A.; Liu, C.; Cooke, K.R. Defibrotide modulates pulmonary endothelial cell activation and protects against lung inflammation in pre-clinical models of LPS-induced lung injury and idiopathic pneumonia syndrome. *Front. Immunol.* **2023**, *14*, 1186422. [CrossRef] [PubMed]
128. Cappell, K.M.; Kochenderfer, J.N. Long-term outcomes following CAR T cell therapy: What we know so far. *Nat. Rev. Clin. Oncol.* **2023**, *20*, 359–371. [CrossRef] [PubMed]
129. Bristol Myers Squibb. ABECMA® (Idecabtagene Vicleucel), Suspension for Intravenous Infusion. Bristol Myers Squibb, 2024. Available online: https://packageinserts.bms.com/pi/pi_abecma.pdf (accessed on 3 May 2025).
130. Janssen Biotech Inc. CARVYKTI® (Ciltacabtagene Autoleucel) Suspension for Intravenous Infusion—Prescribing Information. Janssen Biotech Inc., 2024. Available online: https://www.janssenlabels.com/package-insert/product-monograph/prescribing-information/CARVYKTI-pi.pdf (accessed on 3 May 2025).
131. National Cancer Institute. CAR T Cells: Engineering Patients' Immune Cells to Treat Their Cancers. National Cancer Institute, 2025. Available online: https://www.cancer.gov/about-cancer/treatment/research/car-t-cells (accessed on 3 May 2025).
132. Locke, F.L.; Miklos, D.B.; Jacobson, C.A.; Perales, M.A.; Kersten, M.J.; Oluwole, O.O.; Ghobadi, A.; Rapoport, A.P.; McGuirk, J.; Pagel, J.M.; et al. Axicabtagene Ciloleucel as Second-Line Therapy for Large B-Cell Lymphoma. *N. Engl. J. Med.* **2022**, *386*, 640–654. [CrossRef] [PubMed]
133. Schuster, S.J.; Bishop, M.R.; Tam, C.S.; Waller, E.K.; Borchmann, P.; McGuirk, J.P.; Jager, U.; Jaglowski, S.; Andreadis, C.; Westin, J.R.; et al. Tisagenlecleucel in Adult Relapsed or Refractory Diffuse Large B-Cell Lymphoma. *N. Engl. J. Med.* **2019**, *380*, 45–56. [CrossRef]

134. Jacobson, C.A.; Chavez, J.C.; Sehgal, A.R.; William, B.M.; Munoz, J.; Salles, G.; Munshi, P.N.; Casulo, C.; Maloney, D.G.; de Vos, S.; et al. Axicabtagene ciloleucel in relapsed or refractory indolent non-Hodgkin lymphoma (ZUMA-5): A single-arm, multicentre, phase 2 trial. *Lancet Oncol.* **2022**, *23*, 91–103. [CrossRef]
135. Fowler, N.H.; Dickinson, M.; Dreyling, M.; Martinez-Lopez, J.; Kolstad, A.; Butler, J.; Ghosh, M.; Popplewell, L.; Chavez, J.C.; Bachy, E.; et al. Tisagenlecleucel in adult relapsed or refractory follicular lymphoma: The phase 2 ELARA trial. *Nat. Med.* **2022**, *28*, 325–332. [CrossRef]
136. Wang, M.; Munoz, J.; Goy, A.; Locke, F.L.; Jacobson, C.A.; Hill, B.T.; Timmerman, J.M.; Holmes, H.; Jaglowski, S.; Flinn, I.W.; et al. KTE-X19 CAR T-Cell Therapy in Relapsed or Refractory Mantle-Cell Lymphoma. *N. Engl. J. Med.* **2020**, *382*, 1331–1342. [CrossRef]
137. Bristol Myers Squibb. BREYANZI® (Lisocabtagene Maraleucel) Suspension for Intravenous Infusion. Bristol Myers Squibb, 2024. Available online: https://packageinserts.bms.com/pi/pi_breyanzi.pdf (accessed on 3 May 2025).
138. Shah, B.D.; Ghobadi, A.; Oluwole, O.O.; Logan, A.C.; Boissel, N.; Cassaday, R.D.; Leguay, T.; Bishop, M.R.; Topp, M.S.; Tzachanis, D.; et al. KTE-X19 for relapsed or refractory adult B-cell acute lymphoblastic leukaemia: Phase 2 results of the single-arm, open-label, multicentre ZUMA-3 study. *Lancet* **2021**, *398*, 491–502. [CrossRef] [PubMed]
139. Maude, S.L.; Laetsch, T.W.; Buechner, J.; Rives, S.; Boyer, M.; Bittencourt, H.; Bader, P.; Verneris, M.R.; Stefanski, H.E.; Myers, G.D.; et al. Tisagenlecleucel in Children and Young Adults with B-Cell Lymphoblastic Leukemia. *N. Engl. J. Med.* **2018**, *378*, 439–448. [CrossRef] [PubMed]
140. Roddie, C.; Sandhu, K.S.; Tholouli, E.; Logan, A.C.; Shaughnessy, P.; Barba, P.; Ghobadi, A.; Guerreiro, M.; Yallop, D.; Abedi, M.; et al. Obecabtagene Autoleucel in Adults with B-Cell Acute Lymphoblastic Leukemia. *N. Engl. J. Med.* **2024**, *391*, 2219–2230. [CrossRef]
141. Ailawadhi, S.; Shune, L.; Wong, S.W.; Lin, Y.; Patel, K.; Jagannath, S. Optimizing the CAR T-Cell Therapy Experience in Multiple Myeloma: Clinical Pearls From an Expert Roundtable. *Clin. Lymphoma Myeloma Leuk.* **2024**, *24*, e217–e225. [CrossRef] [PubMed]
142. Janssen Biotech Inc. TECVAYLI® (Teclistamab-Cqyv) Injection, for Subcutaneous Use—Prescribing Information. Janssen Biotech Inc., 2024. Available online: https://www.janssenlabels.com/package-insert/product-monograph/prescribing-information/TECVAYLI-pi.pdf (accessed on 3 May 2025).
143. Pfizer Inc. ELREXFIOTM (Elranatamab-Bcmm) Injection, for Subcutaneous Use—Prescribing Information. Pfizer Inc., 2023. Available online: https://labeling.pfizer.com/ShowLabeling.aspx?id=19669 (accessed on 3 May 2025).
144. Janssen Biotech Inc. TALVEY™ (Talquetamab-Tgvs) Injection, for Subcutaneous Use—Prescribing Information. Janssen Biotech Inc., 2024. Available online: https://www.janssenlabels.com/package-insert/product-monograph/prescribing-information/TALVEY-pi.pdf (accessed on 3 May 2025).
145. Cohen, A.D.; Raje, N.; Fowler, J.A.; Mezzi, K.; Scott, E.C.; Dhodapkar, M.V. How to Train Your T Cells: Overcoming Immune Dysfunction in Multiple Myeloma. *Clin. Cancer Res.* **2020**, *26*, 1541–1554. [CrossRef]
146. Lutfi, F.G.; Ahmed, N.; Hoffmann, M.S.; Tun, A.; McGuirk, J.P. The emergence of bispecific T-cell engagers in the treatment of follicular and large B-cell lymphomas. *Clin. Adv. Hematol. Oncol.* **2024**, *22*, 510–519.
147. Omer, M.H.; Shafqat, A.; Ahmad, O.; Alkattan, K.; Yaqinuddin, A.; Damlaj, M. Bispecific Antibodies in Hematological Malignancies: A Scoping Review. *Cancers* **2023**, *15*, 4550. [CrossRef]
148. Gritti, G.; Belousov, A.; Relf, J.; Dixon, M.; Tandon, M.; Komanduri, K. Predictive model for the risk of cytokine release syndrome with glofitamab treatment for diffuse large B-cell lymphoma. *Blood Adv.* **2024**, *8*, 3615–3618. [CrossRef]
149. Markouli, M.; Ullah, F.; Unlu, S.; Omar, N.; Lopetegui-Lia, N.; Duco, M.; Anwer, F.; Raza, S.; Dima, D. Toxicity Profile of Chimeric Antigen Receptor T-Cell and Bispecific Antibody Therapies in Multiple Myeloma: Pathogenesis, Prevention and Management. *Curr. Oncol.* **2023**, *30*, 6330–6352. [CrossRef]
150. Lee, D.W.; Santomasso, B.D.; Locke, F.L.; Ghobadi, A.; Turtle, C.J.; Brudno, J.N.; Maus, M.V.; Park, J.H.; Mead, E.; Pavletic, S.; et al. ASTCT Consensus Grading for Cytokine Release Syndrome and Neurologic Toxicity Associated with Immune Effector Cells. *Biol. Blood Marrow Transplant.* **2019**, *25*, 625–638. [CrossRef]
151. Neelapu, S.S.; Tummala, S.; Kebriaei, P.; Wierda, W.; Gutierrez, C.; Locke, F.L.; Komanduri, K.V.; Lin, Y.; Jain, N.; Daver, N.; et al. Chimeric antigen receptor T-cell therapy—Assessment and management of toxicities. *Nat. Rev. Clin. Oncol.* **2018**, *15*, 47–62. [CrossRef] [PubMed]
152. Gloude, N.J.; Dandoy, C.E.; Davies, S.M.; Myers, K.C.; Jordan, M.B.; Marsh, R.A.; Kumar, A.; Bleesing, J.; Teusink-Cross, A.; Jodele, S. Thinking Beyond HLH: Clinical Features of Patients with Concurrent Presentation of Hemophagocytic Lymphohistiocytosis and Thrombotic Microangiopathy. *J. Clin. Immunol.* **2020**, *40*, 699–707. [CrossRef] [PubMed]
153. Jacobson, C.A.; Rosenthal, A.C.; Arnason, J.; Agarwal, S.; Zhang, P.; Wu, W.; Amber, V.; Yared, J.A. A phase 2 trial of defibrotide for the prevention of chimeric antigen receptor T-cell-associated neurotoxicity syndrome. *Blood Adv.* **2023**, *7*, 6790–6799. [CrossRef]

154. Teuwen, L.A.; Geldhof, V.; Pasut, A.; Carmeliet, P. COVID-19: The vasculature unleashed. *Nat. Rev. Immunol.* **2020**, *20*, 389–391. [CrossRef]
155. Bednarz, K.; Borek, A.; Drzymala, F.; Rachwal, K.; Gabryel, B. Pharmacological protection of vascular endothelium in acute COVID-19. *J. Physiol. Pharmacol.* **2022**, *73*, 167–177. [CrossRef]
156. García-Bernal, D.; Richardson, E.; Vlodavsky, I.; Carlo-Stella, C.; Iacobelli, M.; Jara, R.; Richardson, P.G.; Moraleda, J.M. Endothelial dysfunction and its critical role in COVID-19-associated coagulopathy: Defibrotide as an endothelium-protective, targeted therapy. *EJHaem* **2021**, *2*, 680–681. [CrossRef] [PubMed]
157. Bonaventura, A.; Vecchie, A.; Dagna, L.; Martinod, K.; Dixon, D.L.; Van Tassell, B.W.; Dentali, F.; Montecucco, F.; Massberg, S.; Levi, M.; et al. Endothelial dysfunction and immunothrombosis as key pathogenic mechanisms in COVID-19. *Nat. Rev. Immunol.* **2021**, *21*, 319–329. [CrossRef]
158. Ahamed, J.; Laurence, J. Long COVID endotheliopathy: Hypothesized mechanisms and potential therapeutic approaches. *J. Clin. Invest.* **2022**, *132*, e161167. [CrossRef]
159. Ariagno, S.; Ragoonanan, D.; Khazal, S.; Mahadeo, K.M.; Cisneros, G.S.; Zinter, M.S.; Blacken, R.A.; Mohan, G.; Lehmann, L.E.; Ferdjallah, A.; et al. Prior COVID-19 infection may increase risk for developing endothelial dysfunction following hematopoietic cell transplantation. *Front. Oncol.* **2022**, *12*, 1000215. [CrossRef]
160. Calabretta, E.; Moraleda, J.M.; Iacobelli, M.; Jara, R.; Vlodavsky, I.; O'Gorman, P.; Pagliuca, A.; Mo, C.; Baron, R.M.; Aghemo, A.; et al. COVID-19-induced endotheliitis: Emerging evidence and possible therapeutic strategies. *Br. J. Haematol.* **2021**, *193*, 43–51. [CrossRef]
161. Moraleda, J.M.; Carlo-Stella, C.; Garcia-Bernal, D.; Rubio, R.J.; Andreu, E.; Calabretta, E.; Aghemo, A.; Diaz-Ricart, M.; Palomo, M.; Carreras, E.; et al. Defibrotide for the Treatment of Endotheliitis Complicating Sars-Cov-2 Infection: Rationale and Ongoing Studies As Part of the International Defacovid Study Group. *Blood* **2020**, *136* (Suppl. S1), 6–8. [CrossRef]
162. Maccio, A.; La Nasa, G.; Oppi, S.; Madeddu, C. Defibrotide Has a Role in COVID-19 Therapy. *Chest* **2022**, *162*, 271–273. [CrossRef]
163. Frame, D.; Scappaticci, G.B.; Braun, T.M.; Maliarik, M.; Sisson, T.H.; Pipe, S.W.; Lawrence, D.A.; Richardson, P.G.; Holinstat, M.; Hyzy, R.C.; et al. Defibrotide Therapy for SARS-CoV-2 ARDS. *Chest* **2022**, *162*, 346–355. [CrossRef] [PubMed]
164. Ruggeri, A.; Corrado, F.; Voza, A.; Wei, L.J.; Catalano, G.; Liberatore, C.; Nitti, R.; Fedeli, C.; Bruno, A.; Calabretta, E.; et al. Use of defibrotide in COVID-19 pneumonia: Comparison of a phase II study and a matched real-world cohort control. *Haematologica* **2024**, *109*, 3261–3268. [CrossRef]
165. Rodríguez-Fortúnez, P.; Martínez-Mellado, A.; Jara-Rubio, R.; Blanquer-Blanquer, M.; Castro-Rebollo, P.; Carrillo-Alcaraz, A.; Rodríguez-Jiménez, C.; Albendin, H.; Solano, E.; Pareja, A.; et al. Defibrotide in the prevention and treatment of acute respiratory distress syndrome in patients with COVID-19. Preliminary safety results update. *Eur. J. Clin. Pharmacol.* **2022**, *78*, S22–S23. [CrossRef]
166. Lang, P.; Eichholz, T.; Bakchoul, T.; Streiter, M.; Petrasch, M.; Bosmuller, H.; Klein, R.; Rabsteyn, A.; Lang, A.M.; Adams, C.; et al. Defibrotide for the Treatment of Pediatric Inflammatory Multisystem Syndrome Temporally Associated With Severe Acute Respiratory Syndrome Coronavirus 2 Infection in 2 Pediatric Patients. *J. Pediatric Infect. Dis. Soc.* **2020**, *9*, 622–625. [CrossRef]
167. Belcaro, G.; Corsi, M.; Agus, G.B.; Cesarone, M.R.; Cornelli, U.; Cotellese, R.; Feragalli, B.; Hu, S. Thrombo-prophylaxis prevents thrombotic events in home-managed COVID patients. A registry study. *Minerva Med.* **2020**, *111*, 366–368. [CrossRef] [PubMed]
168. Kocoglu, M.H.; Richardson, P.G.; Mo, C.C.; Rapoport, A.P.; Atanackovic, D. Defibrotide improves COVID-19-related acute respiratory distress syndrome in myeloma patients after chimeric antigen receptor T-cell treatment without compromising virus-specific and anti-myeloma T-cell responses. *Haematologica* **2024**, *109*, 2372–2377. [CrossRef]
169. Fernandez, S.; Palomo, M.; Molina, P.; Diaz-Ricart, M.; Escolar, G.; Tellez, A.; Segui, F.; Ventosa, H.; Torramade-Moix, S.; Rovira, M.; et al. Progressive endothelial cell damage in correlation with sepsis severity. Defibrotide as a contender. *J. Thromb. Haemost.* **2021**, *19*, 1948–1958. [CrossRef]
170. Richardson, P.G.; Carreras, E.; Iacobelli, M.; Nejadnik, B. The use of defibrotide in blood and marrow transplantation. *Blood Adv.* **2018**, *2*, 1495–1509. [CrossRef]

biomolecules

MDPI

Review

Updated Review on Natural Polyphenols: Molecular Mechanisms, Biological Effects, and Clinical Applications for Cancer Management

Zainab Sabry Othman Ahmed [1,2,*], Elyas Khan [3], Nathan Elias [3], Alhussein Elshebiny [3] and Qingping Dou [3,*]

1 Department of Cytology and Histology, Faculty of Veterinary Medicine, Cairo University, Giza 12211, Egypt
2 Department of Anatomy and Histology, Faculty of Veterinary Medicine, King Salman International University, Ras Sudr 46612, Egypt
3 Departments of Oncology, Pharmacology and Pathology School of Medicine, Barbara Ann Karmanos Cancer Institute, Wayne State University, Detroit, MI 48201, USA; elyaskhan882@gmail.com (E.K.); nathanelias7000@gmail.com (N.E.); alelshebi@gmail.com (A.E.)
* Correspondence: zainab.sabry@cu.edu.eg or zainab.sabry@ksiu.edu.eg (Z.S.O.A.); doup@karmanos.org (Q.D.); Tel.: +1-313-576-8301 (Q.D.)

Abstract: Polyphenols, naturally occurring compounds found exclusively in plants, have gained significant attention for their potential in cancer prevention and treatment. These compounds are known for their antioxidant properties and are abundant in various plant-based foods, such as vegetables, fruits, grains, and beverages. Recent studies have highlighted the broad spectrum of health benefits of polyphenols, including their antiviral, anti-inflammatory, and anticancer properties. In addition, these naturally derived compounds are increasingly important for drug discovery due to their high molecular diversity and novel biofunctionalities. This review provides an in-depth analysis of the current research and knowledge on the potential use of dietary polyphenols as bioactive compounds for the prevention and treatment of various cancers. This review aims to provide valuable insights into the mechanisms underlying the anticancer properties of phenolic compounds in both laboratory and clinical settings. Furthermore, this review highlights the positive clinical outcomes associated with the use of polyphenols as anticancer agents and offers guidance for future research to advance this promising field.

Keywords: dietary sources; polyphenols; phenolic acids; anticancer; cell cycle arrest; apoptosis

1. Introduction

Cancer is a broad category of illnesses in which aberrant cells multiply uncontrollably, starting in nearly any body part and possibly moving to adjacent or distant regions. It is one of the leading causes of death globally, and its prevalence is continuously increasing. According to research by the World Health Organization (WHO), cancer is a major cause of mortality, contributing to almost 10 million deaths in 2020, or about one out of every six deaths.

Maintaining wellness and preventing illness requires a suitable diet and lifestyle. Polyphenols are naturally occurring chemicals produced exclusively by plants and possess potent antioxidant properties [1]. Owing to their prevalence in plant-based foods such as vegetables, fruits, and grains and their antioxidant action, polyphenols have been extensively researched in recent years as adjuvants in reducing the risk factors for debilitating diseases such as diabetes, cancer and cardiovascular disease (CVD) [2]. Examining the role of polyphenols in important signalling pathways could help clarify how a diet rich in polyphenols affects cancer outcomes [3].

Numerous studies have shown that the consumption of polyphenols yields various health benefits, including antiviral, antioxidant, anti-inflammatory, anti-thrombogenic, anti-allergic, antihyperlipidemic, anti-diabetic, anti-asthma, and anticancer effects [4]. Therefore, in response to the growing need for the development of new natural-based therapies, polyphenols have garnered significant scientific attention and have been the subject of substantial investigation in recent years [5].

These molecules exert anticancer effects by targeting different checkpoints in malignant cells and have a high specificity for inducing cell cycle arrest, autophagy, and apoptosis [6]. They exert these anticancer effects by inhibiting telomere expression, angiogenesis, and metastasis, in addition to lowering the expression of transcription factors that regulate the expression of cytoprotective genes, lowering p53 activation, reducing Bcl-2 expression and mitochondrial membrane potential, and decreasing the expression of HIF-1α while increasing cellular apoptosis via downregulation of p-Akt expression [7,8].

The objective of this review is to provide an overview of the current research and knowledge regarding the potential use of dietary polyphenols as naturally occurring bioactive compounds for the prevention and treatment of various malignancies. In addition to offering guidance for future research, this review also sheds light on the mechanisms underlying the possible anticancer properties of phenolic compounds in both clinical and laboratory contexts, as well as the ensuing positive clinical benefits of polyphenols as anticancer compounds.

2. Classification of Polyphenols

Natural polyphenols refer to a large group of phenylpropanoids synthesised by plants as secondary metabolites, ranging from small molecules to highly polymerised compounds, mostly in the form of glycosides. At least 10,000 distinct chemicals with one or more aromatic rings and one or more hydroxyl groups are collectively referred to as polyphenols [9]. Polyphenols can be divided into flavonoids and non-flavonoids. Flavonoids can exist as glycosides or aglycones despite their fundamental structures being aglycones (the non-sugar portion of the corresponding glycoside). Anthocyanins, flavonols, flavan-3-ols, flavones, isoflavones, flavanones, and stilbenes are examples of flavonoids typically found in food [10].

3. The Dietary Sources of Different Polyphenol Compounds

Most fruits and vegetables are rich in polyphenols. Certain fruits and vegetables contain higher levels of some polyphenols than others; therefore, identifying the primary dietary sources of each natural polyphenol is important. For example, sources of flavanols include onions and black tea. Moreover, flavanones are sourced from oranges and lemons. Epigallocatechin-3-gallate is predominantly found in green tea. Genistein is an isoflavone primarily found in soybeans. Quercetin is primarily found in onions, specifically red and yellow onions and citrus fruits. Phenolic acids, such as hydroxybenzoic acids, hydroxycinnamic acid, gallic acid, and caffeic acid, are found in nuts, pineapples, green tea, basil, olives, and other common natural substances. Curcumin, the most common source of curcuminoids, is found in turmeric. Resveratrol, a Stilbene, is primarily found in red grapes and red wine. Lignans, which are commonly found in flaxseeds, sesame seeds, and legumes, are present in a wide variety of foods. The general sources of these natural polyphenols (Figure 1) show that they can be easily and are most likely already incorporated into our daily diets.

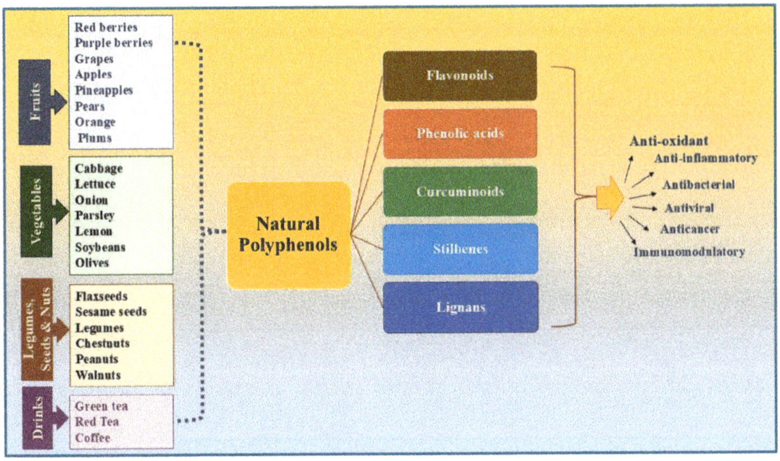

Figure 1. Dietary sources of natural polyphenols.

4. Nutraceuticals and Pharmaceuticals Derived from Dietary Phenolic Compounds

High quantities of polyphenols have been partially linked to the anticancer properties of fruits and vegetables [11]. The well-studied polyphenols present in plants include stilbenes, lignans, phenolic acids, and flavonoids. The consumption of polyphenolic compounds has been linked to several health benefits (Table 1). Polyphenols influence cellular and molecular processes that impede several stages of carcinogenesis, including initiation, promotion, and progression [12]. The antiproliferative properties of polyphenols on various malignant tumours, both in vivo and in vitro, have been the subject of numerous investigations over the past few years [13]. These compounds selectively trigger cell cycle arrest, autophagy, and apoptosis and exert antiproliferative effects on a variety of human cancer cell types both in vivo and in vitro [6].

Table 1. Medicinal properties of natural phenolic compounds.

Class	Compound	Property(s)	Citation
1. Flavonoids	Anthocyanins	Antioxidant and Anticancer	[14]
	Flavanols	Anti-diabetic, Anti-inflammatory and Antioxidant	[15]
	Flavonols: Quercetin	Anti-inflammatory, Antioxidant, Antimicrobial, Anticancer, Antihypertensive, vasodilator, Antiobesity, Antiatherosclerosis	[16]
	Flavonols: Epigallocatechin-3-gallate	Antioxidant Anti-angiogenesis, Anti-inflammation and Anticancer	[17]
	Flavones; Luteolin	Anti-Inflammatory, Antioxidant, Antiallergy, Anticancer and Antibacterial	[18]
	Flavanones; Hesperetin	Anti-inflammatory, Antioxidant, Antibacterial and Anticancer	[18]
	Isoflavones; Genistein	Anti-inflammatory	[19]

Table 1. *Cont.*

Class	Compound	Property(s)	Citation
2. Phenolic acids	Caffeic acid and its derivative caffeic acid phenethyl ester	Antioxidant, anti-inflammatory and anticancer	[20]
	Gallic acids	Anticancer, Antioxidant, and Anti-inflammatory	[21]
	Rosmarinic acid	Anti-inflammatory, Antiviral, Antibacterial, Antidepressant and Anticancer	[22]
	Sinapic acid	Antioxidant, Antimicrobial, Anti-inflammatory, Anticancer, Antianxiety	[23]
	Hydroxy benzoic acid	Antimicrobial	[24]
	Hydroxycinnamic acid	Antioxidant	[25]
	Protocatechuic acid	Anti-inflammatory and Antimicrobial	[26]
	Syringic acid	Antioxidant, antimicrobial, anti-inflammatory, antiendotoxic, neuro and hepatoprotective	[27]
	Protocatechoic acid	Anti-inflammatory and Antimicrobial	[26]
	Synergic acid	Antioxidant, antimicrobial, anti-inflammatory, antiendotoxic, neuro and hepatoprotective	[27]
	Vanillic acid	Antioxidant, Anti-inflammatory and Neuroprotective	[28]
3. Curcuminoids	Curcumin	Anti-inflammatory, Antioxidant and Anticancer	[29]
4. Stilbenes	Resveratrol	Antioxidant, Anti-inflammatory, Immunomodulatory, Neuroprotective, Cardiovascular protective and Anticancer	[30]
5. Lignans	Dibenzocyclooctadiene lignans	Antioxidant, Antiviral, Anti-inflammatory and Anticancer	[31]

5. Polyphenols and Their Anticancer Properties with Insights into Their Molecular Mechanisms, Preclinical Studies, and Clinical Applications

Naturally occurring anticancer chemicals found in dietary phenolic compounds provide a variety of treatment and preventive alternatives for different cancer types. As these compounds can target different checkpoints in malignant cells, investigating their mechanisms of action may increase the effectiveness of treatment [11].

5.1. Flavonoids

5.1.1. Anthocyanins

The most significant class of flavonoids found in plants is anthocyanins (Figure 2), which are water-soluble pigments that have shown antioxidant activity [32]. It has been reported that black elderberries, black chokeberries, and black currants are the richest sources of anthocyanins [33]. However, the bioavailability of anthocyanins is quite poor; only 1–2% of them retain their original structure after consumption [14]. PH, temperature, and solvents are some of the variables that affect the structures and characteristics of anthocyanins and should be managed when conducting investigations on their antioxidant activity [34].

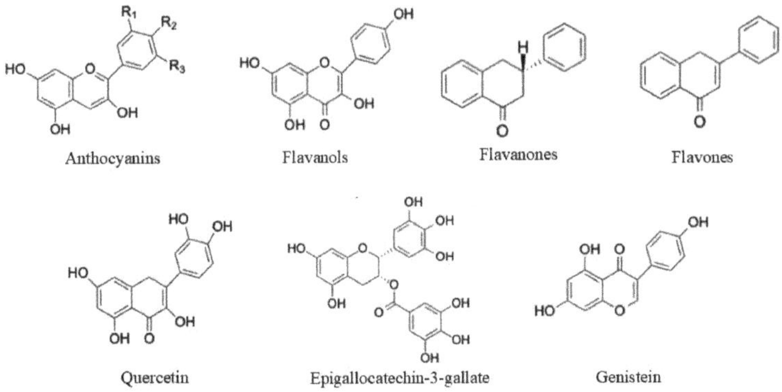

Figure 2. Chemical structures of different classes of flavonoids.

The antimutagenic activity, suppression of oxidative DNA damage and carcinogens, cell cycle arrest, apoptosis, induction of phase II enzymes for detoxification, inhibition of cyclooxygenase-2 enzymes, and anti-angiogenesis are some of the potential mechanisms responsible for the anticancer activity of anthocyanins that have been described in various studies [35]. The downregulation of the pro-survival Sirt1/survivin and Akt/mTOR pathways, anti-proliferation, apoptosis, and decrease in the metastatic markers Sp1, Sp4, and VCAM-1 were confirmed in a variety of cell lines, including MDA-MB-231, MDA-MB-453, BT474, A17, N202/1A, and N202/1E [36]. Another study emphasised the induction of apoptosis via the p38/Fas/FasL/caspase eight and p38/p53/Bax signalling pathways [37]. Moreover, anthocyanins are known to exert strong anti-invasive and antimetastatic properties [14]. For instance, delphinidin treatment causes cell cycle arrest and apoptosis in several cancer types. By specifically inhibiting NF-κB-dependent MMP-9 (matrix metalloproteinase-9) gene expression, delphinidin can function as a potential antimetastatic drug that inhibits PMA-induced cancer cell invasion [38]. In addition, cyanidin-3-glucoside and cyanidin-3-rutinoside, which are extracted from mulberries, inhibited the migration and invasion of A549 lung cancer cells. Moreover, Peonidin-3-glucoside therapy also considerably inhibited lung cancer cell metastasis by downregulating matrix metalloproteinase (MMP) [39]. Furthermore, the growth and proliferation of 22Rv1, PC-3, and C4-2 prostate cancer cell lines were inhibited by anthocyanins [40]. Certain anthocyanins, such as cyanidin and delphinidin, have been shown to be cytotoxicity to colorectal cancer cells through oxidative stress [41].

Laboratory experiments using cell lines from several types of cancer (such as breast, colon, and prostate cancer) have revealed that anthocyanins can strongly inhibit the growth of cancer cells, in addition to inducing apoptosis. Additionally, anthocyanin supplementation decreased the size of tumours and suppressed tumour metastasis in animal experiments using mice [42]. Table 2 summarises the antitumor properties of anthocyanins reported in various malignancies in published studies.

Table 2. The anticancer properties of anthocyanins in different malignancies.

Anticancer Effect of Anthocyanins	Cancer Type	Citation
Anti-invasiveness and inhibition of the proliferation of MDA-MB-231 breast cancer cell lines.	Breast cancer	[43]
Reduction of the viability of breast cancer cell lines MCF-7, MDA-MB-231, and MDA-MB-453, in addition to induction of apoptosis in MDA-MB-453 cells via the intrinsic pathway (caspase cascade activation PARP cleavage and cytochrome C release) and suppression of tumour growth and angiogenesis via inhibiting MMP-9, MMP-2, and uPA expression in BALB/c naked mice with MDA-MB-453 cell xenografts.	Breast cancer	[44]
Inhibition of c-Jun N-terminal kinase, mitogen-activated protein kinase, and fibrosarcoma activation, downregulation of matrix metalloproteinase 2 secretion, and inhibition of cell migration and invasion in MDA-MB-453 breast cancer cells (HER2+).	Breast cancer	[45]
Inhibition of the development of abnormal crypt foci of colon in CF-1 mice.	Colon cancer	[46]
Induction of apoptosis in benign prostatic hyperplasia in rats.	Benign prostate hyperplasia	[47]
Triggering apoptotic factors such as TRAIL in cancer systems and suppression of Akt-mTOR signalling leading to maturation of acute myeloid leukaemia cells.	Leukaemia	[48]

5.1.2. Flavanols

Dark chocolate and cocoa are the primary sources of flavanols [49], which are also found in berries, black chokeberries, blueberries, and blackcurrants. Other significant sources include strawberries, apples, hazelnuts, pecan nuts, pistachios, almonds, red wine, green tea, and black tea [50]. The positive outcomes were mostly linked to monomers/epicatechin/catechin and dimers/procyanidin B2/procyanidin. The bioavailability of procyanidins is approximately 100 times lower than that of their monomers. The monomers created following stomach breakdown that can be quickly absorbed in the gut are typically responsible for the biological effects. The metabolite production process, which is also attributed to the gut microbiota, may have a variety of biological impacts [10]. Isorhamnetin, a derivative of quercetin, exhibits impressive pharmacological properties, such as antioxidant and anti-inflammatory properties [51]. Moreover, the antioxidant properties of epicatechin (EPI), a naturally occurring flavonol, may facilitate the positive effects of natural products like cocoa [52].

5.1.3. Flavonols

Flavonols are mostly found in fruits and vegetables such as cranberries and onions and in some drinks (such as tea and red wine), for which the estimated daily intake ranges from 18 (USA) to 58 mg (Japan) [53]. Nevertheless, these consumption thresholds often only address three main flavonols: kaempferol, myricetin, and quercetin [54]. Studies have investigated the health benefits of flavonol intake, with a focus on antioxidant activity, inflammatory biomarkers, and CVD risk factor reduction, in addition to the effect of quercetin, which has been enzymatically modified, on cognitive function [55].

Quercetin

Quercetin (3,3′,4′,5,7-Pentahydroxyflavone) (Figure 2) is a flavonol [54] with a variety of therapeutic uses [56,57], including cardiovascular protection and antiviral, anti-inflammatory, anti-allergic, and anticancer properties. Additionally, it has been discovered that quercetin is essential for plants [58] as it contributes to photosynthesis, growth, and seed germination due to its antibacterial and antioxidant properties. Moreover, quercetin's presence in different brain regions aids in protection against several neurological disorders, including Parkinson's and Alzheimer's diseases [59].

In vitro studies have demonstrated quercetin's antitumor effect against melanoma [60] and pancreatic [61], breast [62], liver [63,64], and prostate [65,66] cancers. Quercetin's anticancer effect is associated with its ability to control certain enzymatic processes, oxidative stress, and cellular pathways. When applied to tumours with multidrug resistance, quercetin has shown synergistic effects by suppressing the ejection of drugs facilitated by transporter proteins [54]. Furthermore, quercetin has the potential to induce autophagy and is effective in treating breast cancer by inhibiting the Akt-mTOR pathway in glycolysis and cell motility. Moreover, quercetin inhibited the growth and metastasis of breast cancer in mice with MCF-7 tumours. Additionally, it decreased the expression levels of VEGF, p-AKT, and PKM2 in the tumour tissues [67]. Quercetin's antiproliferative effect can be primarily attributed to cell cycle arrest at the G1 phase, which occurs due to the downregulation of cyclin-dependent kinase 1 (CDK1) and cyclin B1 (Figure 3), which are essential components for the progression of the G2/M cell cycle, and the activation of phosphorylation for the retinoblastoma tumour suppressor protein, pRb [68]. In addition, the tumour-suppressing molecules Bax, p21Cip1, p27Kip1, cyt-c, caspase 3, caspase 8, and p53 are upregulated in prostate cancer after treatment with quercetin. Moreover, quercetin inhibited IL-6 and IL-10 cytokine production, resulting in the cytotoxicity of primary effusion lymphoma (PEL). In addition, it downregulated cell survival proteins, such as c-FLIP, cyclin D1, and cMyc, in PEL cells by inhibiting the PI3K/AKT/mTOR and STAT3 pathways [69] (Table 3).

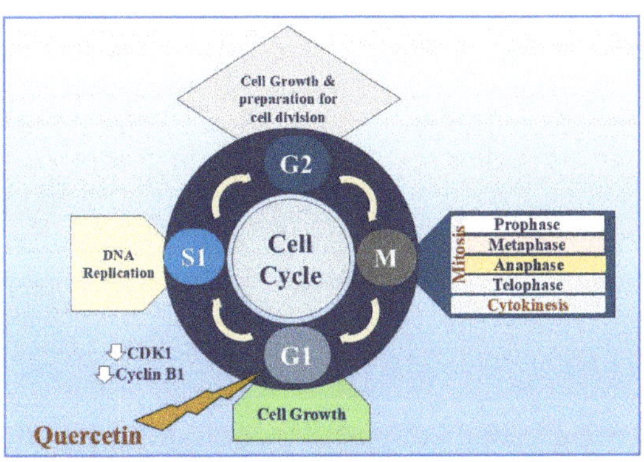

Figure 3. Quercetin induces cell cycle arrest in the G1 phase.

When combined with other drugs, quercetin can increase the apoptotic effect of midkine, elevate caspase 3, and decrease the expression of the survivin gene. It decreased the number of S phase cells and induced G1-phase cell cycle arrest. Furthermore, quercetin increased the expression of PTEN while downregulating the phosphorylation of PI3K, Akt, ERK1/2, p38, NF-κB, and survivin proteins [65]. Using the human gastric cancer cell line AGS, Lei et al. investigated the potential therapeutic benefits of quercetin in combination with irinotecan/SN-38. Quercetin combined with SN-38 synergistically increased apoptosis and anti-proliferation, with alterations in GSK3β/β-catenin signalling. Treatment with quercetin either alone (twice weekly) or combined with irinotecan (10 mg/kg once weekly) resulted in significant inhibition of tumour growth, with lowered levels of COX-2 gene, and downregulation of tumour VEGF-R and VEGF-A [70].

One advantage of quercetin is its low toxicity; however, its shortcomings are low solubility and poor bioavailability, suggesting that nanoparticle encapsulation might improve its efficacy. Polymeric nanoparticles, stimuli-responsive polymeric nanoparticles, and non-responsive polymeric nanoparticles are a few examples. Additionally, quercetin-containing inorganic nanoparticles, including metal oxides, silica, and gold nanoparticles, have been studied [71].

Epigallocatechin-3-Gallate (EGCG)

One of the phenolic components of green tea (*Camellia sinensis*) is epigallocatechin-3-gallate (EGCG) [72]. It is also present in a wide variety of food herbs and plants, including strawberries, blackberries, cranberries, cherries, kiwis, pears, avocados, peaches, apples, pecans, pistachios, and hazelnuts [17]. EGCG, an ester of epigallocatechin and gallic acid, exhibits different biological and pharmacological actions, such as pro-apoptotic, anti-inflammatory, antiangiogenic, antioxidant, and antimetastatic properties [73], and has been used in clinical trials [74]. In addition, EGCG has been reported to have protective potential against neurodegenerative disorders, such as Parkinson's and Alzheimer's diseases [72].

Numerous health benefits, such as reduced circulating cholesterol, weight loss, cardiovascular protection, and inflammatory attenuation, have been demonstrated in different studies [75]. In addition, EGCG's inhibitory effects on the initiation, development, and progression of several tumour types have been shown in several in vitro research using various cancer cell lines, as well as in vivo studies [76–78] (Table 3). Moreover, EGCG has been reported to decrease the side effects linked to chemotherapy and improve the therapeutic efficiency of existing treatments. EGCG has the potential to be a versatile anticancer drug by preventing cell cycle progression, triggering apoptosis, preventing invasion and metastasis, and modifying the tumour microenvironment (TME) [79]. Studies conducted on animals and cells have confirmed these effects and suggested several mechanisms through which EGCG acts as an anticancer agent. The mechanism by which EGCG functions as a potent antioxidant is mediated by reactive oxygen species (ROS). EGCG may also function as a pro-oxidant under specific conditions [80,81]. Despite its natural origin, safety, and affordability, its limited bioavailability is a significant challenge that is being addressed by encapsulating it in nanoparticles for further delivery [79].

In human hepatocellular liver cancer cells (HepG2), EGCG functions as a strong antioxidant that reduces oxidative stress by preventing the generation of ROS and increasing the activity of the antioxidant enzymes glutathione peroxidase and superoxide dismutase. [82]. Moreover, numerous studies have shown that EGCG downregulates MMP1 expression [83]. Considering that MMP1 plays a role in the migration, invasion, and metastasis of cancer cells [84], the anticancer effect of EGCG may be partly attributed to its suppression of MMP1 expression [81]. Furthermore, EGCG's downregulation of MMP2 was suggested to decrease the phosphorylation of PI3K and ERK [81]. Moreover, EGCG inhibits cell division, migration, and Matrigel invasion in TW01 and NA nasopharyngeal cancer cells [85].

In a human pancreatic cancer xenograft model using AsPC-1 cells, EGCG suppressed human umbilical vein endothelial cell (HUVECs') migration, capillary tube formation, and cell proliferation, and these suppressive effects were amplified in the presence of an ERK inhibitor. In addition, EGCG-treated mice tumour samples showed increased p38 and JNK activity and decreased ERK activity. EGCG and catechin gallate inhibited the induction of ConA-mediated MT1-MMP in U87 glioblastoma cells, while EGCG and gallocatechin gallate inhibited the induction of the endoplasmic reticulum stress (ERS) biomarker GRP78 and proMMP2 activation [86].

In cervical cancer HeLa cells, EGCG inhibited proliferation, induced apoptosis, and suppressed cell invasion and migration. Additionally, EGCG decreased MMP9 gene expression while upregulating TIMP1 gene expression [87]. Furthermore, EGCG induced apoptosis in HCT116 cells with wild-type p53 and HT-29 cells with mutant p53 in a manner independent of p53. Regardless of p53 status, EGCG inhibited MMP9 and VEGF expression [88]. HTLV-1-infected cells express Tax oncogenes. In ATL HuT-102 and C91-PL cells that were HTLV-1 positive, EGCG caused cytotoxicity and decreased Tax expression. In these cells, EGCG reduced MMP9 activity, NF-κB activity, and MMP9 mRNA and protein levels [89].

In bladder cancer SW780 cells, EGCG prevented the invasion, migration, and proliferation of cancer cells. Additionally, by activating caspases-8, -9, and -3, Bax, and poly-ADP ribose polymerase, EGCG induced apoptosis in these cells. When EGCG was injected into mice with SW780 tumours, both tumour weight and volume decreased. In SW780 tumours and cells, EGCG suppressed the expression of MMP9 and NF-κB at the protein and mRNA levels. The inhibitory effects of EGCG on the migration and proliferation of SW780 cells were cancelled by the addition of the NF-κB inhibitor SC75741 [90]. According to research about the effects of oral EGCG administration on patients with breast cancer receiving radiation therapy, EGCG decreased the activation of MMP2 and MMP9 in patient sera and decreased serum levels of VEGF and hepatocyte growth factor (HGF) in comparison to patients who were not receiving treatment [91].

Regarding clinical trials, oral delivery of EGCG was safe, practical, and efficient in a phase I clinical trial in unresectable stage III lung cancer when EGCG was combined with standard chemoradiotherapy, with a suggested dose of 440 μmol/L in a phase II clinical trial [92]. Despite the crucial significance of EGCG in cancer prevention, as demonstrated in different phase I and II clinical trials, additional trials are still required to fully comprehend the efficacy of EGCG in cancer treatment [11].

Table 3. The anticancer properties of flavonols in different types of cancer.

Flavonols	Anticancer Effect	Cancer Type	Citation
Quercetin	Increase the expression of PTEN while downregulation of the phosphorylation of PI3K, Akt, ERK1/2, p38, NF-κB, and survivin proteins.	Prostate cancer	[65]
	Induction of autophagy, inhibition of the Akt-mTOR pathway's role in glycolysis and cell motility and reduction of the expression levels of VEGF, p-AKT, and PKM2 in tumour tissue.	Breast cancer	[67]
	Cell cycle arrest at the G1 phase that occurs due to the downregulation of cyclin-dependent kinase 1 (CDK1) and cyclin B1and Upregulation of Bax, Bcl-2, p21Cip1, p27Kip1, cyt-c, caspase 3, caspase 8, and p53.	Prostate cancer	[68]
	Inhibition of IL-6 and IL-10 cytokine production, resulting in cytotoxicity, in addition to downregulation of cell survival proteins, such as c-FLIP, cyclin D1, and cMyc via inhibiting the PI3K/AKT/mTOR and STAT3 pathways.	Primary Effusion Lymphoma	[69]
	Induction of apoptosis and anti-proliferation, with alteration of GSK3β/β-catenin signalling, in addition to reduction of COX-2 level, and downregulation of tumours' VEGF-R and VEGF-A.	Human gastric cancer cell line AGS	[70]

Table 3. *Cont.*

Flavonols	Anticancer Effect	Cancer Type	Citation
Epigallocatechin-3-Gallate (EGCG)	Decreased cell adhesion and downregulated the expression of VEGF, NF-κB, FAK, and MT1-MMP.	Breast cancer MCF-7 cells	[93]
	Inhibition of cell division, migration, and Matrigel invasion.	TW01 and NA nasopharyngeal cancer cells	[85]
	Suppression of human umbilical vein endothelial cells (HUVECs') migration, capillary tube formation, and cell proliferation and downregulation of MMP2, MMP7, MMP9, and MMP12 and reduction of the volume, angiogenesis, and metastasis of tumour, in addition to increased p38 and JNK activity and decreased ERK activity.	Pancreatic cancer xenograft AsPC-1	[94]
	Inhibition of cells proliferation, induction of apoptosis, suppression of cell invasion and migration, in addition to suppression of MMP9 gene expression while upregulation of TIMP1 gene expression.	Cervical cancer HeLa cells	[87]
	Inhibition of the levels of MMP9 and VEGF expression.	Colon cancer HCT116 cells	[88]
	Induction of cytotoxicity, decrease Tax oncogene expression, and reduction of MMP9 activity, NF-κB activity, and MMP9 mRNA and protein levels.	ATL HuT-102 and C91-PL cells	[89]
	Prevention of cell invasion, migration, and proliferation, in addition to activation of caspases-8, -9, and -3, Bax and poly-ADP ribose.	Bladder cancer SW780 cells	[90]
	Reduction of the activation of MMP2 and MMP9 in patient sera in addition to lowering the serum levels of VEGF and hepatocyte growth factor (HGF).	Patients with breast cancer	[91]

5.1.4. Flavones: Luteolin

Luteolin (3,4,5,7-tetrahydroxy flavone), a naturally occurring flavone, is found in fruits and vegetables like celery, sweet bell peppers, chrysanthemum flowers, carrots, onion leaves, parsley, and broccoli [94]. It acts as an antioxidant or pro-oxidant biochemically and has a variety of biological effects, including anti-inflammatory, anti-allergic, and anticancer properties. It is an essential chemopreventive agent for the treatment of different cancers [95]. It can inhibit the proliferation of different types of tumour cells in vitro, with an IC_{50} ranging from roughly 3 to 50 μM [94]. Through a variety of mechanisms, such as kinase suppression, cell cycle regulation, induction of apoptosis, and reduction of transcription factors, luteolin has been shown to impede the progression of carcinogenesis, including cell transformation, metastasis, invasion, and angiogenesis (Table 4). The anticancer properties of luteolin also include DNA damage, redox regulation, and inhibition of cancer cell proliferation, which are linked to apoptosis induction [95].

Luteolin demonstrated cell cycle arrest during the G1 phase in a variety of human cancer cell lines, including gastric, prostate, and melanoma. The induced G1 cell cycle arrest was associated with the inhibition of CDK2 activity in colorectal cancer HT-29 and melanoma OCM-1 cells. In MCF-7 breast cancer cells induced by EGF, luteolin showed significant suppression of the expression of p-STAT3, p-EGFR, p-Akt, and p-Erk1/2 and inhibited cell proliferation. Additionally, it was able to inhibit the EGFR signalling pathway in human breast cancer cell lines. Moreover, previous studies have shown that a moderate dose of luteolin (10 mg/kg) can prevent the establishment of large tumours in a 7,12-

dimethylbenz (a)anthracene-induced tumour model and can dramatically reduce the levels of vascular endothelial growth factor (VEGF) in Sprague-Dawley rats. In addition, luteolin suppressed the growth of MPA-dependent human breast cancer cell xenograft tumours, progestin-dependent VEGF release from breast cancer cells, and tumour cell survival. Furthermore, it reduced blood vessel density and prevented T47-D and BT-474 breast cancer cells from acquiring stem cell-like characteristics [94].

Luteolin inhibited progression of MCF-7 breast cancer cells in the G1 phase and induced sub-G1 cell population, altered the morphology of the nucleus, raised the mRNA levels of death receptors such as DR5 and caspase and inhibited poly-ADP ribose polymerase, a key indicator that helps a cancer cell heal itself in a dose-dependent manner, and increased caspase-9/-8/-3 activity. Furthermore, luteolin induced the release of cytochrome c after impairing the potential of the mitochondrial membrane. As a result, Bcl-2 expression was suppressed, and Bax expression increased [96]. Similarly, by inducing apoptosis, regulating the cell cycle, and inhibiting proliferation, luteolin exhibited anticancer activity against MDA-MB-543 cells. In MDA-MB-231 cells, luteolin caused cell cycle arrest in the S phase by lowering telomerase levels and preventing the phosphorylation of NF-κB inhibitor α. As a result, it lowered the mRNA levels of human telomerase reverse transcriptase, which encodes the catalytic portion of telomerase. Additionally, luteolin inhibits the growth of malignant breast cells and triggers apoptosis, leading to the inhibition of cancer spread [97]. The synergistic effects of luteolin and celecoxib treatment were observed in MCF-7 and MCF7/HER18 cells through Akt inactivation and extracellular signal-regulated kinase (ERK) signalling inhibition [98].

It has been shown that luteolin mono-acylated derivatives exhibit anticancer and antioxidant properties against HCT116 colon cancer cells. The resulting compounds become more lipophilic upon acylation of the -OH groups, increasing their bioavailability [99]. Luteolin inhibited the G2/M phase of the cell cycle and caused colon cancer cells to undergo apoptosis. In a dose- and time-dependent manner, luteolin had an inhibitory effect on cell proliferation in LoVo human colon cells by triggering cell cycle arrest at the G2/M phase and inhibition of cyclin B1/CDC2. Apoptotic protease-activating factor 1 (APAF1) is stimulated by deoxyadenosine triphosphate, which controls these processes [100]. You and colleagues documented how luteolin inhibits colon tumours through apoptosis and autophagy. After treatment with luteolin, HCT116 cells displayed increased p53 phosphorylation and p53 target gene expression, which leads to cell cycle arrest and apoptosis. Thus, luteolin-induced p53 wild-type cells to undergo autophagy. This suggests that autophagy induced by the compound depends on p53 [101].

Through the increased Nrf2 transcription induced by DNA demethylation of its promoter, luteolin exerts anticancer effects on colorectal cancer cells. Furthermore, by strengthening the interaction between Nrf2 and p53, luteolin increases the expression of antioxidant enzymes and apoptotic proteins [102]. Additionally, Lutein has been demonstrated to inhibit the proliferation of colorectal malignant cells, interrupt the cell cycle, damage DNA, and accelerate apoptosis by targeting the MAPK pathway. These findings suggest that luteolin may be a useful adjuvant for the treatment of colorectal cancer in the future [103].

Luteolin has been reported to exert its pharmacological effects by inhibiting the expression of cyclin E, MMP-2, cyclin D1, vimentin, Bcl2, and N-cadherin while promoting the expression of E-cadherin, Bax, and p21. In gastric tumour cells, luteolin's anticancer properties are confirmed by a decrease in the expression of p-PI3K, p-mTOR, p-AKT, p-STAT3, and Notch1, and an increase in the amount of p-P38 signal transduction [104]. Accordingly, luteolin (40 mg/kg) effectively inhibited the growth of cancer in BGC-823 gastric tumour xenografts in experimental mice. According to published studies, luteolin inhibits the activation of the immune system and the expression of MMP-9 and VEGF-A, which prevents cancer growth. Furthermore, in a c-Met-overexpressing individual-derived xenograft model, luteolin significantly suppressed the growth of cancer and decreased the expression of c-Met, ki-67, and MMP-9 in malignant tissues. Moreover, luteolin promoted apoptosis and suppressed invasiveness and proliferation in gastric tumour cells that overexpressed c-Met (SGC7901 and MKN45). Additionally, it downregulated MMP-9 and enhanced the activation of apoptosis-related proteins, such as multi (ADP-ribose) polymerase-1 and caspase-3. Furthermore, luteolin decreased c-Met expression and phosphorylation while knocking down ERK and Akt phosphorylation. It was also found that c-Met was not necessary for the downstream phosphorylated levels of Akt [105].

Luteolin-induced apoptosis in vitro suppressed the growth of tumour cells in vivo by significantly inhibiting the invasion, migration, and proliferation of stomach tumour cells in a time- and dose-dependent manner. In this regard, luteolin therapy caused EMT reversion by shrinking the cytoskeleton and increasing the expression of E-cadherin downstream of mesenchymal markers, such as vimentin, N-cadherin, and Snail. Additionally, it prevents the transduction of Notch1 signals [106]. Moreover, Lutein treatment of GC cells reduced the expression of the target genes Mcl-1 and Bcl-xl and survival while also inhibiting STAT3 phosphorylation. Furthermore, in vivo, research validated luteolin's inhibitory effects on tumour growth and progression [107].

In lung malignant cells, luteolin promotes the production of ROS, which in turn mediates the expression of the tumour necrosis factor-activated cascade. By upregulating c-Jun N-terminal kinase expression and downregulating NF-κB expression, luteolin promoted tumour necrosis factor-induced apoptosis in lung cancer cells. Luteolin also targets a variety of cancer pathways, such as redox stress, ROS formation, cell cycle arrest, autophagy induction, apoptosis initiation, and suppression of cell proliferation, all of which lead to the death of tumour cells [108].

Cai and colleagues suggested that luteolin inhibits the cell cycle and promotes apoptosis by increasing the synthesis of Bax, JNK activation, and enhancing the cleavage of caspase-3 and procaspase-9 in lung cancer cells (A549). Additionally, it inhibits trans-nuclear translocation controlled by TNF-α and NF-kB [109]. Luteolin inhibited cell growth and triggered apoptosis by increasing caspase-9 and -3 activation, decreasing Bcl-2, increasing Bax expression, phosphorylating MEK and its downstream kinase ERK, and activating Akt. Moreover, suppression of MEK-ERK signalling suggests that the pro-apoptotic and anti-migration effects of luteolin are significantly mediated by the MEK-ERK signalling pathway [110].

Through the regulation of both intrinsic and extrinsic cascades, which were suppressed by z-Val-Ala-Asp fluoromethyl ketone, luteolin-induced apoptosis in NCI–H460 human non-small cell lung cancer cells. This suggests that luteolin promotes caspase-dependent apoptosis. Additionally, luteolin-induced autophagy has been discovered to be a mechanism of cell death [111]. Another study showed that luteolin has anticancer effects by increasing Sirt1-regulated apoptotic cell death in NCI-H460 cells [112,113]. Moreover, it increased cleaved caspase-3 levels and reduced cyclin D1 expression by decreasing the mRNA levels of LIM domain kinase signalling-related targets, such as p-cofilin and phosphorylated LIM domain kinase. Furthermore, luteolin reduced phosphorylated LIM domain kinase, Ki-67, and p-cofilin levels, all of which inhibited the development of tumours in a xenograft model of lung tumour patients [113].

Macrophages linked to tumours are essential for the development of cancer [114]. According to Choi et al., luteolin lowers the mRNA levels of M2-associated genes and prevents the attachment of a phosphate group to STAT6, a significant IL-4 downstream signal. Additionally, they found that luteolin inhibited the migration of Lewis lung cancer cell lines in a manner dependent on chemokine (C–C motif) ligand 2 [115].

Ionising radiation and luteolin combination therapy increased programmed cell death in lung cancer cells by downregulating Bcl-2, which in turn stimulated caspase-9, -8, and -3. Additionally, luteolin led to the accumulation of ROS and the addition of phosphate to p38 MAPK. Moreover, in the NCI–H460 cell xenograft mouse model, combined therapy with luteolin and ionising radiation increased programmed cell death and suppressed the progression of cancer. This substance may act as a radiosensitiser, promoting programmed cell death by activating the p38/ROS/cascade pathway [116].

In contrast, luteolin has been reported to significantly reduce the proliferation of oral cancer stem cell lines and the activities of acetaldehyde dehydrogenase and CD44-positive cells. It has also been suggested that luteolin reverses the radiosensitivity of oral tumour cells, as the combined therapy of luteolin and radiation significantly decreases the invasion and spread of oral cancer [117]. Moreover, luteolin demonstrated cytotoxicity against human immortalised keratinocytes (HaCaT) and human melanoma (A375) cells in skin cancer. Furthermore, HaCaT cell lines accumulated cells in the G2/M phase, and A375 cell lines accumulated cells in the G0/G1 phase when luteolin is incubated with cancer cell lines [118].

By reducing miR-301-3p, luteolin decreased the proliferation of pancreatic ductal adenocarcinoma (PDAC) cells and enhanced the antiproliferative effect of TRAIL on tumour cells [119]. In female Syrian golden hamsters, luteolin (100 ppm) reduced carcinogenesis by increasing amylase activity and reducing PDAC incidence and multiplicity, Ki-67 labelling index, pSTAT3 signal transduction, and neoplastic lesion development [120]. Additionally, luteolin (150 and 75 mg/kg) prevented tumour growth in xenografted SCID mice [121].

By inducing apoptosis, decreasing extracellular matrix contraction, and inhibiting growth, luteolin exerts chemopreventive therapeutic effects against prostate cancer. MDM2 was suppressed by luteolin, and luteolin-induced E-cadherin expression decreased due to active Akt overexpression. Therefore, in prostate cancer, luteolin affects E-cadherin expression through the Akt/MDM2 pathway. Furthermore, by suppressing the expression of androgen receptors, luteolin reduced the expression of prostate-specific antigens. It reduced the mRNA levels of numerous genes involved in the cell cycle and epidermal growth factor receptor signal transduction cascades. Luteolin significantly promoted cell cycle arrest at the G2/M phase and triggered the production of p21 RNA and c-FOS. Different studies have revealed that c-FOS or p21 silencing RNAs significantly reduce the expression of RNA of their respective targets, but they have no effect on cell proliferation, and neither double silencing RNA nor single silencing RNA can stop the proliferation of prostate cancer cells [94].

According to Cao et al., luteolin decreased the viability of SMMC-7721 liver cancer cells in a manner that is dependent on both time and dose. Moreover, luteolin decreased Bcl-2 expression at the mRNA and protein levels, increased caspase 8 expression, and caused G0-/G1-phase arrest. Lastly, co-administration of the autophagy inhibitor chloroquine reduced the impact of luteolin on cell death [122].

Nazim and Park [123] showed how luteolin and TRAIL therapy work together, as well as how they affect TRAIL-resistant Huh7 cells. The synergetic effect of luteolin and VV-IL-24 (oncolytic vaccinia virus) to decrease tumour growth through single therapy was validated by Wang et al. [124]. They reported that luteolin inhibited the activation of the PI3K/Akt and NF-κB signalling pathways, which are implicated in the growth and survival of cancer cells. Additionally, it increased the cytotoxicity of chemotherapeutic drugs in kidney cancer cells, indicating that it may be used along with conventional medications to boost their effects. Moreover, luteolin reduces the negative effects of chemotherapy, which makes it a desirable drug for the treatment of kidney cancer.

Nanotechnology is a novel chemoprevention technique for delivering luteolin. To explore its anticancer capabilities against head, neck, and lung cancers, hydrophobic luteolin was synthesised to produce water-soluble polymer-encapsulated nano-luteolin. Nano-luteolin, like luteolin, has been demonstrated to inhibit the growth of lung cancer cells (H292 cell line) and squamous cell carcinoma of the head and neck (SCCHN) cells (Tu212 cell line) in vitro. Using a tumour xenograft mouse model, in vivo experiments comparing nano-luteolin to luteolin revealed that the latter greatly suppressed the growth of SCCHN cancer. This suggests that luteolin may be used for chemoprevention in clinical settings [125].

Table 4. The anticancer properties of luteolin in different tumour cells.

Anticancer Effect of Luteolin	Cancer Type	Citation
Inhibition of CDK2 activity and induction of induced G1 cell cycle arrest.	Colorectal cancer HT-29 and melanoma OCM-1 cells	[94]
Suppression of the expression of p-STAT3, p-EGFR, p-Akt, and p-Erk1/2, as well as inhibition of cell proliferation.	MCF-7 breast cancer cells	[94]
Increase of the mRNA levels of death receptors such as DR5 and caspase-9/-8/-3 activity, in addition to inhibition of poly-ADP ribose polymerase. Moreover, the induction of the release of cytochrome c after impairing the potential of the mitochondrial membrane. As a result, Bcl-2 expression was suppressed, and Bax expression rose.	MCF-7 breast cancer cells	[96]
Induction of cell cycle arrest in the S phase by lowering telomerase levels and preventing the phosphorylation of NF-κB inhibitor α, in addition to inhibition of the growth of breast malignant cells and induction of apoptosis.	MDA-MB-231 breast cancer cells	[97]
Akt inactivation and extracellular signal-regulated kinase (ERK) signalling inhibition.	MCF7/HER18 breast cancer cells	[98]
Induction of cell cycle arrest at the G2/M phase and inhibition of cyclin B1/CDC2.	LoVo human colon cells	[100]
Increased p53 phosphorylation and p53 target gene expression, which leads to cell cycle arrest, apoptosis, and autophagy.	HCT116 colon cells	[101]
Increased Nrf2 transcription by the DNA demethylation of its promoter, in addition to strengthening the interaction between Nrf2 and p53 that results in increased expression of antioxidant enzymes and apoptotic proteins.	Colorectal cancer cells	[102]
Inhibition of the proliferation of colorectal malignant cells, interruption of the cell cycle, damaged DNA and accelerated apoptosis through targeting the MAPK pathway.	Colorectal cancer cells	[103]
Decrease in the expression of p-PI3K, p-mTOR, p-AKT, p-STAT3, and Notch1, and an increase in the amount of p-P38 signal transduction.	Gastric tumour cells	[104]
Inhibition of the immune system and the expression of MMP-9 and VEGF-A, which stops cancer from growing, in addition to suppression of the expression of c-Met, ki-67, and MMP-9 that results in inhibition of tumour cells invasiveness and proliferation and induction of apoptosis.	Gastric tumour cells	[105]

Table 4. *Cont.*

Anticancer Effect of Luteolin	Cancer Type	Citation
EMT reversion by shrinking the cytoskeleton and increasing the expression of E-cadherin downstream of mesenchymal markers such as vimentin, N-cadherin, and Snail, in addition to prevention of the transduction of Notch1 signals.	Gastric tumour cells	[106]
Reduction of the expression of the target genes Mcl-1 and Bcl-xl and survival, in addition to inhibition of STAT3 phosphorylation.	Gastric cancer cells	[107]
Induction of ROS production, which in turn mediates the expression of tumour necrosis factor-activated cascade, in addition to upregulation of c-Jun N-terminal kinase expression and downregulation of NF-κB expression that results in promotion of tumour necrosis factor-induced apoptosis.	Lung cancer cells	[108]
Increase the synthesis of Bax, activation of JNK, and enhancing the cleavage of caspase-3 and procaspase-9, in addition to inhibition of trans-nuclear translocation controlled by TNF-α and NF-kB.	Lung cancer cells (A549)	[109]
Activation of caspase-9 and -3, inhibition of Bcl-2, increasing Bax expressions, phosphorylation of MEK and its downstream kinase ERK, activation of Akt, inhibition of cell growth, and induction of apoptosis.	Lung cancer cells (A549)	[110]
Increased cleaved caspase-3 levels and reduced cyclin D1 expression by decreasing the mRNA levels of LIM domain kinase signalling-related targets, such as p-cofilin and phosphorylated LIM domain kinase, that results into inhibition of tumour development.	Lung tumour xenograft	[111]
Reduction of miR-301-3p level results in inhibition of tumour cells proliferation and enhancing the antiproliferative effect of TRAIL on tumour cells.	Pancreatic ductal adenocarcinoma cells (PDAC)	[119]
Suppression of the expression of androgen receptors and prostate-specific antigen, in addition to the reduction of the mRNA levels of numerous genes involved in the cell cycle, cascades and the epidermal growth factor receptor signal transduction cascade that significantly promoted cell cycle arrest at the G2/M phase and triggered the production of p21 RNA and c-FOS.	Prostate cancer cells	[94]
Decreased Bcl-2 at the mRNA and protein levels, increased caspase 8, and caused G0-/G1-phase arrest. Additionally, it increased Beclin 1 expression, expedited the conversion of LC3B–I to LC3B–II, and increased the number of intracellular autophagosomes.	SMMC-7721 liver cancer cells	[122]
Inhibition of the activation of the PI3K/Akt and NF-κB signalling pathways, which are implicated in the growth and survival of cancer cells.	Kidney malignant cells	[124]

5.1.5. Flavanones

The 40-methoxy derivative of the flavanone eriodictyol is called hesperetin (HSP), and its IUPAC name is 5,7-dihydroxy-2-(3-hydroxy-4 methoxyphenyl)-2,3-dihydrochromen-4-one [126]. HSP, a naturally occurring flavonoid with a variety of pharmacological characteristics, is mostly present in citrus fruits such as Citrus aurantinum, Citrus sinensis, and Citrus limon [127].

HSP may be a promising cancer treatment candidate (Table 5) because it demonstrates a cytotoxic mechanism against a variety of cancer cells, including breast [128], pancreatic [129], prostate [130], glioblastoma [131], liver [132], kidney [133], colon [134], lung [135], oral [136], [137], osteosarcoma [138], ovarian [139], thyroid [140], leukaemia [141], and other cancers [126]. HSP has been shown in numerous studies to be a promising treatment for breast cancer. HSP may promote DNA damage and apoptosis while inhibiting the growth, viability, migration, invasion, mammosphere formation, and colony formation of cancer cells. In MCF-7 breast cancer cells, it increased the mRNA levels of p53, NOTCH1, and PPARG and decreased β-catenin, leading to apoptosis and cell cycle arrest in the G0/G1 phase [142].

In breast cancer, tumour suppressor genes that regulate cell cycle progression are upregulated by HSP. HSP induces both intrinsic and extrinsic pathways that lead to cell death. In addition, HSP can inhibit certain tumor-related growth factors, which will prevent metastasis, inhibit MMP-9 production, and arrest the cell cycle in the Sub G1 phase [126]. According to a recent study, when MCF-7 breast cancer cells were treated with HSP (1–20 μM), aryl hydrocarbon receptor (Ahr) was inhibited, and the expression of CYP1A1, 1A2, and 1B1 was downregulated [143]. Furthermore, HSP inhibited the activity of the aromatase enzyme, cyclin D1, CDK4, Bcl-xL, and pS2, and it increased the expression of CCAAT/C/EBP, pERK-1&-2, and p57Kip2. These actions helped decrease tumour growth in MCF-7 breast cancer cells and female athymic mouse models, both in vitro and in vivo [144]. At a concentration of 95 μM, HSP reduced HER2, MMP-9, and Rac1 expression, lamellipodia formation, and arrested the cell cycle at the G2/M phase, thereby lowering cell viability, invasion and migration, and promoting apoptosis, according to research conducted on HER2 overexpressed breast cancer cells (MCF-7/HER2) and MCF-7/EV cells [145]. In MCF-7, MCF-10A, HMEC, and MDA-MB 231 breast cancer cells, HSP (20–200μM) was found to increase ROS production, cyto-C release, Bax/Bcl-2 ratio, PARP cleavage, caspase-9, -3, -7, JNK, and sk1 activation, in addition to the activation of the ASK1/JNK pathway [146]. In MDA-MB-231 breast cancer cells, HSP suppressed insulin receptor-beta subunit (IR-beta) phosphorylation and Akt, which lowered glucose absorption, leading to decreased cell proliferation [147]. HSP reduced the growth of MDA-MB-231 breast cancer cells by inhibiting HER2-tyrosine Kinase (HER2-TK) activity, causing MMP loss, chromatin condensation, and activating of caspase-8 and-3 [148]. This resulted in cell cycle arrest in the G2 phase and lowered SKBR3. Furthermore, HSP was able to induce apoptosis and prevent metastasis in 4T1 murine breast cancer cells by downregulating MMP-9 production and stopping the cell cycle at the Sub G1 phase [149].

HSP has been known to play a significant role in reducing the risk of prostate cancer and successfully treating it [136]. G0/G1 phase arrest was observed after HSP treatment via increased phosphorylation of the signal transducer and activator of transcription 3 (STAT 3), extracellular signal-regulated kinase $\frac{1}{2}$ (ERK1/2), and AKT signalling pathways, as well as IL-6 gene expression [150]. HSP is also linked to cell cycle arrest at the G1/S phase and elevated mitochondrial membrane depolarisation, leading to apoptosis and decreased cell viability [151].

In H522 lung cancer cells, HSP induced apoptosis by initiating the Fas death receptor/extrinsic pathway, which led to the dose-dependent upregulation of Bax, caspase-3, and caspase-9 [135]. Similarly, by blocking transforming growth factor β and decreasing glucose uptake in cancer cells by downregulating glucose transporter expression, HSP demonstrated strong antiproliferative effects in H441 lung cancer cells [152]. When used with copper, HSP was able to inhibit angiogenesis via the mitochondria-mediated pathway by activating the TRIAL cytotoxic protein, which triggers many mechanisms of apoptosis [153]. Additionally, HSP inhibits IL-1β, which reduces COX-2 expression and PGE2 generation in A549 lung cancer cells [154]. By reducing LPO and altering antioxidant enzymes such as NF-kB, PCNA, and CYP1A1, HSP was able to inhibit the development of cancer in Swiss albino mice. This investigation demonstrates that HSP's free radical-scavenging, antioxidant, anti-inflammatory, and antiproliferative properties of HSPs have the potential to prevent B[a] P-induced lung cancer [155]. Through the activation of mitochondrial-mediated pathways, HSP suppresses cell viability and proliferation while enhancing apoptosis. In a HSP dose-dependent manner, Bcl2 was downregulated while ROS, ATP, Ca^{2+}, Cyto-C, AIF, and Apaf-1 were increased [156].

In the human cancer cell line HCT-116, HSP therapy activates the c-Jun-N-Terminal kinase (JNK) pathway, which reduced cell viability and induced apoptosis [157]. By elevating Bax and Caspase 3 levels and concurrently downregulating the anti-apoptotic protein BCL-2, HSP demonstrated an inhibitory effect on human cancer cell HT-29 through the induction of mitochondrial-mediated apoptosis [158].

HSP has the potential to treat liver cancer by inducing apoptosis and damage to cancer cells and reducing liver injury, liver enlargement, and hepatic fibrosis [126]. Miler et al. [159] found that oral administration of HSP at a dose of 15 mg/kg enhanced the death of cancer cells in male Wistar rats. It also increased the activity of antioxidant enzymes like catalase (CAT), glutathione reductase (GR), and superoxide dismutase (SOD). In Sprague-Dawley rats, HSP (50 mg/kg/day) induced apoptosis and controlled oxidative stress by increasing the expression of Fas/FasL and caspase-8, -3, and albumin levels while decreasing the levels of hepatic glutathione (GSH), hepatic malondialdehyde (MDA), and Bcl-2 [132]. Furthermore, Kong et al. discovered that HSP reduced liver fibrosis and induced apoptosis in HSC-T6 cells and male C57 mice. It inhibited the TGF-β1/Smad pathway and reduced the levels of AST, ALT, hydroxyproline (Hyp), HA, LN, TNF-α, IL-6, extracellular matrix (ECM) production, and Smad2/3 phosphorylation. In Littermate male C57BL/6J mice, HSP derivative reduced the levels of ALP, ALT, AST, TGF-β1, HA, Hyp, F4/80þ macrophage infiltration, MCP-1, TNF-α, IL-1β, IL-6, TNF-α, and IL-1β, Gli-1, and Shh expression at concentrations ranging from 25 to 100 mg/kg [160]. Additionally, HSP induced apoptosis in LX-2 liver cells by decreasing the expression of α-SMA, Col1α1, Col3α1, TIMP-1, PAI-1, and Gli-1 and increasing the levels of Bax and Caspase-3 [161]. In addition, HSP decreases Bcl-2, mitochondrial AIF, mitochondrial Apaf-1, and mitochondrial cyt-c, which drive cancer cell apoptosis, while upregulating a few intracellular ROS, ATP, Ca^{2+}, and cytosolic components such as AIF, Apaf-1, cyt-c, caspase-3, caspase-9, and Bax [156].

In recent years, HSP have shown promise in the treatment of pancreatic cancer [126]. In a study using Miapaca-2, Panc-1, and SNU-213 cell lines at different doses, Lee and his colleagues discovered that HSP (0–20 μM) inhibited the migration of the treated cells. Furthermore, HSP treatment at a dosage of 2.5 μM significantly decreased the viability of Panc-1 pancreatic cancer cells. Additionally, apoptosis was induced because HSP obstructed intracellular signalling, including focal adhesion kinase (FAK), p38 phosphorylation, and caspase-3 activation. Furthermore, HSP at 30 mg/kg exhibited anti-growth properties through the activation of caspase-3 in a Panc-1 xenograft model in BALB/c nude mice [129].

Additionally, HSP may be used to treat renal cancer [126]. HSP decreased oxidative stress, lipid peroxidation, MDA, TNF-α, IL-1β, and IL-6 levels, thereby reducing cisplatin-induced nephrotoxicity in rats [162]. Moreover, HSP reduced renal fibrosis, normalised renal function in the (NRK)-52E cell line and UUO-mouse model, and decreased the expression of fibronectin (FN), Collagen I, α-SMA, EMT, Shh, Gli-1, and E-cadherin [163]. Furthermore, HSP control signalling pathways, metastasis, and some inflammatory indicators, in addition to activating genes linked to antioxidant enzymes. Thus, HSP may exhibit anticancer activity in kidney cancer [126].

In vitro and in xenograft tumours, HSP inhibited the proliferation of gastric cancer cells by inducing apoptosis through a dose-dependent increase in the Bax/Bcl-2 ratio, cyt-c, caspase-3, caspase-9, AIF, and Apaf-1 via a mitochondrial-dependent mechanism [156]. Moreover, HSP reduces cell migration and invasion in gastric cancer cells by inhibiting the expression of genes linked to metastasis and lowering disruptor of telemetric silencing 1-

like (DOT1L) and histone H3K79 methylation by controlling CBP activity [164]. In addition, HSP can elevate the production of ROS to induce apoptosis in SK-OV-3 ovarian cancer cells [139]. By changing the endoplasmic reticulum stress signalling pathway, hesperidin (the aglycone form of HSP) inhibits the growth of (A2780) ovarian cancer cells and triggers apoptosis [165]. HSP also demonstrated cytotoxicity against ovarian cancer cells. These phytochemicals activate cleaved caspase-3 in ovarian cancer cells, promoting antioxidant activity and inducing apoptosis [166].

In U251 and U87 glioblastoma cells, HSP reduced cell viability by decreasing Bcl-2 and increasing Bax protein expression, thus inducing apoptosis in a dose-dependent manner. Moreover, it also caused cell cycle arrest by decreasing cyclin B1 CDK1 and enhancing tumour suppressor gene p21 activation via p38 MAPK, which arrests the G2/M phase [167]. Moreover, HSP controls apoptosis and cell division by generating ROS and activating the SOD enzyme [168]. When C6 glioma cells were implanted in Wistar rats, HSP prevented tumour growth by activating caspase-3 and -9, raising the Bax/Bcl-2 ratio, which caused apoptosis, and downregulating the HIF-1α, VEGF, and VEGFR2 signalling pathways. It also decreased the expression of cyclin B1 and D1 while increasing the expression of Claudin-1 and ZO-1, which decreased the growth of cancer cells [169].

HSP exhibited an anticancer effect on leukaemia HL60 cell lines by inducing apoptosis via increasing caspase-3 activity, MMP loss, and cell cycle arrest in the G2/M and G0/G1 phases [141]. In K562 leukaemia cells, HSP induced apoptosis, arrested the G0/G1 phase, and increased the expression of the DUSP1 (dual specificity phosphatase 1), DUSP3, DUSP5, CDK1A, CDK1B, GADD45B, SPRR2D, MT1F, MT1A, p27Kip1, CASP4, and NFKBIA genes [170]. Furthermore, HSP increased the production of BAD, caspase-3, luciferase activity, PARP cleavage, and Notch 1 signalling [140]. Additionally, HSP increased ROS generation and JNK1/2, p38, Bax, and p21 expression in A431 human cancer cells while suppressing ERK1/2, cyclin B1, D1, D3, and E1 expression, which resulted in apoptosis and decreased cell viability [171].

By stimulating PI3K-Akt signalling, cytotoxic T lymphocytes, and the tolerogenic T cell response, HSP can suppress melanogenic tumour growth [172]. HSP caused oesophageal Eca-109 cancer cells to undergo apoptosis both in vitro (Eca-109) and in vivo (female BALB/c nude mice). Furthermore, the HSP-treated Eca-109 cell line showed decreased PI3K/AKT signalling pathway, cyclin D1, MMP-2,9, and PI3K-p85 expression, as well as increased PTEN phosphorylation and p21 expression, which results in cell cycle arrest at the G0/G1 phase [173].

Hyperplasia, dysplasia, and increased cell proliferation in squamous cell carcinoma (SCC) are abnormalities induced by DMBA-induced oral tumour development in the buccal pouches. HSP treatment has been linked to anticancer effects by mediating apoptotic and antiproliferative effects. HSP inhibits cell proliferation in the buccal mucosa of DMBA-treated animals by downregulating vascular endothelial growth factor (VGEF) in DMBA-treated tissue [174].

It was reported that Hesperetin and Naringenin (Nar), two flavanones, target the mitochondrial fission pathway to exert anticancer effects. They correct abnormal mitochondrial dynamics and lipid metabolism by targeting Drp1, which causes ER stress and apoptosis [175]. The combination of these natural compounds decreased the adverse effects of several medications and showed great benefits against cancer [126]. Phosphorylation of FAK and the p38 signalling pathway were downregulated when HSP was administered in combination with naringin and naringenin, although this was not the case with either of the two treatments [176]. In a xenograft model, Wang et al. found that co-treatment of HSP with platinum caused apoptosis-related cell death by downregulating UGT1A3 and concurrently increasing ROS [177]. Moreover, HSP and luteolin together enhanced the death of MCF-7 breast cancer cells [178]. Dextran and HSP combination enhanced the HSP's antioxidant activity and had a greater cytotoxic effect on MCF and AGS than when HSP was used alone [179]. Additionally, the combination of HSP and doxorubicin arrested the cell cycle in the G2/M phase and prevented metastasis by suppressing the production of MMP-9 in 4T1 cells [149]. Moreover, HSP and 5-FU (fluorouracil) together suppressed cell proliferation in oesophageal cancer Eca-109 cells by downregulating Bcl-2 and increasing cleaved caspase-3 and caspase-9 more efficiently than either drug alone [173].

HSP can also be used as an adjuvant treatment for multidrug resistance. Excessive doxorubicin use results in drug resistance by increasing drug efflux and overexpressing P-glycoprotein (P-gp). However, by reducing the optimum concentrations of both HSP and doxorubicin, HSP combined with doxorubicin therapy inhibits P-gp expression in MCF-7 and MCF-7/DOX cells [180]. Additionally, certain anticancer medications increase the sensitivity of resistant cell lines when NF-κB and IGF1R expressions are inhibited [181]. When HSP was administered to A549/DDP cells, P-gp-mediated MDR was reversed by lowering P-gp expression, which was directly associated with the suppression of the transcription factor NF-κB signalling pathway [182].

Interestingly, Eudragit-E nanoparticles loaded with HSP (HETNPs) showed anticancer efficacy in oral carcinoma (KB) cells. HETNPs more successfully demonstrate elevated ROS levels, loss of mitochondrial membrane potential, and apoptotic morphological alterations than native HSP [183]. Moreover, collagen, nicotinamide adenine dinucleotide (NAD), and flavin adenine dinucleotide (FAD) emissions were reduced in oral carcinoma caused by 7,12-Dimethylbenz[a]anthracene (DMBA); however, oral administration of HSP and its nanoparticles restored the endogenous fluorophore emission and increased the redox ratio in the buccal mucosa of DMBA animals [184].

Table 5. Anticancer effects of hesperetin in different tumour cells.

Anticancer Effect of Hesperetin (HSP)	Cancer Type	Citation
Increased the mRNA levels of p53, NOTCH1, and PPARG and decreased β-catenin, leading to apoptosis and cell cycle arrest in the G0/G1 phase.	MCF-7 breast cancer cells	[142]
Upregulation of tumour suppressor genes that can regulate cell cycle progression, induction of both intrinsic and extrinsic pathways that lead to cell death, inhibition of certain tumour-related growth factors which will prevent metastases, inhibition of MMP-9 production and induction of cell cycle arrest in the Sub G1 phase.	Breast cancer cells	[126]
Inhibition of the aryl hydrocarbon receptor (Ahr) and downregulation of the expression of CYP1A1, 1A2, and 1B1.	MCF-7 breast cancer cells	[143]
Inhibition of the activity of the aromatase enzyme, cyclin D1, CDK4, Bcl-xL, and pS2, while increasing the expression of CCAAT/C/EBP, pERK-1&-2, and p57Kip2 that results in decrease the tumour growth.	MCF-7 breast cancer cells and female athymic mice models	[144]
Reduction of HER2, MMP-9, Rac1 expression, lamellipodia formation, and induction of cell cycle at the G2/M phase, thereby lowering cell viability, invasion, migration, and promoting apoptosis.	HER2 overexpressed breast cancer cells (MCF-7/HER2) and MCF-7/EV cells	[145]

Table 5. *Cont.*

Anticancer Effect of Hesperetin (HSP)	Cancer Type	Citation
Increase ROS production, cyto-C release, the Bax/Bcl-2 ratio, PARP cleavage, caspase-9, -3, -7, JNK, and activation of sk1 and the ASK1/JNK pathway.	MCF-7, MCF-10A, HMEC, and MDA-MB 231 breast cancer cells	[146]
Suppression of insulin receptor-beta subunit (IR-beta) phosphorylation and Akt, which lowers glucose absorption, leading to decreased cell proliferation.	MDA-MB-231 breast cancer cells	[147]
Inhibition of HER2 Tyrosine Kinase (HER2-TK) activity, leading to MMP loss, chromatin condensation, and activating caspase-8 and-3 that resulted in cell cycle arrest at the G2 phase and lowered SKBR3.	MDA-MB-231 breast cancer cells	[148]
Induction of apoptosis and prevention of metastasis of tumour cells by downregulating MMP-9 production and stopping the cell cycle at the Sub G1 phase.	4T1 murine breast cancer cells	[149]
G0/G1 phase arrest via increasing phosphorylation of the signal transducer and activator of transcription 3 (STAT 3), extracellular signal-regulated kinase $\frac{1}{2}$ (ERK1/2), and AKT signalling pathways, as well as IL-6 gene expression.	PC-3 cells	[150]
Induction of apoptosis via initiating the Fas death receptor/extrinsic pathway, which led to the dose-dependent upregulation of Bax, caspase-3, and caspase-9.	H522 lung cancer cells	[151]
Blocking transforming growth factor β and decreasing glucose uptake in a cancer cell by downregulating glucose transporter expression.	H441 lung cancer cells	[152]
Activation of the c-Jun-N-Terminal kinase (JNK) pathway, leading to reduction of cell viability and induction of apoptosis.	Human cancer cell line HCT-116	[157]
Elevation of Bax and caspase3, downregulation of the anti-apoptotic protein BCL-2, and induction of mitochondrial-mediated apoptosis.	Human cancer cell HT-29	[158]
Repression of the TGF-β1/Smad pathway and reduction of the levels of AST, ALT, hydroxyproline (Hyp), HA, LN, TNF-α, IL-6, extracellular matrix (ECM) production, and Smad2/3 phosphorylation.	HSC-T6 cells and male C57 mice	[160]
Reduction of the expression levels of ALP, ALT, AST, TGF-β1, HA, Hyp, F4/80þ macrophage infiltration, MCP-1, TNF-α, IL-1β, IL-6, TNF-α, and IL-1β, Gli-1, and Shh.	Littermate male C57BL/6J mice	[160]
Decreasing the expression of α-SMA, Col1α1, Col3α1, TIMP-1, PAI-1, and Gli-1 and increasing the levels of Bax and Caspase-3 that results in apoptosis.	LX-2 liver cells	[161]
Inhibition of cells migration, decrease cell viability, and induction of apoptosis via obstructing the intracellular signalling, including focal adhesion kinase (FAK), p38 phosphorylation, and caspase-3 activation.	Miapaca-2, Panc-1, and SNU-213 pancreatic cancer cell lines	[129]
Reduction of renal fibrosis, normalising renal function, and decreasing the expression of fibronectin (FN), Collagen I, α-SMA, EMT, Shh, Gli-1, and E-cadherin.	Renal cancer (NRK)-52E cell line and UUO-mouse model	[163]
Induction of apoptosis through a dose-dependent increase in the Bax/Bcl-2 ratio, cyt-c, caspase-3, caspase-9, AIF, and Apaf-1 via a mitochondrial-dependent mechanism.	Gastric cancer cells	[156]
Reduction of cell migration and invasion by inhibiting the expression of genes linked to the metastasis and lowering disruptor of telemetric silencing 1-like (DOT1L) and histone H3K79 methylation by controlling CBP activity.	Gastric cancer cells	[164]

Table 5. *Cont.*

Anticancer Effect of Hesperetin (HSP)	Cancer Type	Citation
Reduction of cell viability by decreasing Bcl-2 and raising Bax protein expression, thus inducing apoptosis, in addition to cell cycle arrest by decreasing cyclin B1 CDK1 and enhancing tumour suppressor gene p21 activation p38 MAPK, which arrests the G2/M phase.	U251 and U87 glioblastoma cells	[131]
Prevention of tumour growth via activating caspase-3 and -9, raising the Bax/Bcl-2 ratio, which caused apoptosis, and downregulation of the HIF-1α, VEGF, and VEGFR2 signalling pathway. In addition, decreasing the expression of cyclin B1 and D1 while increasing the expression of Claudin-1 and ZO-1 decreased the growth of cancer cells.	C6 glioma cells implanted in Wister rats	[169]
Induction of apoptosis via raising caspase-3 activity, MMP loss, and cell cycle arrest in the G2/M and G0/G1 phases.	HL60 leukaemia cell lines	[141]
Induction of apoptosis and cell cycle arrest at the G0/G1 phase, in addition to increasing the expression of the DUSP1 (dual specificity phosphatase 1), DUSP3, DUSP5, CDK1A, CDK1B GADD45B, SPRR2D, MT1F, MT1A, p27Kip1, CASP4, and NFKBIA genes. Moreover, elevation of the production of BAD, caspase-3, luciferase activity, PARP cleavage, and Notch 1 signalling.	K562 leukaemia cells	[140,170]
Increasing ROS generation, JNK1/2, p38, Bax, and p21 expression, while suppressing ERK1/2, cyclin B1, D1, D3, and E1 expression, which resulted in apoptosis and decreased cell viability.	A431 human cancer cells	[171]
Reduction of the expression of GSH, Bcl-2, and survivin while increasing the generation of ROS, cyt-c, caspase-9, -3, Apaf-1, Bax, and Sufu (suppressor of fused protein), in addition to decreasing PI3K/AKT signalling pathway, cyclin D1, MMP-2,9, and PI3K-p85 expression, as well as increased PTEN phosphorylation and p21 expression, which results in cell cycle arrest at the G0/G1 phase.	Oesophageal Eca-109 cancer cells	[173]

5.1.6. Isoflavones

Isoflavones are among the most prevalent categories of phytoestrogens. It is mostly found in the Fabaceae family. These secondary plant metabolites are structurally identical to 17β-estradiol and are typically conjugated to it. They are hydrolysed into aglycones before being metabolised by the enzymes of the gastrointestinal tract or the microbiota found in the digestive tracts of humans and animals. Soy and its derived products are the primary sources of isoflavones [185]. Chickpeas and beans are additional dietary sources of isoflavones, and other plant products, such as fruits, vegetables, and nuts, also contain trace levels of isoflavones [186]. Genistein (Figure 2) is a naturally occurring phytoestrogen and isoflavone found in soybeans. Genistein has been shown to have numerous biological effects, including anti-oxidation, anti-proliferation, and tumoricidal properties [187] (Table 6). Isoflavones have significant antioxidant activity in addition to their oestrogenic activity. Two hydroxyl groups, such as those found in daidzein, must be present for antioxidant activity to occur (in the C-4 and C-7 positions). Compared to glycosides, aglycon molecules have higher activity [188]. Genistein exhibits strong antiproliferative properties against different cancer cells in vitro, inhibits the growth of tumours and shows an antimetastatic effect in vivo [189,190].

Compared to Western countries, where the average daily consumption of isoflavones is less than 2 mg, the incidence of breast cancer is lower in Asia, where the average daily intake of isoflavones approaches 25–50 mg [190]. Increased soy intake is associated with a lower risk of breast cancer [191].

Genistein inhibits breast cancer cell growth by promoting apoptosis [192,193]. Genistein is a prospective treatment for breast cancer since it acts as a weak oestrogen by binding to the oestrogen receptor. This could prevent the effects of natural oestrogens and slow the growth of breast cancer without causing any noticeable adverse effects [192–194]. α-ER activation promotes cell proliferation in breast tissue, whereas β-ER induces apoptosis and inhibits of cell proliferation. The precise ratio of α-ER to β-ER in cells determines how isoflavones affect the suppression or activation of cell growth [195].

In MDA-MB-231 cells, genistein has been shown to have antiproliferative properties, including the inhibition of NF-kB pathways and the subsequent inhibition of NF-κB [196]. Modification of the EGFR/Akt/NFκB pathway contributes to cell differentiation, ultimately resulting in the death of cancer cells. Genistein suppresses Akt activity, which encourages the deactivation of downstream signalling pathways such as NF-κB [197]. Moreover, it has been observed that genistein therapy reduces the expression of MMPs 2, 3, 3, and 15 in T47D cells, inhibiting angiogenesis and metastasis [198].

Genistein treatment of MCF-7-C3 and T47D breast cancer cells resulted in dysregulation of the human oncoprotein known as the carcinogenic inhibitor of protein phosphatase 2A (CIP2A), suggesting that CIP2A is a target of genistein responsible for inducing apoptosis and growth suppression [192].

In LNCaP and DU145 prostate cancer cell lines, both genistein and daidzein reduced cell growth and triggered apoptosis. The cuprous chelator neocuproine and other ROS scavengers, such as superoxide dismutase, catalase, and thiourea, dramatically reduced cell death induced by isoflavone. Copper chelation suppressed ROS production, supporting the idea that isoflavone-induced intracellular copper mobilisation leads to the production of ROS, which causes pro-oxidant cell death [199].

Phase I and II clinical trials on different cancers, including prostate cancer revealed that genistein inhibits metastasis of malignancies [200]. In addition, a significant decrease in serum PSA levels was observed in a prostate cancer trial [201].

Table 6. The anticancer effect of genistein in different tumour cells.

Anticancer Effect of Genistein	Cancer Type	Citation
Inhibition of NF-kB pathways and cell proliferation.	MDA-MB-231	[196]
Suppression of Akt activity, which encourages the deactivation of downstream signalling pathways, such as NF-κB.	MDA-MB-231	[197]
Reduction of the expression of MMPs 2, 3, 3, and 15, in addition to inhibition of angiogenesis and metastasis.	T47D cells	[198]
Dysregulation of the human oncoprotein, known as the carcinogenic inhibitor of protein phosphatase 2A (CIP2A), results in the induction of apoptosis and growth suppression.	MCF-7-C3 and T47D breast cancer cells	[192]
Induction of intracellular copper mobilisation leads to the production of ROS, which causes pro-oxidant cell death.	Prostate cancer cell lines LNCaP and DU145	[199]
Reduction of serum levels of prostate-specific antigen (PSA).	Prostate cancer cell lines	[201]

5.2. Phenolic Acids

One of the most prevalent non-flavonoid plant phenolic components is phenolic acid (Figure 4), which can be found as glycosides or aglycones (free form) [202]. Phenolic acids are secondary metabolites that are widely distributed in plants [203] and are also present in oilseeds, grains, legumes, fruits, vegetables, herbs, and drinks [202]. Based on the C1–C6 and C3–C6 skeletons, phenolic acids are classified into two classes: hydroxybenzoic and hydroxycinnamic acids [204]. Gallic, Protocatechuic, *p*-Hydroxybenzoic, Syringic, and Vanillic acids are examples of hydroxybenzoic acids [202]. They mostly occur as conjugates. The Apiaceae family of spices and herbs has been found to have the highest concentration (fresh weight) of benzoic acids: anise (730–1080 mg kg^{-1}), cumin (42 mg kg^{-1}), fennel (106 mg kg^{-1}), and parsley (30 mg kg^{-1}) [203]. In contrast, P-coumaric, ferulic, caffeic, cinnamic, chlorogenic, and sinapic acids are examples of hydroxycinnamic acids. Plants contain large amounts of cinnamic acid in the form of amides or esters. Cereals, coffee, tea, wine, chocolate, fruits, and vegetables contain high levels of cinnamic acid. Wild blueberries (1470 mg kg^{-1}), coffee (870 mg kg^{-1}), carrots (260 mg kg^{-1}), plums (234 mg kg^{-1}), and eggplant (210 mg kg^{-1}) are among the most significant sources of caffeic acid. Caftaric acid, a characteristic polyphenol found in wine, is one of the most significant derivatives of caffeic acid, and coffee contains significant amounts of chlorogenic acid [205,206]. By modulating several signalling pathways, hydroxycinnamic and hydroxybenzoic acids, as well as their derivatives, exhibit strong antioxidant and antiproliferative properties both in vitro and in vivo [207,208].

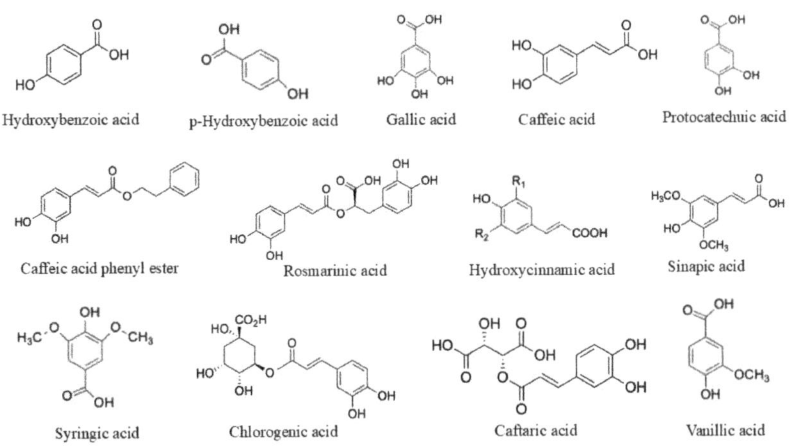

Figure 4. Structural identification of phenolic acids.

5.2.1. Gallic Acid

Gallic acid (GA) (3,4,5-Trihydroxybenzoic acid) is a phenol obtained from plants that can inhibit the development and progression of different malignancies [209]. The strong anticancer effect of gallic acid may be due to its remarkable antioxidant activity. In addition, GA inhibits cancer cell invasion by inducing apoptosis in cancer cells [210]. Because of its antioxidant properties, GA exerts strong anticancer effects. It has been reported it improved the anticancer efficacy of docetaxel, cisplatin, doxorubicin, 5-FU, and paclitaxel in combination with gamma irradiation in vitro in a recent study using oral squamous cell carcinoma (FaDu and Cal33) cell lines. This was achieved through the superoxide-mediated apoptosis pathway, which is powered by lipophagy inhibition via the NRF2-dependent signalling pathway [211]. In a dose- and time-dependent manner, GA suppressed the growth and induction of non-small cell lung carcinoma (NSCLC) A549 cell line, which was linked to downregulated B-cell lymphoma 2 (Bcl-2) and increased (Bcl-2)-associated X protein (Bax) [212]. Upregulation of p53 (tumour suppressor protein) caused suppression of the PI3K/Akt pathway. This, in turn, regulated intrinsic apoptotic proteins like Bcl-2 and Bax and cleaved caspase-3 and cell cycle-related proteins like p27, p21, Cyclin E1, and Cyclin D1.

One of the main bioactive compounds in *Dovyalis caffra* (*D. caffra*) is GA. Specifically, it was discovered that at 1000 μg/mL, the plant's methanol extract had 58.90% toxicity against HepG2 cells, suggesting potential anticancer properties [213]. Interestingly, GA played a crucial role in inducing ferroptosis in HepG2 cells. By preventing β-catenin transport from the nucleus to the cytoplasm, GA inhibits the production of ferroptosis-related proteins SLC7A11 and GPX4 in HepG2 cells. Thus, GA is a novel HC ferroptosis inducer, implying that GA may be a good candidate for the clinical treatment of hepatocellular carcinoma (HCC) [209]. By interacting with G-quadruplexes, GA has the potential to be a promising agent for cancer prevention, as demonstrated by Sanchez-Martin et al. [214]. Moreover, the findings demonstrated that nucleolar stress and the downregulation of G4-containing genes were caused by GA-induced cell cycle arrest in the S and G2/M phases.

However, GA has limited therapeutic utility because of its low oral permeability. 4-methoxybenzenesulfonamide (MBS), 3,4-dimethoxybenzenesulfonamide (DMBS), and 3,4,5-trimethoxybenzenesulfonamide (TMBS) were synthesised as GA analogues. To improve oral permeability and hydrophobicity, different quantities of methoxy groups, which are stronger electron-donating groups, were substituted for hydroxyl groups in these new compounds. Furthermore, a sulfonyl group, a more potent electron-withdrawing group, was substituted for the carboxylic group to boost the molecular polarity and antioxidative properties of the compounds. Compared to GA, TMBS was more successful in reducing DNA damage in lung cancer patients and PBMCs (peripheral blood mononuclear cells) healthy donors. Moreover, TMBS was more cytotoxic to A549 cells, while it did not cause cytotoxicity in healthy PBMCs, indicating that TMBS may have therapeutic value in cancer treatment [210].

5.2.2. Caffeic Acid

It is commonly known that natural caffeic acid (E)-3-(3,4-Dihydroxyphenyl) prop-2-enoic acid has several biological characteristics, including anticancer effects. Min et al. found that caffeic acid (CA) causes apoptosis, which dramatically reduces the proliferation of H1299 NSCLC cells [215]. When CA and paclitaxel (PTX) were combined, they exhibited a synergistic anticancer effect on H1299 cells. This combination inhibited the proliferation of H1299 NSCLC cells. CA administration enhanced H1299 cell apoptosis, caspase-3, and caspase-9 activity, and sub-G1 phase arrest. It also enhanced PTX-induced activation of Bid, Bax, and downstream of PARP cleavage. Moreover, it elevated the phosphorylation of kinase1/2 and c-Jun NH2-terminal protein kinase1/2. Lipid hydroperoxides, reactive thiobarbituric acid substances, and connective dienes are indicators of lipid peroxidation that are elevated by CA. Additionally, it enhanced morphological alterations, changed the potential of the mitochondrial membrane, and increased ROS levels in cells treated with CA, indicating that CA has anticancer activity because of its pro-oxidant function [216]. A study conducted by Rosendahl et al. revealed that CA mimics anti-oestrogen action and modifies important growth regulatory signals, including ER/cyclin D1 and IGF-IR/p-Akt, resulting in cell cycle damage and reduced cell proliferation [217].

To assess the effect of CA on the toll-like receptor 4 (TLR4) signalling pathway, Chen et al. conducted research that showed that CA lowered the production of IL-12 and NF-κB activation. However, by changing the TLR4/MD2 complex, the TLR4 pathway was hindered. These findings demonstrated that apoptosis in breast tumours is caused by the downregulation of TLR4, TRIF, and IRAK4 expression [218].

Furthermore, by inhibiting ERK phosphorylation, CA dramatically reduced lung metastasis caused by CT-26 colon cancer cells. Additionally, CA is directly bound to MEK1 or TOPK and significantly suppresses mitogen-activated MEK1 and TOPK activities. CA inhibited the neoplastic transformation of JB6 P+ cells, AP-1 and NF-κB transactivation, and ERK phosphorylation induced by EGF and H-Ras [219]. Moreover, Yang et al. reported that both in vitro and in vivo models showed that CA efficiently decreases tumour incidence and volume as well as colony formation. The CA-treated mouse skin cancer xenograft model showed a reduction in MAPK phosphorylation. Furthermore, CA directly interfered with ERK1/2 and inhibited ERK1/2 activity in vitro. It also interacts with ERK2 at the Q105, D106, and M108 amino acid residues [220]. High chemopreventive effects against A549 lung adenocarcinoma cells were shown for CA phenethyl ester, a CA derivative that represents a naturally occurring phenolic chemical that is found abundantly in plants and propolis extract, in the context of lung cancer. It was extremely important in reducing TGF-β-promoted cell motility and changing the growth factor-β (TGF-β)-induced activation of Akt (protein kinase β) and blocking the phosphatidyl inositol 3-kinase (PI3K)/Akt pathway.

Research on prostate cancer has revealed that CA phenethyl ester can prevent NF-κB activation in prostate cancer-3 (PC-3) cells by preventing TNF-α and Paclitaxel from activating NF-κB. Nevertheless, this action is also associated with decreased levels of apoptosis-inhibiting proteins (cIAP1, cIAP-2, and XIAP) in cells, as well as the downregulation of elevated levels of spontaneous apoptosis and cIAP-1 expression [221]. Moreover, by controlling Skp2, p53, p21Cip1, and p27Kip1, CA phenethyl ester has been reported to cause cell cycle arrest and growth suppression in castration-resistant prostate cancer cells [222].

Amorim et al. showed that while ROS elimination was unsuccessful, AntiOxCIN6 (a mitochondria-targeted antioxidant) enhanced the antioxidant defence system in HepG2 cells. AntiOxCIN6 markedly affected mitochondrial structure and function, which led to a reduced ability to produce complex I-driven ATP without affecting cell viability. Glycolytic flux increases in tandem with these changes [223]. They also mentioned that AntiOxCIN6 appears to produce metabolic alterations or redox pre-conditioning in lung MRC-5 fibroblasts, protecting cisplatin, while it sensitises A549 adenocarcinoma cells for CIS-induced apoptotic cell death. They suggested that the length and hydrophobicity of the C10-TPP+ alkyl linker are important factors in causing cellular and mitochondrial toxicity, whereas the antioxidant caffeic acid seems to oversee triggering cytoprotective mechanisms.

When paired with anticancer medications, caffeic acid increased apoptosis and suppressed the growth and clonogenicity of acid-adapted cancer cells by inhibiting the hyperactivation of the PI3K/Akt and ERK1/2 signalling pathways linked to drug resistance. Thus, its potential for overcoming drug resistance in cancer therapy is highlighted by its capacity to suppress proliferation, sensitise cells to apoptosis, and alter the signalling pathways [224].

In contrast, a combination of gamma-cyclodextrin (γCD) with CA phenethyl ester was also found to exert cytotoxic effects on several cancer cells [225]. The strong anticancer and antimetastatic effects of this complex were suggested to occur via the disruption of mortalin-p53 complexes, resulting in p53 nuclear translocation and activation, leading to the growth arrest of cancer cells [226].

Further research was performed on pancreatic ductal adenocarcinoma (PDAC) cell lines to investigate the pre-sensitising effects of CA in combination with several medications. In Panc-1 and Mia-PaCa-2 PDAC cell lines, CA pre-sensitisation decreased the doxorubicin IC50 concentration, which also caused ROS production. Following CA treatment, differential gene expression analysis revealed that distinct genes were affected in both cell lines, including p53 and Pi3K/Akt/mTOR in Mia-PaCa-2 cells and N-Cad and Caspase-9 in Panc-1 cells [227].

Comparing the effect of CA phenethyl ester (IC25 = 1.0 μM/IC50 = 2.5 μM) to CA alone (IC25 = 91.0 μM/IC50 = 120.0 μM), it was found that CA phenethyl ester reduced mitochondrial ROS generation, cell migration, and cell survival in murine osteosarcoma UMR-106 [228].

All proposed derivatives of caffeic acid were subjected to molecular docking investigation, which focused on the three-dimensional coordinates of human DHFR (PDB ID 1U72) co-crystallised with methotrexate (MTX). In addition to their anticancer and antibacterial properties, a new series of 1,2,4-triazole analogues of caffeic acid were developed, synthesised, characterised, and evaluated for their ability to inhibit DHFR [229].

5.2.3. Rosmarinic Acid

Rosmarinic acid (RA) is present in 39 plant families [230]. It is particularly present in many species of the Nepetoideae subfamily of the Lamiaceae family and the Boraginaceae family [231,232]. Despite being widely found in the plant kingdom, rosmarinic acid is the only significant chemotaxonomic marker of the Lamiaceae family [233]. Plants in the Lamiaceae family, including rosemary, produce rosmarinic acid (RA) as a secondary metabolite. Rosmarinic Acid (2R)-3-(3,4-Dihydroxyphenyl)-2-{[(E)-3-(3,4-dihydroxyphenyl) prop-2- enoyl]oxy} propanoic acid is one of the esters of caffeic acid [234].

Numerous pharmacological properties of RA have been identified, including antibacterial, antiviral, antimutagenic, antioxidant, and anti-inflammatory activities [232,235]. Furthermore, RA functions as a neuroprotective and immunomodulatory factor [236]. Additionally, its ability to inhibit tumour growth has been observed in a variety of cancer types, including colon, breast, liver, stomach, lung, melanoma, and leukaemia [233,237].

In human oral cancer cells, reduction of cancer cell migratory capacity, activated of apoptosis, induced cell cycle arrest at the G2/M phase, and inhibiting of cell proliferation in a dose-dependent manner [238]. In gastric adenocarcinoma cells, RA reduced the activity of MMP-9, which is crucial for cancer spread because it breaks down collagen and other extracellular matrix proteins [239]. In the WiDr colon cancer cell line, apoptosis was activated, and RA showed antiproliferative effects. Caspase 1 and Caspase 7, which are essential for apoptotic pathways, were upregulated, while BCL2 was downregulated [240]. Additionally, through the regulation of the Nrf2/Keap1 pathway and modulation of miR-1225-5p, RA was able to prevent the migration and invasion of HT-29 colorectal cancer cells. Cellular defence against oxidative stress (OS) is significantly influenced by this pathway, and RA's capacity of RA to inhibit p38/AP-1 signalling through IL-17RA targeting offers more proof of its anticancer potential [241]. Through its suppression of TLR4-mediated NF-κB-STAT3 signalling, which is essential for colon carcinogenesis and inflammation, RA decreased tumour incidence and inflammation in a mouse model [242]. In addition, by suppressing miR-155 and inhibiting hypoxia-inducible factor-1 alpha (HIF-1α), RA influences the IL-6/STAT3 pathway, reducing inflammation and encouraging cancer cell death [243]. In OC3 and DU145 prostate cancer cell lines, RA suppressed colony and spheroid formation, as well as cell proliferation. Moreover, RA therapy was able to successfully inhibit a histone deacetylase enzyme that controls the expression of mitochondrial intrinsic apoptotic pathway genes, such as Bcl-2, Bax, caspase-3, and poly (ADP-ribose) polymerase-1 (PARP-1). When RA downregulates HDAC2, a tumour suppressor protein called p53 is activated, which causes prostate cancer cells to undergo apoptosis. In addition, RA upregulates p21 expression and downregulates proliferating cell nuclear antigen, cyclin D1, and cyclin E1, resulting in apoptosis [244]. In addition to inducing apoptosis, RA dramatically reduced cell invasion, migration, and proliferation in DU-145 prostate cancer cell lines. RA shows promise in preventing the spread of cancer to other organs and improving patient outcomes in advanced prostate cancer by preventing the migration and invasion of cancer cells [245].

In contrast, RA is considered one of the important polyphenolic elements of *Glechoma hederacea* L.'s ethyl acetate fractionated extract (EAFE), which has been shown to preventing HepG2 cell proliferation, leading to apoptosis and cell arrest in the S phase. This EAFE's apoptogenic activity involves Ca^{2+} buildup, ROS generation, MMP disruption, caspase 3, 9 activation, elevated Bax/Bcl-2 ratio, and glutathione depletion [246]. Under the same conditions, RA significantly inhibited the proliferation of SMMC-7721 cells and increased G1 arrest and apoptosis in a concentration-dependent manner. Furthermore, RA was able to inhibit cell invasion by controlling epithelial-mesenchymal transition and suppressing the PI3K/AKT/mTOR signalling cascade [247].

In U2OS and MG63 osteosarcoma cells, RA was able to inhibit DJ-1 expression by regulating the PTEN/PI3K/Akt signalling pathway. Moreover, DJ-1 has been suggested to be a biological target of RA in osteosarcoma cells. RA induced apoptosis by upregulating the cleavage rates of caspase-8, caspase-9, and caspase-3, thus enhancing the Bax/Bcl-2 ratio, which resulted in ROS generation and decreased matrix metalloproteinase (MMP) [248]. Due to its ability to alter many signalling pathways that result in the growth of tumour tissue, there is compelling evidence that RA may be a potential therapy for several BC types [230,249]. Potent antiproliferative effects and cytotoxicity of RA have been reported in a dose- and time-dependent manner in breast cancer cell lines. MDA-MB-231 cell underwent apoptosis and cell cycle arrest in the G0/G1 phase after RA treatment. RA significantly upregulated the expression of tumour necrosis factor receptor superfamily 25 (TNFRSF25), harakiri (HRK), and BCL-2 interacting protein 3 (BNIP3) while inhibiting the expression of TNF superfamily 11B receptor (TNFRSF11B). Moreover, RA was able to markedly activate TNF transcription and cause growth inhibition and DNA damage-inducible 45 alpha (GADD45A) and BNIP3 [236]. Furthermore, in MDA-MB-231 breast cancer cells, RA suppressed MARK4 (microtubule affinity-regulating kinase 4) activity, which led to dose-dependent apoptosis. RA successfully targeted MARK4, a kinase implicated in the progression of cancer, indicating that it may be a suitable therapeutic target for breast cancer. The MARK4 protein, which is closely linked to breast cancer, has a high affinity for RA. Their 500 ns all-atom simulations and molecular docking showed that RA forms stable non-covalent interactions with important residues in the MARK4 active site, indicating that RA may prevent MARK4 from playing a role in the development of cancer [235]. Additionally, RA has been reported to induce both apoptosis and autophagy and to show dose-dependent suppression of breast cancer cell proliferation, especially in oestrogen-dependent MCF7 cells [250]. Moreover, RA promotes apoptosis by upregulating Bax and downregulating Bcl-2 expression. This lends more credence to the theory that important apoptotic proteins, like Bcl-2 and Bax, are regulated by RA to cause apoptotic effects in breast cancer cells [251]. In addition, by decreasing matrix metalloproteinase-9 (MMP-9) activity, RA successfully prevented the invasion and migration of cancer cells. MMP-9 is a proteolytic enzyme essential for the disintegration of the extracellular matrix, which promotes the spread of cancer [252].

In contrast, the RA-loaded microemulsions exhibited superior antioxidant activity compared to free RA in breast cancer cells (T47D and MDA-MB-231). Furthermore, they induced cell cycle arrest apoptosis, and inhibited cell growth. The greater bioavailability and stability of RA when administered via microemulsions have been credited with this increased therapeutic efficacy [253].

Regarding in vivo research on breast cancer models in mice has shown that RA exhibits antitumor activity alone or in combination with paclitaxel. VEGF, TNF-α, and NF-kB were repressed after RA treatment, while Bcl-2, p53, Bax, and caspase-3 were restored, leading to apoptosis. In addition, inhibition of tumour growth with elevated p53 and caspase-3 and repressed Bcl2/Bax ratio was observed in Ehrlich tumours in mice after RA administration, either alone or in combination with paclitaxel [254].

By blocking the ADAM17/EGFR/AKT/GSK3β pathway, RA prevents invasive proliferation and migration of human melanoma A375 cells, induces apoptosis, and increases the susceptibility of melanoma cells to cisplatin [255]. In pancreatic cancer cell lines (Panc-1 and SW1990), RA promoted apoptosis and inhibited cell viability, motility, and invasion. The study found that RA suppresses epithelial-mesenchymal transition (EMT), a crucial step in cancer metastasis, by upregulating miR-506 and suppressing MMP2 and MMP16 [256]. According to Zhou et al., RA was able to inhibit Gli1 nuclear translocation and induce Gli1 degradation by proteasomes. By blocking MMP-9 and E-cadherin, RA prevents cell invasion and migration [257]. Through the regulation of matrix metalloproteinases (MMP-2 and MMP-9) and the upregulation of E-cadherin expression, while downregulating N-cadherin and vimentin, RA also prevented the growth, invasion, and metastasis of hepatocellular carcinoma (HCC) cells in male BALB/c nude mice. By blocking the PI3K/AKT/mTOR signalling pathway, which is necessary for cancer cell survival and proliferation, RA decreases tumour volume and increases apoptosis rates [247].

In liver tumours, RA inhibits NF-κB signalling, thereby decreasing inflammation-related cytokines and angiogenesis. This suggests that RA acts similarly to how it inhibits HSP90AA1 in liver cancer [258].

A crucial chaperone protein, heat shock protein 90 (HSP90) interacts with oncogenic client proteins and co-chaperones to regulate signaling cascades and fix misfolded proteins in cancer cells [259]. The relationship between RA and HSP90AA1, a protein essential for the survival and growth of cancer cells in liver cancer, was examined using molecular docking and dynamics simulations. According to this study, RA generates strong hydrogen bonds at the active site of HSP90AA1 and binds to it with high affinity. This implies that RA may reduce cancer cell survival by blocking HSP90AA1's carcinogenic activity [15]. Furthermore, RA therapy targeted genes implicated in tumour progression and aberrant cell proliferation and downregulated the oncogenic transcription factor forkhead box M1 (FOXM1). FOXM1 was also inhibited in ovarian cancer cells after treatment with a combination of cisplatin and RA methyl ester, which reversed cisplatin resistance [260]. In A549 lung adenocarcinoma cells, RA decreased OS, inflammation, and metastasis, involving pathways such as Akt, P-65-NF-κB, and c-Jun [261]. By inhibiting NF-κB activation and ROS production, RA made human leukaemia U937 cells more sensitive to TNF-α-induced apoptosis. In addition to reducing ROS levels, RA's suppression of NF-κB increases caspase-dependent apoptosis activation, which lowers cancer cell survival [262]. By activating MAPK and inhibiting the expression of P-gp and MDR1, RA can inhibit the growth and cell colony formation, induce apoptosis and cell cycle arrest of non-small cell lung cancer (NSCLC) in a dose-dependent manner, in addition to elevating the sensitivity of cisplatin-resistant cells [263]. In non-small cell lung cancer (NSCLC), RA targets focal adhesion kinase (FAK). RA can bind to FAK to form stable complexes that block the signalling pathways linked to metastasis [264].

Bone metastases can be inhibited by RA. Therefore, RA could be a promising option for a new therapeutic approach for breast cancer bone metastases [265]. Additionally, ST-2 murine bone marrow stromal cells cultured with RA showed a significant and dose-dependent increase in alkaline phosphatase activity, in addition to an increase in the quantity and size of mineralised nodules. RA may prevent bone metastasis from breast cancer by simultaneously reducing the synthesis of interleukin-8 and the receptor activator of NF kappaB ligand (RANKL/RANK/osteoprotegerin) pathway [266]. As osteoprotegerin is a pro-angiogenic factor, its inhibition may help stop the spread of cancer cells. Moreover, increased IL-8 expression by breast cancer cells has been associated with osteolysis in metastatic breast cancer [254]. Furthermore, the decrease in human umbilical vein endothelial cell proliferation, adhesion, migration, and tube formation has demonstrated that RA has antiangiogenic properties, which aid in preventing cancer development and metastasis [267]. In human HaCaT keratinocytes, a model for UV-induced skin cancer, RA in conjunction with fucoxanthin showed that by upregulating Nrf2 and HO-1 and down-regulating inflammasome components such as NLRP3 and Caspase-1, RA and fucoxanthin decreased UVB-induced apoptosis and inflammation. These results suggest that RA may have a preventive function against skin cancer, especially by reducing the negative effects of UV exposure [268]. RA reduces oxidative imbalance and mitochondrial fragmentation caused by UVB rays, which are two major factors in the development of skin cancer. By modifying mitochondrial dynamics and ROS levels, RA appears to shield skin cells from the harmful effects of UV radiation, suggesting a possible protective role for RA against sun-induced skin cancer [269]. Additionally, by activating the Nrf2 pathway, a crucial regulator of the cellular antioxidant response, RA increases the activity of antioxidant enzymes like SOD, CAT, and heme oxygenase-1 (HO-1) [270].

5.2.4. Sinapic Acid

Sinapic acid (5-dimethoxy-4-hydroxycinnamic acid) can be extracted from different vegetables, fruits, cereals, spices, and oilseed crops [271]. Sinapic acid (SA) has anti-inflammatory and anti-apoptotic effects, neuroprotective, anti-inflammatory, anti-nociceptive, anti-allergic, and antioxidant properties [272], and the ability to scavenge free radicals [273]. Additionally, SA reduced intestinal inflammation in a mouse model of colitis, attenuated chemical reagent-induced clinical symptoms such as 2,4,6-trinitrobenzenesulfonic acid (TNBS) and DSS [274,275], and exhibited potent efficacy against bleomycin-induced pulmonary fibrosis in rats [276]. Furthermore, SA preserves epithelial homeostasis in lipopolysaccharide-induced Caco-2 cells and prevents inflammation-induced intercellular hyperpermeability [272].

SA was reported to inhibit histone deacetylase (HDAC), which resulted in ROS release, oxidative stress, apoptosis, cell cycle arrest, and autophagy [277]. Moreover, by down-regulating the AKT/Gsk-3β pathway, SA was able to inhibit the proliferation, migration, and invasion of pancreatic malignant cells [278]. Furthermore, in vitro studies on lung cancer cells revealed elevated ROS and caspase-3 levels by SA, resulting in cytotoxicity and apoptosis, while lung cancer in vivo studies exhibited a reduction in IgG and IgM, leucocytic count, and tumour markers, with improved phagocytic activity and enzymes involved in antioxidant defence [279]. In addition, elevated ROS production, apoptosis, and cell cycle arrest at the G0/G1 phase were observed after exposure of the HEp-2 human laryngeal carcinoma cell line to SA [280].

Exposure of prostate cancer cell lines to SA revealed its antiproliferative and cytotoxic effects. LNCap cells showed significantly elevated levels of caspase-3, caspase-9, CYCS, and Bax, with a marked reduction in CDH2, MMP-2, and MMP-9. PC-3 cells exhibited caspase-3, caspase-8, Bax, CYCS, TMP-1, FAS, and CDH1 expression after SA treatment, with a significantly lower level of MMP-9 [281]. In vivo, research on DMH-induced colon carcinogenesis showed an increase in antioxidant defence through elevated superoxide dismutase (SOD), catalase (CAT), and glutathione peroxidase (GPX) [282].

In contrast, the combination of SA with cisplatin strongly inhibited the migration and proliferation of hepatocellular carcinoma cells by inducing apoptosis and autophagy [283]. The complex of SA with CYP3A4, CYP1A1, and SIRT1 proteins was studied using molecular dynamic simulations and MMPBSA, which showed a stable complex over the course of the simulations. According to these predictions, the mechanism of SA in breast cancer may involve the regulation of several proteins, including cytochrome enzymes (CYP1A1 and CYP3A4), PRKCA, CASP8, SIRT1, and CTNNB1. Interestingly, MDA levels were significantly elevated, SOD activities were significantly reduced, and reduced glutathione (GSH) and catalase levels were elevated in MCF-7 cells treated with sinapic acid-loaded poly (lactic-co-glycolic acid) (PLGA) nanoparticles (SaNPs) at concentrations of 150 and 200 µg/mL for 24 h compared to the control groups [284].

5.3. *Curcuminoids: Curcumin*

Curcumin (Figure 5) is the active ingredient in turmeric. Demethoxycurcumin and Bis-demethoxycurcumin, two curcuminoids found in the yellow-pigmented fraction of Curcuma longa, are chemically related to curcumin, the plant's main ingredient. In aqueous solutions, curcumin (CUR) dissolves very little or not at all; nevertheless, it dissolves in organic solvents such as acetone, ethanol, methanol, and dimethyl sulfoxide (DMSO) [285]. Curcumin has a symmetrical structure with four chemical entities: aryl side chains joined by a linker, two double bonds, an active methylene moiety, and a diketo functional group. Each of these sites has been studied to identify possible locations for appropriate changes to enhance the solubility, bioavailability, and effectiveness of curcumin [286].

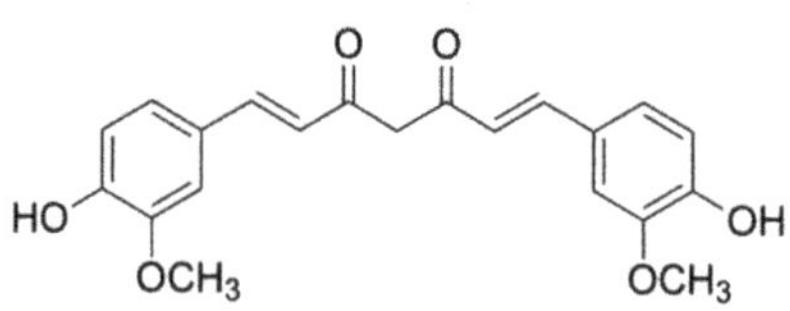

Figure 5. Chemical structure of curcumin.

Curcumin exhibits potent antiproliferative and antimetastatic properties, in addition to its antioxidant and pro-apoptotic activities, in various cancer cell lines [287]. Several in vitro studies have repeatedly demonstrated curcumin's strong antioxidant properties [288–290]. By scavenging free radicals and boosting endogenous antioxidant defences, it can reduce oxidative stress [290]. Additionally, curcumin's anti-inflammatory properties have been clarified by in vitro research, which has shown that it inhibits the generation of proinflammatory mediators and modifies important inflammatory pathways. Curcumin and its analogues have demonstrated therapeutic potential in numerous preclinical models of inflammation-associated diseases, such as arthritis, colitis, and neuroinflammatory disorders [285]. Furthermore, curcumin's anticancer properties have been supported by both in vitro and in vivo research. These effects include inhibition of angiogenesis, induction of apoptosis, suppression of tumour growth, and modification of several signalling pathways implicated in carcinogenesis [291,292]. Furthermore, curcumin exhibits synergistic effects with some anticancer medications, providing a promising approach to treating cancer [293].

It has been reported that curcumin exerts an anticancer effect in human and mouse MM cells in a dose- and time-dependent manner by increasing ROS production, inducing DNA damage and apoptosis, and inhibiting cell survival and growth. PARP-1 cleavage, p53, Caspase-9, and Bax/Bcl-2 ratios were also elevated. Moreover, it stimulated ERK1/2 and P38 MAPK phosphorylation and increased c-Jun expression and phosphorylation while inhibiting P54JNK and AKT phosphorylation and the nuclear translocation of NF-kB [294]. In addition, curcumin can reduce the activity of IkB kinase, thus retarding the degradation of IkBα, which leads to the inhibition of the nuclear translocation of NF-kB [295].

In contrast, curcumin was able to suppress different inflammatory cytokines, including TNF-α, IL-6, IL-8, and IKKβ kinase in head and neck squamous cell carcinoma, in addition to its ability to inhibit protein kinases such as PKA, mTOR, PhK, and MAPK that are essential for regulating cell survival, proliferation, and growth [296]. Moreover, curcumin was able to induce oxidative stress, apoptosis, autophagy, and cell cycle arrest in human glioblastoma by modulating different pathways [297]. In a dose-dependent manner, curcumin inhibited the transcription factor NF-kB in GBM8401 cells via the induction of caspase and mitochondrial-dependent apoptosis [298].

Gefitinib's antitumor activity in the xenografted NSCLC cell lines and mouse model was enhanced by curcumin, which suppressed NSCLC proliferation, EGFR phosphorylation, EGFR ubiquitination, and induced apoptosis [299]. Furthermore, curcumin inhibited JAK2 activity in A549 human lung cancer cell lines, downregulating NF-κB activity and acting on the JAK2/STAT3 signalling pathway. Curcumin is effective in treating lung cancer [300].

Curcumin was able to suppress clonogenicity cell proliferation and induce cell cycle arrest at the G2/M phase in leukaemic cell lines via dose-dependent inhibition of Wilms tumour protein 1 (WT1). Furthermore, by inhibiting EZH2 expression in RPMI8226 and U266 cell lines, curcumin potently suppressed MM cell growth by inducing apoptosis in a time- and dose-dependent manner [301].

Many signalling pathways linked to breast cancer, such as JAK-STA, Hedgehog, Notch, PI3K/mTOR, and Wnt/β-Catenin, were successfully targeted by curcumin [302]. Moreover, it can inhibit the growth of breast cancer cells via DNA methylation and epigenetic changes [303]. Curcumin inhibited the proliferation of MCF-7 breast cancer cells by arresting them in the G1 phase of the cell cycle. This cell cycle arrest was due to the overexpression of CDK inhibitors p21, p53, and p27, as well as increased cyclin E proteasomal degradation [304]. Moreover, by suppressing EZH2 and re-establishing DLC1 expression, CUR suppressed the growth of TNBC and enhanced apoptosis in MDA-MB-231 cells during the G2 phase [305].

Curcumin enhanced the cytoprotective effect and stability against HepG2 cell death induced by tert-butyl hydroperoxide (t-BHP) and facilitated the nuclear translocation of transcription factor Nrf-2, which regulates the antioxidant signalling pathway [306]. Curcumin also targeted and disrupted the intracellular Notch domain of the Notch-1 signalling pathway in HEP3B, SK-Hep-1, and SNU449 cell lines. Additionally, curcumin prevents diethylnitrosamine (DENA)-induced hyperplasia and HCC in animals by lowering the expression of p53, NF-κB, and p21-Ras [307]. A combination of arabinogalactan and curcumin significantly inhibited the proliferation of breast cancer cells without affecting normal cells. This combination induces cell apoptosis by altering membrane potential, increasing ROS levels, and lowering glutathione levels. Additionally, by overexpressing p53 in mice, the combination of curcumin and arabinogalactan prevented the growth of breast tumours [308]. Interestingly, several studies have suggested that the co-treatment of CUR with conventional chemotherapy drugs produces varying levels of efficacy in BC cells compared to normal epithelial cells [309]. In MDA-MB-231, MCF-7, and MCF10A cells, CUR and doxorubicin (DOX) treatments caused G2/M arrest; however, in MCF10A cells, CUR caused S phase arrest [310].

Luo et al. investigated the antiproliferative effects of four curcumin analogues on human gliomas. Curcumin (IC50 = 4.19 μM), bisdemethoxycurcumin (IC50 = 29.15 μM), demethoxycurcumin (IC50 = 30.03 μM), and dimethoxycurcumin (IC50 = 29.55 μM) were the four analogues that were most promising in promoting sub-G1 phase, G2/M arrest, apoptosis, and ROS production in human glioma cells. Dimethoxycurcumin inhibited migration, colony formation, and cell viability; it enhanced LC3B-II expression to trigger autophagy, a natural process that preserves cellular health by dissolving and recycling damaged or unnecessary components; and it caused sub-G1, G2/M arrest, apoptosis, and ROS production. They also examined several curcumin analogues for antitumor properties against the human breast cancer cell (Bcap-37), prostate cancer cells (PC3), and gastric cancer cell line (MGC-803). One of the chemicals was less harmful to NIH3T3 normal cells, while dramatically reducing the development of cancer cells and inducing apoptosis in MGC-803 cells [311].

Due to their high methylation, unsaturated diketone moiety, and low hydrogenation, several curcumin derivatives have demonstrated improved anticancer and anti-inflammatory properties compared to curcumin. Furthermore, numerous analogues of hydrogenated curcumin have also demonstrated strong antioxidant activity [312]. Moreover, novel drug delivery systems have been investigated to improve the stability of curcumin by increasing its absorption and bioavailability. These systems include nanoparticles, metal complexes, liposomes, solid dispersion, microemulsion, micelles, nanogels, and dendrimers [313]. Curcumin-containing catanionic lipid nanosystems (CLNs) have been observed to exhibit improved cytotoxicity against Lewis lung cancer cells [314].

It has been proposed that curcumin and its derivatives enhance BC cells' defence mechanisms by stopping the cell cycle. Every phase contributes to the development of cancer; however, because of its beginning position and function—cell duplication—the G1 phase is frequently seen as being especially important in fostering the development of cancer. Solid lipid nanoparticles (SLNs) loaded with CUR stop the cell cycle at G1/S and reduce the production of CDK4 and cyclin D1 (CCND1), which potently triggers ROS responses and apoptosis [315]. Moreover, the CUR analogue B14 modifies the expression of cyclin D1 (CCD1), cyclin E1, and cyclin-dependent kinase 2 (CDK2), causing G1 phase cell cycle arrest and initiating the mitochondrial apoptotic pathway [316]. Nonetheless, the G2/M phase is where most CUR combinations stop the cell cycle. When CUR was combined with a layered polyelectrolyte capsule, the number of cells in G2 increased significantly. Consequently, the proportion of apoptotic cells increased noticeably [317]. Additionally, MCF7 cells were more significantly affected by CUR, berberine, and a combination of 5-FU loaded into nano micellar particles at lower doses [318]. Moreover, gemcitabine and CUR can be combined as a nanosuspension to increase their anticancer potentiality in a synergistic manner [319].

Disruption of microtubule assembly by mesoporous silica nanoparticles can affect the cell cycle. CUR-MSN-polyethyleneimine (PEI)-FA was more successful in causing the G2/M phase cell cycle arrest by comparing the efficiency of CUR-MSN-HA and CUR-MSN-PEI-FA in MDA-MB-231 and MCF-7 cell lines [320]. In MCF-7 cells, it was demonstrated that the CUR analogue (2E,6E)-2,6-bis-(4-hydroxy-3 methoxybenzylidene)-cyclohexanone (BHMC) stimulates apoptosis and G2/M cell cycle arrest [321].

Co-encapsulation of doxorubicin and curcumin in chitosanpoly (butyl cyanoacrylate) nanoparticles has been shown to reverse multidrug resistance (MDR) [322]. Solid lipid nanoparticles (SLNs) loaded with CUR avoided P-pg MDR in TNBC cells [323]. To reverse multidrug resistance in breast cancer, doxorubicin, and curcumin were delivered using amphiphilic copolymeric micelles [324]. Curcumin reduces doxorubicin resistance in doxorubicin-resistant breast cancer cells by blocking ABCB4's efflux activity [325]. Moreover, the combination of CUR with DOX markedly enhanced apoptosis in proliferative MCF7 cells compared to DOX treatment alone [326].

According to Liu et al., curcumin has been shown to have beneficial effects in clinical trials; nevertheless, it may also have adverse effects on the heart, liver, kidneys, blood, reproductive system, and immune system [327]. Curcumin's efficacy in treating a variety of malignancies, including colorectal cancer, myeloma, oral submucosal fibrosis, and skin lesions, has been investigated in clinical trials [285]. Salehi et al. reported histological improvements and clinical alleviation in individuals with skin lesions and oral submucosal fibrosis after curcumin treatment. Additionally, in individuals with various malignancies, curcumin showed tumour growth reduction and downregulation of inflammatory markers; nevertheless, its effectiveness in treating advanced pancreatic cancer remains restricted [328].

Preclinical research on curcumin's antioxidant, anti-inflammatory, and anticancer properties, as well as those of its analogues, has yielded intriguing therapeutic prospects for a range of diseases. However, further clinical research is required to validate curcumin's toxicity, bioavailability, and effectiveness [285].

5.4. *Stilbenes*

Resveratrol (RSV) (Figure 6) is a naturally occurring phytoalexin produced by plants to defend against pathogenic invasion and environmental stress, and it may help treat cancer and signal advances in cancer treatment. Although it can be found in over 70 plant species, the most significant sources are grapes and wines [329].

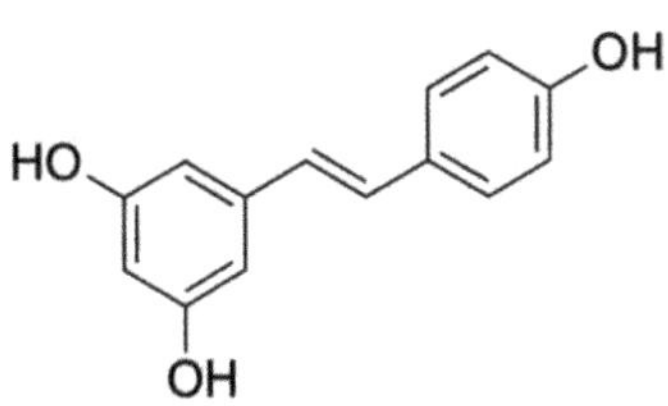

Figure 6. Chemical structure of resveratrol.

There are two isomers of resveratrol. Due to its instability, the cis-isomer is not marketed. Although its trans-isomer is more stable, it degrades more quickly when exposed to high pH or UV light, changing into the cis-isomer. The trans-isomer is thought to be the cause of resveratrol's anticancer and health effects [330]. Resveratrol is crucial for increasing immunity, delaying the ageing process, and imitating the effect that prevents or lessens diseases such as diabetes, in addition to cardiovascular and neurodegenerative diseases [331]. Resveratrol's direct antitumor, anti-inflammatory, and antioxidant properties make it a promising agent for conventional chemotherapy [332].

It has demonstrated effectiveness against lung, skin, and haematological cancers, as well as obesity-related cancers like hepatic, pancreatic, postmenopausal breast, prostate, and colorectal cancers [333].

It has been demonstrated that resveratrol inhibits the plasma levels of insulin-like growth factor-1 (IGF-1) and insulin-like growth factor-binding protein-3 (IGFBP-3), two proteins involved in the insulin signalling pathway that cause carcinogenesis [334]. Additionally, its therapy decreased prostaglandin-E2 (PGE2) production and Ras-association domain family-1a (RASSF-1a) methylation, both of which are associated with antiproliferative and anti-inflammatory effects [335]. Moreover, when resveratrol was administered, immunomodulatory T cell levels were significantly upregulated, proinflammatory cytokines like monocyte chemoattractant protein-1 (MCP1) and tumour necrosis factor-alpha (TNF-α) were downregulated, and plasma antioxidant activity was higher than the baseline [336]. Furthermore, resveratrol demonstrated an antiproliferative effect by blocking wingless-related integration site (Wnt)signalling [337]. Overall, low dosages of resveratrolseem to have chemopreventive ability based on its effects on specific tumour markers [330].

In contrast, pterostilbene (PTE), a naturally occurring resveratrol analogue [338], is abundant in blueberries [339]. Pterostilbene has attracted considerable interest because of its potential medical applications in the treatment of cancer and inflammatory diseases [338]. In vitro and in vivo studies have demonstrated that PTE can inhibit the growth of tumour cells and induce apoptosis by affecting several signalling pathways, such as the PI3K/Akt, MAPK, and NFκB pathways [340–342]. Leukaemia cells undergo apoptosis via the MAPK pathway [341].

Lung cancer has been associated with increased cyclooxygenase-2 (COX-2) activity, and PTE has been shown to control NSCLC cell proliferation and apoptosis by targeting COX-2 [343]. By altering the PTEN/Akt pathway, PTE prevents prostate cancer cells from proliferating [344]. Furthermore, PTE works by increasing the tumour suppressor gene PTEN's acetylation and reactivation. This effect is achieved by suppressing the MTA1/HDAC complex, which usually deacetylates proteins, thereby altering their function. PTE maintains PTEN activity by blocking this complex, which is crucial for controlling the Akt pathway. The Akt pathway is involved in cell growth and survival, and its overactivity can lead to cancer. Therefore, Pterostilbene's capacity to reactivate PTEN and control the Akt pathway shows its promise as a therapeutic agent in cancer treatment [338]. PTE contributes to the anticancer effect of endoplasmic reticulum stress (ERS) in human oesophageal cancer cells by inducing the ROS-mitochondrial-dependent apoptosis mechanism, which inhibits cell adhesion, invasion, and proliferation [345]. Additionally, recent studies have revealed that PTE improves the sensitivity of triple-negative breast cancer cells to TRAIL-driven apoptosis by triggering the ROS/ERS signalling pathway and boosting DR4 and DR5 expression [346]. It exhibits anticancer effectiveness by inducing autophagy; thus, it may be a promising anticancer drug, as highlighted by its safety profile [347]. Additionally, PTE has been reported to reduce mitochondrial membrane potential, induce cell apoptosis, and suppress cancer cell proliferation in a dose-dependent manner [348,349]. Additionally, PTE showed a dose-dependent suppression of cancer stem cell gene expression and self-renewal capabilities in lung cancer cells co-cultured with M2-TAMs (tumour-associated macrophages with the M2 phenotype). According to Huang et al., this impact seems to be mediated by the downregulation of the cancer-promoting gene MUC1, which inhibits polarisation towards M2 and reduces the accumulation of cancer stem cells (CSCs) [350]. RSV and PTE can target CSCs in a variety of cancers, such as breast cancer colorectal cancer, via a variety of signalling pathways [351].

Furthermore, when combined with sunitinib (SUN), PTE exhibits synergistic antitumor activity against gastric cancer (GC) cell lines [352]. SUN causes mitochondrial iron (II) (mtFe) deposition by suppressing the expression of PDZ domain-containing 8 (PDZD8) [353]. As an iron–sulfur complex, mitochondrial iron is crucial for energy production and other functions; however, it also contributes to cell death through ferroptosis [354]. Inhibiting the recruitment of the iron transporters mtNEET and ABCB8 to PDZD8 by DOX in combination with PTE led to the accumulation of mtFe and an increase in mitochondrial ROS generation, which in turn activated HIF1α, further inducing ER stress and apoptosis [355]. The mechanisms include activation of c-Jun N-terminal kinase and p38 pathways [356,357] with elevated BAX and lower BCL2 expression levels [358,359].

Pterostilbene exhibits greater absorption and corresponding plasma levels than its parent compound, resveratrol [360]. Its anticancer effect has been demonstrated to surpass that of resveratrol in vivo [361], and its additional methyl groups reduce its susceptibility to conjugation metabolism [362]. Through a unique mechanism, trans-3,5,4′-trimethoxystilbene promotes cell cycle arrest and apoptosis with improved potency [363,364], with IC50 values 100–200 times lower than those of resveratrol. Greater pharmacokinetic and tumour-suppressive qualities are possessed by trans-3,4,5,4′-tetramethoxystilbene (DMU212/TMS), and its metabolite has demonstrated increased preclinical cytotoxicity in prostate and ovarian cancers [365]. Moreover, trans-2,4,3′,4′,5′-pentamethoxystilbene was more effective than resveratrol in preventing the growth of breast [366] cancer cells.

Piceatannol, the hydroxylated counterpart, demonstrated direct pro-apoptotic, antimetastatic [367], and tyrosine kinase-inhibiting properties [368] along with equipotency for anti-inflammatory, immunomodulatory, and antiproliferative effects [367].

Lowering the levels of circulating cancer biomarkers, such as insulin-like growth factor 1 and insulin-like growth factor-binding protein 3, was detected [334]. Administration of 0.5 g and 1.0 g of resveratrol resulted in significant inhibition of colorectal cancer cell proliferation [369]. Furthermore, elevated caspase-3 levels were recorded in a phase I clinical trial of hepatic malignancy after administration of resveratrol [370]. A trial of the derivative resveratrol-triphosphatase demonstrated that the compound evoked a better reduction in oxidative stress in obese participants [371].

5.5. Lignans

The genus Schisandra (family Schisandraceae) contains a significant group of chemicals known as dibenzocyclooctadiene lignans (DBCLS) (Figure 7), which are distinguished by their distinct chemical structures and wide range of biological activities. This category comprises 40 identified secondary metabolites with hepatoprotective and hepatoregenerative properties, including gomisin A, schisandrin B, schisandrin C, and γ-schisandrin. Additionally, it has been confirmed that deoxyschisandrin and γ-schisandrin have antiviral properties, that schisandrin and schisandrin B have antioxidant properties, and that schisandrin C and gomisin A have anti-inflammatory properties. Recent studies have focused on investigating DBCLS's antitumor capability and function in preventing the proliferation of cancer cells [31].

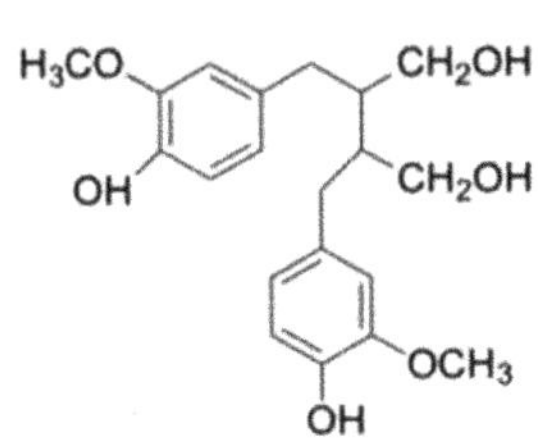

Figure 7. Chemical structure of lignans.

The ability of DBCLS to trigger apoptosis is a key characteristic of its anticancer potential. DBCLS can cause apoptosis in a variety of cancer cell lines, mostly via the mitochondrial apoptotic pathway, according to preclinical research. This entails caspase activation and cytochrome c (cyt-c) release. In addition to their role in apoptosis, DBCLS have cytotoxic effects, including preventing the invasion, migration, and proliferation of cancer cells [372–374]. In GBC-SD and NOZ gallbladder cancer cell lines, schisandrin B was able to effectively inhibit the viability and proliferation of cancer cells and induce apoptosis. Moreover, in vivo investigation conducted on nude mice with subcutaneously placed NOZ tumour xenografts revealed that schisandrin B induced apoptosis in cancer cells by controlling the expression of proteins linked to apoptosis [373]. Furthermore, DBCLS can induce oxidative stress in cancer cells by boosting the generation of ROS. Cell death can result from excessive ROS damage to lipids, proteins, DNA, and other biological macromolecules [31]. By controlling the intracellular generation of reactive oxygen species and blocking NADPH oxidase, gamisin L1 exerts potent cytotoxic effects on cancer cells, inducing apoptosis in A2780 and SKOV3 ovarian cancer cell lines [374]. Additionally, gomisin J was reported to have a unique capacity to alter autophagy in cancer cells, especially in MDA-MB-231 and MCF7 cell lines. It initially causes autophagy in a survival form, but after more exposure, autophagy-mediated cell death takes over. The mTOR pathway, which is a component of the PI3K/Akt/mTOR signalling axis and is frequently used by cancer cells to develop drug resistance, appears to be inhibited in this result. Consequently, gomisin J is a promising therapeutic agent, particularly for the treatment of malignancies that have become resistant to standard therapies [375].

In contrast, DBCLS compounds are able to cause cell cycle arrest, especially during the G1/S and G2/M phases, which inhibits the proliferation of cancer cells [31]. Li et al. used in vitro studies on gastric cancer cell lines to investigate the anticancer mechanism of the combination of schisandrin B and the cytotoxic medication apatinib. This synergistic combination arrested the cell cycle in the G0/G1 phase, which inhibited the division of cancer cells. Schisandrin B enhances cytotoxic drug-induced apoptosis of cancer cells and boost the invasion and migration of apatinib in cancer cells [376]. In melanoma cell lines, gomisin A influenced cell cycle arrest and inhibited cell proliferation. Furthermore, it has been demonstrated that gomisin A decreased melanoma cell viability via blocking the cell cycle, and inhibition of cyclin D1, AMPK, ERK, and JNK phosphorylation which is followed by cell cycle arrest in the G0/G1 phase. Additionally, it has been shown to decrease cell invasion and migration and exert antiproliferative effects. Moreover, gomisin A suppresses epithelial-mesenchymal transition, preventing lung metastasis [377]. Together with TNF-α, this compound also inhibited the activity of signal transducer and transcription activation 1 (STAT1) [378].

Gomisin G suppressed the proliferation of MDA-MB-231 and MDA-MB-468 breast cancer cell lines. Gomisin G's mode of action was predicated on a very significant inhibitory activity on AKT phosphorylation, as well as a reduction in the quantity of phosphorylated retinoblastoma tumour and retinoblastoma tumour suppressor protein. Gomisin G's activity was also centred on reducing the quantity of cyclin D1 and stopping the cell cycle during the G1 phase [379]. By inducing cell cycle arrest in the G2/M phase, causing tumour cell death, and preventing cancer cell trafficking, Schisantherin A showed a strong antiproliferative effect on MKN45 and SGC-7901 gastric cancer cell lines. Furthermore, schisantherin A increases the generation of reactive oxygen species, which are necessary for JNK phosphorylation [380]. Due to their capacity to alter matrix metalloproteinase (MMP) activity, which is essential for cancer metastasis, several studies have indicated that DBCLS compounds can prevent cancer cell migration and invasion. By boosting the number of heat shock proteins, schisandrin B also prevents the development, migration, and invasion of cancer cells (c). In the MDA-MB-435S breast cancer cell line, schisandrin B (encapsulated in liposomes) combined with PFV-modified epirubicin cytotoxic medication increased cytotoxicity, influenced the formation of vascular mimicry, and prevented tumour invasion and spread. The control of vimentin, E-cadherin, vascular endothelial growth factor (VEGF), and matrix metalloproteinase-9 (MMP-9) expression forms the basis of the mechanism of action against cancer cells. Mice have also been used in tests that showed an increase in cancer cell apoptosis [381]. Additionally, schisandrin B completely prevented the growth, proliferation, and invasion of gastric cancer cells in both in vitro and in vivo experiments. Schisandrin B was also shown to suppress STAT3 phosphorylation, induces apoptosis, and boost the effectiveness of 5-fuorouracil [382]. Schisandrin B was shown to successfully suppress cell migration, invasion, and multiplication in both in vitro and in vivo investigations. It has also been reported to increase apoptosis in MG63, Saos2, and U2OS osteosarcoma cell lines. Additionally, studies on healthy cells have demonstrated that schisandrin B has no detrimental effect on the viability of these cells [383].

The ability of lignans to induce oxidative stress in cancer cells may be the cause of schisandrin B's ability to inhibit the proliferation and induce apoptosis of DU145 and LNCaP prostate cancer cell lines. It also inhibits the cell cycle in the S phase [384]. Moreover, schisandrin B inhibits cell division, migration, and invasion, according to in vivo experiments conducted on animal models. Schisandrin B also induces apoptosis by inhibiting the PI3K/Akt and Wnt/β-catenin signalling pathways, although it has no detrimental effects on healthy cells [383]. In order to investigate the action against BGC-823 human gastric cancer cells, Li et al. employed liposomes to encapsulate the cytotoxic medication vinoreblin, R8 peptide, and schisandrin B. Schisandrin B was utilised to suppress metastasis, vinoreblin was employed as a chemotherapeutic drug, and peptide R8 was utilised for its ability to increase cellular absorption. The liposomal complex inhibited invasion and metastasis by lowering the levels of VE-Cadherin, PI3K, VEGF, HIF-1α, FAK, and MMP-2, thereby inducing apoptosis in cancer cells. Furthermore, in vivo experiments have demonstrated that liposomes specifically gather in cancer cell-occupied areas and cause cell death [385].

Breast and ovarian cancer cell lines were used to assess the ability of schisandrin B to treat doxorubicin-resistant malignancies. According to previous studies, schisandrin B significantly increases the intracellular accumulation of doxorubicin by inhibiting the expression and activity of P-glycoprotein. Schisandrin B also decreased the expression of the anti-apoptotic protein survivin [386]. Additionally, schisandrin B and docetaxel together decreased the viability of Caski cervical cancer cells, prevented colony formation, induced apoptosis, and prevented the invasion of tumour cells. Furthermore, BALB/c nude mice xenografted with Caski cells were used for in vivo investigations. These outcomes demonstrate this combination's synergistic effect [387]. Despite the fact that DBCLS from the Schisandra genus have demonstrated encouraging anticancer properties in both in vitro and in vivo models, no research has progressed to clinical trials. This limits the comprehension of the safety, dose, and therapeutic effectiveness of lignans in humans. Furthermore, a thorough knowledge of the metabolism, distribution, excretion, and mode of action of these lignans in the human body is currently lacking, which necessitates further research [31].

6. Conclusions and Future Directions

New approaches to cancer prevention and treatment have been made possible by natural polyphenols that have drawn a lot of interest in response to the increasing demand for the creation of novel therapeutic and preventive solutions derived from natural sources. Compared to the previous literature, this review offers a broader scope by examining not only individual polyphenol classes but also their dietary sources, bioavailability, and recent advances in nanoformulation strategies in clinical settings. By integrating mechanistic insights with translational findings, this review underscores the therapeutic potential of polyphenols in cancer management. These findings support ongoing efforts to incorporate natural compounds into cancer prevention and treatment strategies, offering a safer and potentially cost-effective alternative to conventional therapies.

Although the molecular mechanisms underlying their wide range of biological activities are not completely understood, they are likely linked to cell cycle disruption, apoptosis induction, and improved the efficacy of currently available cytotoxic medications. In recent years, many studies have examined the antiproliferative effects of polyphenols on a variety of malignant tumours, both in vivo and in vitro, while polyphenols causing normal cells little to no harm. Furthermore, natural polyphenols enhance the therapeutic effectiveness of current treatments and reduce the adverse effects associated with chemotherapy. Thus, the promising potential of natural polyphenols makes them appealing candidates for future oncological applications. Despite phase I and II clinical trials, polyphenols have shown their critical importance in cancer prevention; however, further clinical research is still needed to completely understand the effectiveness of natural polyphenols in cancer treatment. The journey from the laboratory to the clinic remains essential. Future efforts should focus on human clinical studies to assess the safety and efficacy of these compounds and close the large gap between preclinical discoveries and clinical applications.

Author Contributions: Conceptualisation, Q.D. and Z.S.O.A.; writing—original draft preparation, Z.S.O.A., E.K., N.E. and A.E.; writing—review, editing, and supervision; Q.D. All authors have read and agreed to the published version of the manuscript.

Funding: This research received no external funding.

Institutional Review Board Statement: Not applicable.

Informed Consent Statement: Not applicable.

Conflicts of Interest: The authors declare no conflicts of interest.

Abbreviations

ABCB8	ATP-Binding Cassette subfamily B member 8
ADAM17	A Disintegrin And Metalloprotease 17
ADP	Adenosine Diphosphate
AgNPs	Silver nanoparticles
Ahr	Aryl hydrocarbon receptor
AIF	Apoptosis Inducing Factor
ALP	Alkaline Phosphatase
ALT	Alanine Transaminase
Antioxcin6	Mitochondrial Targeted Antioxidant
AP-1	Activator Protein-1
APAF1	Apoptotic Protease-Activating Factor 1
ASK-1	Apoptosis signal-regulating kinase 1
AST	Aspartate aminotransferase
ATL	Adult T Cell Leukaemia
ATP	Adenosine Triphosphate
BAX	Bcl-2 Associated X Protein
B-CLL	B-cell Chronic Lymphocytic Leukaemia
Bcl-2	B-Cell Lymphoma 2
BNIP3	BCL-2 Interacting Protein 3
CA	Caffeic Acid
CASP	Caspase
CAT	Catalase
CCND1	Cyclin D1
CD44	Cell surface adhesion receptor
CDC2	Cell division cycle protein 2

CDH1	Cadherin 1
CDH2	Cadherin 2
CDK1	Cyclin-Dependent Kinase 1
CDK2	Cyclin-Dependent Kinase 2
CDK4	Cyclin-Dependent Kinase 4
CDK1A	Cyclin-Dependent Kinase Inhibitor 1A
CDK1B	Cyclin-Dependent Kinase Inhibitor 1B
CEA	Carcinoembryonic Antigen
C/EBP	CCAAT/Enhancer Binding Protein
c-FLIP	Cellular FLICE (FADD-like IL-1β-converting enzyme)-Inhibitory Protein
cIAP1	Cellular Inhibitor of Apoptosis Protein-1
cIAP2	Cellular Inhibitor of Apoptosis Protein-2
CIS	Cisplatin
CIP2A	Carcinogenic Inhibitor of Protein Phosphatase 2A
CLNs	Curcumin-containing catanionic Lipid Nanosystems
Col1α1	Collagen, type I, alpha 1
Col3α1	Collagen type III alpha 1 chain
ConA	Concanavalin A
COX-2	Cyclooxygenase-2
CSCs	Cancer Stem Cells
CTNNB	Catenin Beta 1
CUR	Curcumin
CVD	Cardiovascular Disease
CYCS	Cytochrome C, Somatic
CYP1A1	Cytochrome P450, family 1, subfamily A, Polypeptide 1
CYP3A4	Cytochrome P450 family 3, subfamily A, polypeptide 4
Cyt-c	Cytochrome complex
DBCLS	Dibenzocyclooctadiene lignans
DENA	Diethylnitrosamine
DLC1	DLC1 Rho GTPase Activating Protein
DMBA	7,12-Dimethylbenz[a]anthracene
DMBS	3,4-Dimethoxybenzenesulfonamide
DMH	1,2-Dimethylhydrazine
DMSO	Dimethylsulfoxide
DNA	Deoxyribonucleic Acid
DOT1L	Disruptor of Telomeric Silencing 1-Like
DOX	Doxorubicin
DR4	Death receptor 4
DR5	Death receptor 5
DSS	Dextran Sulfate Sodium
DUSP1	Dual specificity protein phosphatase 1
DUSP3	Dual specificity protein phosphatase 3
DUSP5	Dual specificity protein phosphatase 5
D. caffra	*Dovyalis caffra*
EAFE	Ethyl Acetate Fractionated Extract
ECM	Extracellular Matrix
EGCG	Epigallocatechin-3-Gallate
EGF	Epidermal Growth Factor
EGFR	Epidermal Growth Factor Receptor
EMT	Epithelial-Mesenchymal Transition
EPI	Epicatechin
ER	Endoplasmic Reticulum
ER	Estrogen Receptor

α-ER	Estrogen Receptor Alpha
β-ER	Estrogen Receptor Beta
ERK	Extracellular Signal-Regulated Proteinase
ERS	Endoplasmic Reticulum Stress
EZH2	Enhancer of Zeste Homologue 2
FAD	Flavin Adenine Dinucleotide
FAK	Focal Adhesion Kinase
FasL	Fas Ligand
FIH-1	Factor-Inhibiting HIF-1
FN	Fibronectin
FOXM1	Forkhead Box M1
5-FU	5-Fluorouracil
GA	Gallic Acid
GADD45A	Growth Inhibition And DNA-Damage-Inducible 45 Alpha
GADD45B	Growth Arrest and DNA-Damage Inducible Beta
GC	Gastric Cancer
Gli-1	Glioma-associated oncogene
GSK3β	Glycogen Synthase Kinase-3 beta
GPx	Glutathione Peroxidase
GPx4	Glutathione Peroxidase 4
GRP78	78 kDa Glucose-Regulated Protein
GR	Glutathione Reductase
GSH	Glutathione
HA	Hyaluronic Acid
HC	Hepatic Cells
HCC	Hepatocellular Carcinoma
HDAC	Histone Deacetylase
HEPG2	Human Hepatocellular Liver Cancer Cells
HER2	Human Epidermal Growth Factor Receptor 2
HER2-TK	Human Epidermal Growth Factor Receptor 2-Tyrosine Kinase
HETNPs	Hesperetin-loaded Nanoparticles
HGF	Hepatocyte Growth Factor
HIF-α	Hypoxia-Inducible Factor 1-Alpha
HO-1	Heme Oxygenase-1
Hrk	HaraKiri
HSCs	Hepatic Stellate cells
HSP	Hesperetin
HTLV-1	Human T-Lymphotropic Virus Type 1
HUVEC	Human Umbilical Vein Endothelial Cells
Hyp	Hydroxyproline
IC50:	Half maximal inhibitory concentration
IGF1R	Type 1 Insulin-Like Growth Factor Receptor
IgG	Immunoglobulin G
IgM	Immunoglobulin M
IKKβ	Inhibitory kappa B Kinase Beta
IL-6	Interleukin 6
IL-10	Interleukin 10
IL-12	Interleukin 12
IRAK4	Interleukin 1 Receptor Associated Kinase 4
IR-beta	Insulin Receptor-Beta Subunit
IUPAC	International Union of Pure and Applied Chemistry
JAK	Janus Kinase

JNK	Jun N-Terminal Kinases
LC3B	Light chain 3B
LPO	Lipid Peroxidation
67LR	67 kDa Laminin Receptor
MAPK	Mitogen-Activated Protein Kinase
MARK4	Microtubule Affinity-Regulating Kinase 4
MBS	4-Methoxybenzenesulfon
MD-2	Myeloid Differentiation Factor 2
MCL-1	Myeloid Cell Leukaemia-1
MCP-1	Monocyte Chemoattractant Protein-1
MDM2	Mouse Double Minute 2 Homologue
MDR	MultiDrug Resistance
MEK	Mitogen-Activated protein kinase Kinse
MiR	MicroRNA
MMPBSA	Molecular Mechanics Poisson-Boltzmann Surface Area
MMP-2	Matrix Metalloproteinase-2
MMP-9	Matrix Metalloproteinase-9
MMP	Matrix Metalloproteinase
MM	Multiple myeloma
MPA	Medroxy Progesterone Acetate
mRNA	Messenger ribonucleic acid
MT1A	Metallothionein 1A
MT1F	Metallothionein 1F
mtFe	Mitochondrial iron (II)
mtNEET	Mitochondrial protein mitoNEET
MTX	Methotrexate
mTOR	Mammalian Target Of Rapamycin
M2-TAMs	Tumour-Associated Macrophages with the M2 phenotype
NAD	Nicotinamide Adenine Dinucleotide
NADPH	Reduced Nicotinamide Adenine Dinucleotide Phosphate
Nar	Naringenin
NFKBIA	Nuclear Factor-Kappa-B-Inhibitor Alpha
NF-Kb	Nuclear Factor Kappa B
NLRP3	NLR Family Pyrin Domain-containing 3
Nrf2	Nuclear Factor Erythroid Related Factor 2
NSCLC	Non-Small Cell Lung Carcinoma
OS	Oxidative Stress
P-AKT	Phospho-Akt
PAI	Plasminogen Activator Inhibitor-1
PARP-1	Poly-ADP-Ribose Polymerase 1
PCNA	Proliferating Cell Nuclear Antigen
PC-3	Prostate Cancer-3
PBMC	Peripheral Blood Mononuclear Cells
PDAC	Pancreatic Ductal Adenocarcinoma
PDZD8	PDZ domain-containing 8
PEGFR	Phosphorylated-epidermal growth factor receptor
PEL	Primary Effusion Lymphoma
PERK	Phosphorylation of Extracellular Signal-Related Kinase
PGE2	Prostaglandin E2
P-gp	P-glycoprotein
PhK	Phosphorylase Kinase
PI3K	Phosphatidylinositol 3-kinase

PKA	Protein Kinase A
PKB	Protein Kinase B
PKM2	Pyruvate Kinase Isozymes M1/M2
PLGA	Poly (Lactic-co-Glycolic Acid)
PMA	Phorbol Myristate Acetate
PPARG	Peroxisome Proliferator-Activated Receptor Gamma
PRKCA	Protein Kinase C Alpha Gene
Prp4B	Pre-RNA processing factor 4B
PSA	Prostate-Specific Antigen
PTEN	Phosphatase and Tensin Homologue
PTE	Pterostilbene
PTX	Paclitaxel
RA	Rosmarinic Acid
RANK	Receptor Activator of Nuclear Factor Kappa-B
RANKL	Receptor Activator of Nuclear Factor Kappa-B Ligand
RASSF-1a	Ras-Association Domain Family-1a
Rb	Retinoblastoma protein
RNA	Ribonucleic Acid
ROS	Reactive Oxygen Species
RSV	Resveratrol
SA	Sinapic Acid
SaNPs	Sinapic acid-loaded Nanoparticles
SCC	Squamous Cell Carcinoma
SCCHN	Squamous Cell Carcinoma of The Head and Neck
SCID	Severe Combined Immunodeficiency
Shh	Sonic hedgehog protein
SIRT1	Sirtuin 1
SKP2	S Phase Kinase Associated Protein 2
SLC7A11	Solute Carrier Family 7 Member 11
SLNs	Solid Lipid Nanoparticles
SOD	Superoxide Dismutase
SPRR2D	Small Proline Rich Protein 2D
STAT	Signal Transducer and Activator of Transcription
Sufu	Suppressor of Fused Protein
t-BHP	tert-Butyl Hydroperoxide
TGF-β1	Transforming Growth Factor Beta 1
TIMP-1	Tissue Inhibitor of Metalloproteinases 1
TIMP-2	Tissue Inhibitor of Metalloproteinases 2
TLR4	Toll-Like Receptor 4
TMBS	3,4,5-Trimethoxybenzenesulfonamide
TME	Tumour Microenvironment
TNBS	2,4,6-Trinitrobenzene Sulfonic Acid
TNFRSF11B	TNF Superfamily 11B Receptor
TNFRSF25	Tumour Necrosis Factor Receptor Superfamily 25
TNF-α	Tumour Nerosis Factor-Alpha
TOPK	T-LAK Cell-Originated Protein Kinase
TRAIL	Tumour Necrosis Factor (TNF)- Related Apoptosis-Inducing Ligand
TRIF	TIR-Domain-Containing Adapter-Inducing Interferon-β
UGT1A3	UDP Glucuronosyltransferase Family 1 Member A3
UPA	U-Plasminogen Activator
UV	Ultraviolet rays
UVB	Ultraviolet rays type B

VCAM-1	Vascular Cell Adhesion Molecule-1
VEGF	Vascular Endothelial Growth Factor
VV-IL-24	Vaccinia Virus that encoded the IL-24 gene
WHO	World Health Organization
Wnt	Wingless-Related Integration Site
WT1	Wilms Tumour Protein
XIAP	X-Linked Inhibitor of Apoptosis
α-SMA	Alpha-Smooth Muscle Actin
γCD	Gamma-Cyclodextrin

References

1. Singla, R.K.; Dubey, A.K.; Garg, A.; Sharma, R.K.; Fiorino, M.; Ameen, S.M.; Haddad, M.A.; Al-Hiary, M. Natural polyphenols: Chemical classification, definition of classes, subcategories, and structures. *J. AOAC Int.* **2019**, *102*, 1397–1400. [CrossRef] [PubMed]
2. Tufarelli, V.; Casalino, E.; D'Alessandro, A.G.; Laudadio, V. Dietary phenolic compounds: Biochemistry, metabolism and significance in animal and human health. *Curr. Drug Metab.* **2017**, *18*, 905–913. [CrossRef]
3. Konstantinou, E.K.; Panagiotopoulos, A.A.; Argyri, K.; Panoutsopoulos, G.I.; Dimitriou, M.; Gioxari, A. Molecular pathways of rosmarinic acid anticancer activity in triple-negative breast cancer cells: A literature review. *Nutrients* **2023**, *16*, 2. [CrossRef] [PubMed]
4. Maruca, A.; Catalano, R.; Bagetta, D.; Mesiti, F.; Ambrosio, F.A.; Romeo, I.; Moraca, F.; Rocca, R.; Ortuso, F.; Artese, A.; et al. The Mediterranean Diet as source of bioactive compounds with multi-targeting anti-cancer profile. *Eur. J. Med. Chem.* **2019**, *181*, 111579. [CrossRef]
5. Kumari, M.; Siddiqui, M.A.; Gupta, A. Recent Advancement and Novel Application of Natural Polyphenols for the Treatment of Allergy Asthma: From Phytochemistry to Biological Implications. *Crit. Rev. Immunol.* **2023**, *43*, 29–41. [CrossRef]
6. Rauf, A.; Shariati, M.A.; Imran, M.; Bashir, K.; Khan, S.A.; Mitra, S.; Emran, T.B.; Badalova, K.; Uddin, M.S.; Mubarak, M.S.; et al. Comprehensive review on naringenin and naringin polyphenols as a potent anticancer agent. *Environ. Sci. Pollut. Res.* **2022**, *29*, 31025–31041. [CrossRef]
7. Yoganathan, S.; Alagaratnam, A.; Acharekar, N.; Kong, J. Ellagic acid and schisandrins: Natural biaryl polyphenols with therapeutic potential to overcome multidrug resistance in cancer. *Cells* **2021**, *10*, 458. [CrossRef] [PubMed]
8. Mottaghi, S.; Abbaszadeh, H. Natural lignans honokiol and magnolol as potential anticarcinogenic and anticancer agents. A comprehensive mechanistic review. *Nutr. Cancer* **2022**, *74*, 761–778. [CrossRef]
9. Di Sotto, A.; Di Giacomo, S. Plant polyphenols and human health: Novel findings for future therapeutic developments. *Nutrients* **2023**, *15*, 3764. [CrossRef]
10. Di Lorenzo, C.; Colombo, F.; Biella, S.; Stockley, C.; Restani, P. Polyphenols and human health: The role of bioavailability. *Nutrients* **2021**, *13*, 273. [CrossRef]
11. Bakrim, S.; El Omari, N.; El Hachlafi, N.; Bakri, Y.; Lee, L.H.; Bouyahya, A. Dietary phenolic compounds as anticancer natural drugs: Recent update on molecular mechanisms and clinical trials. *Foods* **2022**, *11*, 3323. [CrossRef]
12. Alam, M.N.; Almoyad, M.; Huq, F. Polyphenols in colorectal cancer: Current state of knowledge including clinical trials and molecular mechanism of action. *BioMed Res. Int.* **2018**, *2018*, 4154185. [CrossRef] [PubMed]
13. Niedzwiecki, A.; Roomi, M.W.; Kalinovsky, T.; Rath, M. Anticancer efficacy of polyphenols and their combinations. *Nutrients* **2016**, *8*, 552. [CrossRef] [PubMed]
14. Tena, N.; Martín, J.; Asuero, A.G. State of the art of anthocyanins: Antioxidant activity, sources, bioavailability, and therapeutic effect in human health. *Antioxidants* **2020**, *9*, 451. [CrossRef] [PubMed]
15. Li, L.; Mohammed, A.H.; Auda, N.A.; Alsallameh, S.M.; Albekairi, N.A.; Muhseen, Z.T.; Butch, C.J. Network Pharmacology, Molecular Docking, and Molecular Dynamics Simulation Analysis Reveal Insights into the Molecular Mechanism of Cordia myxa in the Treatment of Liver Cancer. *Biology* **2024**, *13*, 315. [CrossRef]
16. David, A.V.; Arulmoli, R.; Parasuraman, S. Overviews of biological importance of quercetin: A bioactive flavonoid. *Pharmacogn. Rev.* **2016**, *10*, 84–89.
17. Kong, M.; Xie, K.; Lv, M.; Li, J.; Yao, J.; Yan, K.; Wu, X.; Xu, Y.; Ye, D. Anti-inflammatory phytochemicals for the treatment of diabetes and its complications: Lessons learned and future promise. *Biomed. Pharmacother.* **2021**, *133*, 110975. [CrossRef]
18. Chagas, M.D.; Behrens, M.D.; Moragas-Tellis, C.J.; Penedo, G.X.; Silva, A.R.; Gonçalves-de-Albuquerque, C.F. Flavonols and flavones as potential anti-inflammatory, antioxidant, and antibacterial compounds. *Oxidative Med. Cell. Longev.* **2022**, *2022*, 9966750. [CrossRef]

19. Goh, Y.X.; Jalil, J.; Lam, K.W.; Husain, K.; Premakumar, C.M. Genistein: A review on its anti-inflammatory properties. *Front. Pharmacol.* **2022**, *13*, 820969. [CrossRef]
20. Espíndola, K.M.; Ferreira, R.G.; Narvaez, L.E.; Silva Rosario, A.C.; Da Silva, A.H.; Silva, A.G.; Vieira, A.P.; Monteiro, M.C. Chemical and pharmacological aspects of caffeic acid and its activity in hepatocarcinoma. *Front. Oncol.* **2019**, *9*, 541. [CrossRef]
21. Hadidi, M.; Liñán-Atero, R.; Tarahi, M.; Christodoulou, M.C.; Aghababaei, F. The potential health benefits of gallic acid: Therapeutic and food applications. *Antioxidants* **2024**, *13*, 1001. [CrossRef]
22. Nunes, S.; Madureira, A.R.; Campos, D.; Sarmento, B.; Gomes, A.M.; Pintado, M.; Reis, F. Therapeutic and nutraceutical potential of rosmarinic acid—Cytoprotective properties and pharmacokinetic profile. *Crit. Rev. Food Sci. Nutr.* **2017**, *57*, 1799–1806. [CrossRef]
23. Diederich, M.; Giblin, L.; Malki, M.C. Natural Products and the Hallmarks of Chronic Diseases. In Proceedings of the COST Action 16112—Personalized Nutrition in Ageing Society: Redox Control of Major Age-Related Diseases, Luxembourg, 25–27 March 2019.
24. Kalinowska, M.; Gołębiewska, E.; Świderski, G.; Męczyńska-Wielgosz, S.; Lewandowska, H.; Pietryczuk, A.; Cudowski, A.; Astel, A.; Świsłocka, R.; Samsonowicz, M.; et al. Plant-derived and dietary hydroxybenzoic acids—A comprehensive study of structural, anti-/pro-oxidant, lipophilic, antimicrobial, and cytotoxic activity in MDA-MB-231 and MCF-7 cell lines. *Nutrients* **2021**, *13*, 3107. [CrossRef] [PubMed]
25. Martinez, K.B.; Mackert, J.D.; McIntosh, M.K. Polyphenols and intestinal health. In *Nutrition and Functional Foods for Healthy Aging*; Academic Press: Cambridge, MA, USA, 2017; pp. 191–210.
26. Mahfuz, S.; Mun, H.S.; Dilawar, M.A.; Ampode, K.M.; Yang, C.J. Potential role of protocatechuic acid as natural feed additives in farm animal production. *Animals* **2022**, *12*, 741. [CrossRef]
27. Srinivasulu, C.; Ramgopal, M.; Ramanjaneyulu, G.; Anuradha, C.M.; Kumar, C.S. Syringic acid (SA)—A review of its occurrence, biosynthesis, pharmacological and industrial importance. *Biomed. Pharmacother.* **2018**, *108*, 547–557. [CrossRef] [PubMed]
28. Ullah, R.; Ikram, M.; Park, T.J.; Ahmad, R.; Saeed, K.; Alam, S.I.; Rehman, I.U.; Khan, A.; Khan, I.; Jo, M.G.; et al. Vanillic acid, a bioactive phenolic compound, counteracts LPS-induced neurotoxicity by regulating c-Jun N-terminal kinase in mouse brain. *Int. J. Mol. Sci.* **2020**, *22*, 361. [CrossRef]
29. Hewlings, S.J.; Kalman, D.S. Curcumin: A review of its effects on human health. *Foods* **2017**, *6*, 92. [CrossRef] [PubMed]
30. Meng, X.; Zhou, J.; Zhao, C.N.; Gan, R.Y.; Li, H.B. Health benefits and molecular mechanisms of resveratrol: A narrative review. *Foods* **2020**, *9*, 340. [CrossRef]
31. Jafernik, K.; Motyka, S.; Calina, D.; Sharifi-Rad, J.; Szopa, A. Comprehensive review of dibenzocyclooctadiene lignans from the Schisandra genus: Anticancer potential, mechanistic insights and future prospects in oncology. *Chin. Med.* **2024**, *19*, 17. [CrossRef]
32. Lu, Z.; Wang, X.; Lin, X.; Mostafa, S.; Zou, H.; Wang, L.; Jin, B. Plant anthocyanins: Classification, biosynthesis, regulation, bioactivity, and health benefits. *Plant Physiol. Biochem.* **2024**, *217*, 109268. [CrossRef]
33. Kent, K.; Charlton, K.E.; Lee, S.; Mond, J.; Russell, J.; Mitchell, P.; Flood, V.M. Dietary flavonoid intake in older adults: How many days of dietary assessment are required and what is the impact of seasonality? *Nutr. J.* **2018**, *17*, 7. [CrossRef] [PubMed]
34. Dangles, O.; Fenger, J.A. The chemical reactivity of anthocyanins and its consequences in food science and nutrition. *Molecules* **2018**, *23*, 1970. [CrossRef]
35. Riaz, M.; Zia-Ul-Haq, M.; Saad, B. *Anthocyanins and Human Health: Biomolecular and Therapeutic Aspects*; Springer: Cham, Switzerland, 2016.
36. Lage, N.N.; Layosa, M.A.; Arbizu, S.; Chew, B.P.; Pedrosa, M.L.; Mertens-Talcott, S.; Talcott, S.; Noratto, G.D. Dark sweet cherry (Prunus avium) phenolics enriched in anthocyanins exhibit enhanced activity against the most aggressive breast cancer subtypes without toxicity to normal breast cells. *J. Funct. Foods* **2020**, *64*, 103710. [CrossRef]
37. Kowalczyk, T.; Muskała, M.; Merecz-Sadowska, A.; Sikora, J.; Picot, L.; Sitarek, P. Anti-Inflammatory and Anticancer Effects of Anthocyanins in In Vitro and In Vivo Studies. *Antioxidants* **2024**, *13*, 1143. [CrossRef] [PubMed]
38. Hoshyar, R.; Mahboob, Z.; Zarban, A. The antioxidant and chemical properties of Berberis vulgaris and its cytotoxic effect on human breast carcinoma cells. *Cytotechnology* **2016**, *68*, 1207–1213. [CrossRef]
39. Li, G.; Ding, K.; Qiao, Y.; Zhang, L.; Zheng, L.; Pan, T.; Zhang, L. Flavonoids regulate inflammation and oxidative stress in cancer. *Molecules* **2020**, *25*, 5628. [CrossRef]
40. Eskra, J.N.; Schlicht, M.J.; Bosland, M.C. Effects of black raspberries and their ellagic acid and anthocyanin constituents on taxane chemotherapy of castration-resistant prostate cancer cells. *Sci. Rep.* **2019**, *9*, 4367. [CrossRef]
41. Mazewski, C.; Kim, M.S.; Gonzalez de Mejia, E. Anthocyanins, delphinidin-3-O-glucoside and cyanidin-3-O-glucoside, inhibit immune checkpoints in human colorectal cancer cells in vitro and in silico. *Sci. Rep.* **2019**, *9*, 11560. [CrossRef]
42. Kowalczyk, A.; Tuberoso, C.I.; Jerković, I. The Role of Rosmarinic Acid in Cancer Prevention and Therapy: Mechanisms of Antioxidant and Anticancer Activity. *Antioxidants* **2024**, *13*, 1313. [CrossRef]

43. Faria, A.; Pestana, D.; Teixeira, D.; de Freitas, V.; Mateus, N.; Calhau, C. Blueberry anthocyanins and pyruvic acid adducts: Anticancer properties in breast cancer cell lines. *Phytother. Res.* **2010**, *24*, 1862–1869. [CrossRef]
44. Hui, C.; Bin, Y.; Xiaoping, Y.; Long, Y.; Chunye, C.; Mantian, M.; Wenhua, L. Anticancer activities of an anthocyanin-rich extract from black rice against breast cancer cells in vitro and in vivo. *Nutr. Cancer* **2010**, *62*, 1128–1136. [CrossRef] [PubMed]
45. Chen, X.Y.; Zhou, J.; Luo, L.P.; Han, B.; Li, F.; Chen, J.Y.; Zhu, Y.F.; Chen, W.; Yu, X.P. Black rice anthocyanins suppress metastasis of breast cancer cells by targeting RAS/RAF/MAPK pathway. *BioMed Res. Int.* **2015**, *2015*, 414250. [CrossRef]
46. Lim, S.; Xu, J.; Kim, J.; Chen, T.Y.; Su, X.; Standard, J.; Carey, E.; Griffin, J.; Herndon, B.; Katz, B.; et al. Role of anthocyanin-enriched purple-fleshed sweet potato p40 in colorectal cancer prevention. *Mol. Nutr. Food Res.* **2013**, *57*, 1908–1917. [CrossRef]
47. Jang, H.; Ha, U.S.; Kim, S.J.; Yoon, B.I.; Han, D.S.; Yuk, S.M.; Kim, S.W. Anthocyanin extracted from black soybean reduces prostate weight and promotes apoptosis in the prostatic hyperplasia-induced rat model. *J. Agric. Food Chem.* **2010**, *58*, 12686–12691. [CrossRef] [PubMed]
48. Bontempo, P.; De Masi, L.; Carafa, V.; Rigano, D.; Scisciola, L.; Iside, C.; Grassi, R.; Molinari, A.M.; Aversano, R.; Nebbioso, A.; et al. Anticancer activities of anthocyanin extract from genotyped Solanum tuberosum L."Vitelotte". *J. Funct. Foods* **2015**, *19*, 584–593. [CrossRef]
49. Martin, M.Á.; Ramos, S. Impact of cocoa flavanols on human health. *Food Chem. Toxicol.* **2021**, *151*, 112121. [CrossRef]
50. Joshi, R.; Kulkarni, Y.A.; Wairkar, S. Pharmacokinetic, pharmacodynamic and formulations aspects of Naringenin: An update. *Life Sci.* **2018**, *215*, 43–56. [CrossRef]
51. Wang, L.; Xie, Y.; Xiao, B.; He, X.; Ying, G.; Zha, H.; Yang, C.; Jin, X.; Li, G.; Ping, L.; et al. Isorhamnetin alleviates cisplatin-induced acute kidney injury via enhancing fatty acid oxidation. *Free Radic. Biol. Med.* **2024**, *212*, 22–33. [CrossRef]
52. Ismaeel, A.; McDermott, M.M.; Joshi, J.K.; Sturgis, J.C.; Zhang, D.; Ho, K.J.; Sufit, R.; Ferrucci, L.; Peterson, C.A.; Kosmac, K. Cocoa flavanols, Nrf2 activation, and oxidative stress in peripheral artery disease: Mechanistic findings in muscle based on outcomes from a randomized trial. *Am. J. Physiol.-Cell Physiol.* **2024**, *326*, C589–C605. [CrossRef]
53. Vogiatzoglou, A.; Mulligan, A.A.; Luben, R.N.; Lentjes, M.A.; Heiss, C.; Kelm, M.; Merx, M.W.; Spencer, J.P.; Schroeter, H.; Kuhnle, G.G. Assessment of the dietary intake of total flavan-3-ols, monomeric flavan-3-ols, proanthocyanidins and theaflavins in the European Union. *Br. J. Nutr.* **2014**, *111*, 1463–1473. [CrossRef]
54. Georgiou, N.; Kakava, M.G.; Routsi, E.A.; Petsas, E.; Stavridis, N.; Freris, C.; Zoupanou, N.; Moschovou, K.; Kiriakidi, S.; Mavromoustakos, T. Quercetin: A potential polydynamic drug. *Molecules* **2023**, *28*, 8141. [CrossRef]
55. Bondonno, N.P.; Bondonno, C.P.; Ward, N.C.; Woodman, R.J.; Hodgson, J.M.; Croft, K.D. Enzymatically modified isoquercitrin improves endothelial function in volunteers at risk of cardiovascular disease. *Br. J. Nutr.* **2020**, *123*, 182–189. [CrossRef]
56. Kim, J.K.; Park, S.U. Quercetin and its role in biological functions: An updated review. *EXCLI J.* **2018**, *17*, 856. [PubMed]
57. Yang, D.; Wang, T.; Long, M.; Li, P. Quercetin: Its main pharmacological activity and potential application in clinical medicine. *Oxidative Med. Cell. Longev.* **2020**, *2020*, 8825387. [CrossRef] [PubMed]
58. Singh, P.; Arif, Y.; Bajguz, A.; Hayat, S. The role of quercetin in plants. *Plant Physiol. Biochem.* **2021**, *166*, 10–19. [CrossRef] [PubMed]
59. Amanzadeh, E.; Esmaeili, A.; Rahgozar, S.; Nourbakhshnia, M. Application of quercetin in neurological disorders: From nutrition to nanomedicine. *Rev. Neurosci.* **2019**, *30*, 555–572. [CrossRef]
60. Sturza, A.; Pavel, I.; Ancușa, S.; Danciu, C.; Dehelean, C.; Duicu, O.; Muntean, D. Quercetin exerts an inhibitory effect on cellular bioenergetics of the B164A5 murine melanoma cell line. *Mol. Cell. Biochem.* **2018**, *447*, 103–109. [CrossRef]
61. Pham, T.N.; Stempel, S.; Shields, M.A.; Spaulding, C.; Kumar, K.; Bentrem, D.J.; Matsangou, M.; Munshi, H.G. Quercetin enhances the anti-tumor effects of BET inhibitors by suppressing hnRNPA1. *Int. J. Mol. Sci.* **2019**, *20*, 4293. [CrossRef]
62. Niazvand, F.; Orazizadeh, M.; Khorsandi, L.; Abbaspour, M.; Mansouri, E.; Khodadadi, A. Effects of quercetin-loaded nanoparticles on MCF-7 human breast cancer cells. *Medicina* **2019**, *55*, 114. [CrossRef]
63. Wu, L.; Li, J.; Liu, T.; Li, S.; Feng, J.; Yu, Q.; Zhang, J.; Chen, J.; Zhou, Y.; Ji, J.; et al. Quercetin shows anti-tumor effect in hepatocellular carcinoma LM3 cells by abrogating JAK2/STAT3 signaling pathway. *Cancer Med.* **2019**, *8*, 4806–4820. [CrossRef]
64. Hisaka, T.; Sakai, H.; Sato, T.; Goto, Y.; Nomura, Y.; Fukutomi, S.; Fujita, F.; Mizobe, T.; Nakashima, O.; Tanigawa, M.; et al. Quercetin suppresses proliferation of liver cancer cell lines in vitro. *Anticancer Res.* **2020**, *40*, 4695–4700. [CrossRef]
65. Erdogan, S.; Turkekul, K.; Dibirdik, I.; Doganlar, O.; Doganlar, Z.B.; Bilir, A.; Oktem, G. Midkine downregulation increases the efficacy of quercetin on prostate cancer stem cell survival and migration through PI3K/AKT and MAPK/ERK pathway. *Biomed. Pharmacother.* **2018**, *107*, 793–805. [CrossRef] [PubMed]
66. Ward, A.B.; Mir, H.; Kapur, N.; Gales, D.N.; Carriere, P.P.; Singh, S. Quercetin inhibits prostate cancer by attenuating cell survival and inhibiting anti-apoptotic pathways. *World J. Surg. Oncol.* **2018**, *16*, 108. [CrossRef]
67. Jia, L.; Huang, S.; Yin, X.; Zan, Y.; Guo, Y.; Han, L. Quercetin suppresses the mobility of breast cancer by suppressing glycolysis through Akt-mTOR pathway mediated autophagy induction. *Life Sci.* **2018**, *208*, 123–130. [CrossRef] [PubMed]

68. Reyes-Farias, M.; Carrasco-Pozo, C. The anti-cancer effect of quercetin: Molecular implications in cancer metabolism. *Int. J. Mol. Sci.* **2019**, *20*, 3177. [CrossRef]
69. Granato, M.; Rizzello, C.; Montani, M.S.; Cuomo, L.; Vitillo, M.; Santarelli, R.; Gonnella, R.; D'Orazi, G.; Faggioni, A.; Cirone, M. Quercetin induces apoptosis and autophagy in primary effusion lymphoma cells by inhibiting PI3K/AKT/mTOR and STAT3 signaling pathways. *J. Nutr. Biochem.* **2017**, *41*, 124–136. [CrossRef]
70. Lei, C.S.; Hou, Y.C.; Pai, M.H.; Lin, M.T.; Yeh, S.L. Effects of quercetin combined with anticancer drugs on metastasis-associated factors of gastric cancer cells: In vitro and in vivo studies. *J. Nutr. Biochem.* **2018**, *51*, 105–113. [CrossRef] [PubMed]
71. Caro, C.; Pourmadadi, M.; Eshaghi, M.M.; Rahmani, E.; Shojaei, S.; Paiva-Santos, A.C.; Rahdar, A.; Behzadmehr, R.; García-Martín, M.L.; Díez-Pascual, A.M. Nanomaterials loaded with Quercetin as an advanced tool for cancer treatment. *J. Drug Deliv. Sci. Technol.* **2022**, *78*, 103938. [CrossRef]
72. Bakun, P.; Mlynarczyk, D.T.; Koczorowski, T.; Cerbin-Koczorowska, M.; Piwowarczyk, L.; Kolasiński, E.; Stawny, M.; Kuźmińska, J.; Jelińska, A.; Goslinski, T. Tea-break with epigallocatechin gallate derivatives–Powerful polyphenols of great potential for medicine. *Eur. J. Med. Chem.* **2023**, *261*, 115820. [CrossRef]
73. Zhang, B.Y.; Shi, Y.Q.; Chen, X.; Dai, J.; Jiang, Z.F.; Li, N.; Zhang, Z.B. Protective effect of curcumin against formaldehyde-induced genotoxicity in A549 Cell Lines. *J. Appl. Toxicol.* **2013**, *33*, 1468–1473. [CrossRef]
74. Lee, J.Y.; Kim, H.S.; Song, Y.S. Genistein as a potential anticancer agent against ovarian cancer. *J. Tradit. Complement. Med.* **2012**, *2*, 96–104. [CrossRef] [PubMed]
75. Legeay, S.; Rodier, M.; Fillon, L.; Faure, S.; Clere, N. Epigallocatechin gallate: A review of its beneficial properties to prevent metabolic syndrome. *Nutrients* **2015**, *7*, 5443–5468. [CrossRef] [PubMed]
76. Fujiki, H.; Sueoka, E.; Rawangkan, A.; Suganuma, M. Human cancer stem cells are a target for cancer prevention using (−)-epigallocatechin gallate. *J. Cancer Res. Clin. Oncol.* **2017**, *143*, 2401–2412. [CrossRef]
77. Yu, C.; Jiao, Y.; Xue, J.; Zhang, Q.; Yang, H.; Xing, L.; Chen, G.; Wu, J.; Zhang, S.; Zhu, W.; et al. Metformin sensitizes non-small cell lung cancer cells to an epigallocatechin-3-gallate (EGCG) treatment by suppressing the Nrf2/HO-1 signaling pathway. *Int. J. Biol. Sci.* **2017**, *13*, 1560. [CrossRef] [PubMed]
78. Wei, R.; Cortez Penso, N.E.; Hackman, R.M.; Wang, Y.; Mackenzie, G.G. Epigallocatechin-3-gallate (EGCG) suppresses pancreatic cancer cell growth, invasion, and migration partly through the inhibition of Akt pathway and epithelial–mesenchymal transition: Enhanced efficacy when combined with gemcitabine. *Nutrients* **2019**, *11*, 1856. [CrossRef]
79. Aggarwal, V.; Tuli, H.S.; Tania, M.; Srivastava, S.; Ritzer, E.E.; Pandey, A.; Aggarwal, D.; Barwal, T.S.; Jain, A.; Kaur, G.; et al. Molecular mechanisms of action of epigallocatechin gallate in cancer: Recent trends and advancement. In *Seminars in Cancer Biology*; Academic Press: Cambridge, MA, USA, 2022; Volume 80, pp. 256–275.
80. Hayakawa, S.; Ohishi, T.; Miyoshi, N.; Oishi, Y.; Nakamura, Y.; Isemura, M. Anti-cancer effects of green tea epigallocatchin-3-gallate and coffee chlorogenic acid. *Molecules* **2020**, *25*, 4553. [CrossRef]
81. Ohishi, T.; Hayakawa, S.; Miyoshi, N. Involvement of microRNA modifications in anticancer effects of major polyphenols from green tea, coffee, wine, and curry. *Crit. Rev. Food Sci. Nutr.* **2023**, *63*, 7148–7179. [CrossRef]
82. Bimonte, S.; Albino, V.; Piccirillo, M.; Nasto, A.; Molino, C.; Palaia, R.; Cascella, M. Epigallocatechin-3-gallate in the prevention and treatment of hepatocellular carcinoma: Experimental findings and translational perspectives. *Drug Des. Dev. Ther.* **2019**, *12*, 611–621. [CrossRef]
83. Won, H.R.; Lee, P.; Oh, S.R.; Kim, Y.M. Epigallocatechin-3-gallate suppresses the expression of TNF-α-induced MMP-1 via MAPK/ERK signaling pathways in human dermal fibroblasts. *Biol. Pharm. Bull.* **2021**, *44*, 18–24. [CrossRef]
84. Wang, X.; Liu, Y.; Ding, Y.; Feng, G. CAMSAP2 promotes colorectal cancer cell migration and invasion through activation of JNK/c-Jun/MMP-1 signaling pathway. *Sci. Rep.* **2022**, *12*, 16899. [CrossRef]
85. Fang, C.Y.; Wu, C.C.; Hsu, H.Y.; Chuang, H.Y.; Huang, S.Y.; Tsai, C.H.; Chang, Y.; Tsao, G.S.; Chen, C.L.; Chen, J.Y. EGCG inhibits proliferation, invasiveness and tumor growth by up-regulation of adhesion molecules, suppression of gelatinases activity, and induction of apoptosis in nasopharyngeal carcinoma cells. *Int. J. Mol. Sci.* **2015**, *16*, 2530–2558. [CrossRef]
86. Djerir, D.; Iddir, M.; Bourgault, S.; Lamy, S.; Annabi, B. Biophysical evidence for differential gallated green tea catechins binding to membrane type-1 matrix metalloproteinase and its interactors. *Biophys. Chem.* **2018**, *234*, 34–41. [CrossRef] [PubMed]
87. Sharma, C.; Nusri Qel, A.; Begum, S.; Javed, E.; Rizvi, T.A.; Hussain, A. (-)-Epigallocatechin-3-gallate induces apoptosis and inhibits invasion and migration of human cervical cancer cells. *Asian Pac. J. Cancer Prev.* **2012**, *13*, 4815–4822. [CrossRef] [PubMed]
88. Park, S.Y.; Jung, C.H.; Song, B.; Park, O.J.; Kim, Y.M. Pro-apoptotic and migration-suppressing potential of EGCG, and the involvement of AMPK in the p53-mediated modulation of VEGF and MMP-9 expression. *Oncol. Lett.* **2013**, *6*, 1346–1350. [CrossRef]
89. Harakeh, S.; Diab-Assaf, M.; Azar, R.; Hassan, H.M.; Tayeb, S.; Abou-El-Ardat, K.; Damanhouri, G.A.; Qadri, I.; Abuzenadah, A.; Chaudhary, A.; et al. Epigallocatechin-3-gallate inhibits tax-dependent activation of nuclear factor kappa B and of matrix

metalloproteinase 9 in human T-cell lymphotropic virus-1 positive leukemia cells. *Asian Pac. J. Cancer Prev.* **2014**, *15*, 1219–1225. [CrossRef] [PubMed]
90. Luo, K.W.; Chen, W.; Lung, W.Y.; Wei, X.Y.; Cheng, B.H.; Cai, Z.M.; Huang, W.R. EGCG inhibited bladder cancer SW780 cell proliferation and migration both in vitro and in vivo via down-regulation of NF-κB and MMP-9. *J. Nutr. Biochem.* **2017**, *41*, 56–64. [CrossRef]
91. Zhang, G.; Wang, Y.; Zhang, Y.; Wan, X.; Li, J.; Liu, K.; Wang, F.; Liu, K.; Liu, Q.; Yang, C.; et al. Anti-cancer activities of tea epigallocatechin-3-gallate in breast cancer patients under radiotherapy. *Curr. Mol. Med.* **2012**, *12*, 163–176. [CrossRef]
92. Zhao, H.; Zhu, W.; Xie, P.; Li, H.; Zhang, X.; Sun, X.; Yu, J.; Xing, L. A phase I study of concurrent chemotherapy and thoracic radiotherapy with oral epigallocatechin-3-gallate protection in patients with locally advanced stage III non-small-cell lung cancer. *Radiother. Oncol.* **2014**, *110*, 132–136. [CrossRef] [PubMed]
93. Tanabe, H.; Suzuki, T.; Ohishi, T.; Isemura, M.; Nakamura, Y.; Unno, K. Effects of epigallocatechin-3-gallate on matrix metalloproteinases in terms of its anticancer activity. *Molecules* **2023**, *28*, 525. [CrossRef]
94. Imran, M.; Rauf, A.; Abu-Izneid, T.; Nadeem, M.; Shariati, M.A.; Khan, I.A.; Imran, A.; Orhan, I.E.; Rizwan, M.; Atif, M.; et al. Luteolin, a flavonoid, as an anticancer agent: A review. *Biomed. Pharmacother.* **2019**, *112*, 108612. [CrossRef]
95. Rauf, A.; Wilairatana, P.; Joshi, P.B.; Ahmad, Z.; Olatunde, A.; Hafeez, N.; Hemeg, H.A.; Mubarak, M.S. Revisiting luteolin: An updated review on its anticancer potential. *Heliyon* **2024**, *10*, e26701. [CrossRef] [PubMed]
96. Hussain, Y.; Cui, J.H.; Khan, H.; Aschner, M.; Batiha, G.E.; Jeandet, P. Luteolin and cancer metastasis suppression: Focus on the role of epithelial to mesenchymal transition. *Med. Oncol.* **2021**, *38*, 66. [CrossRef] [PubMed]
97. Huang, L.; Jin, K.; Lan, H. Luteolin inhibits cell cycle progression and induces apoptosis of breast cancer cells through downregulation of human telomerase reverse transcriptase. *Oncol. Lett.* **2019**, *17*, 3842–3850. [CrossRef] [PubMed]
98. Sui, J.Q.; Xie, K.P.; Xie, M.J. Inhibitory effect of luteolin on the proliferation of human breast cancer cell lines induced by epidermal growth factor. *Sheng Li Xue Bao [Acta Physiol. Sin.]* **2016**, *68*, 27–34.
99. Lo, S.; Leung, E.; Fedrizzi, B.; Barker, D. Syntheses of mono-acylated luteolin derivatives, evaluation of their antiproliferative and radical scavenging activities and implications on their oral bioavailability. *Sci. Rep.* **2021**, *11*, 12595. [CrossRef]
100. Chen, Z.; Zhang, B.; Gao, F.; Shi, R. Modulation of G2/M cell cycle arrest and apoptosis by luteolin in human colon cancer cells and xenografts. *Oncol. Lett.* **2018**, *15*, 1559–1565. [CrossRef]
101. Yoo, H.S.; Won, S.B.; Kwon, Y.H. Luteolin induces apoptosis and autophagy in HCT116 colon cancer cells via p53-dependent pathway. *Nutr. Cancer* **2022**, *74*, 677–686. [CrossRef]
102. Kang, K.A.; Piao, M.J.; Hyun, Y.J.; Zhen, A.X.; Cho, S.J.; Ahn, M.J.; Yi, J.M.; Hyun, J.W. Luteolin promotes apoptotic cell death via upregulation of Nrf2 expression by DNA demethylase and the interaction of Nrf2 with p53 in human colon cancer cells. *Exp. Mol. Med.* **2019**, *51*, 1–14. [CrossRef]
103. Song, Y.; Yu, J.; Li, L.; Wang, L.; Dong, L.; Xi, G.; Lu, Y.J.; Li, Z. Luteolin impacts deoxyribonucleic acid repair by modulating the mitogen-activated protein kinase pathway in colorectal cancer. *Bioengineered* **2022**, *13*, 10998–11011. [CrossRef]
104. Pu, Y.; Zhang, T.; Wang, J.; Mao, Z.; Duan, B.; Long, Y.; Xue, F.; Liu, D.; Liu, S.; Gao, Z. Luteolin exerts an anticancer effect on gastric cancer cells through multiple signaling pathways and regulating miRNAs. *J. Cancer* **2018**, *9*, 3669. [CrossRef]
105. Lu, J.; Li, G.; He, K.; Jiang, W.; Xu, C.; Li, Z.; Wang, H.; Wang, W.; Wang, H.; Teng, X.; et al. Luteolin exerts a marked antitumor effect in cMet-overexpressing patient-derived tumor xenograft models of gastric cancer. *J. Transl. Med.* **2015**, *13*, 42. [CrossRef] [PubMed]
106. Zang, M.D.; Hu, L.; Fan, Z.Y.; Wang, H.X.; Zhu, Z.L.; Cao, S.; Wu, X.Y.; Li, J.F.; Su, L.P.; Li, C.; et al. Luteolin suppresses gastric cancer progression by reversing epithelial-mesenchymal transition via suppression of the Notch signaling pathway. *J. Transl. Med.* **2017**, *15*, 52. [CrossRef]
107. Song, S.; Su, Z.; Xu, H.; Niu, M.; Chen, X.; Min, H.; Zhang, B.; Sun, G.; Xie, S.; Wang, H.; et al. Luteolin selectively kills STAT3 highly activated gastric cancer cells through enhancing the binding of STAT3 to SHP-1. *Cell Death Dis.* **2017**, *8*, e2612. [CrossRef]
108. Prasher, P.; Sharma, M.; Singh, S.K.; Gulati, M.; Chellappan, D.K.; Zacconi, F.; De Rubis, G.; Gupta, G.; Sharifi-Rad, J.; Cho, W.C.; et al. Luteolin: A flavonoid with a multifaceted anticancer potential. *Cancer Cell Int.* **2022**, *22*, 386. [CrossRef]
109. Cai, X.; Ye, T.; Liu, C.; Lu, W.; Lu, M.; Zhang, J.; Wang, M.; Cao, P. Luteolin induced G2 phase cell cycle arrest and apoptosis on non-small cell lung cancer cells. *Toxicol. Vitr.* **2011**, *25*, 1385–1391. [CrossRef]
110. Meng, G.; Chai, K.; Li, X.; Zhu, Y.; Huang, W. Luteolin exerts pro-apoptotic effect and anti-migration effects on A549 lung adenocarcinoma cells through the activation of MEK/ERK signaling pathway. *Chem.-Biol. Interact.* **2016**, *257*, 26–34. [CrossRef] [PubMed]
111. Park, S.H.; Park, H.S.; Lee, J.H.; Chi, G.Y.; Kim, G.Y.; Moon, S.K.; Chang, Y.C.; Hyun, J.W.; Kim, W.J.; Choi, Y.H. Induction of endoplasmic reticulum stress-mediated apoptosis and non-canonical autophagy by luteolin in NCI-H460 lung carcinoma cells. *Food Chem. Toxicol.* **2013**, *56*, 100–109. [CrossRef]

112. Ma, L.; Peng, H.; Li, K.; Zhao, R.; Li, L.; Yu, Y.; Wang, X.; Han, Z. Luteolin exerts an anticancer effect on NCI-H460 human non-small cell lung cancer cells through the induction of Sirt1-mediated apoptosis. *Mol. Med. Rep.* **2015**, *12*, 4196–4202. [CrossRef] [PubMed]
113. Zhang, M.; Wang, R.; Tian, J.; Song, M.; Zhao, R.; Liu, K.; Zhu, F.; Shim, J.H.; Dong, Z.; Lee, M.H. Targeting LIMK1 with luteolin inhibits the growth of lung cancer in vitro and in vivo. *J. Cell. Mol. Med.* **2021**, *25*, 5560–5571. [CrossRef]
114. Dallavalasa, S.; Beeraka, N.M.; Basavaraju, C.G.; Tulimilli, S.V.; Sadhu, S.P.; Rajesh, K.; Aliev, G.; Madhunapantula, S.V. The role of tumor associated macrophages (TAMs) in cancer progression, chemoresistance, angiogenesis and metastasis-current status. *Curr. Med. Chem.* **2021**, *28*, 8203–8236. [CrossRef]
115. Choi, H.J.; Choi, H.J.; Chung, T.W.; Ha, K.T. Luteolin inhibits recruitment of monocytes and migration of Lewis lung carcinoma cells by suppressing chemokine (C–C motif) ligand 2 expression in tumor-associated macrophage. *Biochem. Biophys. Res. Commun.* **2016**, *470*, 101–106. [CrossRef] [PubMed]
116. Cho, H.J.; Ahn, K.C.; Choi, J.Y.; Hwang, S.G.; Kim, W.J.; Um, H.D.; Park, J.K. Luteolin acts as a radiosensitizer in non-small cell lung cancer cells by enhancing apoptotic cell death through activation of a p38/ROS/caspase cascade. *Int. J. Oncol.* **2015**, *46*, 1149–1158. [CrossRef]
117. Tu, D.G.; Lin, W.T.; Yu, C.C.; Lee, S.S.; Peng, C.Y.; Lin, T.; Yu, C.H. Chemotherapeutic effects of luteolin on radio-sensitivity enhancement and interleukin-6/signal transducer and activator of transcription 3 signaling repression of oral cancer stem cells. *J. Formos. Med. Assoc.* **2016**, *115*, 1032–1038. [CrossRef] [PubMed]
118. George, V.C.; Kumar, D.R.; Suresh, P.K.; Kumar, S.; Kumar, R.A. Comparative studies to evaluate relative in vitro potency of luteolin in inducing cell cycle arrest and apoptosis in HaCaT and A375 cells. *Asian Pac. J. Cancer Prev.* **2013**, *14*, 631–637. [CrossRef] [PubMed]
119. Moeng, S.; Son, S.W.; Seo, H.A.; Lee, J.S.; Kim, C.K.; Kuh, H.J.; Park, J.K. Luteolin-regulated MicroRNA-301-3p targets caspase-8 and modulates TRAIL sensitivity in PANC-1 cells. *Anticancer Res.* **2020**, *40*, 723–731. [CrossRef]
120. Kato, H.; Naiki-Ito, A.; Suzuki, S.; Inaguma, S.; Komura, M.; Nakao, K.; Naiki, T.; Kachi, K.; Kato, A.; Matsuo, Y.; et al. DPYD, down-regulated by the potentially chemopreventive agent luteolin, interacts with STAT3 in pancreatic cancer. *Carcinogenesis* **2021**, *42*, 940–950. [CrossRef]
121. Li, Z.; Zhang, Y.; Chen, L.; Li, H. The dietary compound luteolin inhibits pancreatic cancer growth by targeting BCL-2. *Food Funct.* **2018**, *9*, 3018–3027. [CrossRef]
122. Cao, Z.; Zhang, H.; Cai, X.; Fang, W.; Chai, D.; Wen, Y.; Chen, H.; Chu, F.; Zhang, Y. Luteolin promotes cell apoptosis by inducing autophagy in hepatocellular carcinoma. *Cell. Physiol. Biochem.* **2018**, *43*, 1803–1812. [CrossRef]
123. Nazim, U.M.; Park, S.Y. Luteolin sensitizes human liver cancer cells to TRAIL-induced apoptosis via autophagy and JNK-mediated death receptor 5 upregulation. *Int. J. Oncol.* **2019**, *54*, 665–672. [CrossRef]
124. Wang, C.; Li, Q.; Xiao, B.; Fang, H.; Huang, B.; Huang, F.; Wang, Y. Luteolin enhances the antitumor efficacy of oncolytic vaccinia virus that harbors IL-24 gene in liver cancer cells. *J. Clin. Lab. Anal.* **2021**, *35*, e23677. [CrossRef]
125. Majumdar, D.; Jung, K.H.; Zhang, H.; Nannapaneni, S.; Wang, X.; Amin, A.R.; Chen, Z.; Chen, Z.G.; Shin, D.M. Luteolin nanoparticle in chemoprevention: In vitro and in vivo anticancer activity. *Cancer Prev. Res.* **2014**, *7*, 65–73. [CrossRef]
126. Sohel, M.; Sultana, H.; Sultana, T.; Al Amin, M.; Aktar, S.; Ali, M.C.; Rahim, Z.B.; Hossain, M.A.; Al Mamun, A.; Amin, M.N.; et al. Chemotherapeutic potential of hesperetin for cancer treatment, with mechanistic insights: A comprehensive review. *Heliyon* **2022**, *8*, e08815. [CrossRef] [PubMed]
127. Babukumar, S.; Vinothkumar, V.; Ramachandhiran, D. Modulating effect of hesperetin on the molecular expression pattern of apoptotic and cell proliferative markers in 7, 12-dimethylbenz (a) anthracene-induced oral carcinogenesis. *Arch. Physiol. Biochem.* **2020**, *126*, 430–439. [CrossRef] [PubMed]
128. Chai, T.; Mohan, M.; Ong, H.; Wong, F. Antioxidant, iron-chelating and anti-glucosidase activities of Typha domingensis Pers (Typhaceae). *Trop. J. Pharm. Res.* **2014**, *13*, 67–72. [CrossRef]
129. Lee, J.; Kim, D.H.; Kim, J.H. Combined administration of naringenin and hesperetin with optimal ratio maximizes the anti-cancer effect in human pancreatic cancer via down regulation of FAK and p38 signaling pathway. *Phytomedicine* **2019**, *58*, 152762. [CrossRef]
130. Shirzad, M.; Beshkar, P.; Heidarian, E. The effects of hesperetin on apoptosis induction and inhibition of cell proliferation in the prostate cancer PC3 cells. *J. HerbMed Pharmacol.* **2015**, *4*, 121–124.
131. Li, Q.; Miao, Z.; Wang, R.; Yang, J.; Zhang, D. Hesperetin induces apoptosis in human glioblastoma cells via p38 MAPK activation. *Nutr. Cancer* **2020**, *72*, 538–545. [CrossRef]
132. Aboismaiel, M.G.; El-Mesery, M.; El-Karef, A.; El-Shishtawy, M.M. Hesperetin upregulates Fas/FasL expression and potentiates the antitumor effect of 5-fluorouracil in rat model of hepatocellular carcinoma. *Egypt. J. Basic Appl. Sci.* **2020**, *7*, 20–34. [CrossRef]

133. Chen, X.; Wei, W.; Li, Y.; Huang, J.; Ci, X. Hesperetin relieves cisplatin-induced acute kidney injury by mitigating oxidative stress, inflammation and apoptosis. *Chem.-Biol. Interact.* **2019**, *308*, 269–278. [CrossRef]
134. Nalini, N.; Aranganathan, S.; Kabalimurthy, J. Chemopreventive efficacy of hesperetin (citrus flavonone) against 1, 2-dimethylhydrazine-induced rat colon carcinogenesis. *Toxicol. Mech. Methods* **2012**, *22*, 397–408. [CrossRef]
135. Elango, R.; Athinarayanan, J.; Subbarayan, V.P.; Lei, D.K.; Alshatwi, A.A. Hesperetin induces an apoptosis-triggered extrinsic pathway and a p53-independent pathway in human lung cancer H522 cells. *J. Asian Nat. Prod. Res.* **2018**, *20*, 559–569. [CrossRef] [PubMed]
136. Gohulkumar, M.; Gurushankar, K.; Prasad, N.R.; Krishnakumar, N. Enhanced cytotoxicity and apoptosis-induced anticancer effect of silibinin-loaded nanoparticles in oral carcinoma (KB) cells. *Mater. Sci. Eng. C* **2014**, *41*, 274–282. [CrossRef]
137. Wu, D.; Zhang, J.; Wang, J.; Li, J.; Liao, F.; Dong, W. Hesperetin induces apoptosis of esophageal cancer cells via mitochondrial pathway mediated by the increased intracellular reactive oxygen species. *Tumor Biol.* **2016**, *37*, 3451–3459. [CrossRef] [PubMed]
138. Coutinho, L.; Oliveira, H.; Pacheco, A.R.; Almeida, L.; Pimentel, F.; Santos, C.; de Oliveira, J.M. Hesperetin-etoposide combinations induce cytotoxicity in U2OS cells: Implications on therapeutic developments for osteosarcoma. *DNA Repair* **2017**, *50*, 36–42. [CrossRef]
139. Mistry, B.; Patel, R.V.; Keum, Y.S. Access to the substituted benzyl-1, 2, 3-triazolyl hesperetin derivatives expressing antioxidant and anticancer effects. *Arab. J. Chem.* **2017**, *10*, 157–166. [CrossRef]
140. Patel, P.N.; Yu, X.M.; Jaskula-Sztul, R.; Chen, H. Hesperetin activates the Notch1 signaling cascade, causes apoptosis, and induces cellular differentiation in anaplastic thyroid cancer. *Ann. Surg. Oncol.* **2014**, *21*, 497–504. [CrossRef] [PubMed]
141. Adan, A.; Baran, Y. The pleiotropic effects of fisetin and hesperetin on human acute promyelocytic leukemia cells are mediated through apoptosis, cell cycle arrest, and alterations in signaling networks. *Tumor Biol.* **2015**, *36*, 8973–8984. [CrossRef]
142. Hermawan, A.; Ikawati, M.; Khumaira, A.; Putri, H.; Jenie, R.I.; Angraini, S.M.; Muflikhasari, H.A. Bioinformatics and in vitro studies reveal the importance of p53, PPARG and notch signaling pathway in inhibition of breast cancer stem cells by hesperetin. *Adv. Pharm. Bull.* **2020**, *11*, 351. [CrossRef]
143. Tan, Y.Q.; Chiu-Leung, L.C.; Lin, S.M.; Leung, L.K. The citrus flavonone hesperetin attenuates the nuclear translocation of aryl hydrocarbon receptor. *Comp. Biochem. Physiol. Part C Toxicol. Pharmacol.* **2018**, *210*, 57–64. [CrossRef]
144. Ye, L.; Chan, F.L.; Chen, S.; Leung, L.K. The citrus flavonone hesperetin inhibits growth of aromatase-expressing MCF-7 tumor in ovariectomized athymic mice. *J. Nutr. Biochem.* **2012**, *23*, 1230–1237. [CrossRef]
145. Nurhayati, I.P.; Khumaira, A.; Ilmawati, G.P.; Meiyanto, E.; Hermawan, A. Cytotoxic and antimetastatic activity of hesperetin and doxorubicin combination toward Her2 expressing breast cancer cells. *Asian Pac. J. Cancer Prev. APJCP* **2020**, *21*, 1259. [CrossRef] [PubMed]
146. Palit, S.; Kar, S.; Sharma, G.; Das, P.K. Hesperetin induces apoptosis in breast carcinoma by triggering accumulation of ROS and activation of ASK1/JNK pathway. *J. Cell. Physiol.* **2015**, *230*, 1729–1739. [CrossRef] [PubMed]
147. Yang, Y.; Wolfram, J.; Boom, K.; Fang, X.; Shen, H.; Ferrari, M. Hesperetin impairs glucose uptake and inhibits proliferation of breast cancer cells. *Cell Biochem. Funct.* **2013**, *31*, 374–379. [CrossRef] [PubMed]
148. Chandrika, B.B.; Steephan, M.; Kumar, T.S.; Sabu, A.; Haridas, M. Hesperetin and naringenin sensitize HER2 positive cancer cells to death by serving as HER2 tyrosine kinase inhibitors. *Life Sci.* **2016**, *160*, 47–56. [CrossRef]
149. Yunita, E.; Muflikhasari, H.A.; Ilmawati, G.P.; Meiyanto, E.; Hermawan, A. Hesperetin alleviates doxorubicin-induced migration in 4T1 breast cancer cells. *Future J. Pharm. Sci.* **2020**, *6*, 23. [CrossRef]
150. Shirzad, M.; Heidarian, E.; Beshkar, P.; Gholami-Arjenaki, M. Biological effects of hesperetin on interleukin-6/phosphorylated signal transducer and activator of transcription 3 pathway signaling in prostate cancer PC3 cells. *Pharmacogn. Res.* **2017**, *9*, 188.
151. Arya, A.; Khandelwal, K.; Ahmad, H.; Laxman, T.S.; Sharma, K.; Mittapelly, N.; Agrawal, S.; Bhatta, R.S.; Dwivedi, A.K. Co-delivery of hesperetin enhanced bicalutamide induced apoptosis by exploiting mitochondrial membrane potential via polymeric nanoparticles in a PC-3 cell line. *RSC Adv.* **2016**, *6*, 5925–5935. [CrossRef]
152. Wolfram, J.; Scott, B.; Boom, K.; Shen, J.; Borsoi, C.; Suri, K.; Grande, R.; Fresta, M.; Celia, C.; Zhao, Y.; et al. Hesperetin liposomes for cancer therapy. *Curr. Drug Deliv.* **2016**, *13*, 711–719. [CrossRef]
153. Tamayo, L.V.; Gouvea, L.R.; Sousa, A.C.; Albuquerque, R.M.; Teixeira, S.F.; de Azevedo, R.A.; Louro, S.R.; Ferreira, A.K.; Beraldo, H. Copper (II) complexes with naringenin and hesperetin: Cytotoxic activity against A 549 human lung adenocarcinoma cells and investigation on the mode of action. *Biometals* **2016**, *29*, 39–52. [CrossRef]
154. Ramteke, P.; Yadav, U. Hesperetin, a Citrus bioflavonoid, prevents IL-1β-induced inflammation and cell proliferation in lung epithelial A549 cells. *Indian J. Exp. Biol.* **2019**, *57*, 7–14.
155. Bodduluru, L.N.; Kasala, E.R.; Barua, C.C.; Karnam, K.C.; Dahiya, V.; Ellutla, M. Antiproliferative and antioxidant potential of hesperetin against benzo (a) pyrene-induced lung carcinogenesis in Swiss albino mice. *Chem.-Biol. Interact.* **2015**, *242*, 345–352. [CrossRef] [PubMed]

156. Zhang, J.; Song, J.; Wu, D.; Wang, J.; Dong, W. Hesperetin induces the apoptosis of hepatocellular carcinoma cells via mitochondrial pathway mediated by the increased intracellular reactive oxygen species, ATP and calcium. *Med. Oncol.* **2015**, *32*, 101, Erratum in: *Med. Oncol.* **2019**, *36*, 38. https://doi.org/10.1007/s12032-019-1258-0. [CrossRef] [PubMed]
157. Jung, K.Y.; Park, J.; Han, Y.S.; Lee, Y.H.; Shin, S.Y.; Lim, Y. Synthesis and biological evaluation of hesperetin derivatives as agents inducing apoptosis. *Bioorg. Med. Chem.* **2017**, *25*, 397–407. [CrossRef] [PubMed]
158. Sivagami, G.; Vinothkumar, R.; Preethy, C.P.; Riyasdeen, A.; Akbarsha, M.A.; Menon, V.P.; Nalini, N. Role of hesperetin (a natural flavonoid) and its analogue on apoptosis in HT-29 human colon adenocarcinoma cell line–A comparative study. *Food Chem. Toxicol.* **2012**, *50*, 660–671. [CrossRef]
159. Miler, M.; Živanović, J.; Ajdžanović, V.; Oreščanin-Dušić, Z.; Milenković, D.; Konić-Ristić, A.; Blagojević, D.; Milošević, V.; Šošić-Jurjević, B. Citrus flavanones naringenin and hesperetin improve antioxidant status and membrane lipid compositions in the liver of old-aged Wistar rats. *Exp. Gerontol.* **2016**, *84*, 49–60. [CrossRef]
160. Kong, R.; Wang, N.; Luo, H.; Lu, J. Hesperetin mitigates bile duct ligation-induced liver fibrosis by inhibiting extracellular matrix and cell apoptosis via the TGF-β1/Smad pathway. *Curr. Mol. Med.* **2018**, *18*, 15–24. [CrossRef] [PubMed]
161. Chen, X.; Li, X.F.; Chen, Y.; Zhu, S.; Li, H.D.; Chen, S.Y.; Wang, J.N.; Pan, X.Y.; Bu, F.T.; Huang, C.; et al. Hesperetin derivative attenuates CCl4-induced hepatic fibrosis and inflammation by Gli-1-dependent mechanisms. *Int. Immunopharmacol.* **2019**, *76*, 105838. [CrossRef]
162. Kumar, M.; Dahiya, V.; Kasala, E.R.; Bodduluru, L.N.; Lahkar, M. The renoprotective activity of hesperetin in cisplatin induced nephrotoxicity in rats: Molecular and biochemical evidence. *Biomed. Pharmacother.* **2017**, *89*, 1207–1215. [CrossRef]
163. Wang, H.W.; Shi, L.; Xu, Y.P.; Qin, X.Y.; Wang, Q.Z. Hesperetin alleviates renal interstitial fibrosis by inhibiting tubular epithelial-mesenchymal transition in vivo and in vitro. *Exp. Ther. Med.* **2017**, *14*, 3713–3719. [CrossRef]
164. Wang, S.W.; Sheng, H.; Zheng, F.; Zhang, F. Hesperetin promotes DOT1L degradation and reduces histone H3K79 methylation to inhibit gastric cancer metastasis. *Phytomedicine* **2021**, *84*, 153499. [CrossRef]
165. Zhao, J.; Li, Y.; Gao, J.; De, Y. Hesperidin inhibits ovarian cancer cell viability through endoplasmic reticulum stress signaling pathways. *Oncol. Lett.* **2017**, *14*, 5569–5574. [CrossRef] [PubMed]
166. Purushothaman, B.K. KMMS Magnetic casein-CaFe2O4 nanohybrid carrier conjugated with progesterone for enhanced cytotoxicity of citrus peel derived hesperidin drug towards breast and ovarian cancer. *Int. J. Biol. Macromol.* **2020**, *151*, 293–304. [CrossRef]
167. Lin, C.Y.; Chen, Y.H.; Huang, Y.C. Hesperetin induces autophagy and delayed apoptosis by modulating the AMPK/Akt/mTOR pathway in human leukemia cells in vitro. *Curr. Issues Mol. Biol.* **2023**, *45*, 1587–1600. [CrossRef] [PubMed]
168. Ersoz, M.; Erdemir, A.; Duranoglu, D.; Uzunoglu, D.; Arasoglu, T.; Derman, S.; Mansuroglu, B. Comparative evaluation of hesperetin loaded nanoparticles for anticancer activity against C6 glioma cancer cells. *Artif. Cells Nanomed. Biotechnol.* **2019**, *47*, 319–329. [CrossRef] [PubMed]
169. Mohi-Ud-Din, R.; Mir, R.H.; Sabreen, S.; Jan, R.; Pottoo, F.H.; Singh, I.P. Recent insights into therapeutic potential of plant-derived flavonoids against cancer. *Anti-Cancer Agents Med. Chem. (Former. Curr. Med. Chem.-Anti-Cancer Agents)* **2022**, *22*, 3343–3369. [CrossRef]
170. Adan, A.; Baran, Y. Fisetin and hesperetin induced apoptosis and cell cycle arrest in chronic myeloid leukemia cells accompanied by modulation of cellular signaling. *Tumor Biol.* **2016**, *37*, 5781–5795. [CrossRef]
171. Smina, T.P.; Mohan, A.; Ayyappa, K.A.; Sethuraman, S.; Krishnan, U.M. Hesperetin exerts apoptotic effect on A431 skin carcinoma cells by regulating mitogen activated protein kinases and cyclins. *Cell. Mol. Biol.* **2015**, *61*, 92–99.
172. Jiang, S.; Wang, S.; Zhang, L.; Tian, L.; Li, L.; Liu, Z.; Dong, Q.; Lv, X.; Mu, H.; Zhang, Q.; et al. Hesperetin as an adjuvant augments protective anti-tumour immunity responses in B16F10 melanoma by stimulating cytotoxic CD8+ T cells. *Scand. J. Immunol.* **2020**, *91*, e12867. [CrossRef]
173. Wu, D.; Li, J.; Hu, X.; Ma, J.; Dong, W. Hesperetin inhibits Eca-109 cell proliferation and invasion by suppressing the PI3K/AKT signaling pathway and synergistically enhances the anti-tumor effect of 5-fluorouracil on esophageal cancer in vitro and in vivo. *RSC Adv.* **2018**, *8*, 24434–24443. [CrossRef]
174. Gurushankar, K.; Nazeer, S.S.; Gohulkumar, M.; Jayasree, R.S.; Krishnakumar, N. Endogenous porphyrin fluorescence as a biomarker for monitoring the anti-angiogenic effect in antitumor response to hesperetin loaded nanoparticles in experimental oral carcinogenesis. *RSC Adv.* **2014**, *4*, 46896–46906.
175. Torcasio, R.; Gallo Cantafio, M.E.; Veneziano, C.; De Marco, C.; Ganino, L.; Valentino, I.; Occhiuzzi, M.A.; Perrotta, I.D.; Mancuso, T.; Conforti, F.; et al. Targeting of mitochondrial fission through natural flavanones elicits anti-myeloma activity. *J. Transl. Med.* **2024**, *22*, 208. [CrossRef] [PubMed]
176. Guo, B.; Zhang, Y.; Hui, Q.; Wang, H.; Tao, K. Naringin suppresses the metabolism of A375 cells by inhibiting the phosphorylation of c-Src. *Tumor Biol.* **2016**, *37*, 3841–3850. [CrossRef]

177. Wang, Y.; Liu, S.; Dong, W.; Qu, X.; Huang, C.; Yan, T.; Du, J. Combination of hesperetin and platinum enhances anticancer effect on lung adenocarcinoma. *Biomed. Pharmacother.* **2019**, *113*, 108779. [CrossRef] [PubMed]
178. Magura, J.; Moodley, R.; Mackraj, I. The effect of hesperidin and luteolin isolated from Eriocephalus africanus on apoptosis, cell cycle and miRNA expression in MCF-7. *J. Biomol. Struct. Dyn.* **2022**, *40*, 1791–1800. [CrossRef]
179. Zare, M.; Norouzi Sarkati, M.; Tashakkorian, H.; Partovi, R.; Rahaiee, S.; Rezaei, P.; Razavi, S.A. Dextran–hesperetin conjugate as a novel biocompatible medicine for antimicrobial and anticancer applications. *J. Polym. Environ.* **2021**, *29*, 811–820. [CrossRef]
180. Susidarti, R.A.; Nugroho, A.E.; Meiyanto, E. Increasing sensitivity of MCF-7/DOX cells towards doxorubicin by hesperetin through suppression of P-glycoprotein expression. *Indones. J. Pharm.* **2014**, *25*, 84.
181. Vidal, S.J.; Rodriguez-Bravo, V.; Quinn, S.A.; Rodriguez-Barrueco, R.; Lujambio, A.; Williams, E.; Sun, X.; de la Iglesia-Vicente, J.; Lee, A.; Readhead, B.; et al. A targetable GATA2-IGF2 axis confers aggressiveness in lethal prostate cancer. *Cancer Cell* **2015**, *27*, 223–239. [CrossRef]
182. Kong, W.; Ling, X.; Chen, Y.; Wu, X.; Zhao, Z.; Wang, W.; Wang, S.; Lai, G.; Yu, Z. Hesperetin reverses P-glycoprotein-mediated cisplatin resistance in DDP-resistant human lung cancer cells via modulation of the nuclear factor-κB signaling pathway. *Int. J. Mol. Med.* **2020**, *45*, 1213–1224. [CrossRef]
183. Gurushankar, K.; Gohulkumar, M.; Prasad, N.R.; Krishnakumar, N. Synthesis, characterization and in vitro anti-cancer evaluation of hesperetin-loaded nanoparticles in human oral carcinoma (KB) cells. *Adv. Nat. Sci. Nanosci. Nanotechnol.* **2013**, *5*, 015006. [CrossRef]
184. Gurushankar, K.; Nazeer, S.S.; Jayasree, R.S.; Krishnakumar, N. Evaluation of antitumor activity of hesperetin-loaded nanoparticles against DMBA-induced oral carcinogenesis based on tissue autofluorescence spectroscopy and multivariate analysis. *J. Fluoresc.* **2015**, *25*, 931–939. [CrossRef]
185. Křížová, L.; Dadáková, K.; Kašparovská, J.; Kašparovský, T. Isoflavones. *Molecules* **2019**, *24*, 1076. [CrossRef] [PubMed]
186. Bustamante-Rangel, M.; Delgado-Zamarreño, M.M.; Pérez-Martín, L.; Rodríguez-Gonzalo, E.; Domínguez-Álvarez, J. Analysis of isoflavones in foods. *Compr. Rev. Food Sci. Food Saf.* **2018**, *17*, 391–411. [CrossRef]
187. Kim, S.H.; Kim, C.W.; Jeon, S.Y.; Go, R.E.; Hwang, K.A.; Choi, K.C. Chemopreventive and chemotherapeutic effects of genistein, a soy isoflavone, upon cancer development and progression in preclinical animal models. *Lab. Anim. Res.* **2014**, *30*, 143–150. [CrossRef] [PubMed]
188. Ko, K.P. Isoflavones: Chemistry, analysis, functions and effects on health and cancer. *Asian Pac. J. Cancer Prev.* **2014**, *15*, 7001–7010. [CrossRef]
189. Wang, G.; Zhang, D.; Yang, S.; Wang, Y.; Tang, Z.; Fu, X. Co-administration of genistein with doxorubicin-loaded polypeptide nanoparticles weakens the metastasis of malignant prostate cancer by amplifying oxidative damage. *Biomater. Sci.* **2018**, *6*, 827–835. [CrossRef] [PubMed]
190. Khongsti, K.; Das, K.B.; Das, B. MAPK pathway and SIRT1 are involved in the down-regulation of secreted osteopontin expression by genistein in metastatic cancer cells. *Life Sci.* **2021**, *265*, 118787. [CrossRef]
191. Shin, H.R.; Joubert, C.; Boniol, M.; Hery, C.; Ahn, S.H.; Won, Y.J.; Nishino, Y.; Sobue, T.; Chen, C.J.; You, S.L.; et al. Recent trends and patterns in breast cancer incidence among Eastern and Southeastern Asian women. *Cancer Causes Control* **2010**, *21*, 1777–1785. [CrossRef]
192. Zhao, Q.; Zhao, M.; Parris, A.B.; Xing, Y.; Yang, X. Genistein targets the cancerous inhibitor of PP2A to induce growth inhibition and apoptosis in breast cancer cells. *Int. J. Oncol.* **2016**, *49*, 1203–1210. [CrossRef]
193. Bhat, S.S.; Prasad, S.K.; Shivamallu, C.; Prasad, K.S.; Syed, A.; Reddy, P.; Cull, C.A.; Amachawadi, R.G. Genistein: A potent anti-breast cancer agent. *Curr. Issues Mol. Biol.* **2021**, *43*, 1502–1517. [CrossRef]
194. Tuli, H.S.; Tuorkey, M.J.; Thakral, F.; Sak, K.; Kumar, M.; Sharma, A.K.; Sharma, U.; Jain, A.; Aggarwal, V.; Bishayee, A. Molecular mechanisms of action of genistein in cancer: Recent advances. *Front. Pharmacol.* **2019**, *10*, 1336. [CrossRef]
195. Islam, M.A.; Bekele, R.; vanden Berg, J.H.; Kuswanti, Y.; Thapa, O.; Soltani, S.; van Leeuwen, F.R.; Rietjens, I.M.; Murk, A.J. Deconjugation of soy isoflavone glucuronides needed for estrogenic activity. *Toxicol. Vitr.* **2015**, *29*, 706–715. [CrossRef] [PubMed]
196. Pavese, J.M.; Farmer, R.L.; Bergan, R.C. Inhibition of cancer cell invasion and metastasis by genistein. *Cancer Metastasis Rev.* **2010**, *29*, 465–482. [CrossRef]
197. Mukund, V. Genistein: Its role in breast cancer growth and metastasis. *Curr. Drug Metab.* **2020**, *21*, 6–10. [CrossRef] [PubMed]
198. Latocha, M.; Płonka, J.; Kuśmierz, D.; Jurzak, M.; Polaniak, R.; Nowosad, A. Transcripional activity of genes encoding MMPs and TIMPs in breast cancer cells treated by genistein and in normal cancer-associated fibroblasts--in vitro studies. *Acta Pol. Pharm.* **2014**, *71*, 1095–1102.
199. Farhan, M.; El Oirdi, M.; Aatif, M.; Nahvi, I.; Muteeb, G.; Alam, M.W. Soy isoflavones induce cell death by copper-mediated mechanism: Understanding its anticancer properties. *Molecules* **2023**, *28*, 2925. [CrossRef] [PubMed]

200. Pavese, J.M.; Krishna, S.N.; Bergan, R.C. Genistein inhibits human prostate cancer cell detachment, invasion, and metastasis. *Am. J. Clin. Nutr.* **2014**, *100*, 431S–436S. [CrossRef]
201. Lazarevic, B.; Boezelijn, G.; Diep, L.M.; Kvernrod, K.; Ogren, O.; Ramberg, H.; Moen, A.; Wessel, N.; Berg, R.E.; Egge Jacobsen, W.; et al. Efficacy and safety of short-term genistein intervention in patients with localized prostate cancer prior to radical prostatectomy: A randomized, placebo-controlled, double-blind Phase 2 clinical trial. *Nutr. Cancer* **2011**, *63*, 889–898. [CrossRef]
202. Chandrasekara, A.; Shahidi, F. Herbal beverages: Bioactive compounds and their role in disease risk reduction-A review. *J. Tradit. Complement. Med.* **2018**, *8*, 451–458. [CrossRef]
203. Bento-Silva, A.; Koistinen, V.M.; Mena, P.; Bronze, M.R.; Hanhineva, K.; Sahlstrøm, S.; Kitrytė, V.; Moco, S.; Aura, A.M. Factors affecting intake, metabolism and health benefits of phenolic acids: Do we understand individual variability? *Eur. J. Nutr.* **2020**, *59*, 1275–1293. [CrossRef]
204. Kumar, N.; Goel, N. Phenolic acids: Natural versatile molecules with promising therapeutic applications. *Biotechnol. Rep.* **2019**, *24*, e00370. [CrossRef]
205. El-Seedi, H.R.; El-Said, A.M.; Khalifa, S.A.; Goransson, U.; Bohlin, L.; Borg-Karlson, A.K.; Verpoorte, R. Biosynthesis, natural sources, dietary intake, pharmacokinetic properties, and biological activities of hydroxycinnamic acids. *J. Agric. Food Chem.* **2012**, *60*, 10877–10895. [CrossRef] [PubMed]
206. Ludwig, I.A.; Mena, P.; Calani, L.; Cid, C.; Del Rio, D.; Lean, M.E.; Crozier, A. Variations in caffeine and chlorogenic acid contents of coffees: What are we drinking? *Food Funct.* **2014**, *5*, 1718–1726. [CrossRef]
207. Süntar, I.; Yakıncı, Ö.F. Potential risks of phytonutrients associated with high-dose or long-term use. In *Phytonutrients in Food*; Woodhead Publishing: Sawston, UK, 2020; pp. 137–155.
208. Thakur, M.; Singh, K.; Khedkar, R. Phytochemicals: Extraction process, safety assessment, toxicological evaluations, and regulatory issues. In *Functional and Preservative Properties of Phytochemicals*; Academic Press: Cambridge, MA, USA, 2020; pp. 341–361.
209. Xie, J.; Wang, H.; Xie, W.; Liu, Y.; Chen, Y. Gallic acid promotes ferroptosis in hepatocellular carcinoma via inactivating Wnt/β-catenin signaling pathway. *Naunyn-Schmiedeberg's Arch. Pharmacol.* **2024**, *397*, 2437–2445. [CrossRef]
210. Jafarinejad, S.; Martin, W.H.; Ras, B.A.; Isreb, M.; Jacob, B.; Aziz, A.; Adoul, Z.; Lagnado, R.; Bowen, R.D.; Najafzadeh, M. The anticancer/cytotoxic effect of a novel gallic acid derivative in non-small cell lung carcinoma A549 cells and peripheral blood mononuclear cells from healthy individuals and lung cancer patients. *BioFactors* **2024**, *50*, 201–213. [CrossRef]
211. Patra, S.; Bhol, C.S.; Panigrahi, D.P.; Praharaj, P.P.; Pradhan, B.; Jena, M.; Bhutia, S.K. Gamma irradiation promotes chemo-sensitization potential of gallic acid through attenuation of autophagic flux to trigger apoptosis in an NRF2 inactivation signalling pathway. *Free Radic. Biol. Med.* **2020**, *160*, 111–124. [CrossRef]
212. Ko, E.B.; Jang, Y.G.; Kim, C.W.; Go, R.E.; Lee, H.K.; Choi, K.C. Gallic acid hindered lung cancer progression by inducing cell cycle arrest and apoptosis in a549 lung cancer cells via PI3K/Akt pathway. *Biomol. Ther.* **2021**, *30*, 151. [CrossRef] [PubMed]
213. Qanash, H.; Yahya, R.; Bakri, M.M.; Bazaid, A.S.; Qanash, S.; Shater, A.F. TMA Anticancer, antioxidant, antiviral and antimicrobial activities of Kei Apple (*Dovyalis caffra*) fruit. *Sci. Rep.* **2022**, *12*, 5914. [CrossRef]
214. Sanchez-Martin, V.; Plaza-Calonge, M.D.; Soriano-Lerma, A.; Ortiz-Gonzalez, M.; Linde-Rodriguez, A.; Perez-Carrasco, V.; Ramirez-Macias, I.; Cuadros, M.; Gutierrez-Fernandez, J.; Murciano-Calles, J.; et al. Gallic acid: A natural phenolic compound exerting antitumoral activities in colorectal cancer via interaction with g-quadruplexes. *Cancers* **2022**, *14*, 2648. [CrossRef] [PubMed]
215. Min, J.; Shen, H.; Xi, W.; Wang, Q.; Yin, L.; Zhang, Y.; Yu, Y.; Yang, Q.; Wang, Z.N. Synergistic anticancer activity of combined use of caffeic acid with paclitaxel enhances apoptosis of non-small-cell lung cancer H1299 cells in vivo and in vitro. *Cell. Physiol. Biochem.* **2018**, *48*, 1433–1442. [CrossRef]
216. Kanimozhi, G.; Prasad, N.R. Anticancer effect of caffeic acid on human cervical cancer cells. In *Coffee in Health and Disease Prevention*; Academic Press: Cambridge, MA, USA, 2015; pp. 655–661.
217. Rosendahl, A.H.; Perks, C.M.; Zeng, L.; Markkula, A.; Simonsson, M.; Rose, C.; Ingvar, C.; Holly, J.M.; Jernström, H. Caffeine and caffeic acid inhibit growth and modify estrogen receptor and insulin-like growth factor I receptor levels in human breast cancer. *Clin. Cancer Res.* **2015**, *21*, 1877–1887. [CrossRef]
218. Chen, L.; Jin, Y.; Chen, H.; Sun, C.; Fu, W.; Zheng, L.; Lu, M.; Chen, P.; Chen, G.; Zhang, Y.; et al. Discovery of caffeic acid phenethyl ester derivatives as novel myeloid differentiation protein 2 inhibitors for treatment of acute lung injury. *Eur. J. Med. Chem.* **2018**, *143*, 361–375. [CrossRef]
219. Kang, K.P.; Park, S.K.; Kim, D.H.; Sung, M.J.; Jung, Y.J.; Lee, A.S.; Lee, J.E.; Ramkumar, K.M.; Lee, S.; Park, M.H.; et al. Luteolin ameliorates cisplatin-induced acute kidney injury in mice by regulation of p53-dependent renal tubular apoptosis. *Nephrol. Dial. Transplant.* **2011**, *26*, 814–822. [CrossRef]

220. Yang, G.; Fu, Y.; Malakhova, M.; Kurinov, I.; Zhu, F.; Yao, K.; Li, H.; Chen, H.; Li, W.; Lim, D.Y.; et al. Caffeic acid directly targets ERK1/2 to attenuate solar UV-induced skin carcinogenesis. *Cancer Prev. Res.* **2014**, *7*, 1056–1066. [CrossRef]
221. Ozturk, G.; Ginis, Z.; Akyol, S.; Erden, G.; Gurel, A.; Akyol, O. The anticancer mechanism of caffeic acid phenethyl ester (CAPE): Review of melanomas, lung and prostate cancers. *Eur. Rev. Med. Pharmacol. Sci.* **2012**, *16*, 2064–2068. [PubMed]
222. Lin, H.P.; Lin, C.Y.; Huo, C.; Hsiao, P.H.; Su, L.C.; Jiang, S.S.; Chan, T.M.; Chang, C.H.; Chen, L.T.; Kung, H.J.; et al. Caffeic acid phenethyl ester induced cell cycle arrest and growth inhibition in androgen-independent prostate cancer cells via regulation of Skp2, p53, p21Cip1 and p27Kip1. *Oncotarget* **2015**, *6*, 6684. [CrossRef]
223. Amorim, R.; Magalhães, C.C.; Benfeito, S.; Cagide, F.; Tavares, L.C.; Santos, K.; Sardão, V.A.; Datta, S.; Cortopassi, G.A.; Baldeiras, I.; et al. Mitochondria dysfunction induced by decyl-TPP mitochondriotropic antioxidant based on caffeic acid AntiOxCIN6 sensitizes cisplatin lung anticancer therapy due to a remodeling of energy metabolism. *Biochem. Pharmacol.* **2024**, *219*, 115953. [CrossRef] [PubMed]
224. Lim, J.; Kim, K.; Kwon, D.Y.; Kim, J.K.; Sathasivam, R.; Park, S.U. Effects of Different Solvents on the Extraction of Phenolic and Flavonoid Compounds, and Antioxidant Activities, in Scutellaria baicalensis Hairy Roots. *Horticulturae* **2024**, *10*, 160. [CrossRef]
225. Ishida, Y.; Gao, R.; Shah, N.; Bhargava, P.; Furune, T.; Kaul, S.C.; Terao, K.; Wadhwa, R. Anticancer activity in honeybee propolis: Functional insights to the role of caffeic acid phenethyl ester and its complex with γ-cyclodextrin. *Integr. Cancer Ther.* **2018**, *17*, 867–873. [CrossRef]
226. Wadhwa, R.; Nigam, N.; Bhargava, P.; Dhanjal, J.K.; Goyal, S.; Grover, A.; Sundar, D.; Ishida, Y.; Terao, K.; Kaul, S.C. Molecular characterization and enhancement of anticancer activity of caffeic acid phenethyl ester by γ cyclodextrin. *J. Cancer* **2016**, *7*, 1755. [CrossRef]
227. Gupta, S.; Tak, H.; Rathore, K.; Banavath, H.N.; Tejavath, K.K. Caffeic acid, a dietary polyphenol, pre-sensitizes pancreatic ductal adenocarcinoma to chemotherapeutic drug. *J. Biomol. Struct. Dyn.* **2024**, 1–15, *Epub ahead of print*. [CrossRef] [PubMed]
228. Pagnan, A.L.; Pessoa, A.S.; Tokuhara, C.K.; Fakhoury, V.S.; Oliveira, G.S.; Sanches, M.L.; Inacio, K.K.; Ximenes, V.F.; Oliveira, R.C. Anti-tumour potential and selectivity of caffeic acid phenethyl ester in osteosarcoma cells. *Tissue Cell* **2022**, *74*, 101705. [CrossRef] [PubMed]
229. Sehrawat, R.; Rathee, P.; Rathee, P.; Khatkar, S.; Küpelli Akkol, E.; Khatkar, A. In silico and in vitro analysis of phenolic acids for identification of potential dhfr inhibitors as antimicrobial and anticancer agents. *Curr. Protein Pept. Sci.* **2024**, *25*, 44–58. [CrossRef]
230. Trócsányi, E.; György, Z.; Zámboriné-Németh, É. New insights into rosmarinic acid biosynthesis based on molecular studies. *Curr. Plant Biol.* **2020**, *23*, 100162. [CrossRef]
231. Levsh, O.; Pluskal, T.; Carballo, V.; Mitchell, A.J.; Weng, J.K. Independent evolution of rosmarinic acid biosynthesis in two sister families under the Lamiids clade of flowering plants. *J. Biol. Chem.* **2019**, *294*, 15193–15205. [CrossRef]
232. Khojasteh, A.; Mirjalili, M.H.; Alcalde, M.A.; Cusido, R.M.; Eibl, R.; Palazon, J. Powerful plant antioxidants: A new biosustainable approach to the production of rosmarinic acid. *Antioxidants* **2020**, *9*, 1273. [CrossRef] [PubMed]
233. Hossan, M.S.; Rahman, S.; Bashar, A.B.; Jahan, R.; Al-Nahain, A.; Rahmatullah, M. Rosmarinic acid: A review of its anticancer action. *World J. Pharm. Pharm. Sci.* **2014**, *3*, 57–70.
234. Tsimogiannis, D.; Oreopoulou, V. Classification of phenolic compounds in plants. In *Polyphenols in Plants*; Academic Press: Cambridge, MA, USA, 2019; pp. 263–284.
235. Anwar, S.; Shamsi, A.; Shahbaaz, M.; Queen, A.; Khan, P.; Hasan, G.M.; Islam, A.; Alajmi, M.F.; Hussain, A.; Ahmad, F.; et al. Rosmarinic acid exhibits anticancer effects via MARK4 inhibition. *Sci. Rep.* **2020**, *10*, 10300. [CrossRef]
236. Messeha, S.S.; Zarmouh, N.O.; Asiri, A.; Soliman, K.F. Rosmarinic acid-induced apoptosis and cell cycle arrest in triple-negative breast cancer cells. *Eur. J. Pharmacol.* **2020**, *885*, 173419. [CrossRef]
237. Yesil-Celiktas, O.; Sevimli, C.; Bedir, E.; Vardar-Sukan, F. Inhibitory effects of rosemary extracts, carnosic acid and rosmarinic acid on the growth of various human cancer cell lines. *Plant Foods Hum. Nutr.* **2010**, *65*, 158–163. [CrossRef]
238. Luo, C.; Zou, L.; Sun, H.; Peng, J.; Gao, C.; Bao, L.; Ji, R.; Jin, Y.; Sun, S. A review of the anti-inflammatory effects of rosmarinic acid on inflammatory diseases. *Front. Pharmacol.* **2020**, *11*, 153. [CrossRef]
239. Radziejewska, I.; Supruniuk, K.; Nazaruk, J.; Karna, E.; Popławska, B.; Bielawska, A.; Galicka, A. Rosmarinic acid influences collagen, MMPs, TIMPs, glycosylation and MUC1 in CRL-1739 gastric cancer cell line. *Biomed. Pharmacother.* **2018**, *107*, 397–407. [CrossRef] [PubMed]
240. Laila, F.; Fardiaz, D.; Yuliana, N.D.; Damanik, M.R.; Nur Annisa Dewi, F. Methanol extract of Coleus amboinicus (Lour) exhibited antiproliferative activity and induced programmed cell death in colon cancer cell WiDr. *Int. J. Food Sci.* **2020**, *2020*, 9068326. [CrossRef]
241. Yang, K.; Shen, Z.; Zou, Y.; Gao, K. Rosmarinic acid inhibits migration, invasion, and p38/AP-1 signaling via miR-1225-5p in colorectal cancer cells. *J. Recept. Signal Transduct.* **2021**, *41*, 284–293. [CrossRef]

242. Jin, B.R.; Chung, K.S.; Hwang, S.; Hwang, S.N.; Rhee, K.J.; Lee, M.; An, H.J. Rosmarinic acid represses colitis-associated colon cancer: A pivotal involvement of the TLR4-mediated NF-κB-STAT3 axis. *Neoplasia* **2021**, *23*, 561–573. [CrossRef] [PubMed]
243. Xu, W.; Yang, F.; Zhang, Y.; Shen, X. Protective effects of rosmarinic acid against radiation-induced damage to the hematopoietic system in mice. *J. Radiat. Res.* **2016**, *57*, 356–362. [CrossRef] [PubMed]
244. Jang, Y.G.; Hwang, K.A.; Choi, K.C. Rosmarinic acid, a component of rosemary tea, induced the cell cycle arrest and apoptosis through modulation of HDAC2 expression in prostate cancer cell lines. *Nutrients* **2018**, *10*, 1784. [CrossRef]
245. García-Sarrió, M.J.; Sanz, M.L.; Palá-Paúl, J.; Díaz, S.; Soria, A.C. Optimization of a Green Microwave-Assisted Extraction Method to Obtain Multifunctional Extracts of *Mentha* sp. *Foods* **2023**, *12*, 2039. [CrossRef]
246. Chao, W.W.; Liou, Y.J.; Ma, H.T.; Chen, Y.H.; Chou, S.T. *The Antitumor Mechanism of the Polyphenol-Enriched Ethyl Acetate Fraction Extract of Glechoma Hederacea (Lamiaceae) Against HepG2 Cells Involves Apoptosis Pathways*; ResearchGate: Berlin, Germany, 2020.
247. Wang, L.; Yang, H.; Wang, C.; Shi, X.; Li, K. Rosmarinic acid inhibits proliferation and invasion of hepatocellular carcinoma cells SMMC 7721 via PI3K/AKT/mTOR signal pathway. *Biomed. Pharmacother.* **2019**, *120*, 109443. [CrossRef]
248. Ma, Z.; Yang, J.; Yang, Y.; Wang, X.; Chen, G.; Shi, A.; Lu, Y.; Jia, S.; Kang, X.; Lu, L. Rosmarinic acid exerts an anticancer effect on osteosarcoma cells by inhibiting DJ-1 via regulation of the PTEN-PI3K-Akt signaling pathway. *Phytomedicine* **2020**, *68*, 153186. [CrossRef]
249. Moore, J.; Yousef, M.; Tsiani, E. Anticancer effects of rosemary (*Rosmarinus officinalis* L.) extract and rosemary extract polyphenols. *Nutrients* **2016**, *8*, 731. [CrossRef]
250. Ghiulai, R.; Avram, S.; Stoian, D.; Pavel, I.Z.; Coricovac, D.; Oprean, C.; Vlase, L.; Farcas, C.; Mioc, M.; Minda, D.; et al. Lemon balm extracts prevent breast cancer progression in vitro and in ovo on chorioallantoic membrane assay. *Evid.-Based Complement. Altern. Med.* **2020**, *2020*, 6489159. [CrossRef] [PubMed]
251. Li, H.; Zhang, Y.; Chen, H.H.; Huang, E.; Zhuang, H.; Li, D.; Ni, F. Rosmarinic acid inhibits stem-like breast cancer through hedgehog and Bcl-2/Bax signaling pathways. *Pharmacogn. Mag.* **2019**, *15*, 600–606. [CrossRef]
252. Khanaree, C.; Punfa, W.; Tantipaiboonwong, P.; Nuntaboon, P.; Suttajit, M.; Topanurak, S.; Dukaew, N.; Mon, M.T.; Hu, R.; Pintha, K. In vitro anti-metastasis of Perilla frutescens leaf water extract on aggressive human breast cancer cells. *J. Assoc. Med. Sci.* **2022**, *55*, 51–59. [CrossRef]
253. Cristy, G.P.; Liana, D.; Chatwichien, J.; Aonbangkhen, C.; Srisomsap, C.; Phanumartwiwath, A. Breast Cancer Prevention by Dietary Polyphenols: Microemulsion Formulation and In vitro Studies. *Sci. Pharm.* **2024**, *92*, 25. [CrossRef]
254. Mahmoud, M.A.; Okda, T.M.; Omran, G.A.; Abd-Alhaseeb, M.M. Rosmarinic acid suppresses inflammation, angiogenesis, and improves paclitaxel induced apoptosis in a breast cancer model via NF3 κB-p53-caspase-3 pathways modulation. *J. Appl. Biomed.* **2021**, *19*, 202–209. [CrossRef] [PubMed]
255. Huang, L.; Chen, J.; Quan, J.; Xiang, D. Rosmarinic acid inhibits proliferation and migration, promotes apoptosis and enhances cisplatin sensitivity of melanoma cells through inhibiting ADAM17/EGFR/AKT/GSK3β axis. *Bioengineered* **2021**, *12*, 3065–3076. [CrossRef]
256. Han, Y.; Ma, L.; Zhao, L.E.; Feng, W.; Zheng, X. Rosmarinic inhibits cell proliferation, invasion and migration via up-regulating miR-506 and suppressing MMP2/16 expression in pancreatic cancer. *Biomed. Pharmacother.* **2019**, *115*, 108878. [CrossRef]
257. Zhou, X.; Wang, W.; Li, Z.; Chen, L.; Wen, C.; Ruan, Q.; Xu, Z.; Liu, R.; Xu, J.; Bai, Y.; et al. Rosmarinic acid decreases the malignancy of pancreatic cancer through inhibiting Gli1 signaling. *Phytomedicine* **2022**, *95*, 153861. [CrossRef]
258. Cao, W.; Hu, C.; Wu, L.; Xu, L.; Jiang, W. Rosmarinic acid inhibits inflammation and angiogenesis of hepatocellular carcinoma by suppression of NF-κB signaling in H22 tumor-bearing mice. *J. Pharmacol. Sci.* **2016**, *132*, 131–137. [CrossRef]
259. Li, Z.N.; Luo, Y. HSP90 inhibitors and cancer: Prospects for use in targeted therapies. *Oncol. Rep.* **2022**, *49*, 6. [CrossRef]
260. Lim, S.H.; Nam, K.H.; Kim, K.; Yi, S.A.; Lee, J.; Han, J.W. Rosmarinic acid methyl ester regulates ovarian cancer cell migration and reverses cisplatin resistance by inhibiting the expression of Forkhead Box M1. *Pharmaceuticals* **2020**, *13*, 302. [CrossRef]
261. Pintha, K.; Chaiwangyen, W.; Yodkeeree, S.; Suttajit, M.; Tantipaiboonwong, P. Suppressive effects of rosmarinic acid rich fraction from perilla on oxidative stress, inflammation and metastasis ability in A549 cells exposed to PM via C-Jun, P-65-Nf-Kb and Akt signaling pathways. *Biomolecules* **2021**, *11*, 1090. [CrossRef] [PubMed]
262. Moon, D.O.; Kim, M.O.; Lee, J.D.; Choi, Y.H.; Kim, G.Y. Rosmarinic acid sensitizes cell death through suppression of TNF-α-induced NF-κB activation and ROS generation in human leukemia U937 cells. *Cancer Lett.* **2010**, *288*, 183–191. [CrossRef]
263. Liao, X.Z.; Gao, Y.; Sun, L.L.; Liu, J.H.; Chen, H.R.; Yu, L.; Chen, Z.Z.; Chen, W.H.; Lin, L.Z. Rosmarinic acid reverses non-small cell lung cancer cisplatin resistance by activating the MAPK signaling pathway. *Phytother. Res.* **2020**, *34*, 1142–1153. [CrossRef]
264. Highland, H.; Thakur, M.; Pandya, P.; Mankad, A.; George, L.B. Molecular Dynamics of A Biglycan-Rosmarinic Acid Complex with Focal Adhesion Kinase for Possible Arrest of Metastasis in Non-Small Cell Lung Cancer (NSCLC): An In-Silico Study. *J. Drug Deliv. Ther.* **2019**, *9*, 159–166. [CrossRef]

265. Xu, Y.; Jiang, Z.; Ji, G.; Liu, J. Inhibition of bone metastasis from breast carcinoma by rosmarinic acid. *Planta Med.* **2010**, *76*, 956–962. [CrossRef] [PubMed]
266. Lee, J.W.; Asai, M.; Jeon, S.K.; Iimura, T.; Yonezawa, T.; Cha, B.Y.; Woo, J.T.; Yamaguchi, A. Rosmarinic acid exerts an antiosteoporotic effect in the RANKL-induced mouse model of bone loss by promotion of osteoblastic differentiation and inhibition of osteoclastic differentiation. *Mol. Nutr. Food Res.* **2015**, *59*, 386–400. [CrossRef]
267. Pagano, K.; Carminati, L.; Tomaselli, S.; Molinari, H.; Taraboletti, G.; Ragona, L. Molecular Basis of the Antiangiogenic Action of Rosmarinic Acid, a Natural Compound Targeting Fibroblast Growth Factor-2/FGFR Interactions. *ChemBioChem* **2021**, *22*, 160–169. [CrossRef]
268. Rodríguez-Luna, A.; Ávila-Román, J.; Oliveira, H.; Motilva, V.; Talero, E. Fucoxanthin and rosmarinic acid combination has anti-inflammatory effects through regulation of NLRP3 inflammasome in UVB-exposed HaCaT keratinocytes. *Mar. Drugs* **2019**, *17*, 451. [CrossRef]
269. Gupta, D.; Archoo, S.; Naikoo, S.H.; Abdullah, S.T. Rosmarinic acid: A naturally occurring plant based agent prevents impaired mitochondrial dynamics and apoptosis in ultraviolet-B-irradiated human skin cells. *Photochem. Photobiol.* **2022**, *98*, 925–934. [CrossRef]
270. Fernando, P.M.; Piao, M.J.; Kang, K.A.; Ryu, Y.S.; Hewage, S.R.; Chae, S.W.; Hyun, J.W. Rosmarinic acid attenuates cell damage against UVB radiation-induced oxidative stress via enhancing antioxidant effects in human HaCaT cells. *Biomol. Ther.* **2016**, *24*, 75. [CrossRef] [PubMed]
271. Pandi, A.; Kalappan, V.M. Pharmacological and therapeutic applications of Sinapic acid—An updated review. *Mol. Biol. Rep.* **2021**, *48*, 3733–3745. [CrossRef]
272. Lan, H.; Zhang, L.Y.; He, W.; Li, W.Y.; Zeng, Z.; Qian, B.; Wang, C.; Song, J.L. Sinapic acid alleviated inflammation-induced intestinal epithelial barrier dysfunction in lipopolysaccharide-(LPS-) treated Caco-2 cells. *Mediat. Inflamm.* **2021**, *2021*, 5514075. [CrossRef]
273. Bin Jardan, Y.A.; Ansari, M.A.; Raish, M.; Alkharfy, K.M.; Ahad, A.; Al-Jenoobi, F.I.; Haq, N.; Khan, M.R.; Ahmad, A. Sinapic acid ameliorates oxidative stress, inflammation, and apoptosis in acute doxorubicin-induced cardiotoxicity via the NF-κB-mediated pathway. *BioMed Res. Int.* **2020**, *2020*, 3921796. [CrossRef] [PubMed]
274. Lee, J.Y. Anti-inflammatory effects of sinapic acid on 2, 4, 6-trinitrobenzenesulfonic acid-induced colitis in mice. *Arch. Pharmacal Res.* **2018**, *41*, 243–250. [CrossRef]
275. Qian, B.; Wang, C.; Zeng, Z.; Ren, Y.; Li, D.; Song, J.L. Ameliorative effect of Sinapic acid on dextran sodium sulfate-(DSS-) induced ulcerative colitis in Kunming (KM) mice. *Oxidative Med. Cell. Longev.* **2020**, *2020*, 8393504. [CrossRef] [PubMed]
276. Raish, M.; Ahmad, A.; Ansari, M.A.; Ahad, A.; Al-Jenoobi, F.I.; Al-Mohizea, A.M.; Khan, A.; Ali, N. Sinapic acid ameliorates bleomycin-induced lung fibrosis in rats. *Biomed. Pharmacother.* **2018**, *108*, 224–231. [CrossRef]
277. Singh, A.K.; Bishayee, A.; Pandey, A.K. Targeting histone deacetylases with natural and synthetic agents: An emerging anticancer strategy. *Nutrients* **2018**, *10*, 731. [CrossRef]
278. Huang, Z.W.; Tan, P.; Yi, X.K.; Chen, H.; Sun, B.; Shi, H.; Mou, Z.Q.; Cheng, Y.L.; Li, T.X.; Li, Q.; et al. Sinapic Acid Alleviates Acute Pancreatitis in Association with Attenuation of Inflammation, Pyroptosis, and the AMPK/NF-κ B Signaling Pathway. *Am. J. Chin. Med.* **2022**, *50*, 2185–2197. [CrossRef]
279. Hu, X.; Geetha, R.V.; Surapaneni, K.M.; Veeraraghavan, V.P.; Chinnathambi, A.; Alahmadi, T.A.; Manikandan, V.; Manokaran, K. Lung cancer induced by Benzo (A) Pyrene: ChemoProtective effect of sinapic acid in swiss albino mice. *Saudi J. Biol. Sci.* **2021**, *28*, 7125–7133. [CrossRef]
280. Janakiraman, K.; Kathiresan, S.; Mariadoss, A.V. Influence of sinapic acid on induction of apoptosis in human laryngeal carcinoma cell line. *Int. J. Mod. Res. Rev.* **2014**, *2*, 165–170.
281. Eroğlu, C.; Avcı, E.; Vural, H.; Kurar, E. Anticancer mechanism of Sinapic acid in PC-3 and LNCaP human prostate cancer cell lines. *Gene* **2018**, *671*, 127–134. [CrossRef] [PubMed]
282. Balaji, C.; Muthukumaran, J.; Nalini, N. Chemopreventive effect of sinapic acid on 1, 2-dimethylhydrazine-induced experimental rat colon carcinogenesis. *Hum. Exp. Toxicol.* **2014**, *33*, 1253–1268. [CrossRef] [PubMed]
283. Zhao, J.; Li, H.; Li, W.; Wang, Z.; Dong, Z.; Lan, H.; Wang, C.; Song, J.L. Effects of Sinapic acid combined with cisplatin on the apoptosis and autophagy of the hepatoma cells HepG2 and SMMC-7721. *Evid.-Based Complement. Altern. Med.* **2021**, *2021*, 6095963. [CrossRef]
284. Poyraz, F.S.; Akbaş, G.; Duranoğlu, D.; Acar, S.; Mansuroğlu, B.; Ersöz, M. Sinapic-Acid-Loaded Nanoparticles Optimized via Experimental Design Methods: Cytotoxic, Antiapoptotoic, Antiproliferative, and Antioxidant Activity. *ACS Omega* **2024**, *9*, 40329–40345. [CrossRef]
285. Kaur, K.; Al-Khazaleh, A.K.; Bhuyan, D.J.; Li, F.; Li, C.G. A review of recent curcumin analogues and their antioxidant, anti-inflammatory, and anticancer activities. *Antioxidants* **2024**, *13*, 1092. [CrossRef]

286. Vyas, A.; Dandawate, P.; Padhye, S.; Ahmad, A.; Sarkar, F. Perspectives on new synthetic curcumin analogs and their potential anticancer properties. *Curr. Pharm. Des.* **2013**, *19*, 2047–2069.
287. Haghi, A.; Azimi, H.; Rahimi, R. A comprehensive review on pharmacotherapeutics of three phytochemicals, curcumin, quercetin, and allicin, in the treatment of gastric cancer. *J. Gastrointest. Cancer* **2017**, *48*, 314–320. [CrossRef]
288. Engwa, G.A. Free radicals and the role of plant phytochemicals as antioxidants against oxidative stress-related diseases. Phytochemicals: Source of antioxidants and role in disease prevention. *BoD—Books Demand* **2018**, *7*, 49–74.
289. Farzaei, M.H.; Zobeiri, M.; Parvizi, F.; El Senduny, F.F.; Marmouzi, I.; Coy-Barrera, E.; Naseri, R.; Nabavi, S.M.; Rahimi, R.; Abdollahi, M. Curcumin in liver diseases: A systematic review of the cellular mechanisms of oxidative stress and clinical perspective. *Nutrients* **2018**, *10*, 855. [CrossRef]
290. Kotha, R.R.; Tareq, F.S.; Yildiz, E.; Luthria, D.L. Oxidative stress and antioxidants—A critical review on in vitro antioxidant assays. *Antioxidants* **2022**, *11*, 2388. [CrossRef] [PubMed]
291. Allegra, A.; Innao, V.; Russo, S.; Gerace, D.; Alonci, A.; Musolino, C. Anticancer activity of curcumin and its analogues: Preclinical and clinical studies. *Cancer Investig.* **2017**, *35*, 1–22. [CrossRef]
292. Kunnumakkara, A.B.; Bordoloi, D.; Harsha, C.; Banik, K.; Gupta, S.C.; Aggarwal, B.B. Curcumin mediates anticancer effects by modulating multiple cell signaling pathways. *Clin. Sci.* **2017**, *131*, 1781–1799. [CrossRef] [PubMed]
293. Chen, D.; Dai, F.; Chen, Z.; Wang, S.; Cheng, X.; Sheng, Q.; Lin, J.; Chen, W. Dimethoxy curcumin induces apoptosis by suppressing survivin and inhibits invasion by enhancing E-cadherin in colon cancer cells. *Med. Sci. Monit. Int. Med. J. Exp. Clin. Res.* **2016**, *22*, 3215. [CrossRef] [PubMed]
294. Masuelli, L.; Benvenuto, M.; Di Stefano, E.; Mattera, R.; Fantini, M.; De Feudis, G.; De Smaele, E.; Tresoldi, I.; Giganti, M.G.; Modesti, A.; et al. Curcumin blocks autophagy and activates apoptosis of malignant mesothelioma cell lines and increases the survival of mice intraperitoneally transplanted with a malignant mesothelioma cell line. *Oncotarget* **2017**, *8*, 34405. [CrossRef]
295. Weng, W.; Goel, A. Curcumin and colorectal cancer: An update and current perspective on this natural medicine. In *Seminars in Cancer Biology*; Academic Press: Cambridge, MA, USA, 2022; Volume 80, pp. 73–86.
296. Ghosh, S.; Banerjee, S.; Sil, P.C. The beneficial role of curcumin on inflammation, diabetes and neurodegenerative disease: A recent update. *Food Chem. Toxicol.* **2015**, *83*, 111–124. [CrossRef]
297. Wong, S.C.; Kamarudin, M.N.; Naidu, R. Anticancer mechanism of curcumin on human glioblastoma. *Nutrients* **2021**, *13*, 950. [CrossRef] [PubMed]
298. Huang, T.Y.; Tsai, T.H.; Hsu, C.W.; Hsu, Y.C. Curcuminoids suppress the growth and induce apoptosis through caspase-3-dependent pathways in glioblastoma multiforme (GBM) 8401 cells. *J. Agric. Food Chem.* **2010**, *58*, 10639–10645. [CrossRef]
299. Lee, J.Y.; Lee, Y.M.; Chang, G.C.; Yu, S.L.; Hsieh, W.Y.; Chen, J.J.; Chen, H.W.; Yang, P.C. Curcumin induces EGFR degradation in lung adenocarcinoma and modulates p38 activation in intestine: The versatile adjuvant for gefitinib therapy. *PLoS ONE* **2011**, *6*, e23756. [CrossRef]
300. Sultana, S.; Munir, N.; Mahmood, Z.; Riaz, M.; Akram, M.; Rebezov, M.; Kuderinova, N.; Moldabayeva, Z.; Shariati, M.A.; Rauf, A.; et al. Molecular targets for the management of cancer using Curcuma longa Linn. phytoconstituents: A Review. *Biomed. Pharmacother.* **2021**, *135*, 111078. [CrossRef]
301. Li, Y.; Wang, J.; Li, X.; Jia, Y.; Huai, L.; He, K.; Yu, P.; Wang, M.; Xing, H.; Rao, Q.; et al. Role of the Wilms' tumor 1 gene in the aberrant biological behavior of leukemic cells and the related mechanisms. *Oncol. Rep.* **2014**, *32*, 2680–2686. [CrossRef] [PubMed]
302. Dandawate, P.R.; Subramaniam, D.; Jensen, R.A.; Anant, S. Targeting cancer stem cells and signaling pathways by phytochemicals: Novel approach for breast cancer therapy. In *Seminars in Cancer Biology*; Academic Press: Cambridge, MA, USA, 2016; Volume 40, pp. 192–208.
303. Fabianowska-Majewska, K.; Kaufman-Szymczyk, A.; Szymanska-Kolba, A.; Jakubik, J.; Majewski, G.; Lubecka, K. Curcumin from turmeric rhizome: A potential modulator of DNA methylation machinery in breast cancer inhibition. *Nutrients* **2021**, *13*, 332. [CrossRef] [PubMed]
304. Song, X.; Zhang, M.; Dai, E.; Luo, Y. Molecular targets of curcumin in breast cancer. *Mol. Med. Rep.* **2019**, *19*, 23–29. [CrossRef]
305. Zhou, X.; Jiao, D.; Dou, M.; Zhang, W.; Lv, L.; Chen, J.; Li, L.; Wang, L.; Han, X. Curcumin inhibits the growth of triple-negative breast cancer cells by silencing EZH2 and restoring DLC1 expression. *J. Cell. Mol. Med.* **2020**, *24*, 10648–10662. [CrossRef]
306. Tu, Z.S.; Wang, Q.; Sun, D.D.; Dai, F.; Zhou, B. Design, synthesis, and evaluation of curcumin derivatives as Nrf2 activators and cytoprotectors against oxidative death. *Eur. J. Med. Chem.* **2017**, *134*, 72–85. [CrossRef] [PubMed]
307. SDarvesh, A.; BAggarwal, B.; Bishayee, A. Curcumin and liver cancer: A review. *Curr. Pharm. Biotechnol.* **2012**, *13*, 218–228. [CrossRef]
308. Moghtaderi, H.; Sepehri, H.; Attari, F. Combination of arabinogalactan and curcumin induces apoptosis in breast cancer cells in vitro and inhibits tumor growth via overexpression of p53 level in vivo. *Biomed. Pharmacother.* **2017**, *88*, 582–594. [CrossRef]

309. Zhu, J.; Li, Q.; Wu, Z.; Xu, Y.; Jiang, R. Curcumin for Treating Breast Cancer: A Review of Molecular Mechanisms, Combinations with Anticancer Drugs, and Nanosystems. *Pharmaceutics* **2024**, *16*, 79. [CrossRef]
310. Kong, W.Y.; Yee, Z.Y.; Mai, C.W.; Fang, C.M.; Abdullah, S.; Ngai, S.C. Zebularine and trichostatin A sensitized human breast adenocarcinoma cells towards tumor necrosis factor-related apoptosis inducing ligand (TRAIL)-induced apoptosis. *Heliyon* **2019**, *5*, e02468. [CrossRef]
311. Luo, L.; Wu, Y.; Liu, C.; Zou, Y.; Huang, L.; Liang, Y.; Ren, J.; Liu, Y.; Lin, Q. Elaboration and characterization of curcumin-loaded soy soluble polysaccharide (SSPS)-based nanocarriers mediated by antimicrobial peptide nisin. *Food Chem.* **2021**, *336*, 127669. [CrossRef]
312. Rodrigues, F.C.; Kumar, N.A.; Thakur, G. Developments in the anticancer activity of structurally modified curcumin: An up-to-date review. *Eur. J. Med. Chem.* **2019**, *177*, 76–104. [CrossRef]
313. Feng, T.; Wei, Y.; Lee, R.J.; Zhao, L. Liposomal curcumin and its application in cancer. *Int. J. Nanomed.* **2017**, *12*, 6027–6044. [CrossRef] [PubMed]
314. Li, S.; Fang, C.; Zhang, J.; Liu, B.; Wei, Z.; Fan, X.; Sui, Z.; Tan, Q. Catanionic lipid nanosystems improve pharmacokinetics and anti-lung cancer activity of curcumin. *Nanomed. Nanotechnol. Biol. Med.* **2016**, *12*, 1567–1579. [CrossRef] [PubMed]
315. Wang, W.; Chen, T.; Xu, H.; Ren, B.; Cheng, X.; Qi, R.; Liu, H.; Wang, Y.; Yan, L.; Chen, S.; et al. Curcumin-loaded solid lipid nanoparticles enhanced anticancer efficiency in breast cancer. *Molecules* **2018**, *23*, 1578. [CrossRef]
316. Shen, H.; Shen, J.; Pan, H.; Xu, L.; Sheng, H.; Liu, B.; Yao, M. Curcumin analog B14 has high bioavailability and enhances the effect of anti-breast cancer cells in vitro and in vivo. *Cancer Sci.* **2021**, *112*, 815–827. [CrossRef]
317. Hanafy, N.A. Optimally designed theranostic system based folic acids and chitosan as a promising mucoadhesive delivery system for encapsulating curcumin LbL nano-template against invasiveness of breast cancer. *Int. J. Biol. Macromol.* **2021**, *182*, 1981–1993. [CrossRef] [PubMed]
318. Ziasarabi, P.; Sahebkar, A.; Ghasemi, F. Evaluation of the effects of nanomicellar curcumin, berberine, and their combination with 5-fluorouracil on breast cancer cells. *Nat. Prod. Hum. Dis. Pharmacol. Mol. Targets Ther. Benefits* **2021**, *1328*, 21–35.
319. Mukhopadhyay, R.; Sen, R.; Paul, B.; Kazi, J.; Ganguly, S.; Debnath, M.C. Gemcitabine co-encapsulated with curcumin in folate decorated PLGA nanoparticles; a novel approach to treat breast adenocarcinoma. *Pharm. Res.* **2020**, *37*, 56. [CrossRef]
320. Li, N.; Wang, Z.; Zhang, Y.; Zhang, K.; Xie, J.; Liu, Y.; Li, W.; Feng, N. Curcumin-loaded redox-responsive mesoporous silica nanoparticles for targeted breast cancer therapy. *Artif. Cells Nanomed. Biotechnol.* **2018**, *46* (Suppl. S2), 921–935. [CrossRef]
321. Yeap, S.K.; Mohd Ali, N.; Akhtar, M.N.; Razak, N.A.; Chong, Z.X.; Ho, W.Y.; Boo, L.; Zareen, S.; Kurniawan, T.A.; Avtar, R.; et al. Induction of apoptosis and regulation of microRNA expression by (2 E, 6 E)-2, 6-bis-(4-hydroxy-3-methoxybenzylidene)-cyclohexanone (BHMC) treatment on MCF-7 breast cancer cells. *Molecules* **2021**, *26*, 1277. [CrossRef]
322. Duan, J.; Mansour, H.M.; Zhang, Y.; Deng, X.; Chen, Y.; Wang, J.; Pan, Y.; Zhao, J. Reversion of multidrug resistance by co-encapsulation of doxorubicin and curcumin in chitosan/poly (butyl cyanoacrylate) nanoparticles. *Int. J. Pharm.* **2012**, *426*, 193–201. [CrossRef]
323. Fathy Abd-Ellatef, G.E.; Gazzano, E.; Chirio, D.; Ragab Hamed, A.; Belisario, D.C.; Zuddas, C.; Peira, E.; Rolando, B.; Kopecka, J.; Assem Said Marie, M.; et al. Curcumin-loaded solid lipid nanoparticles bypass p-glycoprotein mediated doxorubicin resistance in triple negative breast cancer cells. *Pharmaceutics* **2020**, *12*, 96. [CrossRef]
324. Lv, L.I.; Qiu, K.; Yu, X.; Chen, C.; Qin, F.; Shi, Y.; Ou, J.; Zhang, T.; Zhu, H.; Wu, J.; et al. Amphiphilic copolymeric micelles for doxorubicin and curcumin co-delivery to reverse multidrug resistance in breast cancer. *J. Biomed. Nanotechnol.* **2016**, *12*, 973–985. [CrossRef] [PubMed]
325. Wen, C.; Fu, L.; Huang, J.; Dai, Y.; Wang, B.; Xu, G.; Wu, L.; Zhou, H. Curcumin reverses doxorubicin resistance via inhibition the efflux function of ABCB4 in doxorubicin-resistant breast cancer cells. *Mol. Med. Rep.* **2019**, *19*, 5162–5168. [CrossRef] [PubMed]
326. El-Far, A.H.; Darwish, N.H.; Mousa, S.A. Senescent colon and breast cancer cells induced by doxorubicin exhibit enhanced sensitivity to curcumin, caffeine, and thymoquinone. *Integr. Cancer Ther.* **2020**, *19*, 1534735419901160. [CrossRef] [PubMed]
327. Liu, S.; Liu, J.; He, L.; Liu, L.; Cheng, B.; Zhou, F.; Cao, D.; He, Y. A comprehensive review on the benefits and problems of curcumin with respect to human health. *Molecules* **2022**, *27*, 4400. [CrossRef]
328. Salehi, B.; Stojanović-Radić, Z.; Matejić, J.; Sharifi-Rad, M.; Kumar, N.V.; Martins, N.; Sharifi-Rad, J. The therapeutic potential of curcumin: A review of clinical trials. *Eur. J. Med. Chem.* **2019**, *163*, 527–545. [CrossRef]
329. Boon, E.A.; Croft, K.D.; Shinde, S.; Hodgson, J.M.; Ward, N.C. The acute effect of coffee on endothelial function and glucose metabolism following a glucose load in healthy human volunteers. *Food Funct.* **2017**, *8*, 3366–3373. [CrossRef]
330. Ren, B.; Kwah, M.X.; Liu, C.; Ma, Z.; Shanmugam, M.K.; Ding, L.; Xiang, X.; Ho, P.C.; Wang, L.; Ong, P.S.; et al. Resveratrol for cancer therapy: Challenges and future perspectives. *Cancer Lett.* **2021**, *515*, 63–72. [CrossRef]
331. Wahab, A.; Gao, K.; Jia, C.; Zhang, F.; Tian, G.; Murtaza, G.; Chen, J. Significance of resveratrol in clinical management of chronic diseases. *Molecules* **2017**, *22*, 1329. [CrossRef]

332. Harikumar, K.B.; Kunnumakkara, A.B.; Sethi, G.; Diagaradjane, P.; Anand, P.; Pandey, M.K.; Gelovani, J.; Krishnan, S.; Guha, S.; Aggarwal, B.B. Resveratrol, a multitargeted agent, can enhance antitumor activity of gemcitabine in vitro and in orthotopic mouse model of human pancreatic cancer. *Int. J. Cancer* **2010**, *127*, 257–268. [CrossRef] [PubMed]
333. Carter, L.G.; D'Orazio, J.A.; Pearson, K.J. Resveratrol and cancer: Focus on in vivo evidence. *Endocr.-Relat. Cancer* **2014**, *21*, R209–R225. [CrossRef]
334. Brown, V.A.; Patel, K.R.; Viskaduraki, M.; Crowell, J.A.; Perloff, M.; Booth, T.D.; Vasilinin, G.; Sen, A.; Schinas, A.M.; Piccirilli, G.; et al. Repeat dose study of the cancer chemopreventive agent resveratrol in healthy volunteers: Safety, pharmacokinetics, and effect on the insulin-like growth factor axis. *Cancer Res.* **2010**, *70*, 9003–9011. [CrossRef] [PubMed]
335. Zhu, W.; Qin, W.; Zhang, K.; Rottinghaus, G.E.; Chen, Y.C.; Kliethermes, B.; Sauter, E.R. Trans-resveratrol alters mammary promoter hypermethylation in women at increased risk for breast cancer. *Nutr. Cancer* **2012**, *64*, 393–400. [CrossRef]
336. Espinoza, J.L.; Trung, L.Q.; Inaoka, P.T.; Yamada, K.; An, D.T.; Mizuno, S.; Nakao, S.; Takami, A. The repeated administration of resveratrol has measurable effects on circulating T-cell subsets in humans. *Oxidative Med. Cell. Longev.* **2017**, *2017*, 6781872. [CrossRef] [PubMed]
337. Holcombe, R.F.; Martinez, M.; Planutis, K.; Planutiene, M. Effects of a grape-supplemented diet on proliferation and Wnt signaling in the colonic mucosa are greatest for those over age 50 and with high arginine consumption. *Nutr. J.* **2015**, *14*, 62. [CrossRef]
338. Liu, P.; Tang, W.; Xiang, K.; Li, G. Pterostilbene in the treatment of inflammatory and oncological diseases. *Front. Pharmacol.* **2024**, *14*, 1323377. [CrossRef]
339. Daniel, M.; Tollefsbol, T.O. Pterostilbene down-regulates hTERT at physiological concentrations in breast cancer cells: Potentially through the inhibition of cMyc. *J. Cell. Biochem.* **2018**, *119*, 3326–3337. [CrossRef]
340. Mak, K.K.; Wu, A.T.; Lee, W.H.; Chang, T.C.; Chiou, J.F.; Wang, L.S.; Wu, C.H.; Huang, C.Y.; Shieh, Y.S.; Chao, T.Y.; et al. Pterostilbene, a bioactive component of blueberries, suppresses the generation of breast cancer stem cells within tumor microenvironment and metastasis via modulating NF-κ B/microRNA 448 circuit. *Mol. Nutr. Food Res.* **2013**, *57*, 1123–1134. [CrossRef]
341. Hsiao, P.C.; Chou, Y.E.; Tan, P.; Lee, W.J.; Yang, S.F.; Chow, J.M.; Chen, H.Y.; Lin, C.H.; Lee, L.M.; Chien, M.H. Pterostilbene simultaneously induced G0/G1-phase arrest and MAPK-mediated mitochondrial-derived apoptosis in human acute myeloid leukemia cell lines. *PLoS ONE* **2014**, *9*, e105342. [CrossRef]
342. Tong, C.; Wang, Y.; Li, J.; Cen, W.; Zhang, W.; Zhu, Z.; Yu, J.; Lu, B. Pterostilbene inhibits gallbladder cancer progression by suppressing the PI3K/Akt pathway. *Sci. Rep.* **2021**, *11*, 4391. [CrossRef]
343. Wang, Z.; Wang, T.; Chen, X.; Cheng, J.; Wang, L. Pterostilbene regulates cell proliferation and apoptosis in non-small-cell lung cancer via targeting COX-2. *Biotechnol. Appl. Biochem.* **2023**, *70*, 106–119. [CrossRef] [PubMed]
344. Dhar, S.; Kumar, A.; Li, K.; Tzivion, G.; Levenson, A.S. Resveratrol regulates PTEN/Akt pathway through inhibition of MTA1/HDAC unit of the NuRD complex in prostate cancer. *Biochim. Biophys. Acta (BBA)-Mol. Cell Res.* **2015**, *1853*, 265–275. [CrossRef]
345. Feng, Y.; Yang, Y.; Fan, C.; Di, S.; Hu, W.; Jiang, S.; Li, T.; Ma, Z.; Chao, D.; Feng, X.; et al. Pterostilbene inhibits the growth of human esophageal cancer cells by regulating endoplasmic reticulum stress. *Cell. Physiol. Biochem.* **2016**, *38*, 1226–1244. [CrossRef] [PubMed]
346. Hung, C.M.; Liu, L.C.; Ho, C.T.; Lin, Y.C.; Way, T.D. Pterostilbene enhances TRAIL-induced apoptosis through the induction of death receptors and downregulation of cell survival proteins in TRAIL-resistance triple negative breast cancer cells. *J. Agric. Food Chem.* **2017**, *65*, 11179–11191. [CrossRef] [PubMed]
347. Mei, H.; Xiang, Y.; Mei, H.; Fang, B.; Wang, Q.; Cao, D.; Hu, Y.; Guo, T. Pterostilbene inhibits nutrient metabolism and induces apoptosis through AMPK activation in multiple myeloma cells. *Int. J. Mol. Med.* **2018**, *42*, 2676–2688. [CrossRef]
348. Moon, D.; McCormack, D.; McDonald, D.; McFadden, D. Pterostilbene induces mitochondrially derived apoptosis in breast cancer cells in vitro. *J. Surg. Res.* **2013**, *180*, 208–215. [CrossRef]
349. Bin, W.H.; Da, L.H.; Xue, Y.; Jing, B. Pterostilbene (3′, 5′-dimethoxy-resveratrol) exerts potent antitumor effects in HeLa human cervical cancer cells via disruption of mitochondrial membrane potential, apoptosis induction and targeting m-TOR/PI3K/Akt signalling pathway. *JBUON* **2018**, *23*, 1384–1389.
350. Huang, W.C.; Chan, M.L.; Chen, M.J.; Tsai, T.H.; Chen, Y.J. Modulation of macrophage polarization and lung cancer cell stemness by MUC1 and development of a related small-molecule inhibitor pterostilbene. *Oncotarget* **2016**, *7*, 39363. [CrossRef]
351. Zhang, L.; Wen, X.; Li, M.; Li, S.; Zhao, H. Targeting cancer stem cells and signaling pathways by resveratrol and pterostilbene. *Biofactors* **2018**, *44*, 61–68. [CrossRef]
352. Hojo, Y.; Kishi, S.; Mori, S.; Fujiwara-Tani, R.; Sasaki, T.; Fujii, K.; Nishiguchi, Y.; Nakashima, C.; Luo, Y.; Shinohara, H.; et al. Sunitinib and pterostilbene combination treatment exerts antitumor effects in gastric cancer via suppression of PDZD8. *Int. J. Mol. Sci.* **2022**, *23*, 4002. [CrossRef] [PubMed]

353. Roemhild, K.; von Maltzahn, F.; Weiskirchen, R.; Knüchel, R.; von Stillfried, S.; Lammers, T. Iron metabolism: Pathophysiology and pharmacology. *Trends Pharmacol. Sci.* **2021**, *42*, 640–656. [CrossRef] [PubMed]
354. Wang, R.; Hussain, A.; Guo, Q.Q.; Jin, X.W.; Wang, M.M. Oxygen and Iron Availability Shapes Metabolic Adaptations of Cancer Cells. *World J. Oncol.* **2024**, *15*, 28. [CrossRef]
355. Nishiguch, Y.; Fujiwara-Tani, R.; Nukaga, S.; Nishida, R.; Ikemoto, A.; Sasaki, R.; Mori, S.; Ogata, R.; Kishi, S.; Hojo, Y.; et al. Pterostilbene induces apoptosis from endoplasmic reticulum stress synergistically with anticancer drugs that deposit iron in mitochondria. *Int. J. Mol. Sci.* **2024**, *25*, 2611. [CrossRef]
356. Kong, Y.; Chen, G.; Xu, Z.; Yang, G.; Li, B.; Wu, X.; Xiao, W.; Xie, B.; Hu, L.; Sun, X.; et al. Pterostilbene induces apoptosis and cell cycle arrest in diffuse large B-cell lymphoma cells. *Sci. Rep.* **2016**, *6*, 37417. [CrossRef]
357. Gao, H.; Liu, Z.; Xu, W.; Wang, Q.; Zhang, C.; Ding, Y.; Nie, W.; Lai, J.; Chen, Y.; Huang, H. Pterostilbene promotes mitochondrial apoptosis and inhibits proliferation in glioma cells. *Sci. Rep.* **2021**, *11*, 6381. [CrossRef]
358. Tan, K.T.; Chen, P.W.; Li, S.; Ke, T.M.; Lin, S.H.; Yang, C.C. Pterostilbene inhibits lung squamous cell carcinoma growth in vitro and in vivo by inducing S phase arrest and apoptosis. *Oncol. Lett.* **2019**, *18*, 1631–1640. [CrossRef] [PubMed]
359. Elsherbini, A.M.; Sheweita, S.A.; Sultan, A.S. Pterostilbene as a phytochemical compound induces signaling pathways involved in the apoptosis and death of mutant P53-breast cancer cell lines. *Nutr. Cancer* **2021**, *73*, 1976–1984. [CrossRef]
360. Kapetanovic, I.M.; Muzzio, M.; Huang, Z.; Thompson, T.N.; McCormick, D.L. Pharmacokinetics, oral bioavailability, and metabolic profile of resveratrol and its dimethylether analog, pterostilbene, in rats. *Cancer Chemother. Pharmacol.* **2011**, *68*, 593–601. [CrossRef]
361. Fulda, S. Resveratrol and derivatives for the prevention and treatment of cancer. *Drug Discov. Today* **2010**, *15*, 757–765. [CrossRef]
362. Yeo, S.C.; Ho, P.C.; Lin, H.S. Pharmacokinetics of pterostilbene in S prague-D awley rats: The impacts of aqueous solubility, fasting, dose escalation, and dosing route on bioavailability. *Mol. Nutr. Food Res.* **2013**, *57*, 1015–1025. [CrossRef] [PubMed]
363. Scherzberg, M.C.; Kiehl, A.; Zivkovic, A.; Stark, H.; Stein, J.; Fürst, R.; Steinhilber, D.; Ulrich-Rückert, S. Structural modification of resveratrol leads to increased anti-tumor activity, but causes profound changes in the mode of action. *Toxicol. Appl. Pharmacol.* **2015**, *287*, 67–76. [CrossRef] [PubMed]
364. Aldawsari, F.S.; Velázquez-Martínez, C.A. 3, 4′, 5-trans-Trimethoxystilbene; a natural analogue of resveratrol with enhanced anticancer potency. *Investig. New Drugs* **2015**, *33*, 775–786. [CrossRef] [PubMed]
365. Piotrowska-Kempisty, H.; Ruciński, M.; Borys, S.; Kucińska, M.; Kaczmarek, M.; Zawierucha, P.; Wierzchowski, M.; Łażewski, D.; Murias, M.; Jodynis-Liebert, J. 3′-hydroxy-3, 4, 5, 4′-tetramethoxystilbene, the metabolite of resveratrol analogue DMU-212, inhibits ovarian cancer cell growth in vitro and in a mice xenograft model. *Sci. Rep.* **2016**, *6*, 32627. [CrossRef]
366. Pan, M.H.; Lin, C.L.; Tsai, J.H.; Ho, C.T.; Chen, W.J. 3, 5, 3′, 4′, 5′-pentamethoxystilbene (MR-5), a synthetically methoxylated analogue of resveratrol, inhibits growth and induces G1 cell cycle arrest of human breast carcinoma MCF-7 cells. *J. Agric. Food Chem.* **2010**, *58*, 226–234. [CrossRef]
367. Seyed, M.A.; Jantan, I.; Bukhari, S.N.; Vijayaraghavan, K. A comprehensive review on the chemotherapeutic potential of piceatannol for cancer treatment, with mechanistic insights. *J. Agric. Food Chem.* **2016**, *64*, 725–737. [CrossRef]
368. Choi, K.H.; Kim, J.E.; Song, N.R.; Son, J.E.; Hwang, M.K.; Byun, S.; Kim, J.H.; Lee, K.W.; Lee, H.J. Phosphoinositide 3-kinase is a novel target of piceatannol for inhibiting PDGF-BB-induced proliferation and migration in human aortic smooth muscle cells. *Cardiovasc. Res.* **2010**, *85*, 836–844. [CrossRef]
369. Patel, K.R.; Brown, V.A.; Jones, D.J.; Britton, R.G.; Hemingway, D.; Miller, A.S.; West, K.P.; Booth, T.D.; Perloff, M.; Crowell, J.A.; et al. Clinical pharmacology of resveratrol and its metabolites in colorectal cancer patients. *Cancer Res.* **2010**, *70*, 7392–7399. [CrossRef]
370. Howells, L.M.; Berry, D.P.; Elliott, P.J.; Jacobson, E.W.; Hoffmann, E.; Hegarty, B.; Brown, K.; Steward, W.P.; Gescher, A.J. Phase I randomized, double-blind pilot study of micronized resveratrol (SRT501) in patients with hepatic metastases—Safety, pharmacokinetics, and pharmacodynamics. *Cancer Prev. Res.* **2011**, *4*, 1419–1425. [CrossRef]
371. De Groote, D.; Van Belleghem, K.; Devière, J.; Van Brussel, W.; Mukaneza, A.; Amininejad, L. Effect of the intake of resveratrol, resveratrol phosphate, and catechin-rich grape seed extract on markers of oxidative stress and gene expression in adult obese subjects. *Ann. Nutr. Metab.* **2012**, *61*, 15–24. [CrossRef]
372. Hwang, D.; Shin, S.Y.; Lee, Y.; Hyun, J.; Yong, Y.; Park, J.C.; Lee, Y.H.; Lim, Y. A compound isolated from Schisandra chinensis induces apoptosis. *Bioorg. Med. Chem. Lett.* **2011**, *21*, 6054–6057. [CrossRef] [PubMed]
373. Xiang, S.S.; Wang, X.A.; Li, H.F.; Shu, Y.J.; Bao, R.F.; Zhang, F.; Cao, Y.; Ye, Y.Y.; Weng, H.; Wu, W.G.; et al. Schisandrin B induces apoptosis and cell cycle arrest of gallbladder cancer cells. *Molecules* **2014**, *19*, 13235–13250. [CrossRef]
374. Ko, Y.H.; Jeong, M.; Jang, D.S.; Choi, J.H. Gomisin L1, a lignan isolated from Schisandra berries, induces apoptosis by regulating NADPH oxidase in human ovarian cancer cells. *Life* **2021**, *11*, 858. [CrossRef]

375. Jung, S.; Moon, H.I.; Kim, S.; Quynh, N.T.; Yu, J.; Sandag, Z.; Le, D.D.; Lee, H.; Lee, H.; Lee, M.S. Anticancer activity of gomisin J from Schisandra chinensis fruit. *Oncol. Rep.* **2019**, *41*, 711–717. [CrossRef]
376. Li, Y.J.; Liu, H.T.; Xue, C.J.; Xing, X.Q.; Dong, S.T.; Wang, L.S.; Ding, C.Y.; Meng, L.; Dong, Z.J. The synergistic anti-tumor effect of schisandrin B and apatinib. *J. Asian Nat. Prod. Res.* **2020**, *22*, 839–849. [CrossRef]
377. Han, Y.H.; Mun, J.G.; Jeon, H.D.; Park, J.; Kee, J.Y.; Hong, S.H. Gomisin A ameliorates metastatic melanoma by inhibiting AMPK and ERK/JNK-mediated cell survival and metastatic phenotypes. *Phytomedicine* **2020**, *68*, 153147. [CrossRef] [PubMed]
378. Waiwut, P.; Shin, M.S.; Yokoyama, S.; Saiki, I.; Sakurai, H. Gomisin A enhances tumor necrosis factor-α-induced G1 cell cycle arrest via signal transducer and activator of transcription 1-mediated phosphorylation of retinoblastoma protein. *Biol. Pharm. Bull.* **2012**, *35*, 1997–2003. [CrossRef]
379. Maharjan, S.; Park, B.K.; Lee, S.I.; Lim, Y.; Lee, K.; Kwon, H.J. Gomisin G inhibits the growth of triple-negative breast cancer cells by suppressing AKT phosphorylation and decreasing cyclin D1. *Biomol. Ther.* **2018**, *26*, 322. [CrossRef] [PubMed]
380. Wang, Z.; Yu, K.; Hu, Y.; Su, F.; Gao, Z.; Hu, T.; Yang, Y.; Cao, X.; Qian, F. Schisantherin A induces cell apoptosis through ROS/JNK signaling pathway in human gastric cancer cells. *Biochem. Pharmacol.* **2020**, *173*, 113673. [CrossRef]
381. Jing, M.; Bi, X.J.; Yao, X.M.; Cai, F.; Liu, J.J.; Fu, M.; Kong, L.; Liu, X.Z.; Zhang, L.; He, S.Y.; et al. Enhanced antitumor efficacy using epirubicin and schisandrin B co-delivery liposomes modified with PFV via inhibiting tumor metastasis. *Drug Dev. Ind. Pharm.* **2020**, *46*, 621–634. [CrossRef]
382. He, L.; Chen, H.; Qi, Q.; Wu, N.; Wang, Y.; Chen, M.; Feng, Q.; Dong, B.; Jin, R.; Jiang, L. Schisandrin B suppresses gastric cancer cell growth and enhances the efficacy of chemotherapy drug 5-FU in vitro and in vivo. *Eur. J. Pharmacol.* **2022**, *920*, 174823. [CrossRef] [PubMed]
383. Wang, Y.; Chen, J.; Huang, Y.; Yang, S.; Tan, T.; Wang, N.; Zhang, J.; Ye, C.; Wei, M.; Luo, J.; et al. Schisandrin B suppresses osteosarcoma lung metastasis in vivo by inhibiting the activation of the Wnt/β-catenin and PI3K/Akt signaling pathways. *Oncol. Rep.* **2022**, *47*, 50. [CrossRef] [PubMed]
384. Nasser, M.I.; Han, T.; Adlat, S.; Tian, Y.; Jiang, N. Inhibitory effects of Schisandrin B on human prostate cancer cells. *Oncol. Rep.* **2019**, *41*, 677–685. [CrossRef] [PubMed]
385. Li, X.Y.; Shi, L.X.; Yao, X.M.; Jing, M.; Li, Q.Q.; Wang, Y.L.; Li, Q.S. Functional vinorelbine plus schisandrin B liposomes destroying tumor metastasis in treatment of gastric cancer. *Drug Dev. Ind. Pharm.* **2021**, *47*, 100–112. [CrossRef]
386. Wang, S.; Wang, A.; Shao, M.; Lin, L.; Li, P.; Wang, Y. Schisandrin B reverses doxorubicin resistance through inhibiting P-glycoprotein and promoting proteasome-mediated degradation of survivin. *Sci. Rep.* **2017**, *7*, 8419. [CrossRef]
387. Yan, C.; Gao, L.; Qiu, X.; Deng, C. Schisandrin B synergizes docetaxel-induced restriction of growth and invasion of cervical cancer cells in vitro and in vivo. *Ann. Transl. Med.* **2020**, *8*, 1157. [CrossRef]

biomolecules

MDPI

Review

Triggering Pyroptosis in Cancer

Daniel E. Johnson [1,2] and Zhibin Cui [3,*]

1 Department of Otolaryngology—Head and Neck Surgery, University of California at San Francisco, San Francisco, CA 94143, USA; daniel.johnson@ucsf.edu
2 Helen Diller Family Comprehensive Cancer Center, University of California at San Francisco, San Francisco, CA 94143, USA
3 Department of Oral Biology, School of Dental Medicine, University at Buffalo, The State University of New York, Buffalo, NY 14214, USA
* Correspondence: zhibincu@buffalo.edu

Abstract: Pyroptosis is an inflammatory programmed cell death recently identified as a crucial cellular process in various diseases, including cancers. Unlike other forms of cell death, canonical pyroptosis involves the specific cleavage of gasdermin by caspase-1, resulting in cell membrane damage and the release of the pro-inflammatory cytokines IL-1β and IL-18. Initially observed in innate immune cells responding to external pathogens or internal death signals, pyroptotic cell death has now been observed in numerous cell types. Recent studies have extensively explored different ways to trigger pyroptotic cell death in solid tumors, presenting a promising avenue for cancer treatment. This review outlines the mechanisms of both canonical and noncanonical pyroptosis pertinent to cancer and primarily focuses on various biomolecules that can induce pyroptosis in malignancies. This strategy aims not only to eliminate cancer cells but also to promote an improved tumor immune microenvironment. Furthermore, emerging research indicates that targeting pyroptotic pathways may improve the effectiveness of existing cancer treatments, making them more potent against resistant tumor types, offering new hope for overcoming treatment resistance in aggressive malignancies.

Keywords: cancer; pyroptosis; cell death; chemotherapy drugs; caspase-1; gasdermin; IL-1b; IL-18; *GSDM-E*; natural products; LPS

1. Introduction to Pyroptosis

Homeostasis of cell numbers in eukaryotes is maintained via a balance between cellular proliferation and cell death pathways. In cancer, this balance is lost, either as a result of aberrant proliferation or the loss of normal cell death processes, or both. Multiple forms of cell death have been described, with apoptosis being the most extensively characterized. Numerous studies have shown that apoptotic cell death is frequently abrogated in human tumors, resulting in tumor growth and resistance to anti-cancer drugs and biologics. More recently, additional forms of cell death have been identified, including autophagy, necroptosis, pyroptosis, ferroptosis, cuproptosis, and disulfidptosis. In this review, we summarize the current state of knowledge regarding the mechanisms of pyroptosis and the molecules and drugs that can be used to induce this cell death process.

The term pyroptosis was introduced by Cookson and Brennan in 2001 to describe a novel form of cell death distinct from apoptosis [1]. While the hallmark cell death features of nuclear condensation and DNA fragmentation are observed in both pyroptosis and apoptosis, the process of pyroptosis is uniquely dependent on caspase-1 protease, whereas apoptosis is typically dependent on the caspase-3 or caspase-7 proteases [1–3]. Caspase-1

 https://doi.org/10.3390/biom15030348

is responsible for proteolytic production of the pro-inflammatory cytokines interleukin-1β (IL-1β) and interleukin-18 (IL-18) [4]. Hence, in contrast to apoptosis, which does not induce inflammation, cell death resulting from pyroptosis is pro-inflammatory. Moreover, pyroptosis is associated with rapid damage to the cell membrane, leading to the early release of pro-inflammatory cytokines [3]. This also contrasts with apoptotic cell death, where the loss of cell membrane integrity is a late event in the death process. Collectively, pyroptosis is distinguished from apoptosis by caspase-1 dependency, pro-inflammatory cytokine production, and early cell membrane rupture.

Pyroptosis was initially described as a cellular suicide process that macrophages undergo in response to interaction with Salmonella or other pathogens [2,5,6]. Later, the process was identified in other innate immune cells, including neutrophils and the barrier intestinal epithelial cells [7]. The induction of pyroptosis during an innate immune response is believed to function in fighting infections by promoting the release of pro-inflammatory cytokines (i.e., IL-1β and IL-18), resulting in the attraction of other immune cells to the site of infection and effective pathogen elimination. However, due to its inflammatory-associated properties, pyroptosis has also been found to be associated with several chronic inflammatory diseases, such as arthritis, pneumonia, hepatitis, colitis, and cardiovascular disease [8,9]. In addition, significant associations have been reported between pyroptosis and various cancers, highlighting the prospect of inducing this unique form of cell death as a novel therapeutic approach for cancer. In this review, we discuss the role of pyroptosis in cancer and explore promising biomolecules that might be utilized to induce pyroptosis in human malignancies.

2. Pyroptosis Signaling Pathways

Pyroptosis is characterized by rapid cell membrane rupture, driven by the gasdermin (GSDM) protein family (particularly GSDM-D), along with secretion of the pro-inflammatory cytokines IL-1β and IL-18, which recruit other innate immune cells and enhance the cytolytic activities of NK and T-cells [10,11]. Initially, this process was thought to be mediated specifically by caspase-1, defining what is now known as the canonical pathway. However, in the past decade, additional mechanisms of pyroptosis activation have been identified, leading to the classification of these alternative routes as noncanonical pathways.

2.1. Canonical Pathway of Pyroptosis

The canonical pyroptosis pathway (Figure 1A) was originally identified in innate immune cells or antigen-presenting cells, such as macrophages and monocytes [12,13]. These cells play a defensive role by detecting "danger" signals—including pathogens, cell stress, cytokines, or disruptions to cellular homeostasis—which activate pyroptotic cell death and trigger inflammation [14].

During infection, microorganisms release pathogen-associated molecular patterns (PAMPs) like lipopolysaccharide (LPS), while stressed or damaged host cells release damage-associated molecular patterns (DAMPs), including histones, hyaluronic acid, high-mobility group protein B1 (HMGB1), and fragments of host cell RNA and DNA [15]. PAMPs and DAMPs are recognized by pattern recognition receptors (PRRs), leading to the activation of immune cells. Based on homologies in protein domains, several different PRRs have been defined, including Toll-like receptors (TLRs), nucleotide oligomerization domain (NOD)-like receptors (NLRs), retinoic acid-inducible gene-I (RIG-I)-like receptors (RLRs), C-type lectin receptors (CLRs), and absent in melanoma-2 (AIM2)-like receptors (ALRs). The most extensively studied PRR involved in the canonical pyroptosis pathway is NLRP3 (NOD-, LRR-, and pyrin domain-containing protein 3) [16,17]. Activated NLRP3 induces the assembly of the inflammasome complex by recruiting pro-caspase-1 and the adapter protein ASC (apoptosis-associated speck-like protein containing a caspase recruitment domain). The inflammasome complex recruits multiple pro-caspase-1 molecules, resulting in processing to generate active caspase-1. The activated caspase-1 then cleaves the substrate protein gasdermin D (GSDM-D), generating proteins encompassing the N-terminal domain (N-GSDM-D) and the C-terminal domain (C-GSDM-D). The N-GSDM-D protein is recruited to the cell membrane where it forms pores, resulting in the early rupture of the cell membrane, a defining feature of pyroptosis [18,19]. GSDM-D is an executor of pyroptosis and belongs to the gasdermin (GSDM) protein family, which includes six paralogues: GSDM-A to GSDM-E, and PJVK (DFNB59). In their inactive forms, GSDM proteins are primarily localized in the cytoplasm. Upon activation, the N-terminal domains translocate to the plasma membrane to execute their pore-forming functions. The expression of GSDMs is tissue- and cell-type-specific. For instance, GSDM-A is expressed in the skin, GSDM-A/B/C in the gastrointestinal tract, and GSDMB/C in the lungs and reproductive organs. GSDM-D and GSDM-E, however, are ubiquitously expressed in immune cells such as macrophages and dendritic cells, highlighting their essential role in innate immunity [20].

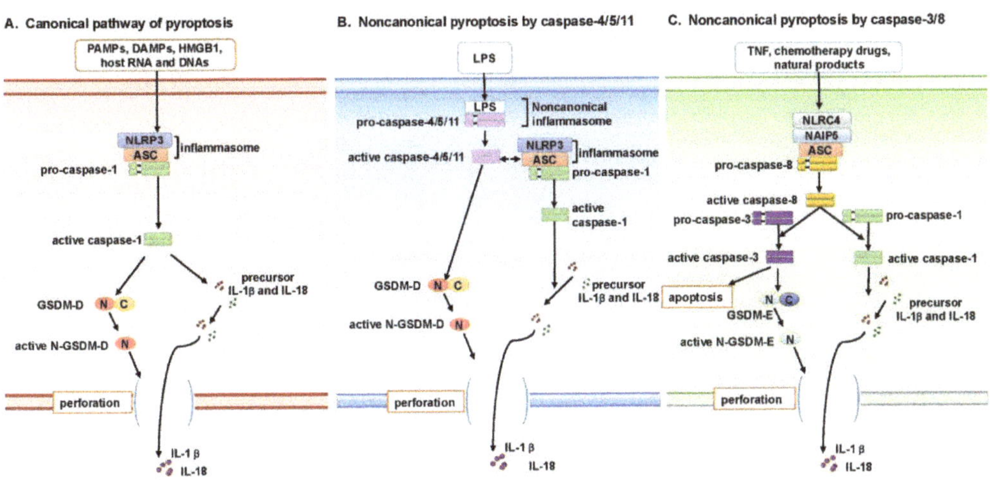

Figure 1. Canonical and noncanonical pyroptosis signaling pathways. (**A**) The canonical pyroptosis pathway in innate immune cells, like macrophages, is initiated by PAMPs and DAMPs that activate pattern recognition receptors (PRRs), primarily the NLRP3 receptor. When NLRP3 is engaged, it forms the inflammasome complex with ASC, which subsequently recruits and activates caspase-1. The active form of caspase-1 cleaves GSDM-D, leading to the formation of active N-terminal fragments (N-GSDM-D) that create pores in the cell membrane, culminating in pyroptotic cell death. Additionally, caspase-1 cleaves and activates the pro-inflammatory cytokines IL-1β and IL-18, which are released upon cell rupture and serve to attract and activate both innate and adaptive immune cells, thereby promoting inflammation. (**B**) Noncanonical pyroptosis pathways operate independently of the NLRP3/ASC/caspase-1 axis and involve the direct detection of LPS by caspase-11 in mice or by caspase-4/5 in humans. This engagement leads to the formation of a "noncanonical inflammasome", activating these caspases and resulting in GSDM-D cleavage and pyroptotic cell death. While caspase-11 does not directly process IL-1β or IL-18, it can activate the canonical NLRP3/caspase-1 inflammasome to enhance cytokine release. (**C**) Apoptotic mediators such as caspase-3 and caspase-8 can also provoke pyroptosis via noncanonical pathways triggered by TNF, chemotherapy agents, or natural products. Caspase-3 cleaves GSDM-E into N-terminal fragments, and caspase-8 can induce pyroptosis in macrophages that lack caspase-1 and -11 through a GSDM-D-independent mechanism that involves the NAIP5/NLRC4/ASC inflammasome.

IL-1β and IL-18 are pro-inflammatory cytokines that belong to the interleukin-1 protein family. Both are synthesized as inactive precursor forms which are converted to active forms upon cleavage by caspase-1 [21]. During canonical pyroptosis signaling, cell membrane rupture mediated by GSDM-D facilitates the release of IL-1b and IL-18, which promotes the recruitment of innate immune cells like neutrophils and macrophages and modulates the activation of adaptive immune cells like CD4$^+$ T-cells or even CD8$^+$ T-cells [22–24].

2.2. Noncanonical Pyroptosis Mediated by Caspase-4/5/11

Following the identification and characterization of the canonical pyroptosis pathway, multiple studies revealed that pyroptosis can also be activated through alternative signaling mechanisms, collectively referred to as noncanonical pathways. One such noncanonical pathway is triggered directly in response to the PAMP molecule LPS and involves the formation of a 'noncanonical inflammasome', distinct from the inflammasome generated via the NLRP3/ASC/caspase-1 axis (Figure 1B). In this noncanonical pathway, LPS binds directly, and with high affinity, to caspase-11 in mouse macrophages, or to its homologs, caspase-4 and caspase-5, in human monocytes, without the need for NLRP3-mediated PAMP sensing [25,26].

The binding of LPS to caspase-11 (or caspase-4/5 in humans) induces the formation of a "noncanonical inflammasome", which is sufficient to induce the oligomerization and activation of these caspases [25]. In mouse macrophages, activated caspase-11 directly cleaves GSDM-D, resulting in cell membrane perforation and pyroptotic cell death. Although caspase-11 does not directly cleave pro-IL-1β or pro-IL-18, evidence suggests it can activate a cascade involving the NLRP3/caspase-1 inflammasome, ultimately leading to IL-1β processing and release [26,27]. However, the precise mechanisms by which caspase-11 interacts with the canonical inflammasome remain unclear.

2.3. Noncanonical Pyroptosis Mediated by Caspase-3/8

In addition to pyroptosis induction by caspase-1 or caspases-4/5/11, the apoptotic mediators caspase-3 and caspase-8 also have been found to play roles in inducing a noncanonical pathway of pyroptosis (Figure 1C). The gasdermin (GSDM) protein family comprises six members (GSDM-A, -B, -C, -D, -E, and PJVK). Cleavage of these proteins generates N-terminal fragments (N-GSDM) that form pores in the cell membrane, leading to pyroptotic cell death [28]. Recent studies have shown that GSDM-E is a specific substrate of caspase-3. In cells expressing GSDM-E, tumor necrosis factor (TNF) can induce caspase-3-mediated cleavage of GSDM-E into active N-GSDM-E fragments, which perforate the cell membrane and induce pyroptosis [29].

A role for caspase-8 in noncanonical pyroptosis has come to light in studies of caspase-1 and caspase-11 double knockout mice. Infection of macrophages from these mice with Legionella pneumophila triggers the recruitment of caspase-8 to a noncanonical inflammasome complex, which consists of neuronal apoptosis inhibitory protein 5 (NAIP5), NLR family CARD-containing protein 4 (NLRC4), and ASC. Formation of this complex results in the activation of caspase-8 and subsequent cell membrane perforation via a GSDM-D-independent process [30]. Key questions regarding this pathway that remain unanswered are as follows: (1) whether caspase-8 activation is necessary to trigger caspase-3-mediated GSDM-E activation; and (2) whether the absence of caspase-1/11 and GSDM-D is essential for caspase-8-mediated membrane rupture. Additional findings reveal complex roles for caspase-8 in inflammasome activity. One study reported that expression of a catalytically inactive caspase-8 mutant (C362S) leads to ASC inflammasome complex formation and results in caspase-1 activation and IL-1β secretion. Embryonic expression of this caspase-8 mutant causes necroptosis-independent embryonic lethality due to pyroptosis [31]. Another study found that Yersinia infection, by inhibiting TGF-β-activated kinase 1 (TAK1) or IKK kinases, induces cell death mediated by both GSDM-D and GSDM-E. This pathway operates independently of caspase-1 and caspase-11 but is dependent on caspase-8 activity [32]. While the role of caspase-8 in this process is significant, it is unclear whether caspase-8 directly cleaves GSDM proteins.

3. Inducing Pyroptosis in Cancer

The role of pyroptosis in cancer development is not fully understood. Pyroptosis occurs in innate immune cells (primarily macrophages and neutrophils) in response to pathogen invasion or cellular stress, leading to the release of cytokines that trigger an inflammatory defense. Chronic inflammation, however, can promote cancer initiation [33]. During cancer progression, pyroptosis-induced immune cell infiltration may contribute to both pro-tumor and anti-tumor effects, with the balance of interactions between cancer cells and immune cells playing a critical role in cancer outcomes.

Pyroptosis has been observed not only in immune and barrier epithelial cells but also in various cancer cell types, including those representing liver cancer [34], breast cancer [35], esophageal cancer [36], lung cancer [37], colon cancer [38], and cervical cancer [39]. Despite the complex nature of pyroptosis within the tumor immune microenvironment, the induction of pyroptosis in cancer cells may present a promising opportunity for cancer therapy. The induction of pyroptosis in cancer cells not only leads to direct cell death but also releases cancer antigens, which enable antigen-presenting cells (APCs) to present these antigens to immune cells, including $CD8^+$ T-cells. Additionally, the secretion of inflammatory cytokines can further support anti-tumor immunity to some extent.

Here, we describe biomolecules that have been shown to induce pyroptosis in cancers and the potential for incorporating these molecules in new therapeutic approaches.

3.1. Chemotherapy Drugs That Induce Pyroptosis

Chemotherapy drugs are primarily known to induce cell death via caspase-3-mediated apoptosis. However, recent studies have shown that certain chemotherapeutic agents can also induce pyroptosis. In particular, topotecan, etoposide, cisplatin, and CPT-11 have been observed to trigger pyroptosis in SH-SY5Y neuroblastoma cells and MeWo skin melanoma cells, both of which exhibit high levels of gasdermin E (GSDM-E) expression [29]. These chemotherapeutic agents promote pyroptosis by inducing the caspase-3-mediated cleavage of GSDM-E. Generally, GSDM-E expression is low in most cancer cells, leading to apoptosis as the primary cell death pathway. However, in cancer cells with elevated GSDM-E expression, the cell death mechanism is shifted towards pyroptosis following chemotherapy treatment. This switch highlights the potential of GSDM-E as a key factor in modulating the response of certain cancers to chemotherapy [29]. As summarized in Table 1, multiple chemotherapy drugs have now been shown to induce pyroptosis in a variety of cancer types via caspase-3-mediated cleavage of GSDM-E.

Table 1. Chemotherapy drugs that induce pyroptosis.

Reagents	Category	Mechanism	Cells	Reference
Topotecan	Topoisomerase 1 inhibitor	caspase-3/GSDM-E	SH-SY5Y, MeWo	[29]
Etoposide	Topoisomerase II inhibitor	caspase-3/GSDM-E	SH-SY5Y, MeWo	[29]
Cisplatin	Alkylating agent	caspase-3/GSDM-E	SH-SY5Y, MeWo, A549	[37,40]
5-Fu	Anti-metabolites	caspase-3/GSDM-E	SGC-7901 MKN-45	[41]

Table 1. *Cont.*

Reagents	Category	Mechanism	Cells	Reference
Paclitaxel	Microtubules interference	caspase 3/GSDM E	A549	[41]
Ru (II) polypyridyl + Paclitaxel		caspase-1/GSDM-D	HeLa	[42]
Lobaplatin	Platinum	caspase-3/GSDM-E	HT-29, HCT116	[38]
Doxorubicin	Anthracycline	caspase-3/GSDM-E	HepG2, Hep 3B	[43]
FL118	Camptothecin	_	Colon cancer cells	[44]

3.2. *Metformin*

Metformin, well known as an anti-diabetic medication, has shown potential as an anti-cancer agent by inducing pyroptosis in cancer cells. Wang et al. [45] demonstrated that metformin induces GSDM-D-mediated pyroptosis in esophageal squamous cell carcinoma (ESCC) following inhibition of the oncogenic scaffold protein PELP1, a biomarker of poor prognosis. However, the mechanisms linking PELP1- and GSDM-D-mediated pyroptosis were not fully elaborated in this study.

Further research has demonstrated that metformin can promote cancer cell pyroptosis through activation of the SIRT1/NF-κB signaling pathway. Metformin activates SIRT1, an NAD^+-dependent deacetylase involved in regulating immunity and inflammatory responses. SIRT1 activation enhances the expression of NF-κB (p65), which then promotes the induction of BAX and the activation of caspase-3, leading to GSDM-E-mediated pyroptosis. This cascade of events underscores metformin's capacity to harness the SIRT1/NF-κB signaling axis to drive pyroptosis and positions metformin as a novel therapeutic option to induce pyroptotic cell death in cancer cells, particularly those resistant to traditional apoptotic pathways [46].

3.3. *Small Molecule Inhibitors That Induce Pyroptosis*

3.3.1. DPP8/9 Inhibitors

Dipeptidyl peptidases (DPPs) are proteases that cleave the N-terminal dipeptide of their substrates. Inhibitors of DPP8 and DPP9 have attracted significant interest due to their ability to induce pyroptosis in immune cells [47]. DPP8/9 interacts with the inflammasome complex containing the PRR protein NLRP1 and CARD8 (caspase recruitment domain-containing 8 protein). When DPP8/9 binds to this inflammasome complex, CARD8 is retained within it. However, upon inhibition of DPP8/9, CARD8 is released from the complex, triggering caspase-1 activation and induction of pyroptosis [48]. The pyroptosis pathway induced by DPP8/9 inhibitors is primarily observed in monocytes and macrophages [47,49].

Several DPP8/9 inhibitors are currently available, including the nonselective DPP inhibitor Val-boroPro, the DPP8/9 selective inhibitor 1G244, as well as CQ31, and compounds 8j and L-allo-Ile-isoindoline [50–53]. Beyond immune cells, DPP8/9 inhibitors have been shown to induce pyroptosis in acute myeloid leukemia (AML) cell lines and primary AML samples [54]. While the effects of these inhibitors on solid tumors are not yet reported, DPP8 and DPP9 expressions have been detected in various normal tissues and a range of tumors, including brain tumors, gynecological malignancies, and liver cancer. Thus, DPP8/9 inhibitors represent a potential strategy for inducing pyroptotic cell death in solid cancers.

3.3.2. BRD4 Inhibitor

BRD4 is a member of the bromodomain and extra-terminal domain (BET) protein family, and it binds to acetylated lysine residues on histone tails, thereby altering chromatin structure. The BRD4 inhibitor JQ1, a thienotriazolodiazepine-based small molecule, competitively inhibits BET-histone binding, preventing BET proteins from interacting with chromatin. Recently, JQ1 has been shown to induce pyroptosis in renal cell carcinoma (RCC) via the activation of NF-κB, which subsequently activates the NLRP3 inflammasome, leading to caspase-1-dependent pyroptosis [55]. While the precise interaction between BRD4 and NLRP3 remains unclear, another study reported that BRD4 is an essential regulator of NLRC4 inflammasome activation in Salmonella-infected bone marrow-derived macrophages [56].

3.3.3. Other Inhibitors

Recent studies have demonstrated that inhibitors targeting EGFR, MEK, and ALK in lung cancer cells that express mutant forms of these proteins induce not only apoptosis but also pyroptosis. These inhibitors activate the intrinsic apoptotic pathway through the BIM-BAX-cytochrome c release signaling axis, which subsequently triggers caspase-3/GSDM-E-mediated pyroptosis. Notably, pyroptosis induced by these inhibitors occurs concurrently with caspase-3-mediated apoptosis, and there is a complex interplay among the proteins mediating both cell death pathways. However, the precise mechanisms underlying this crosstalk are incompletely understood [57].

In melanomas harboring the BRAF V600E/K mutation, the combined use of the BRAF inhibitor PLX4720 and the MEK inhibitor PD0325901 promotes GSDM-E cleavage and the release of HMGB1, along with an upregulation of IL-1β and IL-18, leading to pyroptosis induction [58]. This pyroptosis significantly enhances $CD8^+$ T-cell infiltration into the tumor microenvironment, facilitating the immune-mediated inhibition of tumor growth. Interestingly, melanoma tumors that relapse after this combination therapy remain susceptible to etoposide- and doxorubicin-induced pyroptosis, highlighting the potential of pyroptosis induction as a therapeutic strategy not only for primary tumors but also for recurrent tumors resistant to targeted therapies [58].

Additional small molecule inhibitors that act to induce pyroptosis are summarized in Table 2.

Table 2. Small-molecule inhibitors that induce pyroptosis.

Name	Target	Mechanisms	Cancer Types	Reference
BI2536	PLK1	caspase-3/GSDM-E	Esophageal	[59,60]
Alpha-NETA	Choline acetylcholine transferase	caspase-4/GSDM-D	Ovarian	[61]
BIX-01294 (BIX)	G9a	caspase-3/GSDM-E	Gastric	[62]
NO. 0449-0145	APE1	caspase-4/GSDM-D	Lung	[63]
Famotidine	Histamine H2-receptor	GSDM-E	gastric	[64]
Bexarotene	RXR	caspase-4/GSDM-E	ovarian	[65]
AT7519	CDK	caspase-3/GSDM-E	Glioblastoma	[66]
Elraglusib	GSK-3	GSDM-B	Colon	[67]

3.4. *Natural Products That Induce Pyroptosis*

3.4.1. Curcumin

Curcumin is a natural compound derived from the rhizomes of turmeric (*Curcuma longa*), a hydrophobic polyphenol (Table 3). The United States Food and Drug Administration (FDA) has approved the use of curcumin as a safe flavoring and antioxidant agent in certain foods [68]. Over the past two decades, extensive research has shown that curcumin interferes with multiple cell signaling pathways, impacting cancer by activating apoptosis, inhibiting proliferation, reducing inflammation, and suppressing angiogenesis and metastasis [69–71]. In light of its promising impact in preclinical cancer models, curcumin-containing products have been investigated in numerous clinical trials for treating pancreatic, breast, prostate, and colorectal cancers [72–74].

Table 3. Natural products that induce pyroptosis.

Name	Category	Mechanisms	Cancer Types	Reference
Curcumin	Polyphenol	caspase-3/GSDM-E	Liver	[75]
Anthocyanins	Flavonoid	caspase-1/GSDM-D	Oral	[76]
Miltirone	Abietane-type diterpenoids	caspase-3/GSDM-E	Liver	[77]
Tetraarsenic hexoxide	Arsenic oxide	caspase-3/GSDM-E	Breast	[78]
Triptolide	Diterpenoid epoxide	caspase-3/GSDM-E	Oral	[79]
Trichosanthin	Ribosome-inactivating protein	caspase-1/GSDM-D	Lung	[80]
Aloe-emodin	Anthraquinone	caspase-3/GSDM-E	Cervical	[81]
Myricetin	Flavonoid	caspase-3/GSDM-E	Lung	[82]
Luteolin	Flavonoid	caspase-1/GSDM-D	Colon	[83]
Mallotucin D	Flavonoid glycoside	caspase-3/GSDM-E	Liver	[84]
Shikonin	Naphthoquinone	caspase-3/GSDM-E	Gastric	[85]

Recent studies indicate that curcumin can induce pyroptosis in cancer cells. In HepG2 hepatocellular carcinoma cells, curcumin treatment activates the intrinsic apoptosis pathway, which in turn triggers caspase-3/GSDM-E-mediated pyroptotic cell death. This process results from elevation in the levels of reactive oxygen species (ROS) following curcumin treatment [75]. However, curcumin is far from a specific inducer of pyroptosis, as it also induces other forms of cell death, including apoptosis, necroptosis, and ferroptosis [86,87].

3.4.2. Iron

Iron plays a crucial role in both the formation and destruction of intracellular ROS, which serve as key mediators in multiple forms of cell death. Mitochondrial ROS, for example, drives intrinsic apoptosis, while ROS are also integral to necroptosis, ferroptosis, and, as recently identified, ROS-induced pyroptosis [88]. In melanoma cells, the combination of iron with the ROS-inducing agent carbonyl cyanide m-chlorophenyl hydrazone (CCCP) markedly amplifies ROS signaling, resulting in oxidation and oligomerization of the mitochondrial outer membrane protein TOM20. Oxidized TOM20 subsequently recruits BAX to the mitochondria, facilitating cytochrome c release into the cytosol and caspase-3 activation, which then leads to GSDM-E-mediated pyroptosis, highlighting a mechanistic pathway by which intracellular iron and enhanced ROS drive pyroptosis in melanoma [89].

3.4.3. Cucurbitacin B

Cucurbitacin B (CuB) is a bioactive compound derived from *cucurbitaceae* plants that has shown significant promise as an anti-tumor and anti-inflammatory agent. In non-small cell lung cancer (NSCLC), cucurbitacin B induces pyroptosis by binding directly to Toll-like receptor 4 (TLR4). This binding activates the NLRP3 inflammasome, resulting in the cleavage of GSDM-D and subsequent pyroptotic cell death. Cucurbitacin B further promotes pyroptosis by increasing the production of mitochondrial ROS, leading to the accumulation/oxidation of mitochondrial TOM20, as well as to the release of cytosolic calcium. In mouse models of NSCLC, cucurbitacin B effectively inhibits tumor growth and shows superior anti-tumor effects compared to standard treatments, underscoring its potential as a targeted therapy through the TLR4/NLRP3/GSDMD signaling axis [90].

Beyond its anti-cancer role, cucurbitacin B also exhibits anti-inflammatory properties, particularly in osteoarthritis. In osteoarthritis models, cucurbitacin B interacts with key signaling proteins, including NRF2 and NF-κB, to inhibit the secretion of pro-inflammatory cytokines such as IL-1β and IL-18, thereby reducing inflammation [91]. These effects indicate that the impact of cucurbitacin B treatment may differ according to tissue and disease type, acting as an anti-inflammatory agent in osteoarthritis and triggering pyroptosis in cancer. This adaptability makes cucurbitacin B a strong candidate for distinct therapeutic applications across multiple diseases settings.

Other natural products that induce pyroptosis are summarized in Table 3.

3.5. Pathogen or Pathogen-Derived Compounds That Induce Pyroptosis

Lipopolysaccharide

Lipopolysaccharide (LPS) is a bacterial endotoxin known to trigger pyroptosis in immune cells, particularly in monocytes and macrophages. In HT29 colon cancer cells, LPS exposure induces cleavage/activation of GSDM-D, leading to the secretion of IL-1β and IL-18 and, ultimately, pyroptotic cell death. Furthermore, LPS has been shown to increase the chemosensitivity of HT29 cells to oxaliplatin, and this combined treatment enhances GSDM-D-mediated pyroptosis. These findings suggest that combination of LPS with chemotherapy drugs may provide a strategy to promote pyroptosis and improve therapeutic outcomes in colon cancer [92].

For therapeutic purposes focused on the induction of pyroptosis, researchers have utilized bacterial outer membrane vesicles (OMVs) as natural carriers to deliver LPS into cancer cells. OMV-LPS has been shown to effectively induce GSDM-D-mediated pyroptosis in cancer cells, leading to a significant inhibition of tumor growth in vivo [93]. Additionally, OMV-LPS treatment promoted greater $CD8^+$ T-cell infiltration into the tumor microenvironment, enhancing anti-tumor immunity. This approach represents a promising strategy for treating solid tumors by inducing cancer cell death and stimulating immune responses [93].

Other pathogen or pathogen-derived components that induce pyroptosis in cancer cells are summarized in Table 4.

Table 4. Pathogen or pathogen-derived compound that induce pyroptosis.

Name	Category	Mechanisms	Cancer Types	Reference
4-hydroxybenzoic acid	Benzoic acid derivatives	caspase-1/GSDM-D	Lung	[94]
EV-A71	Enterovirus	caspase-1/GSDM-D	Neuroblastoma	[95]
CagA	Protein from *Helicobacter pylori*	caspase-1/GSDM-D	Gastric	[96]
Apoptin	Protein from chicken anemia virus	caspase-3/GSDM-E	Colon	[97]
Talaromyces marneffei	Fungus	caspase-1/GSDM-D	Liver	[98]
Coxsackievirus group B3	Enterovirus	caspase-3/GSDM-E	Colon	[99]

3.6. Endogenous Metabolites That Induce Pyroptosis

Metabolites play pivotal roles in cancer development, influencing cellular processes like proliferation, survival, and immune evasion within the tumor microenvironment. Recently, several endogenous metabolites—including omega-3 docosahexaenoic acid, citric acid, α, β-unsaturated ketone, and α-ketoglutarate (α-KG)—were identified as inducers of pyroptosis (Table 5). These metabolites can engage inflammasome pathways or enhance ROS production, thereby activating pyroptotic signaling in cancer cells.

Table 5. Endogenous metabolites that induce pyroptosis.

Name	Category	Mechanisms	Cancer Types	Reference
Omega-3 docosahexaenoic acid	Polyunsaturated fatty acid	caspase-1/GSDM-D	Breast	[100]
Citric acid	Tricarboxylic acid	caspase-4/GSDM-D	Ovarian	[101]
α, β-Unsaturated ketone	Unsaturated carbonyl compound	caspase-3/GSDM-E	Lung cancer	[102]

As an example, α-KG is a key metabolite in the tricarboxylic acid (TCA) cycle and is involved in lipid biosynthesis, oxidative stress response, protein modification, autophagy, and cell death. α-KG induces pyroptosis in cancer cells via caspase-8-mediated cleavage of GSDM-C, showing promise for limiting tumor growth and metastasis. Treatment with DM-α-KG, a cell-permeable α-KG derivative, elevates ROS levels, which oxidizes and triggers endocytosis of the plasma membrane receptor DR6. This process recruits pro-caspase-8 and GSDM-C into a "receptosome", where active caspase-8 cleaves GSDM-C, initiating pyroptosis [103]. This mechanism relies on an acidic environment, where α-KG is converted to L-2-hydroxyglutarate (L-2HG) by malate dehydrogenase 1 (MDH1), boosting ROS production. The presence of lactic acid further intensifies this effect, sensitizing even pyroptosis-resistant cancer cells to α-KG-induced pyroptosis [103]. Together, these insights reveal the dual role of metabolites as drivers of tumor progression and as therapeutic agents. Further exploration into metabolite-driven pyroptosis could inspire innovative treatments that reshape the tumor immune microenvironment for more effective cancer therapy.

4. Challenges and Opportunities

The induction of pyroptosis shows great promise as a therapeutic strategy, especially for cancers resistant to conventional chemotherapies that induce apoptosis. Chemotherapy-induced cell death involves not only apoptosis but also necroptosis and pyroptosis, yet clear molecular mechanisms that differentiate these processes or highlight crosstalk are not well defined. Numerous chemotherapeutic agents trigger the intrinsic apoptotic pathway through caspase-3 activation. This, in turn, leads to the cleavage of multiple substrates, such as PARP-1, which promote apoptotic cell death, and GSDM-E, which initiates pyroptosis. Although apoptosis and pyroptosis may occur simultaneously during caspase-3 activation, pyroptosis is more immunogenic and promotes immune cell infiltration, particularly by $CD8^+$ T-cells, thereby boosting anti-tumor responses. Nonetheless, further research is necessary to clarify how apoptosis might be switched toward pyroptosis to enhance therapeutic outcomes.

The role of pyroptosis in various stages of cancer development is still not fully understood. While chronic inflammation can facilitate cellular transformation, cytokine release due to pyroptosis has the potential to attract effector immune cells to combat cancer. A significant challenge is how to effectively induce pyroptosis in cancers with varied genetic backgrounds. For instance, many cancer cells demonstrate low GSDM-E expression, making pyroptosis induction more challenging. Overcoming this limitation requires a deeper understanding of how pyroptosis affects cancer progression across different genetic contexts and tumor types. To date, no clinical trials have been undertaken with the primary goal of targeting pyroptosis in cancer. Combined therapies that utilize agents capable of converting apoptosis into pyroptosis may enhance treatment efficacy, particularly in apoptosis-resistant cancers. Pyroptosis facilitates the conversion of immunologically "cold" tumors into more immunogenic phenotypes, thereby amplifying the advantages of immunotherapy. Merging pyroptosis-inducing strategies with immunotherapeutic methods might significantly advance cancer treatment. It is crucial to comprehend the mechanisms that regulate pyroptosis and its interplay with the immune microenvironment to advance this strategy. Since pyroptosis-induced inflammation may, under certain circumstances, contribute to tumor progression, combining anti-inflammatory drugs may serve as a complementary approach to enhance the efficacy of anti-pyroptosis strategies. This combinatorial approach could help mitigate the pro-tumorigenic effects of inflammation while simultaneously targeting pyroptosis, offering a more comprehensive therapeutic strategy for tumor suppression. By tackling these challenges, pyroptosis-based therapies may become vital for innovative cancer treatments, providing new hope for patients with resistant tumors.

Natural products, primarily derived from plants and microbes, have proven to be valuable sources of anti-cancer agents, including those capable of inducing pyroptosis. These compounds enhance ROS generation and activate pathways comprising signaling molecules like NLRP3, MEK/ERK1/2, and JNK, thereby promoting tumor cell death while fostering an immune-stimulatory tumor microenvironment. However, limitations such as low bioavailability and off-target effects have hindered the clinical application of natural products. Nanotechnology offers a potential solution by improving the solubility, stability, and specificity of natural products through nanosizing and targeted delivery systems. The incorporation of pyroptosis-inducing agents into nanoparticles enhances their efficacy and enables synergy with therapies like chemotherapy and immunotherapy. These advancements position natural product-based pyroptosis inducers as promising tools for precise and potent cancer immunotherapy.

Collectively, despite significant progress, challenges remain in translating pyroptosis-based therapies to clinical settings. Effective screening strategies are essential for identifying potent pyroptosis-inducing agents. Addressing tumor heterogeneity and the pro-inflammatory nature of pyroptosis is also critical to minimizing side effects and increasing treatment specificity. By integrating pyroptosis inducers with other therapeutic modalities, these combinatorial strategies may overcome treatment resistance, stimulate robust immune responses, and establish pyroptosis as a cornerstone in the fight against cancer.

Cancer cells exhibit unique vulnerabilities to cell death, and exploitation of these vulnerabilities has strong potential to lead to tumor-selective anti-cancer therapies. Pyroptosis represents a new area of cell death research that should be explored as a means of provoking cell death in cancer cells and tumors. At present, only a modest number of biomolecules have been identified to promote pyroptosis. More work is needed to identify pyroptosis-inducing agents, and combination of these agents with immune modulating therapies is warranted. Coincident with these investigations, it will be important to closely monitor whether selective induction of pyroptosis leads to adverse toxicities, and whether biomarkers that will help determine which patient populations are most likely to benefit from treatment with pyroptosis inducers can be identified. As research into these questions continues, we will learn the full utility of targeting pyroptosis to improve clinical outcomes in patients suffering from cancer.

Author Contributions: D.E.J.: Manuscript writing and editing, and figure and table designing; Z.C.: Manuscript writing and editing, and figure and table designing. All authors have read and agreed to the published version of the manuscript.

Funding: This work was supported by NIH grant R01 DE028289 (DEJ) and the New York State Foundation (ZC).

Institutional Review Board Statement: Not applicable.

Informed Consent Statement: Not applicable.

Data Availability Statement: Not applicable.

Acknowledgments: The grammar correction tool Grammarly Inc. was used to enhance the language by improving the quality of English and readability, all while being utilized under human oversight and control.

Conflicts of Interest: The authors declare no conflicts of interest.

References

1. Cookson, B.T.; Brennan, M.A. Pro-inflammatory programmed cell death. *Trends Microbiol.* **2001**, *9*, 113–114. [CrossRef] [PubMed]
2. Hersh, D.; Monack, D.M.; Smith, M.R.; Ghori, N.; Falkow, S.; Zychlinsky, A. The Salmonella invasin SipB induces macrophage apoptosis by binding to caspase-1. *Proc. Natl. Acad. Sci. USA* **1999**, *96*, 2396–2401. [CrossRef] [PubMed]
3. Brennan, M.A.; Cookson, B.T. Salmonella induces macrophage death by caspase-1-dependent necrosis. *Mol. Microbiol.* **2000**, *38*, 31–40. [CrossRef] [PubMed]
4. Arend, W.P.; Palmer, G.; Gabay, C. IL-1, IL-18, and IL-33 families of cytokines. *Immunol. Rev.* **2008**, *223*, 20–38. [CrossRef]
5. Zychlinsky, A.; Prevost, M.C.; Sansonetti, P.J. Shigella flexneri induces apoptosis in infected macrophages. *Nature* **1992**, *358*, 167–169. [CrossRef]
6. Man, S.M.; Hopkins, L.J.; Nugent, E.; Cox, S.; Gluck, I.M.; Tourlomousis, P.; Wright, J.A.; Cicuta, P.; Monie, T.P.; Bryant, C.E. Inflammasome activation causes dual recruitment of NLRC4 and NLRP3 to the same macromolecular complex. *Proc. Natl. Acad. Sci. USA* **2014**, *111*, 7403–7408. [CrossRef]
7. Oh, C.; Spears, T.J.; Aachoui, Y. Inflammasome-mediated pyroptosis in defense against pathogenic bacteria. *Immunol. Rev.* **2024**, *329*, e13408. [CrossRef]
8. Wu, Y.; Zhang, J.; Yu, S.; Li, Y.; Zhu, J.; Zhang, K.; Zhang, R. Cell pyroptosis in health and inflammatory diseases. *Cell Death Discov.* **2022**, *8*, 191. [CrossRef]
9. Zhaolin, Z.; Guohua, L.; Shiyuan, W.; Zuo, W. Role of pyroptosis in cardiovascular disease. *Cell Prolif.* **2019**, *52*, e12563. [CrossRef]
10. Dinarello, C.A. Immunological and inflammatory functions of the interleukin-1 family. *Annu. Rev. Immunol.* **2009**, *27*, 519–550. [CrossRef]
11. Jorgensen, I.; Lopez, J.P.; Laufer, S.A.; Miao, E.A. IL-1beta, IL-18, and eicosanoids promote neutrophil recruitment to pore-induced intracellular traps following pyroptosis. *Eur. J. Immunol.* **2016**, *46*, 2761–2766. [CrossRef] [PubMed]
12. Cordoba-Rodriguez, R.; Fang, H.; Lankford, C.S.; Frucht, D.M. Anthrax lethal toxin rapidly activates caspase-1/ICE and induces extracellular release of interleukin (IL)-1beta and IL-18. *J. Biol. Chem.* **2004**, *279*, 20563–20566. [CrossRef] [PubMed]
13. Boyden, E.D.; Dietrich, W.F. Nalp1b controls mouse macrophage susceptibility to anthrax lethal toxin. *Nat. Genet.* **2006**, *38*, 240–244. [CrossRef] [PubMed]
14. Man, S.M.; Kanneganti, T.D. Innate immune sensing of cell death in disease and therapeutics. *Nat. Cell Biol.* **2024**, *26*, 1420–1433. [CrossRef]
15. Tang, D.; Kang, R.; Coyne, C.B.; Zeh, H.J.; Lotze, M.T. PAMPs and DAMPs: Signal 0s that spur autophagy and immunity. *Immunol. Rev.* **2012**, *249*, 158–175. [CrossRef]
16. Wei, X.; Xie, F.; Zhou, X.; Wu, Y.; Yan, H.; Liu, T.; Huang, J.; Wang, F.; Zhou, F.; Zhang, L. Role of pyroptosis in inflammation and cancer. *Cell Mol. Immunol.* **2022**, *19*, 971–992. [CrossRef]
17. Jin, C.; Flavell, R.A. Molecular mechanism of NLRP3 inflammasome activation. *J. Clin. Immunol.* **2010**, *30*, 628–631. [CrossRef]
18. Ding, J.; Wang, K.; Liu, W.; She, Y.; Sun, Q.; Shi, J.; Sun, H.; Wang, D.C.; Shao, F. Pore-forming activity and structural autoinhibition of the gasdermin family. *Nature* **2016**, *535*, 111–116. [CrossRef]
19. Liu, X.; Zhang, Z.; Ruan, J.; Pan, Y.; Magupalli, V.G.; Wu, H.; Lieberman, J. Inflammasome-activated gasdermin D causes pyroptosis by forming membrane pores. *Nature* **2016**, *535*, 153–158. [CrossRef]
20. Chen, K.W.; Broz, P. Gasdermins as evolutionarily conserved executors of inflammation and cell death. *Nat. Cell Biol.* **2024**, *26*, 1394–1406. [CrossRef]
21. Dinarello, C.A. Overview of the IL-1 family in innate inflammation and acquired immunity. *Immunol. Rev.* **2018**, *281*, 8–27. [CrossRef]
22. Evavold, C.L.; Ruan, J.; Tan, Y.; Xia, S.; Wu, H.; Kagan, J.C. The Pore-Forming Protein Gasdermin D Regulates Interleukin-1 Secretion from Living Macrophages. *Immunity* **2018**, *48*, 35–44 e36. [CrossRef] [PubMed]
23. Iwasaki, A.; Medzhitov, R. Control of adaptive immunity by the innate immune system. *Nat. Immunol.* **2015**, *16*, 343–353. [CrossRef] [PubMed]
24. Van Den Eeckhout, B.; Tavernier, J.; Gerlo, S. Interleukin-1 as Innate Mediator of T Cell Immunity. *Front. Immunol.* **2020**, *11*, 621931. [CrossRef]
25. Shi, J.; Zhao, Y.; Wang, Y.; Gao, W.; Ding, J.; Li, P.; Hu, L.; Shao, F. Inflammatory caspases are innate immune receptors for intracellular LPS. *Nature* **2014**, *514*, 187–192. [CrossRef]

26. Kayagaki, N.; Warming, S.; Lamkanfi, M.; Vande Walle, L.; Louie, S.; Dong, J.; Newton, K.; Qu, Y.; Liu, J.; Heldens, S.; et al. Non-canonical inflammasome activation targets caspase-11. *Nature* **2011**, *479*, 117–121. [CrossRef]
27. Kayagaki, N.; Stowe, I.B.; Lee, B.L.; O'Rourke, K.; Anderson, K.; Warming, S.; Cuellar, T.; Haley, B.; Roose-Girma, M.; Phung, Q.T.; et al. Caspase-11 cleaves gasdermin D for non-canonical inflammasome signalling. *Nature* **2015**, *526*, 666–671. [CrossRef]
28. Zhu, C.; Xu, S.; Jiang, R.; Yu, Y.; Bian, J.; Zou, Z. The gasdermin family: Emerging therapeutic targets in diseases. *Signal Transduct. Target. Ther.* **2024**, *9*, 87. [CrossRef]
29. Wang, Y.; Gao, W.; Shi, X.; Ding, J.; Liu, W.; He, H.; Wang, K.; Shao, F. Chemotherapy drugs induce pyroptosis through caspase-3 cleavage of a gasdermin. *Nature* **2017**, *547*, 99–103. [CrossRef]
30. Mascarenhas, D.P.A.; Cerqueira, D.M.; Pereira, M.S.F.; Castanheira, F.V.S.; Fernandes, T.D.; Manin, G.Z.; Cunha, L.D.; Zamboni, D.S. Inhibition of caspase-1 or gasdermin-D enable caspase-8 activation in the Naip5/NLRC4/ASC inflammasome. *PLoS Pathog.* **2017**, *13*, e1006502. [CrossRef]
31. Fritsch, M.; Gunther, S.D.; Schwarzer, R.; Albert, M.C.; Schorn, F.; Werthenbach, J.P.; Schiffmann, L.M.; Stair, N.; Stocks, H.; Seeger, J.M.; et al. Caspase-8 is the molecular switch for apoptosis, necroptosis and pyroptosis. *Nature* **2019**, *575*, 683–687. [CrossRef] [PubMed]
32. Orning, P.; Weng, D.; Starheim, K.; Ratner, D.; Best, Z.; Lee, B.; Brooks, A.; Xia, S.; Wu, H.; Kelliher, M.A.; et al. Pathogen blockade of TAK1 triggers caspase-8-dependent cleavage of gasdermin D and cell death. *Science* **2018**, *362*, 1064–1069. [CrossRef] [PubMed]
33. Pazhouhesh Far, N.; Hajiheidari Varnousafaderani, M.; Faghihkhorasani, F.; Etemad, S.; Abdulwahid, A.R.R.; Bakhtiarinia, N.; Mousaei, A.; Dortaj, E.; Karimi, S.; Ebrahimi, N.; et al. Breaking the barriers: Overcoming cancer resistance by targeting the NLRP3 inflammasome. *Br. J. Pharmacol.* **2024**, *182*, 3–25. [CrossRef]
34. Wu, X.; Cao, J.; Wan, X.; Du, S. Programmed cell death in hepatocellular carcinoma: Mechanisms and therapeutic prospects. *Cell Death Discov.* **2024**, *10*, 356. [CrossRef]
35. Yan, H.; Luo, B.; Wu, X.; Guan, F.; Yu, X.; Zhao, L.; Ke, X.; Wu, J.; Yuan, J. Cisplatin Induces Pyroptosis via Activation of MEG3/NLRP3/caspase-1/GSDMD Pathway in Triple-Negative Breast Cancer. *Int. J. Biol. Sci.* **2021**, *17*, 2606–2621. [CrossRef] [PubMed]
36. Li, R.Y.; Zheng, Z.Y.; Li, Z.M.; Heng, J.H.; Zheng, Y.Q.; Deng, D.X.; Xu, X.E.; Liao, L.D.; Lin, W.; Xu, H.Y.; et al. Cisplatin-induced pyroptosis is mediated via the CAPN1/CAPN2-BAK/BAX-caspase-9-caspase-3-GSDME axis in esophageal cancer. *Chem. Biol. Interact.* **2022**, *361*, 109967. [CrossRef]
37. Zhang, C.C.; Li, C.G.; Wang, Y.F.; Xu, L.H.; He, X.H.; Zeng, Q.Z.; Zeng, C.Y.; Mai, F.Y.; Hu, B.; Ouyang, D.Y. Chemotherapeutic paclitaxel and cisplatin differentially induce pyroptosis in A549 lung cancer cells via caspase-3/GSDME activation. *Apoptosis* **2019**, *24*, 312–325. [CrossRef]
38. Yu, J.; Li, S.; Qi, J.; Chen, Z.; Wu, Y.; Guo, J.; Wang, K.; Sun, X.; Zheng, J. Cleavage of GSDME by caspase-3 determines lobaplatin-induced pyroptosis in colon cancer cells. *Cell Death Dis.* **2019**, *10*, 193. [CrossRef]
39. Chen, J.; Ge, L.; Shi, X.; Liu, J.; Ruan, H.; Heng, D.; Ye, C. Lobaplatin Induces Pyroptosis in Cervical Cancer Cells via the Caspase-3/GSDME Pathway. *Anticancer. Agents Med. Chem.* **2022**, *22*, 2091–2097. [CrossRef]
40. Peng, Z.; Wang, P.; Song, W.; Yao, Q.; Li, Y.; Liu, L.; Li, Y.; Zhou, S. GSDME enhances Cisplatin sensitivity to regress non-small cell lung carcinoma by mediating pyroptosis to trigger antitumor immunocyte infiltration. *Signal Transduct. Target. Ther.* **2020**, *5*, 159. [CrossRef]
41. Wang, Y.; Yin, B.; Li, D.; Wang, G.; Han, X.; Sun, X. GSDME mediates caspase-3-dependent pyroptosis in gastric cancer. *Biochem. Biophys. Res. Commun.* **2018**, *495*, 1418–1425. [CrossRef] [PubMed]
42. Chen, D.; Guo, S.; Tang, X.; Rong, Y.; Bo, H.; Shen, H.; Zhao, Z.; Qiao, A.; Shen, J.; Wang, J. Combination of ruthenium (II) polypyridyl complex Delta-Ru1 and Taxol enhances the anti-cancer effect on Taxol-resistant cancer cells through Caspase-1/GSDMD-mediated pyroptosis. *J. Inorg. Biochem.* **2022**, *230*, 111749. [CrossRef] [PubMed]
43. Xiao, B.; Ying, C.; Chen, Y.; Huang, F.; Wang, B.; Fang, H.; Guo, W.; Liu, T.; Zhou, X.; Huang, B.; et al. Doxorubicin hydrochloride enhanced antitumour effect of CEA-regulated oncolytic virotherapy in live cancer cells and a mouse model. *J. Cell Mol. Med.* **2020**, *24*, 13431–13439. [CrossRef] [PubMed]
44. Tang, Z.; Ji, L.; Han, M.; Xie, J.; Zhong, F.; Zhang, X.; Su, Q.; Yang, Z.; Liu, Z.; Gao, H.; et al. Pyroptosis is involved in the inhibitory effect of FL118 on growth and metastasis in colorectal cancer. *Life Sci.* **2020**, *257*, 118065. [CrossRef]
45. Wang, L.; Li, K.; Lin, X.; Yao, Z.; Wang, S.; Xiong, X.; Ning, Z.; Wang, J.; Xu, X.; Jiang, Y.; et al. Metformin induces human esophageal carcinoma cell pyroptosis by targeting the miR-497/PELP1 axis. *Cancer Lett.* **2019**, *450*, 22–31. [CrossRef]
46. Zheng, Z.; Bian, Y.; Zhang, Y.; Ren, G.; Li, G. Metformin activates AMPK/SIRT1/NF-kappaB pathway and induces mitochondrial dysfunction to drive caspase3/GSDME-mediated cancer cell pyroptosis. *Cell Cycle* **2020**, *19*, 1089–1104. [CrossRef]
47. Okondo, M.C.; Johnson, D.C.; Sridharan, R.; Go, E.B.; Chui, A.J.; Wang, M.S.; Poplawski, S.E.; Wu, W.; Liu, Y.; Lai, J.H.; et al. DPP8 and DPP9 inhibition induces pro-caspase-1-dependent monocyte and macrophage pyroptosis. *Nat. Chem. Biol.* **2017**, *13*, 46–53. [CrossRef]

48. Karakaya, T.; Slaufova, M.; Di Filippo, M.; Hennig, P.; Kundig, T.; Beer, H.D. CARD8: A Novel Inflammasome Sensor with Well-Known Anti-Inflammatory and Anti-Apoptotic Activity. *Cells* **2024**, *13*, 1032. [CrossRef]
49. de Vasconcelos, N.M.; Vliegen, G.; Goncalves, A.; De Hert, E.; Martin-Perez, R.; Van Opdenbosch, N.; Jallapally, A.; Geiss-Friedlander, R.; Lambeir, A.M.; Augustyns, K.; et al. DPP8/DPP9 inhibition elicits canonical Nlrp1b inflammasome hallmarks in murine macrophages. *Life Sci. Alliance* **2019**, *2*, 313. [CrossRef]
50. Coutts, S.J.; Kelly, T.A.; Snow, R.J.; Kennedy, C.A.; Barton, R.W.; Adams, J.; Krolikowski, D.A.; Freeman, D.M.; Campbell, S.J.; Ksiazek, J.F.; et al. Structure-activity relationships of boronic acid inhibitors of dipeptidyl peptidase IV. 1. Variation of the P2 position of Xaa-boroPro dipeptides. *J. Med. Chem.* **1996**, *39*, 2087–2094. [CrossRef]
51. Lankas, G.R.; Leiting, B.; Roy, R.S.; Eiermann, G.J.; Beconi, M.G.; Biftu, T.; Chan, C.C.; Edmondson, S.; Feeney, W.P.; He, H.; et al. Dipeptidyl peptidase IV inhibition for the treatment of type 2 diabetes: Potential importance of selectivity over dipeptidyl peptidases 8 and 9. *Diabetes* **2005**, *54*, 2988–2994. [CrossRef] [PubMed]
52. Van Goethem, S.; Van der Veken, P.; Dubois, V.; Soroka, A.; Lambeir, A.M.; Chen, X.; Haemers, A.; Scharpe, S.; De Meester, I.; Augustyns, K. Inhibitors of dipeptidyl peptidase 8 and dipeptidyl peptidase 9. Part 2: Isoindoline containing inhibitors. *Bioorg. Med. Chem. Lett.* **2008**, *18*, 4159–4162. [CrossRef] [PubMed]
53. Rao, S.D.; Chen, Q.; Wang, Q.; Orth-He, E.L.; Saoi, M.; Griswold, A.R.; Bhattacharjee, A.; Ball, D.P.; Huang, H.C.; Chui, A.J.; et al. M24B aminopeptidase inhibitors selectively activate the CARD8 inflammasome. *Nat. Chem. Biol.* **2022**, *18*, 565–574. [CrossRef]
54. Johnson, D.C.; Taabazuing, C.Y.; Okondo, M.C.; Chui, A.J.; Rao, S.D.; Brown, F.C.; Reed, C.; Peguero, E.; de Stanchina, E.; Kentsis, A.; et al. DPP8/DPP9 inhibitor-induced pyroptosis for treatment of acute myeloid leukemia. *Nat. Med.* **2018**, *24*, 1151–1156. [CrossRef]
55. Tan, Y.F.; Wang, M.; Chen, Z.Y.; Wang, L.; Liu, X.H. Inhibition of BRD4 prevents proliferation and epithelial-mesenchymal transition in renal cell carcinoma via NLRP3 inflammasome-induced pyroptosis. *Cell Death Dis.* **2020**, *11*, 239. [CrossRef]
56. Dong, X.; Hu, X.; Bao, Y.; Li, G.; Yang, X.D.; Slauch, J.M.; Chen, L.F. Brd4 regulates NLRC4 inflammasome activation by facilitating IRF8-mediated transcription of Naips. *J. Cell Biol.* **2021**, *220*, e202005148. [CrossRef]
57. Lu, H.; Zhang, S.; Wu, J.; Chen, M.; Cai, M.C.; Fu, Y.; Li, W.; Wang, J.; Zhao, X.; Yu, Z.; et al. Molecular Targeted Therapies Elicit Concurrent Apoptotic and GSDME-Dependent Pyroptotic Tumor Cell Death. *Clin. Cancer Res.* **2018**, *24*, 6066–6077. [CrossRef]
58. Erkes, D.A.; Cai, W.; Sanchez, I.M.; Purwin, T.J.; Rogers, C.; Field, C.O.; Berger, A.C.; Hartsough, E.J.; Rodeck, U.; Alnemri, E.S.; et al. Mutant BRAF and MEK Inhibitors Regulate the Tumor Immune Microenvironment via Pyroptosis. *Cancer Discov.* **2020**, *10*, 254–269. [CrossRef]
59. Wu, M.; Wang, Y.; Yang, D.; Gong, Y.; Rao, F.; Liu, R.; Danna, Y.; Li, J.; Fan, J.; Chen, J.; et al. A PLK1 kinase inhibitor enhances the chemosensitivity of cisplatin by inducing pyroptosis in oesophageal squamous cell carcinoma. *EBioMedicine* **2019**, *41*, 244–255. [CrossRef]
60. Huo, J.; Shen, Y.; Zhang, Y.; Shen, L. BI 2536 induces gasdermin E-dependent pyroptosis in ovarian cancer. *Front. Oncol.* **2022**, *12*, 963928. [CrossRef]
61. Qiao, L.; Wu, X.; Zhang, J.; Liu, L.; Sui, X.; Zhang, R.; Liu, W.; Shen, F.; Sun, Y.; Xi, X. alpha-NETA induces pyroptosis of epithelial ovarian cancer cells through the GSDMD/caspase-4 pathway. *FASEB J.* **2019**, *33*, 12760–12767. [CrossRef] [PubMed]
62. Deng, B.B.; Jiao, B.P.; Liu, Y.J.; Li, Y.R.; Wang, G.J. BIX-01294 enhanced chemotherapy effect in gastric cancer by inducing GSDME-mediated pyroptosis. *Cell Biol. Int.* **2020**, *44*, 1890–1899. [CrossRef] [PubMed]
63. Long, K.; Gu, L.; Li, L.; Zhang, Z.; Li, E.; Zhang, Y.; He, L.; Pan, F.; Guo, Z.; Hu, Z. Small-molecule inhibition of APE1 induces apoptosis, pyroptosis, and necroptosis in non-small cell lung cancer. *Cell Death Dis.* **2021**, *12*, 503. [CrossRef] [PubMed]
64. Huang, J.; Fan, P.; Liu, M.; Weng, C.; Fan, G.; Zhang, T.; Duan, X.; Wu, Y.; Tang, L.; Yang, G.; et al. Famotidine promotes inflammation by triggering cell pyroptosis in gastric cancer cells. *BMC Pharmacol. Toxicol.* **2021**, *22*, 62. [CrossRef]
65. Kobayashi, T.; Mitsuhashi, A.; Hongying, P.; Shioya, M.; Kojima, K.; Nishikimi, K.; Yahiro, K.; Shozu, M. Bexarotene-induced cell death in ovarian cancer cells through Caspase-4-gasdermin E mediated pyroptosis. *Sci. Rep.* **2022**, *12*, 11123. [CrossRef]
66. Zhao, W.; Zhang, L.; Zhang, Y.; Jiang, Z.; Lu, H.; Xie, Y.; Han, W.; Zhao, W.; He, J.; Shi, Z.; et al. The CDK inhibitor AT7519 inhibits human glioblastoma cell growth by inducing apoptosis, pyroptosis and cell cycle arrest. *Cell Death Dis.* **2023**, *14*, 11. [CrossRef]
67. Huntington, K.E.; Louie, A.D.; Srinivasan, P.R.; Schorl, C.; Lu, S.; Silverberg, D.; Newhouse, D.; Wu, Z.; Zhou, L.; Borden, B.A.; et al. GSK-3 inhibitor elraglusib enhances tumor-infiltrating immune cell activation in tumor biopsies and synergizes with anti-PD-L1 in a murine model of colorectal cancer. *bioRxiv* **2023**. [CrossRef]
68. Prasad, S.; Tyagi, A.K.; Aggarwal, B.B. Recent developments in delivery, bioavailability, absorption and metabolism of curcumin: The golden pigment from golden spice. *Cancer Res. Treat.* **2014**, *46*, 2–18. [CrossRef]
69. Alexandrow, M.G.; Song, L.J.; Altiok, S.; Gray, J.; Haura, E.B.; Kumar, N.B. Curcumin: A novel Stat3 pathway inhibitor for chemoprevention of lung cancer. *Eur. J. Cancer Prev.* **2012**, *21*, 407–412. [CrossRef]
70. Panahi, Y.; Darvishi, B.; Ghanei, M.; Jowzi, N.; Beiraghdar, F.; Varnamkhasti, B.S. Molecular mechanisms of curcumins suppressing effects on tumorigenesis, angiogenesis and metastasis, focusing on NF-kappaB pathway. *Cytokine Growth Factor. Rev.* **2016**, *28*, 21–29. [CrossRef]

71. Cho, J.W.; Lee, K.S.; Kim, C.W. Curcumin attenuates the expression of IL-1beta, IL-6, and TNF-alpha as well as cyclin E in TNF-alpha-treated HaCaT cells; NF-kappaB and MAPKs as potential upstream targets. *Int. J. Mol. Med.* **2007**, *19*, 469–474. [PubMed]
72. Dhillon, N.; Aggarwal, B.B.; Newman, R.A.; Wolff, R.A.; Kunnumakkara, A.B.; Abbruzzese, J.L.; Ng, C.S.; Badmaev, V.; Kurzrock, R. Phase II trial of curcumin in patients with advanced pancreatic cancer. *Clin. Cancer Res.* **2008**, *14*, 4491–4499. [CrossRef] [PubMed]
73. Kanai, M.; Yoshimura, K.; Asada, M.; Imaizumi, A.; Suzuki, C.; Matsumoto, S.; Nishimura, T.; Mori, Y.; Masui, T.; Kawaguchi, Y.; et al. A phase I/II study of gemcitabine-based chemotherapy plus curcumin for patients with gemcitabine-resistant pancreatic cancer. *Cancer Chemother. Pharmacol.* **2011**, *68*, 157–164. [CrossRef] [PubMed]
74. Passildas-Jahanmohan, J.; Eymard, J.C.; Pouget, M.; Kwiatkowski, F.; Van Praagh, I.; Savareux, L.; Atger, M.; Durando, X.; Abrial, C.; Richard, D.; et al. Multicenter randomized phase II study comparing docetaxel plus curcumin versus docetaxel plus placebo in first-line treatment of metastatic castration-resistant prostate cancer. *Cancer Med.* **2021**, *10*, 2332–2340. [CrossRef]
75. Liang, W.F.; Gong, Y.X.; Li, H.F.; Sun, F.L.; Li, W.L.; Chen, D.Q.; Xie, D.P.; Ren, C.X.; Guo, X.Y.; Wang, Z.Y.; et al. Curcumin Activates ROS Signaling to Promote Pyroptosis in Hepatocellular Carcinoma HepG2 Cells. *In Vivo* **2021**, *35*, 249–257. [CrossRef]
76. Yue, E.; Tuguzbaeva, G.; Chen, X.; Qin, Y.; Li, A.; Sun, X.; Dong, C.; Liu, Y.; Yu, Y.; Zahra, S.M.; et al. Anthocyanin is involved in the activation of pyroptosis in oral squamous cell carcinoma. *Phytomedicine* **2019**, *56*, 286–294. [CrossRef]
77. Zhang, X.; Zhang, P.; An, L.; Sun, N.; Peng, L.; Tang, W.; Ma, D.; Chen, J. Miltirone induces cell death in hepatocellular carcinoma cell through GSDME-dependent pyroptosis. *Acta Pharm. Sin. B* **2020**, *10*, 1397–1413. [CrossRef]
78. An, H.; Heo, J.S.; Kim, P.; Lian, Z.; Lee, S.; Park, J.; Hong, E.; Pang, K.; Park, Y.; Ooshima, A.; et al. Tetraarsenic hexoxide enhances generation of mitochondrial ROS to promote pyroptosis by inducing the activation of caspase-3/GSDME in triple-negative breast cancer cells. *Cell Death Dis.* **2021**, *12*, 159. [CrossRef]
79. Cai, J.; Yi, M.; Tan, Y.; Li, X.; Li, G.; Zeng, Z.; Xiong, W.; Xiang, B. Natural product triptolide induces GSDME-mediated pyroptosis in head and neck cancer through suppressing mitochondrial hexokinase-IotaIota. *J. Exp. Clin. Cancer Res.* **2021**, *40*, 190. [CrossRef]
80. Tan, Y.; Xiang, J.; Huang, Z.; Wang, L.; Huang, Y. Trichosanthin inhibits cell growth and metastasis by promoting pyroptosis in non-small cell lung cancer. *J. Thorac. Dis.* **2022**, *14*, 1193–1202. [CrossRef]
81. Li, T.; Shi, L.; Liu, W.; Hu, X.; Hui, Y.; Di, M.; Xue, S.; Zheng, Y.; Yao, M.; Li, C.; et al. Aloe-Emodin Induces Mitochondrial Dysfunction and Pyroptosis by Activation of the Caspase-9/3/Gasdermin E Axis in HeLa Cells. *Front. Pharmacol.* **2022**, *13*, 854526. [CrossRef] [PubMed]
82. Han, J.; Cheng, C.; Zhang, J.; Fang, J.; Yao, W.; Zhu, Y.; Xiu, Z.; Jin, N.; Lu, H.; Li, X.; et al. Myricetin activates the Caspase-3/GSDME pathway via ER stress induction of pyroptosis in lung cancer cells. *Front. Pharmacol.* **2022**, *13*, 959938. [CrossRef] [PubMed]
83. Chen, Y.; Ma, S.; Pi, D.; Wu, Y.; Zuo, Q.; Li, C.; Ouyang, M. Luteolin induces pyroptosis in HT-29 cells by activating the Caspase1/Gasdermin D signalling pathway. *Front. Pharmacol.* **2022**, *13*, 952587. [CrossRef] [PubMed]
84. Dai, X.; Sun, F.; Deng, K.; Lin, G.; Yin, W.; Chen, H.; Yang, D.; Liu, K.; Zhang, Y.; Huang, L. Mallotucin D, a Clerodane Diterpenoid from Croton crassifolius, Suppresses HepG2 Cell Growth via Inducing Autophagic Cell Death and Pyroptosis. *Int. J. Mol. Sci.* **2022**, *23*, 14217. [CrossRef]
85. Ju, X.; Zhang, H.; Wang, J.; Sun, Z.; Guo, L.; Wang, Q. Shikonin triggers GSDME-mediated pyroptosis in tumours by regulating autophagy via the ROS-MAPK14/p38alpha axis. *Phytomedicine* **2023**, *109*, 154596. [CrossRef]
86. Lee, Y.J.; Park, K.S.; Lee, S.H. Curcumin Targets Both Apoptosis and Necroptosis in Acidity-Tolerant Prostate Carcinoma Cells. *Biomed. Res. Int.* **2021**, *2021*, 8859181. [CrossRef]
87. Li, R.; Zhang, J.; Zhou, Y.; Gao, Q.; Wang, R.; Fu, Y.; Zheng, L.; Yu, H. Transcriptome Investigation and In Vitro Verification of Curcumin-Induced HO-1 as a Feature of Ferroptosis in Breast Cancer Cells. *Oxid. Med. Cell Longev.* **2020**, *2020*, 3469840. [CrossRef]
88. Dixon, S.J.; Stockwell, B.R. The role of iron and reactive oxygen species in cell death. *Nat. Chem. Biol.* **2014**, *10*, 9–17. [CrossRef]
89. Zhou, B.; Zhang, J.Y.; Liu, X.S.; Chen, H.Z.; Ai, Y.L.; Cheng, K.; Sun, R.Y.; Zhou, D.; Han, J.; Wu, Q. Tom20 senses iron-activated ROS signaling to promote melanoma cell pyroptosis. *Cell Res.* **2018**, *28*, 1171–1185. [CrossRef]
90. Yuan, R.; Zhao, W.; Wang, Q.Q.; He, J.; Han, S.; Gao, H.; Feng, Y.; Yang, S. Cucurbitacin B inhibits non-small cell lung cancer in vivo and in vitro by triggering TLR4/NLRP3/GSDMD-dependent pyroptosis. *Pharmacol. Res.* **2021**, *170*, 105748. [CrossRef]
91. Lou, C.; Fang, Y.; Mei, Y.; Hu, W.; Sun, L.; Jin, C.; Chen, H.; Zheng, W. Cucurbitacin B attenuates osteoarthritis development by inhibiting NLRP3 inflammasome activation and pyroptosis through activating Nrf2/HO-1 pathway. *Phytother. Res.* **2024**, *38*, 3352–3369. [CrossRef] [PubMed]
92. Wu, L.S.; Liu, Y.; Wang, X.W.; Xu, B.; Lin, Y.L.; Song, Y.; Dong, Y.; Liu, J.L.; Wang, X.J.; Liu, S.; et al. LPS Enhances the Chemosensitivity of Oxaliplatin in HT29 Cells via GSDMD-Mediated Pyroptosis. *Cancer Manag. Res.* **2020**, *12*, 10397–10409. [CrossRef] [PubMed]

93. Chen, L.; Ma, X.; Liu, W.; Hu, Q.; Yang, H. Targeting Pyroptosis through Lipopolysaccharide-Triggered Noncanonical Pathway for Safe and Efficient Cancer Immunotherapy. *Nano Lett.* **2023**, *23*, 8725–8733. [CrossRef] [PubMed]
94. Sannino, F.; Sansone, C.; Galasso, C.; Kildgaard, S.; Tedesco, P.; Fani, R.; Marino, G.; de Pascale, D.; Ianora, A.; Parrilli, E.; et al. Pseudoalteromonas haloplanktis TAC125 produces 4-hydroxybenzoic acid that induces pyroptosis in human A459 lung adenocarcinoma cells. *Sci. Rep.* **2018**, *8*, 1190. [CrossRef]
95. Zhu, X.; Wu, T.; Chi, Y.; Ge, Y.; Wu, B.; Zhou, M.; Zhu, F.; Ji, M.; Cui, L. Pyroptosis induced by enterovirus A71 infection in cultured human neuroblastoma cells. *Virology* **2018**, *521*, 69–76. [CrossRef]
96. Zhang, X.; Li, C.; Chen, D.; He, X.; Zhao, Y.; Bao, L.; Wang, Q.; Zhou, J.; Xie, Y.H. pylori CagA activates the NLRP3 inflammasome to promote gastric cancer cell migration and invasion. *Inflamm. Res.* **2022**, *71*, 141–155. [CrossRef]
97. Liu, Z.; Li, Y.; Zhu, Y.; Li, N.; Li, W.; Shang, C.; Song, G.; Li, S.; Cong, J.; Li, T.; et al. Apoptin induces pyroptosis of colorectal cancer cells via the GSDME dependent pathway. *Int. J. Biol. Sci.* **2022**, *18*, 717–730. [CrossRef]
98. Wang, G.; Wei, W.; Jiang, Z.; Jiang, J.; Han, J.; Zhang, H.; Hu, J.; Zhang, P.; Li, X.; Chen, T.; et al. Talaromyces marneffei activates the AIM2-caspase-1/-4-GSDMD axis to induce pyroptosis in hepatocytes. *Virulence* **2022**, *13*, 963–979. [CrossRef]
99. Zhang, Y.; Xu, T.; Tian, H.; Wu, J.; Yu, X.; Zeng, L.; Liu, F.; Liu, Q.; Huang, X. Coxsackievirus Group B3 Has Oncolytic Activity against Colon Cancer through Gasdermin E-Mediated Pyroptosis. *Cancers* **2022**, *14*, 6206. [CrossRef]
100. Pizato, N.; Luzete, B.C.; Kiffer, L.; Correa, L.H.; de Oliveira Santos, I.; Assumpcao, J.A.F.; Ito, M.K.; Magalhaes, K.G. Omega-3 docosahexaenoic acid induces pyroptosis cell death in triple-negative breast cancer cells. *Sci. Rep.* **2018**, *8*, 1952. [CrossRef]
101. Wang, X.; Yin, Y.; Qian, W.; Peng, C.; Shen, S.; Wang, T.; Zhao, S. Citric acid of ovarian cancer metabolite induces pyroptosis via the caspase-4/TXNIP-NLRP3-GSDMD pathway in ovarian cancer. *FASEB J.* **2022**, *36*, e22362. [CrossRef] [PubMed]
102. Zhu, M.; Wang, J.; Xie, J.; Chen, L.; Wei, X.; Jiang, X.; Bao, M.; Qiu, Y.; Chen, Q.; Li, W.; et al. Design, synthesis, and evaluation of chalcone analogues incorporate alpha,beta-Unsaturated ketone functionality as anti-lung cancer agents via evoking ROS to induce pyroptosis. *Eur. J. Med. Chem.* **2018**, *157*, 1395–1405. [CrossRef] [PubMed]
103. Zhang, J.Y.; Zhou, B.; Sun, R.Y.; Ai, Y.L.; Cheng, K.; Li, F.N.; Wang, B.R.; Liu, F.J.; Jiang, Z.H.; Wang, W.J.; et al. The metabolite alpha-KG induces GSDMC-dependent pyroptosis through death receptor 6-activated caspase-8. *Cell Res.* **2021**, *31*, 980–997. [CrossRef] [PubMed]

biomolecules

MDPI

Review

Ubiquitin-Specific Protease Inhibitors for Cancer Therapy: Recent Advances and Future Prospects

Mohamad Bakkar [1,2], Sara Khalil [3], Komal Bhayekar [1], Narva Deshwar Kushwaha [1], Amirreza Samarbakhsh [1], Sadaf Dorandish [1], Holly Edwards [4,5], Q. Ping Dou [3,4,5], Yubin Ge [3,4,5,*] and Navnath S. Gavande [1,5,*]

1 Department of Pharmaceutical Sciences, Eugene Applebaum College of Pharmacy and Health Sciences (EACPHS), Wayne State University, Detroit, MI 48201, USA; mbakkar@med.wayne.edu (M.B.); hc2647komal@wayne.edu (K.B.); hn3752@wayne.edu (N.D.K.); gk1514@wayne.edu (A.S.); sdorandish@wayne.edu (S.D.)

2 Division of Pediatric Hematology and Oncology, Children's Hospital of Michigan, Detroit, MI 48201, USA

3 Cancer Biology Graduate Program, Wayne State University School of Medicine, Detroit, MI 48201, USA; ho9320@wayne.edu (S.K.); doup@karmanos.org (Q.P.D.)

4 Department of Oncology, Wayne State University School of Medicine, Detroit, MI 48201, USA; pitmanh@karmanos.org

5 Molecular Therapeutics Program, Barbara Ann Karmanos Cancer Institute (KCI), Wayne State University School of Medicine, Detroit, MI 48201, USA

* Correspondence: gey@karmanos.org (Y.G.); ngavande@wayne.edu (N.S.G.); Tel.: +1-313-578-4285 (Y.G.); +1-(313)-577-1523 (N.S.G.); Fax: +1-(313)-577-2033 (N.S.G.)

Abstract: Cancer management has traditionally depended on chemotherapy as the mainstay of treatment; however, recent advancements in targeted therapies and immunotherapies have offered new options. Ubiquitin-specific proteases (USPs) have emerged as promising therapeutic targets in cancer treatment due to their crucial roles in regulating protein homeostasis and various essential cellular processes. This review covers the following: (1) the structural and functional characteristics of USPs, highlighting their involvement in key cancer-related pathways, and (2) the discovery, chemical structures, mechanisms of action, and potential clinical implications of USP inhibitors in cancer therapy. Particular attention is given to the role of USP inhibitors in enhancing cancer immunotherapy, e.g., modulation of the tumor microenvironment, effect on regulatory T cell function, and influence on immune checkpoint pathways. Furthermore, this review summarizes the current progress and challenges of clinical trials involving USP inhibitors as cancer therapy. We also discuss the complexities of achieving target selectivity, the ongoing efforts to develop more specific and potent USP inhibitors, and the potential of USP inhibitors to overcome drug resistance and synergize with existing cancer treatments. We finally provide a perspective on future directions in targeting USPs, including the potential for personalized medicine based on specific gene mutations, underscoring their significant potential for enhancing cancer treatment. By elucidating their mechanisms of action, clinical progress, and potential future applications, we hope that this review could serve as a useful resource for both basic scientists and clinicians in the field of cancer therapeutics.

Keywords: ubiquitin-specific proteases (USPs); deubiquitinating enzymes (DUBs); small molecule inhibitors; cancer therapy

1. Introduction

1.1. Overview of the Ubiquitin–Proteasome Pathway

The ubiquitin–proteasome system (UPS) is a vital pathway regulating protein levels in eukaryotic cells, which is essential for maintaining cellular homeostasis. This pathway consists of several enzymatic steps that ensure the targeted degradation of substrate proteins. The UPS process begins with the activation of ubiquitin (Ub) by the E1 ubiquitin-activating enzyme in an ATP-dependent reaction, which is then transferred to an E2 ubiquitin-conjugating enzyme and finally transferred to a specific lysine residue on the target proteins by an E3 ubiquitin ligase (Figure 1). This process is repeated to form a polyubiquitin chain on the target proteins, which serves as a signal for recognition by the 26S proteasome [1]. The 26S proteasome, a multi-subunit protease complex consisting of a 20S core particle with proteolytic activity and two 19S regulatory particles that recognize ubiquitinated proteins, is then responsible for the degradation of the targeted substrate proteins into small peptides. Deubiquitinating enzymes (DUBs) play a significant role in the regulation of the UPS by removing ubiquitin molecules from ubiquitinated substrate proteins. The UPS is involved in multiple cellular processes, including DNA damage response (DDR), apoptosis, signal transduction, and drug resistance. Dysfunction of the UPS has been implicated in a range of disease types, including cancer [2]. Cancers with higher protein turnover are generally more sensitive to the inhibition of proteasome and deubiquitinases. The importance of the UPS in pathogenesis is underscored by the development of proteasome inhibitors as therapeutic agents. To date, the FDA has approved three 20S proteasome inhibitors, Borezomib, Carfilzomib, and Ixazomib, for the treatment of hematologic malignancies, namely multiple myeloma and mantle cell lymphoma. Bortezomib, which was discovered in 1995 and approved by the US FDA for the treatment of multiple myeloma in 2003 and later for the treatment of mantle cell lymphoma, validated the therapeutic potential of targeting the UPS, particularly in hematologic malignancies [3]. Carfilzomib was subsequently approved in 2012, followed by Ixazomib, both specifically for the treatment of multiple myeloma. These drugs have offered new therapeutic options for patients with these challenging blood cancers [4]. The UPS's roles in a wide range of cellular processes make it an important target of ongoing research.

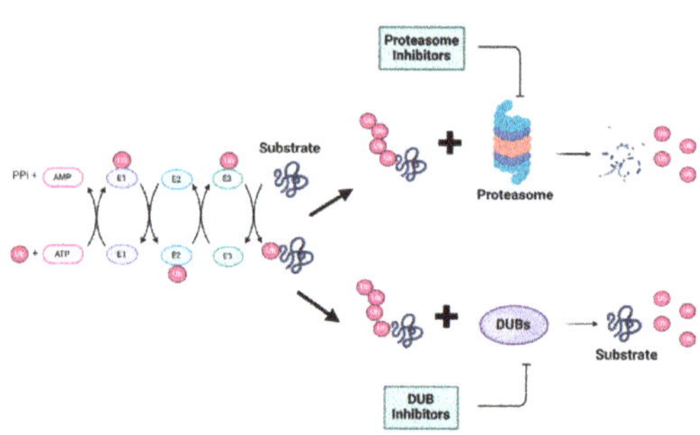

Figure 1. Schematic representation of ubiquitin-tagged protein degradation. This schematic illustrates the UPS, a vital cellular protein degradation mechanism. The process begins with the ATP-dependent activation of ubiquitin, which leads to a cascade of enzymatic reactions that ultimately results in the degradation of the substrates into smaller peptides by the proteasome or stabilization of the substrates by the relevant DUB. In both cases, ubiquitin molecules are released for reuse. Ub: ubiquitin (Ub) molecules (purple); ATP/AMP: energy source for ubiquitin activation; PPi: inorganic pyrophosphate; proteasome: multi-subunit protein degradation complex (blue and pink); degradation products: released peptide fragments and free ubiquitin molecules; DUBs: deubiquitinating enzymes. (Created using BioRender. Bakkar, M. (2025)).

1.2. Role of Deubiquitinating Enzymes

Deubiquitinating enzymes (DUBs) are proteases that cleave ubiquitin from various substrate proteins, playing a crucial role in regulating the ubiquitin signaling pathway. The human genome encodes approximately 100 putative DUBs, which are classified into two main classes: metalloproteases and cysteine (Cys) proteases. These classes are further subclassed based on their domain structures. The first family belongs to the JAB1/MPN/Mov34 metalloenzyme (JAMM) domain zinc-dependent metalloprotease family, while the other five families—the ubiquitin C-terminal hydrolases (UCH), the ovarian tumor proteases (OTU), the Machado–Josephin domain proteases (MJDs), the ubiquitin-specific protease (USP/UBP), and the recently discovered motif interacting with ubiquitin (MIU)-containing novel DUB family (MINDY) and Zinc finger with UFM1-specific peptidase domain protein (ZUFSP)—are papain-like cysteine proteases.

1.2.1. Metalloprotease DUBs

Metalloproteases form one of the two main classes of DUBs, with the primary subclass of metalloproteases being JAMM (JAB1/MPN/MOV34 metalloenzyme) domain proteases. The catalytic site of this subclass of DUBs contains two histidine residues, an aspartate (Asp), and a catalytic serine (Ser). The main unique property of metalloproteases is that they use zinc in their catalytic mechanism, which distinguishes them from the cysteine protease DUBs. The zinc ion plays a role in the generation of a hydroxide ion from water, which acts as a nucleophile to hydrolyze the isopeptide bond between the protein substrate and ubiquitin. A key consequence of this mechanism is that this type of DUB does not form a covalent intermediate with the substrate during catalysis. In contrast to ubiquitin-specific proteases (USPs), which use their catalytic serine to form a covalent intermediate with the substrate during enzyme cleavage, metalloprotease DUBs utilize a nucleophilic hydroxide ion that directly attacks the peptide bond without forming a covalent bond with the enzyme. This characteristic makes them inherently resistant to DUB inhibitors, which often target the formation of such covalent intermediates [5].

1.2.2. Cysteine Protease DUBs

The six subfamilies of cysteine protease DUBs are organized based on their domain architecture. It is important to note that most DUBs in this class utilize a catalytic triad composed of a histidine (His), an active site cysteine (Cys), and in most cases, an asparagine (Asn) or aspartate (Asp).

Ubiquitin C-Terminal Hydrolases (UCHs)

This subfamily of cysteine proteases was one of the first types of DUBs identified with UCHL3 discovered in 1997 [6]. The UCH domain contains three conserved residues, namely His, Cys, and Asp, and there are four known members of this subfamily. These small enzymes preferentially remove small peptides from the C-terminus of ubiquitin. The loop structure covering the active site limits the size of the substrate with which they can interact, unlike USPs, which can handle larger substrates. UCHs have been implicated in playing a role in the oncogenesis of various cancers [7].

Ovarian Tumor Proteases (OTUs)

This subfamily displays specificity for the ubiquitin substrates with which it interacts by utilizing additional ubiquitin interaction sites that can bind to specific linkages on longer protein chains. Another important distinction is the lack of an asparagine or aspartate residue in some members of this subfamily [5]. OTUD5 is an important member of the

OTU subfamily of cysteine proteases and has been implicated in both tumor progression and tumor suppression depending on the disease type. These DUBs play an important role in various cellular processes, including DNA repair and protein quality control [8].

Machado–Josephin Domain Proteases (MJDs)

These DUBs comprise four members in humans, and all contain a catalytic domain known as the Josephin domain. This important feature contains two conserved histidines, a catalytic cysteine, and two ubiquitin-binding sites. This subfamily is named after the Machado–Josephin disease, a neurological disorder caused by a CAG repeat expansion motif producing polyglutamine resulting in protein aggregation and misfolding [9]. Ataxin-3, a critical MJD in humans, has been shown to play a role in the proliferation of testicular cancer, gastric cancer, and lung cancer cells [10].

Motif Interacting with Ubiquitin (MIU)-Containing Novel DUB Family (MINDY) and Zinc Finger with UFM1-Specific Peptidase Domain Protein (ZUFSP)

These two subfamilies of cysteine proteases were the two most recently identified. The MINDY subfamily of DUBs has a unique catalytic triad consisting of cysteine, histidine, and threonine [11]. There are four known members of this subfamily, and these enzymes appear to preferentially cleave long ubiquitin chains starting from the direction of the distal end [12]. ZUFSP contains only one known member in humans, which utilizes multiple ubiquitin-binding domains and possesses highly specific cleavage of K63 ubiquitin linkage [13]. ZUFSP has a modular architecture, with its overall specificity and activity being influenced by its various modular ubiquitin-binding domains [14].

Ubiquitin-Specific Proteases (USPs)

USPs form the largest subclass of DUBs with 58 known members and are the focus of this review article. USPs are generally larger enzymes, ranging from 50 to 300 kDa and typically have N-terminal extensions involved in substrate recognition and protein–protein interactions. USPs can remove ubiquitin from protein conjugates, process ubiquitin precursors, and disassemble ubiquitin chains. Some types of USPs show specificity for certain substrates, while others have broader activity. They play crucial roles in various cellular processes including the cell cycle, protein degradation, signal transduction, and DNA repair. Their substrate specificities make them important potential targets for therapeutic development in the treatment of diseases such as cancer [5]. Given the crucial role USPs play in regulating protein homeostasis as well as their potential as therapeutic targets in cancer therapy, the remainder of this review will focus on the structural and functional characteristics of USPs, their involvement in key cancer-related pathways, and the development of USP inhibitors as promising anticancer agents.

2. USPs: Structure, Biological Function, and Role in Tumorigenesis and Cancer Progression

2.1. The Structure of USPs

USPs represent the most prominent family of DUBs. They represent a sophisticated class of enzymes whose function depends on the intricate interplay between three vital structural elements. These elements—the core catalytic domain, ubiquitin-like domains, and ancillary domains—work congruently to achieve precise protein regulation [15]. Their activities are fine-tuned through post-translational modifications (PTMs), creating a complex regulatory network [16]. Understanding this structural and regulatory framework is crucial for basic research and therapeutic development. As common with signaling proteins, most USP deubiquitinases have a modular architecture with a catalytic domain, putative site of interaction, and localization domains.

Most USP domains have at least two ubiquitin-binding sites, one for the proximal ubiquitin molecules and one for the distal ubiquitin molecules. These sites cleave the isopeptide bond linking two ubiquitin moieties or the proximal ubiquitin with a protein substrate [17]. The USP family is the largest, with more than 50 members. The USP catalytic domain may be flanked at its N- or C-terminus by several additional domains involved in differential substrate recognition, sub-cellular localization, and regulation of enzymatic activity [18]. Among the members of the USP domain family, the majority act upon larger protein substrates in contrast to other subclasses of DUBs [19].

2.1.1. Core Catalytic Domain

At the heart of USP function is the core catalytic domain, whose structure reveals the precise ingenuity of these enzymes. This domain features homolog sequences derived from papain-like cysteine proteases arranged in a distinctive "palm, thumb, and fingers" configuration (Figure 2) that orchestrates substrate binding and cleavage [20,21].

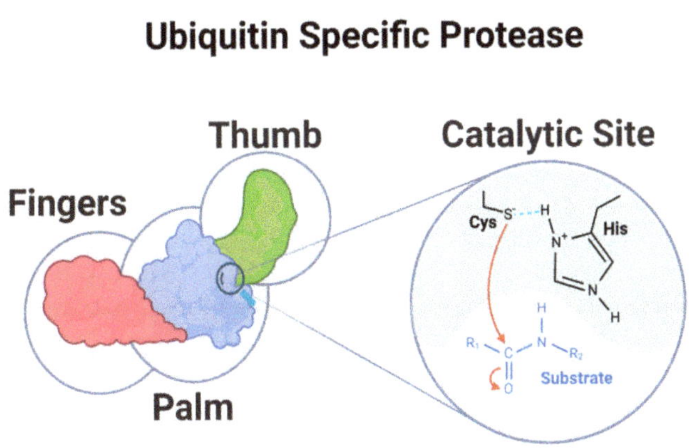

Figure 2. Structure of ubiquitin-specific protease. This diagram depicts the structural organization and catalytic mechanism of a USP. The USP contains a catalytic core domain featuring the characteristic catalytic triad consisting of cysteine (Cys), histidine (His), and, in most cases, an asparagine (Asn) or aspartate (Asp), which interacts with histidine (the cysteine and histidine residues are illustrated in the figure) highlighted in the active site between the palm and thumb sub-domains. The isopeptide bond between ubiquitin and the substrate protein is cleaved resulting in the removal of ubiquitin modifications, which is crucial for protein regulation and cellular homeostasis. (Created using BioRender. Bakkar, M. (2025)).

The catalytic center of USPs is located at the interface between the palm and thumb regions of the USP domain. The USP catalytic core can be thought of as being divided into six conserved boxes (Figure 3).

Box 1:	Box 2:	Box 3:	Box 4:	Box 5:	Box 6:
Catalytic Cys residue	Structural Domain	Cys-X-X-Cys motif	Cys-X-X-Cys motif	Catalytic His residue	Catalytic Asp/Asn residue

Figure 3. Molecular basis of ubiquitin-specific protease catalytic core protein domains.

Box 1 includes a catalytic Cys residue, box 5 includes a catalytic His, and box 6 includes a catalytic Asp/Asn residue. Each of these boxes is characterized by several other conserved features and residues. Boxes 3 and 4 contain a Cys-X-X-Cys motif, which has been reported to form a functional zinc binding motif in some USPs. It is speculated that zinc binding facilitates the folding of the USP core, thus allowing the interaction of sequence motifs that are several hundred residues apart. USP domains exhibit a common conserved fold. The catalytic triad resides between the thumb (Cys) and palm sub-domains (His/Asp) [17]. The principal site of the DUB has a strong interaction with the distal ubiquitin, mainly through the Ile44 patch, with different interacting surfaces within DUB subfamilies. The stretch of amino acids extending from the binding site of the C-terminus of the distal ubiquitin to the DUB catalytic center is responsible for distinguishing ubiquitin from other ubiquitin-like molecules (ULMs) [22,23]. The residues that are responsible for the difference in the C-terminal sequence of ubiquitin (Leu71, Arg72, Leu73, Arg74, Gly75, Gly76) when compared to that of ULMs are vital for identifying ubiquitin by DUBs [22,24]. The catalytic domain is responsible for the isopeptidase activity that cleaves the isopeptide bond between ubiquitin and the target substrate protein [18]. Most USPs contain a core catalytic domain with insertions and terminal extensions bearing additional protein interaction domains [20].

2.1.2. Ubiquitin-like Domains (UBL)

Building upon this catalytic foundation, UBL domains extend the functional repertoire of USPs. Strategically positioned near the distal ubiquitin-binding site, they play a crucial role in substrate recognition and regulation. Their integration within the catalytic core's variable insertions facilitates essential intra-molecular interactions that maintain proper USP conformation. Many USPs contain UBL domains that can bind to ubiquitin or other UBLs, as well as compete with ubiquitin binding and regulate catalysis. These UBL domains may mediate substrate recognition and binding, as well as play a vital role in localization at the proteasome [18].

2.1.3. Ancillary Domains

Complementing these two core components mentioned above, ancillary domains add another layer of functional sophistication through specialized structural elements. These include zinc finger domains, ubiquitin-interacting motifs (UIMs), and SUMO-interacting motifs (SIMs) that regulate their activity, localization, and substrate specificity [18]. These domains have fine-tuned substrate specificity and mediate crucial protein–protein interactions, although the role of these ancillary domains in regulating USP activity is not always consistent [21]. Their structural adaptability, particularly in the substrate-induced ordering of the unstructured region, provides additional regulatory flexibility through PTM-mediated control [17].

2.2. *The Roles of USPs in Tumorigenesis and Cancer Progression*

USPs regulate several critical cancer-related pathways, including NF-κB, Wnt/β-catenin, JAK/STAT, p53 signaling, c-MYC, TGF-β, DNA repair, apoptosis, cell cycle regulation, MAPK, and hypoxia pathways (Table 1). These pathways are involved in various aspects of cancer biology, such as cellular proliferation, survival, metastasis, and therapy resistance. USPs modulate these pathways through deubiquitination of key substrate proteins, which affects their stability and, consequently, their functions. This regulation can either promote or inhibit cancer progression depending on the specific USP and pathway involved. The diverse roles of USPs in these pathways highlight their importance in cancer biology and their potential as therapeutic targets.

Table 1. Cancer-related cellular pathways influenced by USPs.

Pathway	Role of USPs	Cancers	References
NF-κB pathway	Nuclear factor-kappa B (NF-κB) is involved in the regulation of several key biological functions, including cell survival. USPs regulate the NF-κB pathway by modulating the deubiquitination of key signaling proteins. USP21 plays a key role in the downregulation of TNF- α induced NF-κB signaling through deubiquitination of receptor-interacting protein 1 (RIP1), while USP4 promotes TNF-α mediated NF-κB activation by deubiquitinating TGF- β-activated kinase 1, highlighting the diverse roles USPs can play in regulating NF-κB signaling. Downregulation of NF-κB promotes cell death and reduces cell survival.	Breast cancer, prostate cancer, lung cancer, colorectal cancer, and ovarian cancer	[25–27]
Wnt/β-catenin pathway	The Wnt/β-catenin pathway is a crucial signaling mechanism that regulates various biological processes, including cell proliferation, differentiation, and survival. β-catenin is a key signal transducer in this pathway and is positively regulated by USP4. The C-terminal catalytic domain of USP4 is responsible for binding and nuclear transport of β-catenin leading to increased transcription of oncogenes. USP4 knockdown in a cell line of colon cancer has also been found to reduce invasion and migration, while overexpression has been demonstrated to enhance β-catenin-regulated transcription.	Colorectal cancer, gastric cancer, breast cancer, lung cancer, pancreatic cancer, and melanoma	[28,29]
JAK/STAT pathway	USPs regulate the JAK/STAT pathway by modulating the stability and degradation of key signaling components, such as STAT3 which plays a crucial role in transcription. Aberrant activation of STAT3 can enhance cancer cell proliferation, increase cell survival by up-regulating anti-apoptotic genes, and may contribute to drug resistance as well as immune evasion. USP5 plays a significant role in the progression and metastasis of pancreatic cancer by stabilizing STAT3.	Breast cancer, lung cancer, colorectal cancer, gastric cancer, leukemia, and bladder cancer	[30–32]
p53 signaling pathway	Several USPs regulate the p53 signaling pathway. Importantly, USP7 has been demonstrated to regulate p53 by stabilizing MDM2, an E3 ubiquitin ligase that typically promotes p53 degradation. USP7 has been proposed to be a potential therapeutic target given the contribution it has to cancer pathogenesis through the downregulation of p53. This in turn allows damaged cells to evade apoptosis and proliferate abnormally.	Breast cancer, colorectal cancer, ovarian cancer, lung cancer, and head and neck cancer	[33,34]
c-MYC	c-MYC is a well-established oncogene that plays a critical role in the progression of various cancers. This pathway is significantly influenced by various USPs, such as USP16 and USP36, which interact directly with c-MYC by deubiquitinating and stabilizing it, promoting chemoresistance. USP1 has been demonstrated to promote c-MYC pathway activation, leading to increased transcription of MYC target genes. USP7 indirectly affects MYC levels by stabilizing MDM2 and MDMX which regulate p53 that functions as a MYC antagonist.	Colorectal cancer, breast cancer, non-small cell lung cancer, pancreatic ductal adenocarcinoma, glioblastoma, and ovarian cancer	[35–37]
TGF-beta pathway	TGF-beta signaling has been shown to drive epithelial-mesenchymal transition enhancing tumor migration and invasion through pathways regulated by USPs. USP22 is upregulated in many types of cancers. In epithelial ovarian cancer, higher USP22 levels are associated with higher stage and worse clinical outcomes. USP22 facilitates cell proliferation by inducing G1 cell cycle arrest through interaction with oncogenic TGF-β1.	Breast cancer, colorectal cancer, liver cancer, non-small cell lung cancer, pancreatic cancer, and epithelial ovarian cancer	[38–40]
DNA repair pathways	USP21 has shown positive regulatory effects on BRCA2 which plays a critical role in homologous recombination. USP1 also plays a critical role in DNA damage repair by modulating the ubiquitination of key regulators of DNA repair such as PCNA and FANCD2.	Breast cancer, non-small cell lung cancer, colorectal cancer, gastric cancer, ovarian cancer, and hepatocellular carcinoma	[3,41–43]
Apoptotic pathways	Apoptotic signaling pathways are regulated by USPs. BCL-2 family proteins regulate apoptosis by controlling the release of apoptogenic factors, with both pro-apoptotic and anti-apoptotic proteins regulated by this group of proteins. Overexpression of USP9x in certain cancers leads to the overexpression of Mcl-1, an anti-apoptotic protein, by way of deubiquitination. This has been shown to lead to increased resistance to radiation therapy.	Breast cancer, lung cancer, colorectal cancer, ovarian cancer, and hepatocellular carcinoma	[44–46]
Cell cycle regulation pathways	The cell cycle is characterized by stages G1, S, G2, and M. Cyclins, cyclin-dependent kinases (CDKs), cyclin-dependent kinase inhibitors (CDKIs), and aurora kinases function to regulate the cellular transition among these stages. USPs regulate cell cycle progression by stabilizing regulatory proteins such as CDKs. USP2, USP5, USP13, USP20, and USP22 have been found to promote the stability of cyclin D1 which facilitates cell cycle advancement and tumorigenesis.	Breast cancer, lung cancer, colorectal cancer, gastric cancer, and prostate cancer	[16,44,47–49]

Table 1. *Cont.*

Pathway	Role of USPs	Cancers	References
MAPK pathway	The MAPK (mitogen-activated protein kinase) pathway is a system of intracellular communication that transmits signals from the cell surface to the nucleus. USP11 promotes the progression and chemoresistance of various cancers, and overexpression of USP11 has been associated with abnormal activation of the ERK-dependent MAPK pathway during the progression of colorectal carcinoma.	Breast cancer, colorectal cancer, lung cancer, gastric cancer, ovarian cancer, and hepatocellular carcinoma	[44,50]
Hypoxia pathways	Hypoxic conditions in tissues, particularly in cancer, are known to induce a cellular transcriptional response. Transcriptional complex hypoxia-inducible factors (HIF) are the key factors in this signaling pathway and the α-subunit of HIF triggers a cascade that leads to degradation by the 26S proteasome. USP22 has been shown to regulate the hypoxia pathway by stabilizing HIFs such as HIF-1α and enhancing their transcriptional activity, allowing cancer cells to adapt to hypoxic conditions.	Breast cancer, lung cancer, colorectal cancer, pancreatic cancer, cervical cancer, and head and neck cancer	[51,52]

2.3. Unique Structural and Functional Characteristics of Various USPs

USPs are a diverse family of DUBs that play crucial roles in protein homeostasis. A common theme among many USPs is their association with cancer pathogenesis, highlighting the importance of ubiquitin regulation in malignancies. USPs share some common structural features, but they also possess unique elements that contribute to their specific functions [53]. Table 2 provides a description of some of the similarities and differences of USPs.

The mechanism of action of the USP family of DUBs is generally conserved across members. The enzymatic activity relies on nucleophilic substitution of the catalytic cysteine on the isopeptide bond between ubiquitin and its protein substrate. The histidine residue acts as a general base and activates the catalytic cysteine for this reaction by lowering its pKa (shown in Figure 2). This mechanism is crucial for the deubiquitinating activity of USPs, allowing them to remove ubiquitin modifications from target proteins and thereby regulate their stability and functions [18].

Beyond the catalytic domain, USPs often have additional structural elements that regulate their activity and confer specificity. These include ubiquitin-binding domains, zinc finger domains, and other protein–protein interaction motifs that allow USPs to recognize specific substrates or participate in particular cellular pathways [18]. Despite their structural and functional similarities, the development of specific inhibitors for USPs has been uneven. Several USP inhibitors have been developed through various drug discovery approaches, including high-throughput screening and structure-based drug design. However, for many USPs, specific inhibitors have not yet been discovered [54]. This gap represents both a challenge and an opportunity in the field of cancer drug development and research, as targeting specific USPs could lead to novel cancer therapies that can potentially overcome drug resistance, improve efficacy, and reduce side effects.

Table 2. Unique structural and functional characteristics of various USPs.

USP Name	Domains	Catalytic Site	Cancer Association	References
USP1	Conserved USP catalytic domain (thumb, palm, and finger sub-domains).	Three catalytic residues: Cys90, His593, and Asp751.	Osteosarcoma, renal clear cell carcinoma, colorectal carcinoma, non-small cell lung cancer, and gastric cancers	[42]

Table 2. *Cont.*

USP Name	Domains	Catalytic Site	Cancer Association	References
USP3	Zinc finger (ZnF) domain and catalytic domain.	Conserved catalytic domain containing key Cys, Hys, and Asp/Asn residues critical for USP activity.	Gastric cancer and breast cancer	[55,56]
USP5	Catalytic domain flanked by two ubiquitin-associated (UBA) domains and two Zinc finger (ZnF) ubiquitin-binding protein (UBP) domains.	Papain-like proteases have a Cys and a His residue, which form an ion pair with the negatively charged cysteine thiolate functioning as a nucleophile.	Breast, lung, colorectal, hepatocellular, pancreatic, and non-small cell lung cancer	[57–59]
USP7	N-terminal MATH domain, central catalytic domain, and five C-terminal tandem ubiquitin-like domains (UbL).	Cys223, His464, and Asp481 residues are critical for deubiquitination.	Colorectal cancer, osteosarcoma, acute myeloid leukemia, breast cancer, prostate cancer, multiple myeloma, ovarian cancer, bladder cancer, esophageal squamous cell carcinoma, chronic lymphocytic lymphoma, and medulloblastoma	[54,60,61]
USP10	N-terminus region, USP catalytic structure domain, and a smaller C-terminus region.	Conserved catalytic domain containing key Cys, His, and Asp/Asn residues critical for USP activity.	Colorectal cancer, prostate cancer, hepatocellular carcinoma, glioblastoma multiforme, non-small-cell lung cancer, chronic myeloid leukemia, and acute myeloid leukemia	[62]
USP19	Two CHORD-SGT1/P23 domains (CS1 and CS2 at the N-terminus), USP catalytic domain.	Contains Cys and His residues critical for enzymatic activity and auto-inhibition mechanism.	Breast cancer and osteosarcoma	[63,64]
USP20	ZnF-UBP domain, catalytic USP domain, and two domains present in ubiquitin-specific protease (DUSP) domains.	Conserved Cys and His residues that catalyze proteolysis of the isopeptide bond between a target protein lysine residue and a glycine residue of ubiquitin.	Breast cancer, lung cancer, colon cancer, gastric cancer, and adult T-cell leukemia	[44,65]
USP32	USP catalytic domain, Domain in USP (DUSP), two UbL domains, and calcium-binding EF-hand with a signal transduction mechanism.	Conserved catalytic domain containing key Cys and His residues critical for USP activity.	Small cell lung cancer, gastric cancer, breast cancer, epithelial ovarian cancer, glioblastoma, gastrointestinal stromal tumor, AML, and pancreatic adenocarcinoma	[66]

Table 2. *Cont.*

USP Name	Domains	Catalytic Site	Cancer Association	References
USP36	Conserved USP catalytic domain (thumb, palm, and finger sub-domains).	Key residues: Cys223, His464, Asp481 in the catalytic domain.	Colon cancer and breast cancer	[36,67]

2.4. History of the Development of USP Inhibitors

Over the past decade, the development of USP inhibitors has gained significant momentum as a promising area in cancer therapeutics. Discovery of USP inhibitors primarily relied on high-throughput screening prior to 2014. However, the field has since evolved with structure-guided drug design based on co-crystal structure complexes becoming a prominent approach [68,69]. The interest in USP inhibitors stems from the crucial role USPs play in various cellular processes and their involvement in multiple cancer-related pathways [54]. The shift in focus to targeting USPs was partly due to the limitations of existing UPS-targeted inhibitors such as Bortezomib, which showed efficacy primarily in a couple of hematologic malignancies but only marginal effects on solid tumors [68]. More than seventy USP inhibitors have been reported over the past 20 years with six reaching clinical trials, but despite these advancements, no USP inhibitor has yet been approved for clinical use [69]. The development of USP inhibitors faces the challenge of achieving target selectivity, but despite this, there is growing optimism that USPs represent a new reservoir of therapeutic targets [54].

2.5. Therapeutic Implications

The therapeutic significance of the USP structural organization becomes apparent in disease contexts, particularly in cancer biology. This integrated understanding of USP structure and regulation reveals a remarkable system where each domain contributes to a more extensive functional network. PTMs modulate the activity of signaling proteins further, forming an intricate regulatory network evident in several disease processes, particularly in cancer biology. Specifically, these modifications regulate essential cellular processes, including proliferation, immune responses, and apoptosis [70]. Drug discovery efforts over the last decade have capitalized on this knowledge for the development of adequate targeted therapeutic strategies.

3. Small Molecule Inhibitors Targeting USPs

3.1. Chemical Structure, Inhibition Potency, and Mechanism of Action of Small Molecule Inhibitors Targeting USPs

To date, more than seventy small-molecule USP inhibitors have been reported in the literature. The chemical structures, inhibition potency, and mechanisms of action of some selected USP inhibitors are summarized in Table 3. USP inhibitors that are currently being or were previously tested in cancer clinical trials will be discussed in Section 3.3. USP inhibitors can be broadly categorized based on their specificity and the USPs they target. Some USP inhibitors, such as SP-002 and Pimozide (PMZ), are highly selective for USP1, while others like PR619 are pan-inhibitors affecting multiple USPs. The mechanisms of action vary among these compounds, with some binding reversibly to their targets (e.g., IU1), while others form irreversible covalent bonds (e.g., Q29, XL177A). Many of these inhibitors demonstrate promising anticancer effects through various mechanisms. Common themes include inducing cell cycle arrest, enhancing DNA damage, promoting apoptosis, and modulating key signaling pathways, as described in Table 1. For instance,

USP7 inhibitors HBX-19818 and P22077 work by stabilizing p53 and promoting the degradation of oncogenic proteins. USP14 inhibitors like IU1 and its derivatives have shown potential in treating neurodegenerative disorders by modulating the degradation of tau protein. Research on these USP inhibitors spans various stages, from in vitro studies to in vivo animal models, with some compounds showing efficacy in xenograft mouse models. Several inhibitors have been shown to exhibit synergistic effects when combined with other cancer therapies, such as PARP inhibitors or traditional chemotherapeutic agents. While these USP inhibitors show promise as a novel class of anticancer drugs in preclinical studies, the potential of these agents in clinical settings needs to be determined.

Table 3. Small molecule inhibitors targeting USPs.

USP Target	Name of the Inhibitor	Chemical Structure	IC_{50} Values	Mechanism of Action and Anticancer Activity	Refs
USP1	SP-002	Structure not disclosed.	15.7 nM	- Highly selective USP1 inhibitor. - Induces accumulation of Ub-PCNA on chromatin leading to S/G2 cell cycle arrest. - Synergistic with the PARP inhibitor olaparib in malignancies with homologous recombination defects (HRD). - Minimal side effects on hematopoietic stem cells in vitro. - Tumor growth suppressed in in vivo xenograft mice model.	[71,72]
USP1 + UAF1	C527	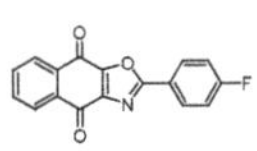	0.88 μM	- A non-covalent inhibitor that binds reversibly to the active site of the USP1/UAF1 complex. - Promotes ID1 degradation which is a transcription factor essential for the proliferation of many types of cancer as well as upregulates cell cycle inhibitor p21. - Enhances cisplatin sensitivity in vitro and in vivo.	[73,74]
	SJB2-043	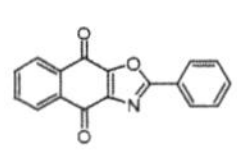	0.544 μM	- More potent derivative of C527. - Reversibly binds to and inhibits the enzymatic activity of the USP1/UAF1. - Leads to degradation of ID1/ID2/ID3 and apoptosis. - Inhibits growth of myeloid leukemia and multiple myeloma cells.	[73,75]
	SJB3-019A	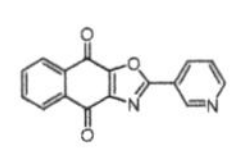	78.1 nM	- Similar mechanism of action as SJB2-043 but more potent. - Increases the level of Ub-FANCD2 and Ub-PCNA and leads to inhibition of DNA repair by decreasing homologous recombination activity.	
	ML323	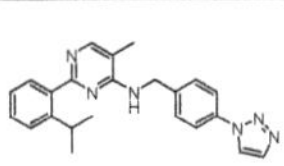	76 nM	- Allosterically leads to changes of the active site and disrupts the hydrophobic core of USP1. - Inhibits PCNA and FANCI-FANCD2 deubiquitination, leading to disruption of DNA replication and repair pathways in cancer cells.	[76–78]
USP1 + UAF1/USP7	Pimozide	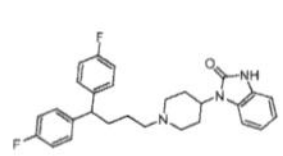	2 μM/47 μM	- A non-covalent and non-competitive inhibitor that binds to the USP1/UAF1 complex and reversibly inhibits its enzymatic activity. - Upregulates cell cycle inhibitory proteins p21 and p27 as well as downregulates cell cycle promoting proteins cyclin D3 and CDK2. - Induces ubiquitination and degradation of Max (a MYC binding protein). - Leads to G0/G1 arrest and apoptosis in diffuse large B-cell lymphoma (DLBCL). - Reduces tumor burden in rituximab-resistant DLBCL cells- and patient-derived xenograft (PDX) mouse models. - Shown to inhibit breast cancer and non-small cell lung cancer (NSCLC) cell proliferation.	[79–81]

Table 3. *Cont.*

USP Target	Name of the Inhibitor	Chemical Structure	IC_{50} Values	Mechanism of Action and Anticancer Activity	Refs
	Trifluoperazine (TFP)	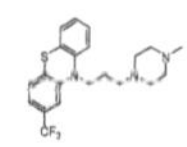	8 μM/9 μM	- A non-covalent inhibitor that binds reversibly to the USP1/UAF1 complex and inhibits its activity. - Classified as a typical antipsychotic agent. - Decreases cyclin D1/CDK4 and cyclin E/CDK2 which stimulates intrinsic apoptosis. - Leads to G0/G1 cell cycle arrest and apoptosis in triple-negative breast cancer (TNBC). - Inhibits TNBC tumor growth.	[79,82]
	Flupenthixol	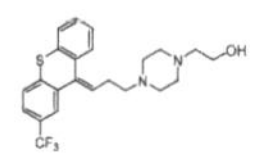	7 μM/13 μM	- A non-covalent non-competitive inhibitor that binds reversibly to the USP1/UAF1 complex and inhibits its activity. - Similar in structure to TFP and classified as a typical antipsychotic medication. Less selective than Pimozide and GW7647.	[79]
USP1 + UAF1/USP2/ USP7/USP8	Rottlerin	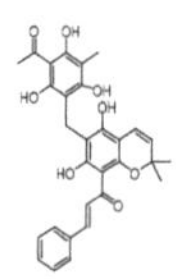	8 μM/34 μM/13 μM/6 μM	- Irreversibly inhibits the USP1/UAF1 complex. - Contains an α, β-unsaturated carbonyl group, which acts as a potential Michael acceptor. - Covalently modifies the cysteines at the active sites. - Causes G0/G1 cell cycle arrest. - Downregulates cyclin D1. - Suppresses NF-kB and its target genes. - Induces antiangiogenic effects and causes apoptosis in breast cancer cells.	[79,83, 84]
USP1 + UAF1/USP12	GW7647	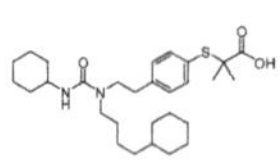	5 μM/44 μM	- A potent non-competitive reversible inhibitor of the USP1/UAF1 complex. - Binds to an allosteric site. - May sensitize cells to DNA-damaging agents as the timely deubiquitination of PCNA and FANCD2 is important for proper DNA repair. - Enhances the cytotoxicity of cisplatin in NSCLC.	[79,85]
USP 1/2/3/4/5/7/8/ 15/20/28/47/ UCHL1/UCHL3/ UCHL5	PR619(Broad range or Pan DUBs inhibitor)	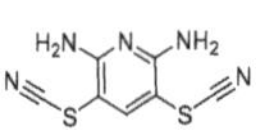	3.9–8.9 μM	- Pan inhibitor of multiple USPs. - Binds reversibly to the active sites of several USPs and interferes with the removal of ubiquitin from the respective substrate. - Triggers apoptosis by activating caspases and PARP cleavage which leads to G0/G1 cell cycle arrest. - In vivo xenograft mouse model demonstrated inhibition of human chondrosarcoma tumor growth.	[54,86, 87]
USP2	Q29	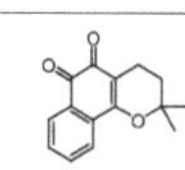	<5 μM	- Selective, irreversible inhibitor of USP2. - Causes oxidation of the catalytic cysteine residue of USP2. - Interferes with cancer cell cycle progression.	[88,89]
	ML364	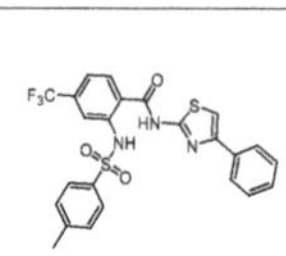	1.1 μM	- Reversibly and non-competitively inhibits USP2 at an allosteric site. - Interrupts cyclinD1-USP2 interaction. - Enhances cyclinD1 degradation. - Causes G0/G1 cell cycle arrest in colorectal cancer cells. - Decreases homologous recombination-mediated DNA repair.	[89,90]
	LCAHA	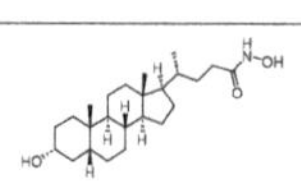	3.7–9.7 μM	- Selective, non-competitive inhibitor. - Destabilizes cyclin D1. - Causes G0/G1 cell cycle arrest in colon cancer cells.	[89,91]
	STD1T	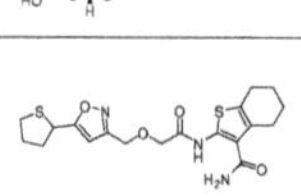	3.3 μM	- Selective, reversible inhibitor of USP2a. - Inhibits deubiquitination of cyclin D1. - Leads to G1 phase cell cycle arrest. - Enhances the cyclin D1degradation in colon cancer and breast cancer cells.	[89,92]

Table 3. *Cont.*

USP Target	Name of the Inhibitor	Chemical Structure	IC_{50} Values	Mechanism of Action and Anticancer Activity	Refs
	6TG	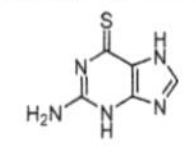	40 μM	- Non-competitive, slow-binding covalent inhibitor of the USP2 catalytic domain. - Disrupts enzyme–substrate interaction and covalently binds to the Cys276 of USP2.	[89,93,94]
USP2/5/8/UCHL1/UCHL3	RA-9	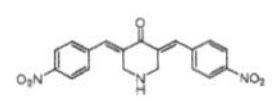	Not determined by enzymatic assay	- Non-specific irreversible inhibitor - Exposes its carbonyl to nucleophilic attack by cysteine - Downregulates cyclin D1, upregulates p53, p27Kip1, and p16Ink4A. - Causes G2-M phase cell cycle arrest and apoptosis. - Increases ER stress markers. - Reduces tumor growth and prolongs survival in ovarian cancer xenograft mouse model.	[54,89,95–97]
	RA-14	Chalcone derivative of RA-9 (structure not disclosed)	Not determined by enzymatic assay	- Derivative of RA-9 - Irreversible USP inhibitor. - Presence of the α, β-unsaturated carbonyl group of RA-14 is susceptible to nucleophilic attack from the sulfhydryl group of the USP catalytic cysteine residues - Causes S-G2/M phase cell cycle arrest. - Downregulates cyclinD1, upregulates p53, p27Kip1, and p16Ink4A. - Induces apoptosis in breast, cervical, and ovarian cancer cells.	[54,95]
	AM146	Chalcone derivative of RA-9 (structure not disclosed)	Not determined by enzymatic assay	- Derivative of RA-9. - Similar mechanism of action as RA-9 - Downregulates cyclin D1. - Upregulates p53.	[54,95]
USP2/7	NSC632839	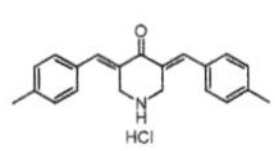	45 μM/37 μM	- Inhibits cleavage of z-LRGG sequence of Ub of USP. - Non-selective USP2 and USP7 isopeptidase inhibitor - Also inhibits deSUMOylase SUMO specific protease 2 (SENP2). - Induces apoptosis by stabilizing the second mitochondrial-derived activator of caspases	[89,98]
	Compound 14	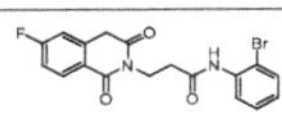	0.25 μM (USP2)	- Reversible inhibitor of USP2 and USP7. - Uncompetitive mechanism of action.	[54,89,99]
USP4/5	Vialinin A	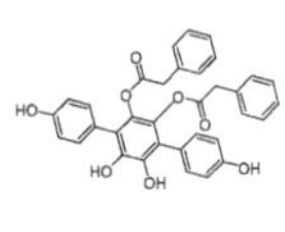	1.5 μM/5.9 μM	- Non-covalent inhibitor. - Binds specifically, to the catalytic domain near Val98 within USP4 and forms H-bonds via residues Asp91, Glu92, and Leu97. - Forms H-bonds via residues Asp91, Glu92, and Leu97 on the USP4. - π-stacking interactions with residues Phe44 and Phe53 contribute to stability. - Isolated from the Chinese mushroom Thelephora vialis. - Inhibits TNF-α translation and enhances its degradation. - Inhibits TGF-β induced cell proliferation.	[100–102]
USP5	USP5-IN-1 (compound 64)	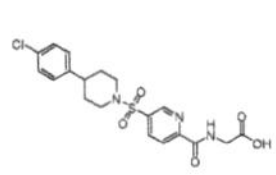	0.8–26 μM	- Non-covalent inhibitor. - Selective inhibitor of USP5. - Binds to the C-terminal ubiquitin-binding site of the ZnF-UBP domain. - Displaces ubiquitin in a concentration-dependent manner.	[103]
USP5/7/8/13/14/15/22	Curcusone D	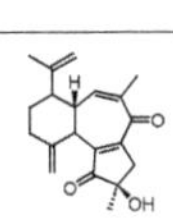	Not determined by enzymatic assay	- Covalent, non-selective inhibitor of USPs. - Induces ROS production. - Inhibits cancer cell growth in combination with Bortezomib in multiple myeloma cells.	[54,104]

Table 3. *Cont.*

USP Target	Name of the Inhibitor	Chemical Structure	IC_{50} Values	Mechanism of Action and Anticancer Activity	Refs
USP5/9X/14/24/37/UCHL5	WP1130 (Degrasyn)	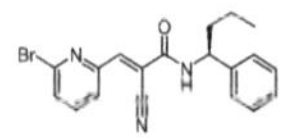	2.5–5 µM	Partly selective non-covalent USP inhibitor. - Inhibition of USP9X leads to downregulation of survivin, c-FLIP, and MCL-1 (anti-apoptotic proteins). - Sensitizes cancer cells to TNF-related apoptosis-inducing ligand mediated apoptosis. - Increases sensitivity of NSCLC cells to cisplatin	[100, 105–108]
	EOAI3402143 (G9)	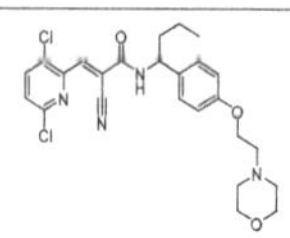	1.6–5 µM	- Reversibly targets multiple USPs. - Downregulates cyclin D1. - Suppresses NSCLC proliferation in a xenograft mouse model.	[49,107, 109,110]
USP7	HBX-41108	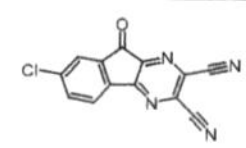	0.424 µM	- Reversible and uncompetitive USP7 inhibitor. - Upregulates p53 and p21. - Induces apoptosis in cancer cells. - Antiproliferative against colon cancer cell line.	[111, 112]
	HBX-28258	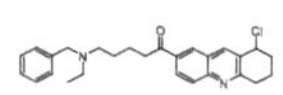	22.6 µM	- Covalent inhibitor that binds to the active site of USP7 at Cys223Promotes MDM2 degradation and p53 activation in colon cancer cells.	[113, 114]
	Spongiacidin C	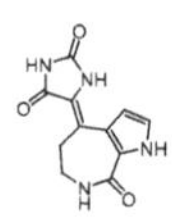	3.8 µM	- Selective USP7 inhibitor - One of the first USP7 inhibitors derived from a natural source. - Pyrrole alkaloid isolated from the marine sponge Stylissa massa. - Inhibition of USP7 leads to MDM2 and MDMX degradation which in turn promotes p53 upregulation. - In vitro studies demonstrated a lack of cytotoxic activity.	[114, 115]
	GNE-6640	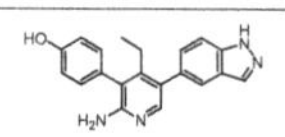	0.75 µM	- Bind to the ubiquitin-binding site within the USP catalytic domain. - Interacts with the acidic residues that mediate hydrogen-bond interactions with the ubiquitin Lys48 side chain. - Increases ubiquitination and degradation of MDM2, resulting in stabilization of p53.	[116–118]
	GNE-6776	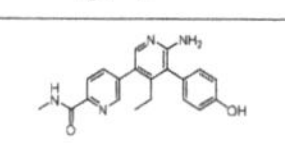	1.34 µM		
	XL188	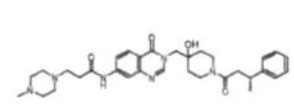	90 nM	- Targets the S4-S5 pocket of catalytic domain ~5 angstroms from the cysteine residue. - The 4-hydroxy-piperidine group of XL188 forms hydrogen bonds with the Gln297 carboxylic group of USP7. - Leads to increased expression of tumor suppressors p21 and p53.	[119, 120]
	ALM4	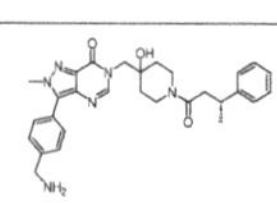	6 nM	- Both ALM4 and ALM46 are highly potent non-covalent inhibitors. - Binds to the USP7 catalytic domain. - Promotes MDM2 and MDMX degradation, which in turn activates p53.	[121, 122]
	ALM46	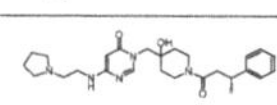	0.09 µM		
	FT671	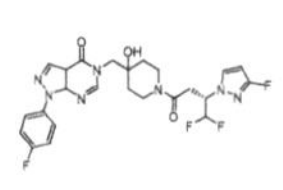	52 nM	- High affinity and specific non-covalent inhibitor. - Targets dynamic pocket near the catalytic center. - Downregulates USP7 substrates (e.g., MDM2). - Increases expression of tumor suppressors p53 and p21. - Reduces TNBC tumor growth in vivo.	[123, 124]
	FT827	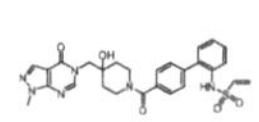	K_i = 66 µM	- Covalent inhibitor of USP7. - The vinylsulfonamide moiety covalently binds at Cys223. - Specifically targets USP7 at the active site. - Similar downstream effects of FT671.	[123]

Table 3. *Cont.*

USP Target	Name of the Inhibitor	Chemical Structure	IC_{50} Values	Mechanism of Action and Anticancer Activity	Refs
	Parthenolide (PTL)	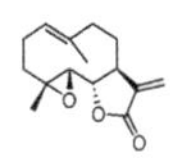	6.58 μM	- Covalent inhibitor. - Contains an α-methylene-γ-butyrolactone moiety, which directly interacts with USP7 in a competitive manner. - Covalently modifies the Cys223 residue. - Induces G2/M phase cell cycle arrest and apoptosis. - Disrupts wnt/β-catenin signaling pathway by enhancing β-catenin ubiquitination and degradation. - Suppresses colorectal cancer cell proliferation in vitro. - Leads to the inhibition of the NF-kB pathway. - Stabilizes p53 by promoting MDM2 ubiquitination and degradation. - Suppresses JAK activity reducing STAT3 activation. - Reduces tumor burden in colon cancer xenograft mouse model.	[125]
	Costunolide	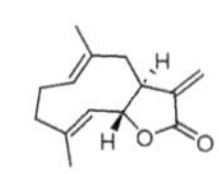	Not determined by enzymatic assay	- Covalent inhibitor. - Contains an α-methylene-γ-butyrolactone moiety, which directly interacts with USP7. - Degradation of β-catenin leads to suppression of the Wnt/β-catenin signaling pathway. - Inhibits growth of colon cancer cells in vitro.	[125]
	YCH2823	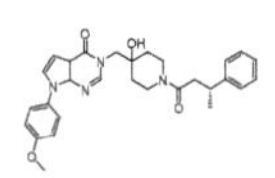	49.6 nM	- Directly interacts with the catalytic domain. - Impedes deubiquitination of substrates such as MDM2. - Increases expression of tumor suppressors p53 and p21 levels. - Causes G1 phase cell cycle arrest.	[126]
	XL177A	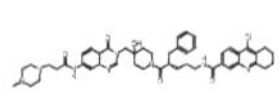	0.34 nM	- Covalent, irreversible, and extremely potent sub-nanomolar inhibitor. - Irreversibly binds to USP7 by forming a covalent bond with the catalytic cysteine. - Induces conformational changes in USP7 protein dynamics. - Degradation of negative regulators of p53 (e.g. MDM2), which leads to G1 cell cycle arrest. - Exerts anti-tumor activity against Ewing sarcoma and malignant rhabdoid tumor.	[120]
USP7/10	HBX-19818	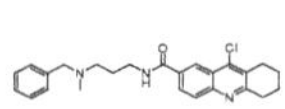	28.1 μM/14 μM/	- Covalent inhibitor. - Irreversibly targets the USP7 catalytic site at Cys223 through nucleophilic attack, resulting in chloride release from the molecule and subsequently covalent binding. - Promotes MDM2 protein degradation and p53 activation. - Induces caspase 3 activity and PARP cleavage. - Induces G1 phase cell cycle arrest. - Reduces colon cancer cell proliferation.	[113, 114,127]
	Compound 3	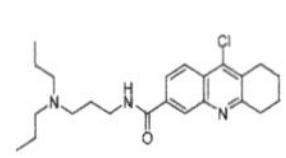	3.6 μM (USP7)	- Non-covalent inhibitor that makes H-bond with Asp295 and hydrophobic contact with Phe409 in the binding pocket. - Structural analog of HBX-19818. - More specific to USP10 than USP7. - Leads to FMS-like tyrosine kinase 3 (FLT3) degradation at lower concentrations.	[54,62, 127,128]

Table 3. *Cont.*

USP Target	Name of the Inhibitor	Chemical Structure	IC_{50} Values	Mechanism of Action and Anticancer Activity	Refs
USP7/47	P22077	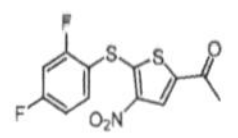	8 μM (USP7)	- Covalent inhibitor. - Binds to the Cys223 residue at the catalytic site which induces the confirmational change in active site rearrangement to inhibit enzymatic activity. - Disrupts the NF-kB and MAPK pathways. - Stabilizes p53 by triggering HDM2 degradation. - Re-sensitizes chemo-resistant neuroblastoma cells. - Reduces tumor burden in orthotopic neuroblastoma xenograft mouse model. - Overcomes cytarabine resistance in AML cells.	[54,86, 129–132]
	P5091	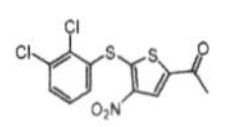	4.2 μM (USP7)	- Selective and potent covalent inhibitor - Enhances β-catenin degradation, which leads to disruption of the wnt/β-catenin signaling pathway and induces apoptosis. - Suppresses colorectal cancer cell proliferation in vitro and in vivo. - Induces apoptosis in multiple myeloma cells that are resistant to Bortezomib.	[54,133–135]
	P50429	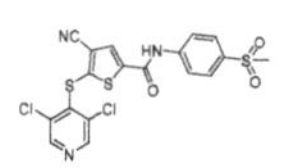	0.42 μM/1 μM	- Covalent analog of P5091. - Covalently modifies the catalytic cysteine residue, Cys223, at the active site of USP7. - Induces conformational switch in USP7 active site. - Inhibits colon cancer cell proliferation.	[54,132, 136]
USP8	HY50736	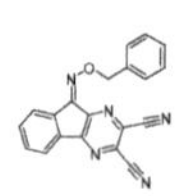	0.24 μM	- Mainly target USP8 with very poor activity against USP7. - Derivatives of HBX-41108. - The introduction of the O-alkyloxime moieties at C-9 of the tricyclic scaffold is important for USP8 inhibition. USP8 plays a key role in the recycling of cell surface receptors such as EGFR, and inhibition leads to decreased cellular survival and proliferation.	[54,137]
	HY50737A	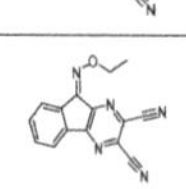	0.28 μM		
USP9x	FT709	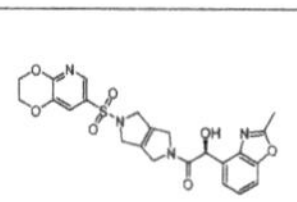	82 nM	- Non-covalent, potent, and highly selective inhibitor. - Competitive USP9x inhibitor. - Destabilizes Makorin and ZNF598 ubiquitin E3 ligases, which are substrates that regulate the ribosomal quality control pathway.	[138]
USP10	Compound 9	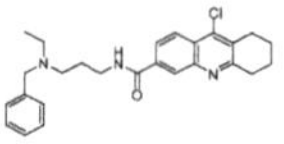	Not determined by enzymatic assay	- Non-covalent inhibitor. - Interacts with Asp295, Val296, and Tyr456 at the catalytic domain. - Structural analog of HBX-19818. - Similar in structure to compound 3 but lacks activity against USP7. - Reversible inhibition of USP10 leads to FLT3 degradation.	[54,62, 127,128]
	Wu-5	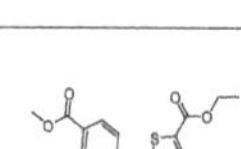	8.3 μM	- Selective covalent USP10 inhibitor. - Induces degradation of FLT3-ITD and shows efficacy against FLT3 inhibitor-resistant AML cells. - Enhances anti-leukemic effects of FLT3 inhibitors (crenolanib).	[139, 140]
USP10/13	Spautin-1	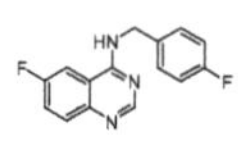	0.6–0.7 μM	- Potent and specific inhibitor. - Leads to beclin-1 degradation, which is important for autophagy. - Reduces cancer cell proliferation in nutrition-deprived conditions. - Prevents induction of UPR-associated proteins and induces cytotoxicity preferentially in glucose-starved cancer cells.	[141–143]

Table 3. *Cont.*

USP Target	Name of the Inhibitor	Chemical Structure	IC_{50} Values	Mechanism of Action and Anticancer Activity	Refs
USP11	Mitoxantrone	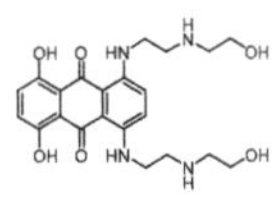	3.15 μM	- Synthetic anthracenedione. - Inhibits the activity of USP11, however, the exact mechanism of action is elusive. - USP11 plays an essential role in DNA repair. - Exhibits anticancer activity through several mechanisms, such as DNA intercalation to produce DNA strand breaks, topoisomerase II inhibition, and interference with RNA synthesis.	[144]
USP12/USP46	Galeterone	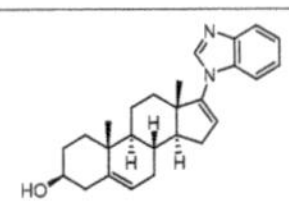	2.1–3.4/3.4–4.2 μM	- Selectively inhibits USP12 and USP46. - Disrupts androgen receptor (AR) stability and the signaling pathway. - Suppresses prostate cancer cell proliferation in vitro.	[145]
USP14	IU1	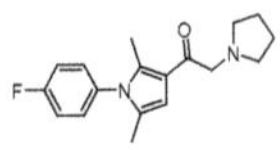	4–5 μM	- Reversible and selective USP14 inhibitor. - Promotes MDM2 protein degradation in cervical cancer cells leading to G0/G1 cell cycle arrest.	[146, 147]
	IU1-47	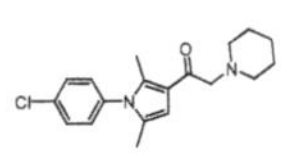	0.6 μM	- More potent derivatives of IU1 with a similar mechanism of action, but several folds more potent. - Prevents USP14 from docking on the proteasome. - Binds to the thumb-palm cleft region of the catalytic domain of USP14 which prevents the C-terminus of ubiquitin from accessing USP14. - Promotion of proteasome substrate degradation of proteins crucial for cancer cell survival. - IU1-248 shows evidence of synergy with the PARG inhibitor COH34 against TNBC in vitro.	[146, 148–150]
	IU1-206	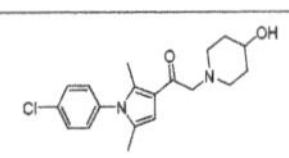	Not determined by enzymatic assay		
	IU1-248	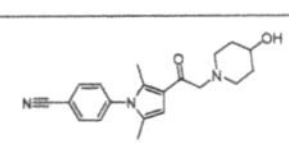	0.83 μM		
USP14/UCHL5	Auranofin	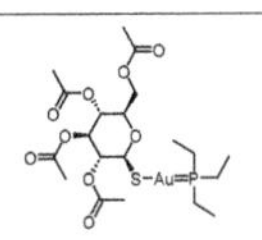	Not determined by enzymatic assay	- Competitively reduces substrate binding (HA-Ub-VS) to the active site Cys of USP14. - FDA approved gold containing disease-modifying antirheumatic drug (DMARD).	[151, 152]
	CuPT	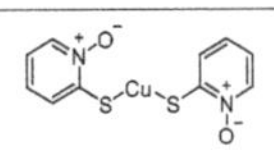	Not determined by enzymatic assay	- These inhibitors compete with Ub-VS for binding to UCHL5 and USP14. - Metal-center containing compounds with an identical organic structure but different metal. - Induce G2 cell cycle arrest. - Lead to accumulation of ubiquitinated proteins inducing G2 cell cycle arrest.	[153–157]
	PtPT	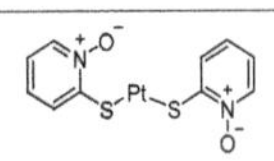	Not determined by enzymatic assay		
	NiPT	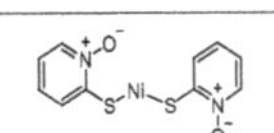	Not determined by enzymatic assay		
	b-AP15	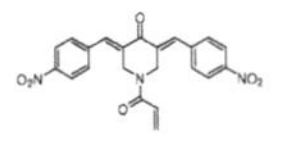	<10 μM	- Covalent inhibitor. - Selectively inhibits the active site at residues Cys203 and Cys257 in the ubiquitin-binding pocket of USP14. - Contains an α, β-Unsaturated carbonyl - Contains electrophilic Michael acceptor motifs which contribute to its ability to form covalent bonds with its targets via Michael addition. - Reaction forms high molecular weight complexes, which can be reduced by thiol-containing reducing agents such as dithiothreitol (DTT) or glutathione. - Alters levels of cell cycle regulators such as cyclin D1, CDKs, and p27 which leads to G0/G1 cell cycle arrest and apoptosis.	[158–161]

Table 3. *Cont.*

USP Target	Name of the Inhibitor	Chemical Structure	IC_{50} Values	Mechanism of Action and Anticancer Activity	Refs
USP22	USP22i-S02	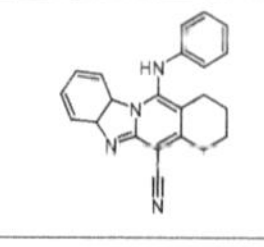	Not determined by enzymatic assay	- Potent, reversible USP22 inhibitor - Leads to the degradation of cell cycle progression proteins including cyclin B1 and cyclin D1. - Synergizes with cisplatin in breast cancer cells. - Suppresses Foxp3 in T regulatory cells (Tregs).	[162–165]
USP24	NCI677397	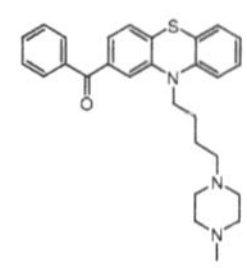	Not determined by enzymatic assay	- Non-covalent inhibitor. - Interacts with the catalytic domain of USP24. - Leads to decreased stability of the ATP-binding cassette (ABC) transporters which function to pump toxins including chemotherapy out of cells. - Potential ferroptosis inducer (FIN) by increasing levels of lipid ROS leading to ferroptotic cellular death.	[118, 166]
	USP24-i-101	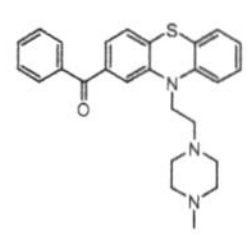	Not determined by enzymatic assay	- Reversible inhibitor. - Increases PD-L1 degradation leading to increased apoptotic signaling. - Activates autophagy by inhibiting E2F4 and TRAF6. - Excessive activation of autophagy paradoxically leads to overcoming chemotherapy resistance in drug-resistant lung cancer cells in vitro, as excessive autophagy induction can lead to cell death when combined with chemotherapy. - Overcomes Taxol- or gefitinib-induced drug resistance in lung cancer.	[167]
USP25/28	AZ1	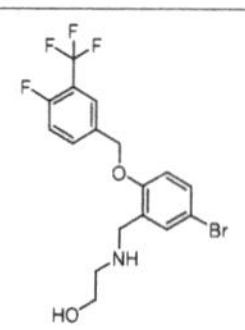	1.08 μM/1.76 μM	- Non-competitive dual inhibitor. - Benzylaminoethanol derivative. - Binds to the conserved site of USP25 and USP28 at the distal portion of the S1 cleft. - Fluorine is essential for binding and allosteric regulatory inhibition. - Several USP25 and USP28 substrates function as oncogenes such as ΔNp63.	[168, 169]
	FT206	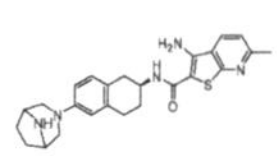	1.01/0.15 μM	- Binds to the same cleft as AZ1. - Higher selectivity for USP28. - Destabilizes c-MYC. - Induces regression of squamous cell lung cancer in vivo.	[37,169]
	Vismodegib (VSM)	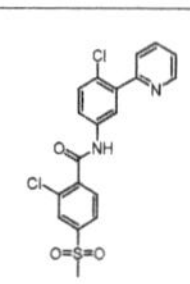	2.92/3.51–4.41 μM	- Interacts with the binding pocket of USP28 composed of two helical structures spanning D255-N278 and N286-Y293. - Leads to downregulation of c-MYC, Notch1, and Tankyrase-1/2 in colorectal cancer cells. - Inhibits the Hedgehog signaling pathway, which has been implicated in the pathogenesis of basal cell carcinoma.	[169–172]
USP30	MF-094	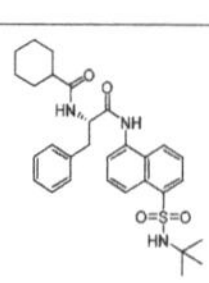	0.12 μM	- Selectively inhibits the active site of USP30, a mitochondrial localized DUB. - Enhances mitophagy by suppressing deubiquitination and enhancing the degradation of damaged mitochondria. - Inhibits oral squamous cell carcinoma in vivo.	[173–175]
	15-oxospiramilactone	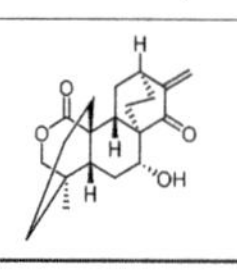	Not determined by enzymatic assay	- Binds to Cys77 of the catalytic domain of USP30. - Semi-synthetic product isolated from the Spiraea japonica plant. - Promotes mitochondrial fusion by enhancing the activity of mitofusin proteins Mfn1 and Mfn2.	[176, 177]

Table 3. *Cont.*

USP Target	Name of the Inhibitor	Chemical Structure	IC_{50} Values	Mechanism of Action and Anticancer Activity	Refs
	Compound 39 (CMPD-39)	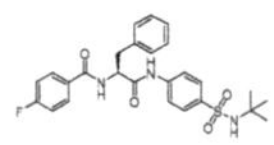	20 nM	- Selective competitive inhibitor. - Binds to the catalytic site of USP30. - Increases ubiquitination of TOMM20 and SYNJ2BP, which leads to increased mitophagy.	[178–180]
USP36	Cinobufotalin	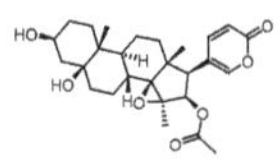	2.75 μM	- Interacts directly with the catalytic domain of USP36. - Is a type of bufadienolide, which is a subclass of cardiac glycosides that are derived from toad venom glands. - Promotes the degradation of c-Myc. - Inhibits proliferation, migration, and invasion of colon cancer cells.	[36]

3.2. USP Inhibitors and Immunotherapy

As mentioned above, the development of USP inhibitors represents a significant advancement in cancer treatment, offering a multifaceted approach to improving patient outcomes. USP inhibitors work by various mechanisms to inhibit cancer cell growth and proliferation, such as destabilizing oncoproteins [181]. USP inhibitors thus offer a novel strategy to overcome drug resistance, a major challenge in cancer therapy, by targeting the mechanisms that cancer cells employ to evade chemotherapy and targeted treatments [182]. Some USP inhibitors also show promise in enhancing cancer immunotherapy by making cancer cells more vulnerable to immune attacks [183]. In this section, we focus on the potential of USP inhibitors in enhancing the immune system and immunotherapy against cancer.

3.2.1. Role of USP1 in Immune Response

USP1 plays a significant role in immunotherapy, particularly in the context of cancer treatment. USP1 has been identified as a critical regulator of T-cell differentiation, which is crucial for effective immunotherapy responses. By deubiquitinating and stabilizing the transcriptional co-activator with PDZ-binding motif (TAZ), USP1 leads to enhanced activity of RORγt, which is an important transcription factor for T helper type 17 (Th17) cell development. Deubiquitination of TAZ also causes decreased acetylation of Foxp3, which promotes its proteasomal degradation leading to reduced Treg cell differentiation [184]. USP1 has also been implicated in resistance to chemotherapy and Rituximab in DLBCL. USP1 deubiquitinates MAX, which is an important MYC-binding protein and promotes MYC gene transcription. High expression of USP1 in DLBCL has been found to be associated with a poorer prognosis, while inhibition with Pimozide or knockdown of USP1 has been shown to reduce cancer cell growth and induce cell cycle arrest. Inhibition of USP1 was also demonstrated to significantly reduce tumor burden in a therapy-resistant DLBCL-engrafted PDX mouse model. Importantly, Pimozide was shown to have evidence of synergy with Etoposide in therapy-resistant DLBCL [80]. The potential of USP1 inhibitors lies in their ability to modulate the tumor microenvironment and enhance the efficacy of immunotherapeutics. Combining USP1 inhibitors with existing immunotherapies may enhance treatment efficacy, particularly in cases of resistance to current therapies.

3.2.2. Role of USP7 in Immune Response

Recent research has substantially advanced our understanding of how USP7 modulates the immune response in cancer patients. Elevated levels of USP7 have been shown to facilitate tumor growth by enhancing the immunosuppressive functions of Foxp3+ Tregs [185,186]. Blocking USP7 hinders the activity of Tip60-dependent Foxp3+ Treg cells, decreasing their inhibitory function and enhancing the body's ability to fight against tumors [186]. Histone acetyltransferase Tat-interactive protein (Tip60) is essential for the survival of Treg cells. Tip60 is a USP7 substrate, thus targeting USP7 could interfere with Tip60-mediated Foxp3 dimer formation, resulting in reduced activation of immunosuppressive molecules such as CTLA4 and IL-10 while promoting the expression of pro-inflammatory cytokines such as IL-2 and IFN-γ [186]. USP7 is important in regulating the equilibrium of M1 (which suppresses tumor growth and promotes an anti-tumor immune response) and M2 (which aids in tumor progression and inhibits the immune system) macrophages. Targeted inhibition of USP7 results in a change in the appearance and behavior of M2 macrophages, leading to a higher growth rate of differentiated CD8+ T cell groups in a laboratory setting [61]. In research on mice with Lewis lung carcinoma, inhibiting USP7 led to decreased tumor growth and increased levels of M1 macrophages and CD8+ T cells that expressed IFN-γ. USP7 also maintains the stability of PD-L1 by inhibiting its degradation [187].

3.2.3. Enhancing Immunotherapy by USP Inhibitors

USP inhibitors can make cancer cells more susceptible to immune attack by preventing them from evading the immune system and enhancing the effectiveness of immunotherapy [183]. Tumors have specific properties that allow them to evade the immune system, hindering an anti-tumor response and promoting their growth. Mechanisms of immune evasion include the production of immunosuppressive factors, down-regulation of major histocompatibility complex molecules, and recruitment of immunoregulatory cells. Additional studies have shown that USPs can influence the effectiveness of immunotherapy through the regulation of immune cell function and the immune response within the tumor microenvironment [157,188].

HBX-19818 and GNE-6776 are covalent USP7 inhibitors that have shown promise in stabilizing p53 and enhancing its activity, which can lead to improved anti-tumor responses [116]. USP22 inhibitors have also shown potential in improving responses to immunotherapy by increasing levels of CD8+ T cells and NK cells, transforming immune-desert tumors into immune-inflamed tumors [189]. In liver tumors, inhibiting USP22 has been shown to boost tumor immunogenicity, enhance T-cell infiltration, and increase responsiveness to anti-PD-L1 immunotherapy [183,190]. Recent research has identified potent macrocycle inhibitors of USP22, including compound S02, which binds tightly to the catalytic domain pocket of USP22. These USP22 inhibitors have demonstrated significant potential in reducing cancer cell growth, promoting apoptosis, and enhancing the efficacy of existing treatments [183]. USP8 inhibitors have shown potential in enhancing anti-tumor activity when combined with anti-PD-1/PD-L1 immunotherapy. USP13 inhibitors may enhance the antitumor effects of DNA damage response inhibitors as well as promote immune cell infiltration and innate immunity [191]. USP14 inhibitors have been shown to sensitize cancer cells to immune checkpoint inhibitors [192]. USP15 inhibitors can disrupt pathways controlling Toll-like receptor, RIG-I, and NF-kB signaling, enhancing the immune response against tumors [183]. USP9X inhibitors like G9 have shown promise in inactivating Notch signaling, reducing proinflammatory cytokines, and enhancing antitumor immune response [183,193].

In summary, USP inhibitors represent a promising avenue for enhancing cancer immunotherapy by modulating various aspects of the immune response as well as the tumor microenvironment. Their ability to target multiple pathways involved in immune evasion and tumor progression and to enhance the efficacy of existing treatments makes them a valuable addition to the arsenal of cancer therapeutics.

3.3. Development of USP Inhibitors for Cancer Therapy: Clinical Trials Status

Regarding the development of USP inhibitors for the treatment of cancer, most of these inhibitors are still in the preclinical stage. However, a few USP inhibitors are being explored in clinical trials for their potential therapeutic effects, given their mechanism of action and possible benefits to patients with resistant or refractory disease. Multiple USP1 inhibitors are currently being tested in phase 1 and phase 2 clinical trials, which aim to evaluate the safety, tolerability, and potential antitumor activity of these compounds. While there are no USP7 inhibitors currently in clinical trials, several compounds have shown promise in preclinical studies. Inhibition of USP7 has long been viewed as a promising anticancer target being the key DUB in regulating p53 levels. FT671 and FT827 are both small-molecule USP7 inhibitors that have shown high affinity and selectivity to USP7 in vitro. Both compounds appear to target the auto-inhibited apo form of USP7 near the catalytic center, which is distinct from other USPs [123]. Most recently, YCH2823 was discovered as a next-generation USP7 inhibitor with enhanced cellular activity compared to FT671, demonstrating about five times the potency. YCH2823 inhibited the growth of *TP53*-wild-type, *TP53*-mutant, and *MYCN*-amplified cell lines with exceptional efficacy by binding directly to the catalytic domain of USP7. A synergy between USP7 and mTOR inhibitors was also observed, demonstrating potential for novel therapeutic strategies [126]. Table 4 provides a comprehensive overview of the USP inhibitors that are currently or were previously investigated in clinical trials. The study design, preliminary/current results, and key findings from these clinical studies are also described where applicable in the table.

3.3.1. USP1 Inhibitors

USP1 inhibitors represent a promising class of drugs targeting DNA damage repair mechanisms, particularly in homologous recombination deficient (HRD) cancers. Several USP1 inhibitor compounds are currently being studied in various stages of clinical development. TNG348, developed by Tango Therapeutics, is an oral, potent, and highly selective allosteric USP1 inhibitor. It showed potential in preclinical models, demonstrating enhanced activity when combined with DNA repair pathway targeted therapies and efficacy in xenografts with both primary and acquired resistance to PARP inhibitors. Malignancies with HRD are generally more sensitive to therapies that target DNA repair. BRCA1-mutated tumor cells are known to be sensitive to PARP inhibitors due to defects in replication fork stability, and they have shown synergy with chemotherapy. HRD cancers represent up to 1 in 2 ovarian cancers, 1 in 4 breast cancers, 1 in 10 prostate cancers, and 1 in 20 pancreatic cancers. USP1 has been found to be upregulated in tumors with BRCA1 mutations, and its knockdown destabilizes replication forks resulting in cell death [194]. Inhibiting USP1 could be a potential therapeutic strategy for cancers with BRCA1-deficient cells, especially those resistant to PARP inhibitors [194]. However, the phase 1/2 trial (NCT06065059) was terminated due to significant liver toxicity, with grade 3/4 hepatotoxicity observed in patients who continued treatment for longer than eight weeks. These results demonstrate the challenges in developing novel therapies and highlight the importance of safety monitoring in early-phase clinical trials.

KSQ-4279, also known as RO7623066, was developed by Roche/KSQ Therapeutics, and it is a potent and highly selective oral USP1 inhibitor. It binds to a cryptic site in USP1, similar to ML323, but with subtle rearrangements in the site folding, accounting for differences in potency and selectivity [195]. Currently, in a phase 1 trial (NCT05240898) for advanced solid tumors, KSQ-4279 is being studied as a single agent and in combination with Carboplatin or Olaparib. The combination of a PARP inhibitor and KSQ-4279 was shown to induce regression of several PDX PARP-resistant tumors in the preclinical phase [196]. Preliminary results showed limited efficacy as monotherapy but suggested potential synergy with Olaparib, particularly in BRCA1-mutated tumors. Pharmacokinetic studies demonstrated that the tested drug becomes saturated at higher doses, suggesting the need for twice-daily dosing. Pharmacodynamic data also appeared to support the theorized mechanism of action of USP1 inhibition.

XL309, also known as ISM3091, was developed using AI technology by Insilico Medicine, Boston, Massachusetts, USA and licensed to Exelixis, Alameda, California, USA, and it is a highly selective and non-covalent oral inhibitor of USP1 [197]. Preclinical data demonstrated high efficacy in BRCA1 mutated triple-negative breast cancer cell lines and synergistic effects with Olaparib in vivo. Interestingly, it also showed efficacy against a lung adenocarcinoma cell line without HRD, suggesting potential applications beyond HRD cancers. This opens the door to consider potentially utilizing XL309 in HR-proficient cancers. A phase 1 trial (NCT05932862) is ongoing to investigate its safety and preliminary antitumor activity.

SIM0501, developed by Simcere Jiangsu Pharmaceutical Co., Nanjing, Jiangsu, China, is another highly selective and non-covalent oral USP1 inhibitor [195]. It has shown evidence of synergy with Olaparib both in vitro and in vivo across various HRD-positive cancers. In vivo efficacy studies have demonstrated a dose-dependent increase of Ub-PCNA, which may be a useful biomarker to track for response, as Ub-PCNA is a direct target of USP1 [196]. A phase 1 trial (NCT06331559) is evaluating its safety and preliminary efficacy in patients with advanced solid tumors.

HSK39775, developed by Tibet Haisco Pharmaceutical Co. Ltd., Zedang Town, Shannan, Tibet, is a highly selective and non-covalent oral USP1 inhibitor. It has been shown to inhibit the USP1/UAF1 complex and exhibit strong dose-dependent tumor response in xenograft-derived BRCA-mutated triple-negative breast cancer with synergy seen when given with a PARP inhibitor. HSK39775 monotherapy was also found to inhibit the growth of BRCA wild-type lung cancer [196]. A phase 1/2 trial (NCT06314373) is underway to evaluate its safety and efficacy in advanced solid tumors. The field of USP1 inhibitors is rapidly evolving, and these clinical trials will provide crucial data for this novel therapeutic approach in cancer management, particularly for patients with HRD-positive tumors and those resistant to PARP inhibitors.

3.3.2. USP14 Inhibitor

VLX1570, a small molecule DUB inhibitor derived from b-AP15 [(3E,5E)-3,5-bis[(4-nitrophenyl)methylidene]-1-(prop-2-enoyl)piperidin-4-one], selectively inhibits USP14 and UCHL5, which are associated with the 19S regulatory subunit of the proteasome. Both of these small molecule inhibitors are α,β-unsaturated carbonyl DUB inhibitors and have similar mechanisms of action. It was studied in a phase 1 clinical trial (NCT02372240) for relapsed or refractory multiple myeloma patients who had developed resistance to Bortezomib [198]. The trial, conducted by Eric K Rowinsky et al., aimed to determine the safety and tolerability of VLX1570. While the compound showed anti-myeloma effects at doses ≥ 0.6 mg/kg, the study was discontinued due to severe respiratory insufficiency in two patients that led to death [199]. Despite these setbacks, the unique mechanism of action of DUB inhibitors in Bortezomib-resistant multiple myeloma has prompted further research, including the development of a rat model to assess VLX1570-induced lung toxicity and the development of specific DUB inhibitors [199].

Table 4. Clinical trials involving USP inhibitors.

Chemical Structure	Type of USP Inhibitor	Clinical Trials.gov ID	Trial Phase	Study Design	Results	References
TNG348 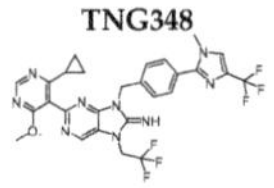	USP1 IC_{50}: 82 nM	NCT06065059	Phase 1 dose escalation and Phase 2 dose expansion trial (terminated)	- Target: BRCA 1/2 mutant tumors or HRD+ solid tumors. - Two arms: TNG348 monotherapy and in combination with Olaparib (PARP inhibitor). - Enrollment: 7 patients. - Study period: December 2023–May 2024.	- Terminated due to significant safety concerns. - Grade 3/4 hepatotoxicity observed in patients after 8 weeks. - No patients received combination therapy with Olaparib. - Highlights safety challenges in early-phase trials.	[194,200–202]
KSQ-4279, also known as RO7623066 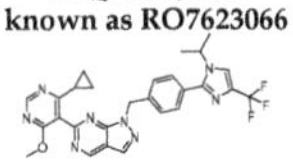	USP1 IC_{50}: 10 nM	NCT05240898	Phase 1 (recruiting)	- Target: Advanced solid tumors, particularly those with HRD. - Three arms: monotherapy (arm 1), combination with Olaparib (arm 2), combination with Carboplatin (arm 3). - Dose escalation followed by dose expansion. - Study period: August 2021–June 2027.	- 64 heavily pretreated patients enrolled. - Disease control rates: 28% (monotherapy), 40% (Olaparib combo), 29% (Carboplatin combo). - One partial response in advanced fallopian tube cancer. - 17% (5/29) achieved stable disease for >16 weeks with monotherapy. - Anti-tumor activity in Olaparib combo linked to BRCA1 mutational status. - Most common side effect: anemia (86.7% in Olaparib combo arm). - No study discontinuations due to adverse events. - Study ongoing to determine maximum tolerated dose.	[71,196,203,204]
XL309, also known as ISM3091 Chemical structure is not publicly disclosed	USP1	NCT05932862	Phase 1 (recruiting)	- Target: Advanced solid tumors, including those with HRD mutations. - Investigating monotherapy and combination with Olaparib. - Estimated enrollment: 377 patients. - Study Period: August 2023–August 2029.	- No reported results as of December 2024. - Study ongoing.	[197,205,206]
SIM0501 Chemical structure is not publicly disclosed.	USP1	NCT06331559	Phase 1 (recruiting)	- Target: Advanced solid tumors, including Olaparib-resistant cases. - Estimated enrollment: 176 patients. - Study period: March 2024–December 2027.	- No reported results as of December 2024. - Study ongoing. - First patient enrolled in March 2024.	[197,207,208]
HSK39775 Chemical structure is not publicly disclosed.	USP1	NCT06314373	Phase 1 dose escalation and Phase 2 dose expansion trial (recruiting)	- Target: Advanced solid tumors. - Investigating monotherapy in two phases. - Phase 1: Safety, efficacy, and RP2D. - Phase 2: To evaluate ORR and further safety. - Estimated enrollment: 243 patients. - Study period: March 2024–September 2028	- No reported results as of December 2024. - Study ongoing. - First patient enrolled in March 2024.	[197,209,210]

Table 4. *Cont.*

Chemical Structure	Type of USP Inhibitor	Clinical Trials.gov ID	Trial Phase	Study Design	Results	References
VLX1570 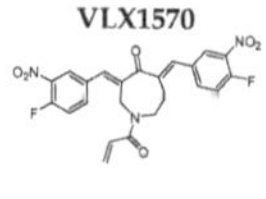	USP14/UCHL5 IC_{50}: ~10 μM	NCT02372240	Phase 1 dose escalation and Phase 2 dose expansion trial (terminated)	- Target. Relapsed/refractory multiple myeloma. - Intravenous administration of compound formulated in polyethylene glycol and polysorbate 80 mixture given the poor aqueous solubility. - 28-day cycle with escalating doses. - Enrollment: 14 patients. - Study period: April 2015–May 2017.	- Terminated due to severe toxicity. - 14 patients enrolled. - Anti-myeloma effects observed at doses ≥ 0.6 mg/kg. - Two patient deaths due to abrupt, severe respiratory insufficiency. Study discontinued due to the severity and abrupt nature of toxicity. - Toxicity comparable to some cases of Bortezomib. - Highlighted the need for further research into safer USP inhibitors.	[198,199, 211,212]

4. Summary and Future Directions

USP inhibitors have shown significant potential in combating drug resistance and enhancing the effectiveness of existing cancer therapies. The UPS is involved in a vast array of cellular functions; thus, developing more potent and selective USP inhibitors remains a primary focus. Current inhibitors often face challenges related to specificity and potency, given the possible off-target effects these USP inhibitors may have. Researchers are currently exploring novel allosteric binding sites within USPs to design inhibitors with improved selectivity and efficacy. Combination therapies involving USP inhibitors are being studied to overcome drug resistance in cancer treatment. Integrating USP inhibitors with existing anticancer therapies, such as chemotherapy and targeted therapy, shows potential for enhancing treatment outcomes [182]. USP inhibitors also have the key benefit of enhancing the effectiveness of immunotherapy by modulating the tumor microenvironment. Tailored approaches to treatment may lead to more effective and personalized therapy. As research progresses, expanding our understanding of the complex roles of USPs in various signaling pathways and cellular processes will be crucial [54]. Ongoing clinical trials suggest that USP inhibitors, particularly those targeting USP1, may have the potential to become valuable components of cancer therapy. While no USP inhibitors have yet been approved for clinical use, they have shown significant promise in preclinical studies. Advancing USP inhibitors from preclinical studies to clinical trials remains a significant goal, with the aim of bringing these promising therapeutic agents closer to clinical application. As these trials continue to expand and progress, more information about their safety and efficacy in patients will emerge, potentially leading to novel treatment strategies for patients with cancer.

5. Conclusions

In conclusion, this review provides an in-depth analysis of USPs' structural and functional characteristics, their roles in various cancer-related pathways, and the mechanisms of action of over seventy different small molecule USP inhibitors. We also discuss how these inhibitors can enhance cancer immunotherapy, overcome drug resistance, and synergize with existing cancer treatments. Additionally, we explore the current progress and challenges in clinical trials involving USP inhibitors, highlighting their growing potential.

Research on USP inhibitors has shown their potential in targeting oncogenic pathways, modulating immune responses, and mitigating neurodegenerative protein aggregation. Targeting USPs by small-molecule inhibitors has demonstrated significant potential in tackling drug resistance and enhancing the effectiveness of existing cancer therapies. Several USP inhibitors, such as VLX1570 and b-AP15, have demonstrated efficacy in preclinical and early clinical trials, particularly in multiple myeloma and other malignancies.

However, challenges remain in developing selective and potent USP inhibitors due to the highly conserved catalytic domains among DUBs and potential off-target effects. The USPs are involved in multiple pathways, and additional studies are required to understand their complex mechanisms. Further advancements in structure-based drug design, high-throughput screening, and combination therapies may enhance their clinical applicability.

Author Contributions: M.B.: conceptualization; writing—original draft; writing—review and editing of the manuscript. S.K.: writing—original draft of USP inhibitors and immunotherapy section; writing—review and editing of the manuscript. K.B.: writing—original draft of USP inhibitor mechanisms of action section included in Table 3; writing—review and editing of the manuscript. N.D.K. and A.S.: writing—original draft of Tables 3 and 4 structures and IC_{50} values; writing—review and editing of the manuscript. S.D.: writing—original draft of USP structure, biological function, and role in tumorigenesis and progression section; writing—review and editing of the manuscript. H.E.: writing—review and editing of the manuscript. Q.P.D.: writing—review and editing of the manuscript. Y.G.: conceptualization; supervision; writing—original draft; writing—review and editing of the manuscript. N.S.G.: conceptualization; supervision; writing—original draft; writing—review and editing of the manuscript. All authors have read and agreed to the published version of the manuscript.

Funding: The research work in the Gavande laboratory is supported by the National Institutes of Health (R01CA247370 and R01AI161570), the Department of Defense (W81XWH-22-1-0369), the VA, KCI's Michigan SPORE, Richard Barber Interdisciplinary Research Program, DMC Foundation, WSU Applebaum Faculty Research Award (FRAP), and Wayne State University. Sadaf Dorandish is supported by the NIH-funded Wayne State University Chemistry Biology Interface (CBI) Program T32GM142519-01.

Institutional Review Board Statement: Not applicable.

Informed Consent Statement: Not applicable.

Data Availability Statement: Not applicable.

Acknowledgments: Y.G. and N.S.G. highly appreciate the Karmanos Cancer Institute's Molecular Therapeutics Program's support for our initial studies on developing USP inhibitors for cancer therapy. The research work in the Ge laboratory is supported by the U Can-Cer Vive Foundation, Kids without Cancer, Children's Foundation, and the Molecular Therapeutics Program of the Karmanos Cancer Institute.

Conflicts of Interest: The authors declare no conflicts of interest.

References

1. Bedford, L.; Lowe, J.; Dick, L.R.; Mayer, R.J.; Brownell, J.E. Ubiquitin-like protein conjugation and the ubiquitin-proteasome system as drug targets. *Nat. Rev. Drug Discov.* **2011**, *10*, 29–46. [CrossRef]
2. Du, W.; Mei, Q.B. Ubiquitin-proteasome system, a new anti-tumor target. *Acta Pharmacol. Sin.* **2013**, *34*, 187–188. [CrossRef] [PubMed]
3. Arkwright, R.; Pham, T.M.; Zonder, J.A.; Dou, Q.P. The preclinical discovery and development of bortezomib for the treatment of mantle cell lymphoma. *Expert. Opin. Drug Discov.* **2017**, *12*, 225–235. [CrossRef]
4. Teicher, B.A.; Tomaszewski, J.E. Proteasome inhibitors. *Biochem. Pharmacol.* **2015**, *96*, 1–9. [CrossRef] [PubMed]
5. Snyder, N.A.; Silva, G.M. Deubiquitinating enzymes (DUBs): Regulation, homeostasis, and oxidative stress response. *J. Biol. Chem.* **2021**, *297*, 101077. [CrossRef]
6. Johnston, S.C.; Larsen, C.N.; Cook, W.J.; Wilkinson, K.D.; Hill, C.P. Crystal structure of a deubiquitinating enzyme (human UCH-L3) at 1.8 A resolution. *EMBO J.* **1997**, *16*, 3787–3796. [CrossRef]

7. Fang, Y.; Fu, D.; Shen, X.Z. The potential role of ubiquitin c-terminal hydrolases in oncogenesis. *Biochim. Biophys. Acta* **2010**, *1806*, 1–6. [CrossRef] [PubMed]
8. Fu, L.; Lu, K.; Jiao, Q.; Chen, X.; Jia, F. The Regulation and Double-Edged Roles of the Deubiquitinase OTUD5. *Cells* **2023**, *12*, 1161. [CrossRef] [PubMed]
9. Kawaguchi, Y.; Okamoto, T.; Taniwaki, M.; Aizawa, M.; Inoue, M.; Katayama, S.; Kawakami, H.; Nakamura, S.; Nishimura, M.; Akiguchi, I.; et al. CAG expansions in a novel gene for Machado-Joseph disease at chromosome 14q32.1. *Nat. Genet.* **1994**, *8*, 221–228. [CrossRef]
10. Shi, Z.; Chen, J.; Zhang, X.; Chu, J.; Han, Z.; Xu, D.; Gan, S.; Pan, X.; Ye, J.; Cui, X. Ataxin-3 promotes testicular cancer cell proliferation by inhibiting anti-oncogene PTEN. *Biochem. Biophys. Res. Commun.* **2018**, *503*, 391–396. [CrossRef] [PubMed]
11. Abdul Rehman, S.A.; Armstrong, L.A.; Lange, S.M.; Kristariyanto, Y.A.; Grawert, T.W.; Knebel, A.; Svergun, D.I.; Kulathu, Y. Mechanism of activation and regulation of deubiquitinase activity in MINDY1 and MINDY2. *Mol. Cell* **2021**, *81*, 4176–4190.e6. [CrossRef]
12. Abdul Rehman, S.A.; Kristariyanto, Y.A.; Choi, S.Y.; Nkosi, P.J.; Weidlich, S.; Labib, K.; Hofmann, K.; Kulathu, Y. MINDY-1 Is a Member of an Evolutionarily Conserved and Structurally Distinct New Family of Deubiquitinating Enzymes. *Mol. Cell* **2016**, *63*, 146–155. [CrossRef]
13. Trulsson, F.; Akimov, V.; Robu, M.; van Overbeek, N.; Berrocal, D.A.P.; Shah, R.G.; Cox, J.; Shah, G.M.; Blagoev, B.; Vertegaal, A.C.O. Deubiquitinating enzymes and the proteasome regulate preferential sets of ubiquitin substrates. *Nat. Commun.* **2022**, *13*, 2736. [CrossRef] [PubMed]
14. Hermanns, T.; Pichlo, C.; Woiwode, I.; Klopffleisch, K.; Witting, K.F.; Ovaa, H.; Baumann, U.; Hofmann, K. A family of unconventional deubiquitinases with modular chain specificity determinants. *Nat. Commun.* **2018**, *9*, 799. [CrossRef]
15. Harper, S.; Gratton, H.E.; Cornaciu, I.; Oberer, M.; Scott, D.J.; Emsley, J.; Dreveny, I. Structure and catalytic regulatory function of ubiquitin specific protease 11 N-terminal and ubiquitin-like domains. *Biochemistry* **2014**, *53*, 2966–2978. [CrossRef] [PubMed]
16. Cruz, L.; Soares, P.; Correia, M. Ubiquitin-Specific Proteases: Players in Cancer Cellular Processes. *Pharmaceuticals* **2021**, *14*, 848. [CrossRef]
17. Ye, Y.; Scheel, H.; Hofmann, K.; Komander, D. Dissection of USP catalytic domains reveals five common insertion points. *Mol. Biosyst.* **2009**, *5*, 1797–1808. [CrossRef]
18. Harrigan, J.A.; Jacq, X.; Martin, N.M.; Jackson, S.P. Deubiquitylating enzymes and drug discovery: Emerging opportunities. *Nat. Rev. Drug Discov.* **2018**, *17*, 57–78. [CrossRef]
19. Reyes-Turcu, F.E.; Wilkinson, K.D. Polyubiquitin binding and disassembly by deubiquitinating enzymes. *Chem. Rev.* **2009**, *109*, 1495–1508. [CrossRef]
20. Reyes-Turcu, F.E.; Ventii, K.H.; Wilkinson, K.D. Regulation and cellular roles of ubiquitin-specific deubiquitinating enzymes. *Annu. Rev. Biochem.* **2009**, *78*, 363–397. [CrossRef]
21. Komander, D. Mechanism, specificity and structure of the deubiquitinases. *Subcell. Biochem.* **2010**, *54*, 69–87. [CrossRef] [PubMed]
22. He, M.; Zhou, Z.; Shah, A.A.; Zou, H.; Tao, J.; Chen, Q.; Wan, Y. The emerging role of deubiquitinating enzymes in genomic integrity, diseases, and therapeutics. *Cell Biosci.* **2016**, *6*, 62. [CrossRef] [PubMed]
23. Dikic, I.; Wakatsuki, S.; Walters, K.J. Ubiquitin-binding domains—From structures to functions. *Nat. Rev. Mol. Cell Biol.* **2009**, *10*, 659–671. [CrossRef]
24. Drag, M.; Mikolajczyk, J.; Bekes, M.; Reyes-Turcu, F.E.; Ellman, J.A.; Wilkinson, K.D.; Salvesen, G.S. Positional-scanning fluorigenic substrate libraries reveal unexpected specificity determinants of DUBs (deubiquitinating enzymes). *Biochem. J.* **2008**, *415*, 367–375. [CrossRef]
25. Fan, Y.H.; Yu, Y.; Mao, R.F.; Tan, X.J.; Xu, G.F.; Zhang, H.; Lu, X.B.; Fu, S.B.; Yang, J. USP4 targets TAK1 to downregulate TNFalpha-induced NF-kappaB activation. *Cell Death Differ.* **2011**, *18*, 1547–1560. [CrossRef] [PubMed]
26. Xu, G.; Tan, X.; Wang, H.; Sun, W.; Shi, Y.; Burlingame, S.; Gu, X.; Cao, G.; Zhang, T.; Qin, J.; et al. Ubiquitin-specific peptidase 21 inhibits tumor necrosis factor alpha-induced nuclear factor kappaB activation via binding to and deubiquitinating receptor-interacting protein 1. *J. Biol. Chem.* **2010**, *285*, 969–978. [CrossRef]
27. Yamamoto, Y.; Gaynor, R.B. Therapeutic potential of inhibition of the NF-kappaB pathway in the treatment of inflammation and cancer. *J. Clin. Investig.* **2001**, *107*, 135–142. [CrossRef]
28. Yun, S.I.; Kim, H.H.; Yoon, J.H.; Park, W.S.; Hahn, M.J.; Kim, H.C.; Chung, C.H.; Kim, K.K. Ubiquitin specific protease 4 positively regulates the WNT/beta-catenin signaling in colorectal cancer. *Mol. Oncol.* **2015**, *9*, 1834–1851. [CrossRef] [PubMed]
29. Krishnamurthy, N.; Kurzrock, R. Targeting the Wnt/beta-catenin pathway in cancer: Update on effectors and inhibitors. *Cancer Treat. Rev.* **2018**, *62*, 50–60. [CrossRef]
30. Hu, X.; Li, J.; Fu, M.; Zhao, X.; Wang, W. The JAK/STAT signaling pathway: From bench to clinic. *Signal Transduct. Target. Ther.* **2021**, *6*, 402. [CrossRef] [PubMed]

31. Lian, J.; Liu, C.; Guan, X.; Wang, B.; Yao, Y.; Su, D.; Ma, Y.; Fang, L.; Zhang, Y. Ubiquitin specific peptidase 5 enhances STAT3 signaling and promotes migration and invasion in Pancreatic Cancer. *J. Cancer* **2020**, *11*, 6802–6811. [CrossRef] [PubMed]
32. Qureshy, Z.; Johnson, D.E.; Grandis, J.R. Targeting the JAK/STAT pathway in solid tumors. *J. Cancer Metastasis Treat* **2020**, *6*, 21. [CrossRef]
33. Lee, J.T.; Gu, W. The multiple levels of regulation by p53 ubiquitination. *Cell Death Differ.* **2010**, *17*, 86–92. [CrossRef] [PubMed]
34. Chen, X.; Zhang, T.; Su, W.; Dou, Z.; Zhao, D.; Jin, X.; Lei, H.; Wang, J.; Xie, X.; Cheng, B.; et al. Mutant p53 in cancer: From molecular mechanism to therapeutic modulation. *Cell Death Dis.* **2022**, *13*, 974. [CrossRef] [PubMed]
35. Ge, J.; Yu, W.; Li, J.; Ma, H.; Wang, P.; Zhou, Y.; Wang, Y.; Zhang, J.; Shi, G. USP16 regulates castration-resistant prostate cancer cell proliferation by deubiquitinating and stablizing c-Myc. *J. Exp. Clin. Cancer Res.* **2021**, *40*, 59. [CrossRef] [PubMed]
36. Hu, Y.; Luo, M. Cinobufotalin regulates the USP36/c-Myc axis to suppress malignant phenotypes of colon cancer cells in vitro and in vivo. *Aging* **2024**, *16*, 5526–5544. [CrossRef]
37. Ruiz, E.J.; Pinto-Fernandez, A.; Turnbull, A.P.; Lan, L.; Charlton, T.M.; Scott, H.C.; Damianou, A.; Vere, G.; Riising, E.M.; Da Costa, C.; et al. USP28 deletion and small-molecule inhibition destabilizes c-MYC and elicits regression of squamous cell lung carcinoma. *Elife* **2021**, *10*, e71596. [CrossRef]
38. Guo, J.; Zhao, J.; Sun, L.; Yang, C. Role of ubiquitin specific proteases in the immune microenvironment of prostate cancer: A new direction. *Front. Oncol.* **2022**, *12*, 955718. [CrossRef]
39. Ji, M.; Shi, H.; Xie, Y.; Zhao, Z.; Li, S.; Chang, C.; Cheng, X.; Li, Y. Ubiquitin specific protease 22 promotes cell proliferation and tumor growth of epithelial ovarian cancer through synergy with transforming growth factor beta1. *Oncol. Rep.* **2015**, *33*, 133–140. [CrossRef] [PubMed]
40. Wang, J.; Xiang, H.; Lu, Y.; Wu, T. Role and clinical significance of TGF-beta1 and TGF-betaR1 in malignant tumors (Review). *Int. J. Mol. Med.* **2021**, *47*, 55. [CrossRef]
41. Liu, J.; Kruswick, A.; Dang, H.; Tran, A.D.; Kwon, S.M.; Wang, X.W.; Oberdoerffer, P. Ubiquitin-specific protease 21 stabilizes BRCA2 to control DNA repair and tumor growth. *Nat. Commun.* **2017**, *8*, 137. [CrossRef] [PubMed]
42. Woo, S.M.; Kim, S.; Seo, S.U.; Kim, S.; Park, J.W.; Kim, G.; Choi, Y.R.; Hur, K.; Kwon, T.K. Inhibition of USP1 enhances anticancer drugs-induced cancer cell death through downregulation of survivin and miR-216a-5p-mediated upregulation of DR5. *Cell Death Dis.* **2022**, *13*, 821. [CrossRef]
43. Foster, B.M.; Wang, Z.; Schmidt, C.K. DoUBLing up: Ubiquitin and ubiquitin-like proteases in genome stability. *Biochem. J.* **2024**, *481*, 515–545. [CrossRef]
44. Li, Q.; Ye, C.; Tian, T.; Jiang, Q.; Zhao, P.; Wang, X.; Liu, F.; Shan, J.; Ruan, J. The emerging role of ubiquitin-specific protease 20 in tumorigenesis and cancer therapeutics. *Cell Death Dis.* **2022**, *13*, 434. [CrossRef] [PubMed]
45. Shimizu, S.; Narita, M.; Tsujimoto, Y. Bcl-2 family proteins regulate the release of apoptogenic cytochrome c by the mitochondrial channel VDAC. *Nature* **1999**, *399*, 483–487. [CrossRef]
46. Trivigno, D.; Essmann, F.; Huber, S.M.; Rudner, J. Deubiquitinase USP9x confers radioresistance through stabilization of Mcl-1. *Neoplasia* **2012**, *14*, 893–904. [CrossRef]
47. Ma, C.; Wang, D.; Tian, Z.; Gao, W.; Zang, Y.; Qian, L.; Xu, X.; Jia, J.; Liu, Z. USP13 deubiquitinates and stabilizes cyclin D1 to promote gastric cancer cell cycle progression and cell proliferation. *Oncogene* **2023**, *42*, 2249–2262. [CrossRef]
48. Li, G.; Yang, T.; Chen, Y.; Bao, J.; Wu, D.; Hu, X.; Feng, C.; Xu, L.; Li, M.; Li, G.; et al. USP5 Sustains the Proliferation of Glioblastoma Through Stabilization of CyclinD1. *Front. Pharmacol.* **2021**, *12*, 720307. [CrossRef]
49. Zhang, Z.; Cui, Z.; Xie, Z.; Li, C.; Xu, C.; Guo, X.; Yu, J.; Chen, T.; Facchinetti, F.; Bohnenberger, H.; et al. Deubiquitinase USP5 promotes non-small cell lung cancer cell proliferation by stabilizing cyclin D1. *Transl. Lung Cancer Res.* **2021**, *10*, 3995–4011. [CrossRef]
50. Sun, H.; Ou, B.; Zhao, S.; Liu, X.; Song, L.; Liu, X.; Wang, R.; Peng, Z. USP11 promotes growth and metastasis of colorectal cancer via PPP1CA-mediated activation of ERK/MAPK signaling pathway. *EBioMedicine* **2019**, *48*, 236–247. [CrossRef]
51. Brahimi-Horn, C.; Pouyssegur, J. When hypoxia signalling meets the ubiquitin-proteasomal pathway, new targets for cancer therapy. *Crit. Rev. Oncol. Hematol.* **2005**, *53*, 115–123. [CrossRef] [PubMed]
52. Yang, Y.C.; Zhao, C.J.; Jin, Z.F.; Zheng, J.; Ma, L.T. Targeted therapy based on ubiquitin-specific proteases, signalling pathways and E3 ligases in non-small-cell lung cancer. *Front. Oncol.* **2023**, *13*, 1120828. [CrossRef]
53. Keijzer, N.; Priyanka, A.; Stijf-Bultsma, Y.; Fish, A.; Gersch, M.; Sixma, T.K. Variety in the USP deubiquitinase catalytic mechanism. *Life Sci. Alliance* **2024**, *7*, e202302533. [CrossRef] [PubMed]
54. Chen, S.; Liu, Y.; Zhou, H. Advances in the Development Ubiquitin-Specific Peptidase (USP) Inhibitors. *Int. J. Mol. Sci.* **2021**, *22*, 4546. [CrossRef]
55. Wang, Y.; Shi, Y.; Niu, K.; Yang, R.; Lv, Q.; Zhang, W.; Feng, K.; Zhang, Y. Ubiquitin specific peptidase 3: An emerging deubiquitinase that regulates physiology and diseases. *Cell Death Discov.* **2024**, *10*, 243. [CrossRef]

56. Wu, X.; Wang, H.; Zhu, D.; Chai, Y.; Wang, J.; Dai, W.; Xiao, Y.; Tang, W.; Li, J.; Hong, L.; et al. USP3 promotes gastric cancer progression and metastasis by deubiquitination-dependent COL9A3/COL6A5 stabilisation. *Cell Death Dis.* **2021**, *13*, 10. [CrossRef] [PubMed]
57. Scott, D.; Layfield, R.; Oldham, N.J. Structural insights into interactions between ubiquitin specific protease 5 and its polyubiquitin substrates by mass spectrometry and ion mobility spectrometry. *Protein Sci.* **2015**, *24*, 1257–1263. [CrossRef] [PubMed]
58. Novinec, M.; Lenarcic, B. Papain-like peptidases: Structure, function, and evolution. *Biomol. Concepts* **2013**, *4*, 287–308. [CrossRef]
59. Yan, B.; Guo, J.; Deng, S.; Chen, D.; Huang, M. A pan-cancer analysis of the role of USP5 in human cancers. *Sci. Rep.* **2023**, *13*, 8972. [CrossRef]
60. Nie, L.; Wang, C.; Liu, X.; Teng, H.; Li, S.; Huang, M.; Feng, X.; Pei, G.; Hang, Q.; Zhao, Z.; et al. USP7 substrates identified by proteomics analysis reveal the specificity of USP7. *Genes. Dev.* **2022**, *36*, 1016–1030. [CrossRef] [PubMed]
61. Korenev, G.; Yakukhnov, S.; Druk, A.; Golovina, A.; Chasov, V.; Mirgayazova, R.; Ivanov, R.; Bulatov, E. USP7 Inhibitors in Cancer Immunotherapy: Current Status and Perspective. *Cancers* **2022**, *14*, 5539. [CrossRef]
62. Tao, L.; Liu, X.; Jiang, X.; Zhang, K.; Wang, Y.; Li, X.; Jiang, S.; Han, T. USP10 as a Potential Therapeutic Target in Human Cancers. *Genes.* **2022**, *13*, 831. [CrossRef] [PubMed]
63. Rossi, F.A.; Rossi, M. Emerging Role of Ubiquitin-Specific Protease 19 in Oncogenesis and Cancer Development. *Front. Cell Dev. Biol.* **2022**, *10*, 889166. [CrossRef] [PubMed]
64. Zhang, J.; van Dinther, M.; Thorikay, M.; Gourabi, B.M.; Kruithof, B.P.T.; Ten Dijke, P. Opposing USP19 splice variants in TGF-beta signaling and TGF-beta-induced epithelial-mesenchymal transition of breast cancer cells. *Cell Mol. Life Sci.* **2023**, *80*, 43. [CrossRef] [PubMed]
65. Komander, D.; Clague, M.J.; Urbe, S. Breaking the chains: Structure and function of the deubiquitinases. *Nat. Rev. Mol. Cell Biol.* **2009**, *10*, 550–563. [CrossRef] [PubMed]
66. Li, S.; Song, Y.; Wang, K.; Liu, G.; Dong, X.; Yang, F.; Chen, G.; Cao, C.; Zhang, H.; Wang, M.; et al. USP32 deubiquitinase: Cellular functions, regulatory mechanisms, and potential as a cancer therapy target. *Cell Death Discov.* **2023**, *9*, 338. [CrossRef]
67. O'Dea, R.; Kazi, N.; Hoffmann-Benito, A.; Zhao, Z.; Recknagel, S.; Wendrich, K.; Janning, P.; Gersch, M. Molecular basis for ubiquitin/Fubi cross-reactivity in USP16 and USP36. *Nat. Chem. Biol.* **2023**, *19*, 1394–1405. [CrossRef] [PubMed]
68. Pal, A.; Young, M.A.; Donato, N.J. Emerging potential of therapeutic targeting of ubiquitin-specific proteases in the treatment of cancer. *Cancer Res.* **2014**, *74*, 4955–4966. [CrossRef] [PubMed]
69. Huang, M.L.; Shen, G.T.; Li, N.L. Emerging potential of ubiquitin-specific proteases and ubiquitin-specific proteases inhibitors in breast cancer treatment. *World J. Clin. Cases* **2022**, *10*, 11690–11701. [CrossRef]
70. Clague, M.J.; Urbe, S.; Komander, D. Breaking the chains: Deubiquitylating enzyme specificity begets function. *Nat. Rev. Mol. Cell Biol.* **2019**, *20*, 338–352. [CrossRef]
71. Zhou, F.; Li, Z.; Zhou, L.; Zhu, W.; Zhang, J.; Yang, W.; Xue, L.; Qin, X.; Chen, P.; Tang, R. Abstract 6201: Identification of SP-002, a highly selective USP1 inhibitor effectively inhibits HRD tumor growth and displays low hematotoxicity risk. *Cancer Res.* **2023**, *83*, 6201. [CrossRef]
72. Gore, S.D.; Carducci, M.A. Modifying histones to tame cancer: Clinical development of sodium phenylbutyrate and other histone deacetylase inhibitors. *Expert. Opin. Investig. Drugs* **2000**, *9*, 2923–2934. [CrossRef] [PubMed]
73. Mistry, H.; Hsieh, G.; Buhrlage, S.J.; Huang, M.; Park, E.; Cuny, G.D.; Galinsky, I.; Stone, R.M.; Gray, N.S.; D'Andrea, A.D.; et al. Small-molecule inhibitors of USP1 target ID1 degradation in leukemic cells. *Mol. Cancer Ther.* **2013**, *12*, 2651–2662. [CrossRef] [PubMed]
74. Shang, K.; Zhang, L.; Yu, Y.; Xiao, H.; Gao, Y.; Yang, L.; Huang, J.; Song, H.; Han, H. Disulfide-containing polymer delivery of C527 and a Platinum(IV) prodrug selectively inhibited protein ubiquitination and tumor growth on cisplatin resistant and patient-derived liver cancer models. *Mater. Today Bio* **2023**, *18*, 100548. [CrossRef]
75. Das, D.S.; Das, A.; Ray, A.; Song, Y.; Samur, M.K.; Munshi, N.C.; Chauhan, D.; Anderson, K.C. Blockade of Deubiquitylating Enzyme USP1 Inhibits DNA Repair and Triggers Apoptosis in Multiple Myeloma Cells. *Clin. Cancer Res.* **2017**, *23*, 4280–4289. [CrossRef]
76. Dexheimer, T.S.; Rosenthal, A.S.; Luci, D.K.; Liang, Q.; Villamil, M.A.; Chen, J.; Sun, H.; Kerns, E.H.; Simeonov, A.; Jadhav, A.; et al. Synthesis and structure-activity relationship studies of N-benzyl-2-phenylpyrimidin-4-amine derivatives as potent USP1/UAF1 deubiquitinase inhibitors with anticancer activity against nonsmall cell lung cancer. *J. Med. Chem.* **2014**, *57*, 8099–8110. [CrossRef] [PubMed]
77. Rennie, M.L.; Arkinson, C.; Chaugule, V.K.; Walden, H. Cryo-EM reveals a mechanism of USP1 inhibition through a cryptic binding site. *Sci. Adv.* **2022**, *8*, eabq6353. [CrossRef] [PubMed]

78. Dexheimer, T.S.; Rosenthal, A.S.; Liang, Q.; Chen, J.; Villamil, M.A.; Kerns, E.H.; Simeonov, A.; Jadhav, A.; Zhuang, Z.; Maloney, D.J. Discovery of ML323 as a Novel Inhibitor of the USP1/UAF1 Deubiquitinase Complex. In *Probe Reports from the NIH Molecular Libraries Program*; National Center for Biotechnology Information: Bethesda, MD, USA, 2010.
79. Chen, J.; Dexheimer, T.S.; Ai, Y.; Liang, Q.; Villamil, M.A.; Inglese, J.; Maloney, D.J.; Jadhav, A.; Simeonov, A.; Zhuang, Z. Selective and cell-active inhibitors of the USP1/ UAF1 deubiquitinase complex reverse cisplatin resistance in non-small cell lung cancer cells. *Chem. Biol.* **2011**, *18*, 1390–1400. [CrossRef]
80. Li, X.Y.; Wu, J.C.; Liu, P.; Li, Z.J.; Wang, Y.; Chen, B.Y.; Hu, C.L.; Fei, M.Y.; Yu, P.C.; Jiang, Y.L.; et al. Inhibition of USP1 reverses the chemotherapy resistance through destabilization of MAX in the relapsed/refractory B-cell lymphoma. *Leukemia* **2023**, *37*, 164–177. [CrossRef] [PubMed]
81. Dakir, E.H.; Pickard, A.; Srivastava, K.; McCrudden, C.M.; Gross, S.R.; Lloyd, S.; Zhang, S.D.; Margariti, A.; Morgan, R.; Rudland, P.S.; et al. The anti-psychotic drug pimozide is a novel chemotherapeutic for breast cancer. *Oncotarget* **2018**, *9*, 34889–34910. [CrossRef] [PubMed]
82. Feng, Z.; Xia, Y.; Gao, T.; Xu, F.; Lei, Q.; Peng, C.; Yang, Y.; Xue, Q.; Hu, X.; Wang, Q.; et al. The antipsychotic agent trifluoperazine hydrochloride suppresses triple-negative breast cancer tumor growth and brain metastasis by inducing G0/G1 arrest and apoptosis. *Cell Death Dis.* **2018**, *9*, 1006. [CrossRef] [PubMed]
83. Valacchi, G.; Pecorelli, A.; Sticozzi, C.; Torricelli, C.; Muscettola, M.; Aldinucci, C.; Maioli, E. Rottlerin exhibits antiangiogenic effects in vitro. *Chem. Biol. Drug Des.* **2011**, *77*, 460–470. [CrossRef]
84. Kumar, D.; Shankar, S.; Srivastava, R.K. Rottlerin-induced autophagy leads to the apoptosis in breast cancer stem cells: Molecular mechanisms. *Mol. Cancer* **2013**, *12*, 171. [CrossRef] [PubMed]
85. McClurg, U.L.; Summerscales, E.E.; Harle, V.J.; Gaughan, L.; Robson, C.N. Deubiquitinating enzyme Usp12 regulates the interaction between the androgen receptor and the Akt pathway. *Oncotarget* **2014**, *5*, 7081–7092. [CrossRef] [PubMed]
86. Altun, M.; Kramer, H.B.; Willems, L.I.; McDermott, J.L.; Leach, C.A.; Goldenberg, S.J.; Kumar, K.G.; Konietzny, R.; Fischer, R.; Kogan, E.; et al. Activity-based chemical proteomics accelerates inhibitor development for deubiquitylating enzymes. *Chem. Biol.* **2011**, *18*, 1401–1412. [CrossRef]
87. Lin, W.C.; Chiu, Y.L.; Kuo, K.L.; Chow, P.M.; Hsu, C.H.; Liao, S.M.; Dong, J.R.; Chang, S.C.; Liu, S.H.; Liu, T.J.; et al. Anti-tumor effects of deubiquitinating enzyme inhibitor PR-619 in human chondrosarcoma through reduced cell proliferation and endoplasmic reticulum stress-related apoptosis. *Am. J. Cancer Res.* **2023**, *13*, 3055–3066. [PubMed]
88. Ohayon, S.; Refua, M.; Hendler, A.; Aharoni, A.; Brik, A. Harnessing the oxidation susceptibility of deubiquitinases for inhibition with small molecules. *Angew. Chem. Int. Ed. Engl.* **2015**, *54*, 599–603. [CrossRef]
89. Zhang, S.; Guo, Y.; Zhang, S.; Wang, Z.; Zhang, Y.; Zuo, S. Targeting the deubiquitinase USP2 for malignant tumor therapy (Review). *Oncol. Rep.* **2023**, *50*, 176. [CrossRef] [PubMed]
90. Davis, M.I.; Pragani, R.; Fox, J.T.; Shen, M.; Parmar, K.; Gaudiano, E.F.; Liu, L.; Tanega, C.; McGee, L.; Hall, M.D.; et al. Small Molecule Inhibition of the Ubiquitin-specific Protease USP2 Accelerates cyclin D1 Degradation and Leads to Cell Cycle Arrest in Colorectal Cancer and Mantle Cell Lymphoma Models. *J. Biol. Chem.* **2016**, *291*, 24628–24640. [CrossRef] [PubMed]
91. Magiera, K.; Tomala, M.; Kubica, K.; De Cesare, V.; Trost, M.; Zieba, B.J.; Kachamakova-Trojanowska, N.; Les, M.; Dubin, G.; Holak, T.A.; et al. Lithocholic Acid Hydroxyamide Destabilizes Cyclin D1 and Induces G(0)/G(1) Arrest by Inhibiting Deubiquitinase USP2a. *Cell Chem. Biol.* **2017**, *24*, 458–470.e18. [CrossRef]
92. Tomala, M.D.; Magiera-Mularz, K.; Kubica, K.; Krzanik, S.; Zieba, B.; Musielak, B.; Pustula, M.; Popowicz, G.M.; Sattler, M.; Dubin, G.; et al. Identification of small-molecule inhibitors of USP2a. *Eur. J. Med. Chem.* **2018**, *150*, 261–267. [CrossRef] [PubMed]
93. Chuang, S.J.; Cheng, S.C.; Tang, H.C.; Sun, C.Y.; Chou, C.Y. 6-Thioguanine is a noncompetitive and slow binding inhibitor of human deubiquitinating protease USP2. *Sci. Rep.* **2018**, *8*, 3102. [CrossRef]
94. Lin, H.C.; Kuan, Y.; Chu, H.F.; Cheng, S.C.; Pan, H.C.; Chen, W.Y.; Sun, C.Y.; Lin, T.H. Disulfiram and 6-Thioguanine synergistically inhibit the enzymatic activities of USP2 and USP21. *Int. J. Biol. Macromol.* **2021**, *176*, 490–497. [CrossRef]
95. Issaenko, O.A.; Amerik, A.Y. Chalcone-based small-molecule inhibitors attenuate malignant phenotype via targeting deubiquitinating enzymes. *Cell Cycle* **2012**, *11*, 1804–1817. [CrossRef]
96. Coughlin, K.; Anchoori, R.; Iizuka, Y.; Meints, J.; MacNeill, L.; Vogel, R.I.; Orlowski, R.Z.; Lee, M.K.; Roden, R.B.; Bazzaro, M. Small-molecule RA-9 inhibits proteasome-associated DUBs and ovarian cancer in vitro and in vivo via exacerbating unfolded protein responses. *Clin. Cancer Res.* **2014**, *20*, 3174–3186. [CrossRef] [PubMed]
97. Mirzapoiazova, T.; Pozhitkov, A.; Nam, A.; Mambetsariev, I.; Nelson, M.S.; Tan, Y.C.; Zhang, K.; Raz, D.; Singhal, S.; Nasser, M.W.; et al. Effects of selected deubiquitinating enzyme inhibitors on the proliferation and motility of lung cancer and mesothelioma cell lines. *Int. J. Oncol.* **2020**, *57*, 80–86. [CrossRef]

98. Nicholson, B.; Leach, C.A.; Goldenberg, S.J.; Francis, D.M.; Kodrasov, M.P.; Tian, X.; Shanks, J.; Sterner, D.E.; Bernal, A.; Mattern, M.R.; et al. Characterization of ubiquitin and ubiquitin-like-protein isopeptidase activities. *Protein Sci.* **2008**, *17*, 1035–1043. [CrossRef]
99. Vamisetti, G.B.; Meledin, R.; Gopinath, P.; Brik, A. Halogen Substituents in the Isoquinoline Scaffold Switches the Selectivity of Inhibition between USP2 and USP7. *Chembiochem* **2019**, *20*, 282–286. [CrossRef]
100. Okada, K.; Ye, Y.Q.; Taniguchi, K.; Yoshida, A.; Akiyama, T.; Yoshioka, Y.; Onose, J.; Koshino, H.; Takahashi, S.; Yajima, A.; et al. Vialinin A is a ubiquitin-specific peptidase inhibitor. *Bioorg. Med. Chem. Lett.* **2013**, *23*, 4328–4331. [CrossRef]
101. Xu, J.; Chen, D.; Jin, L.; Chen, Z.; Tu, Y.; Huang, X.; Xue, F.; Xu, J.; Chen, M.; Wang, X.; et al. Ubiquitously specific protease 4 inhibitor-Vialinin A attenuates inflammation and fibrosis in S100-induced hepatitis mice through Rheb/mTOR signalling. *J. Cell Mol. Med.* **2021**, *25*, 1140–1150. [CrossRef]
102. Bailly, C.; Vergoten, G. Binding of Vialinin A and p-Terphenyl Derivatives to Ubiquitin-Specific Protease 4 (USP4): A Molecular Docking Study. *Molecules* **2022**, *27*, 5909. [CrossRef]
103. Mann, M.K.; Zepeda-Velázquez, C.A.; González-Álvarez, H.; Dong, A.; Kiyota, T.; Aman, A.M.; Loppnau, P.; Li, Y.; Wilson, B.; Arrowsmith, C.H.; et al. Structure-Activity Relationship of USP5 Inhibitors. *J. Med. Chem.* **2021**, *64*, 15017–15036. [CrossRef] [PubMed]
104. Cao, M.N.; Zhou, Y.B.; Gao, A.H.; Cao, J.Y.; Gao, L.X.; Sheng, L.; Xu, L.; Su, M.B.; Cao, X.C.; Han, M.M.; et al. Curcusone D, a novel ubiquitin-proteasome pathway inhibitor via ROS-induced DUB inhibition, is synergistic with bortezomib against multiple myeloma cell growth. *Biochim. Biophys. Acta* **2014**, *1840*, 2004–2013. [CrossRef]
105. Kapuria, V.; Peterson, L.F.; Fang, D.; Bornmann, W.G.; Talpaz, M.; Donato, N.J. Deubiquitinase inhibition by small-molecule WP1130 triggers aggresome formation and tumor cell apoptosis. *Cancer Res.* **2010**, *70*, 9265–9276. [CrossRef]
106. Kim, S.; Woo, S.M.; Min, K.J.; Seo, S.U.; Lee, T.J.; Kubatka, P.; Kim, D.E.; Kwon, T.K. WP1130 Enhances TRAIL-Induced Apoptosis through USP9X-Dependent miR-708-Mediated Downregulation of c-FLIP. *Cancers* **2019**, *11*, 344. [CrossRef] [PubMed]
107. Li, P.; Liu, H.M. Recent advances in the development of ubiquitin-specific-processing protease 7 (USP7) inhibitors. *Eur. J. Med. Chem.* **2020**, *191*, 112107. [CrossRef]
108. D'Arcy, P.; Wang, X.; Linder, S. Deubiquitinase inhibition as a cancer therapeutic strategy. *Pharmacol. Ther.* **2015**, *147*, 32–54. [CrossRef] [PubMed]
109. Peterson, L.F.; Sun, H.; Liu, Y.; Potu, H.; Kandarpa, M.; Ermann, M.; Courtney, S.M.; Young, M.; Showalter, H.D.; Sun, D.; et al. Targeting deubiquitinase activity with a novel small-molecule inhibitor as therapy for B-cell malignancies. *Blood* **2015**, *125*, 3588–3597. [CrossRef]
110. Zheng, Y.; Wang, L.; Niu, X.; Guo, Y.; Zhao, J.; Li, L.; Zhao, J. EOAI, a ubiquitin-specific peptidase 5 inhibitor, prevents non-small cell lung cancer progression by inducing DNA damage. *BMC Cancer* **2023**, *23*, 28. [CrossRef] [PubMed]
111. Colland, F.; Formstecher, E.; Jacq, X.; Reverdy, C.; Planquette, C.; Conrath, S.; Trouplin, V.; Bianchi, J.; Aushev, V.N.; Camonis, J.; et al. Small-molecule inhibitor of USP7/HAUSP ubiquitin protease stabilizes and activates p53 in cells. *Mol. Cancer Ther.* **2009**, *8*, 2286–2295. [CrossRef] [PubMed]
112. Pal, A.; Donato, N.J. Ubiquitin-specific proteases as therapeutic targets for the treatment of breast cancer. *Breast Cancer Res.* **2014**, *16*, 461. [CrossRef] [PubMed]
113. Reverdy, C.; Conrath, S.; Lopez, R.; Planquette, C.; Atmanene, C.; Collura, V.; Harpon, J.; Battaglia, V.; Vivat, V.; Sippl, W.; et al. Discovery of specific inhibitors of human USP7/HAUSP deubiquitinating enzyme. *Chem. Biol.* **2012**, *19*, 467–477. [CrossRef] [PubMed]
114. Qi, S.M.; Cheng, G.; Cheng, X.D.; Xu, Z.; Xu, B.; Zhang, W.D.; Qin, J.J. Targeting USP7-Mediated Deubiquitination of MDM2/MDMX-p53 Pathway for Cancer Therapy: Are We There Yet? *Front. Cell Dev. Biol.* **2020**, *8*, 233. [CrossRef]
115. Yamaguchi, M.; Miyazaki, M.; Kodrasov, M.P.; Rotinsulu, H.; Losung, F.; Mangindaan, R.E.; de Voogd, N.J.; Yokosawa, H.; Nicholson, B.; Tsukamoto, S. Spongiacidin C, a pyrrole alkaloid from the marine sponge Stylissa massa, functions as a USP7 inhibitor. *Bioorg Med. Chem. Lett.* **2013**, *23*, 3884–3886. [CrossRef] [PubMed]
116. Kategaya, L.; Di Lello, P.; Rouge, L.; Pastor, R.; Clark, K.R.; Drummond, J.; Kleinheinz, T.; Lin, E.; Upton, J.P.; Prakash, S.; et al. USP7 small-molecule inhibitors interfere with ubiquitin binding. *Nature* **2017**, *550*, 534–538. [CrossRef]
117. Kategaya, L.; Di Lello, P.; Rougé, L.; Pastor , R.; Clark, K.R.; Drummond, J. Identified Selective USP7 Inhibitors Compete with Ubiquitin Binding. *Cancer Discov.* **2017**, *7*, 1365. [CrossRef]
118. Wang, S.A.; Young, M.J.; Wang, Y.C.; Chen, S.H.; Liu, C.Y.; Lo, Y.A.; Jen, H.H.; Hsu, K.C.; Hung, J.J. USP24 promotes drug resistance during cancer therapy. *Cell Death Differ.* **2021**, *28*, 2690–2707. [CrossRef]
119. Lamberto, I.; Liu, X.; Seo, H.S.; Schauer, N.J.; Iacob, R.E.; Hu, W.; Das, D.; Mikhailova, T.; Weisberg, E.L.; Engen, J.R.; et al. Structure-Guided Development of a Potent and Selective Non-covalent Active-Site Inhibitor of USP7. *Cell Chem. Biol.* **2017**, *24*, 1490–1500 e1411. [CrossRef] [PubMed]

120. Schauer, N.J.; Liu, X.; Magin, R.S.; Doherty, L.M.; Chan, W.C.; Ficarro, S.B.; Hu, W.; Roberts, R.M.; Iacob, R.E.; Stolte, B.; et al. Selective USP7 inhibition elicits cancer cell killing through a p53-dependent mechanism. *Sci. Rep.* **2020**, *10*, 5324. [CrossRef] [PubMed]
121. Gavory, G.; O'Dowd, C.R.; Helm, M.D.; Flasz, J.; Arkoudis, E.; Dossang, A.; Hughes, C.; Cassidy, E.; McClelland, K.; Odrzywol, E.; et al. Discovery and characterization of highly potent and selective allosteric USP7 inhibitors. *Nat. Chem. Biol.* **2018**, *14*, 118–125. [CrossRef]
122. O'Dowd, C.R.; Helm, M.D.; Rountree, J.S.S.; Flasz, J.T.; Arkoudis, E.; Miel, H.; Hewitt, P.R.; Jordan, L.; Barker, O.; Hughes, C.; et al. Identification and Structure-Guided Development of Pyrimidinone Based USP7 Inhibitors. *ACS Med. Chem. Lett.* **2018**, *9*, 238–243. [CrossRef]
123. Turnbull, A.P.; Ioannidis, S.; Krajewski, W.W.; Pinto-Fernandez, A.; Heride, C.; Martin, A.C.L.; Tonkin, L.M.; Townsend, E.C.; Buker, S.M.; Lancia, D.R.; et al. Molecular basis of USP7 inhibition by selective small-molecule inhibitors. *Nature* **2017**, *550*, 481–486. [CrossRef] [PubMed]
124. Kim, A.; Benavente, C. Abstract PR04: USP7 inhibitors prevent triple negative breast cancer metastasis by inducing UHRF1 protein degradation. *Cancer Res.* **2024**, *84*, PR04. [CrossRef]
125. Li, X.; Kong, L.; Yang, Q.; Duan, A.; Ju, X.; Cai, B.; Chen, L.; An, T.; Li, Y. Parthenolide inhibits ubiquitin-specific peptidase 7 (USP7), Wnt signaling, and colorectal cancer cell growth. *J. Biol. Chem.* **2020**, *295*, 3576–3589. [CrossRef]
126. Cheng, Y.J.; Zhuang, Z.; Miao, Y.L.; Song, S.S.; Bao, X.B.; Yang, C.H.; He, J.X. Identification of YCH2823 as a novel USP7 inhibitor for cancer therapy. *Biochem. Pharmacol.* **2024**, *222*, 116071. [CrossRef] [PubMed]
127. Weisberg, E.L.; Schauer, N.J.; Yang, J.; Lamberto, I.; Doherty, L.; Bhatt, S.; Nonami, A.; Meng, C.; Letai, A.; Wright, R.; et al. Inhibition of USP10 induces degradation of oncogenic FLT3. *Nat. Chem. Biol.* **2017**, *13*, 1207–1215. [CrossRef] [PubMed]
128. Javaid, S.; Zadi, S.; Awais, M.; Wahab, A.T.; Zafar, H.; Maslennikov, I.; Choudhary, M.I. Identification of new leads against ubiquitin specific protease-7 (USP7): A step towards the potential treatment of cancers. *RSC Adv.* **2024**, *14*, 33080–33093. [CrossRef]
129. Lee, G.; Oh, T.I.; Um, K.B.; Yoon, H.; Son, J.; Kim, B.M.; Kim, H.I.; Kim, H.; Kim, Y.J.; Lee, C.S.; et al. Small-molecule inhibitors of USP7 induce apoptosis through oxidative and endoplasmic reticulum stress in cancer cells. *Biochem. Biophys. Res. Commun.* **2016**, *470*, 181–186. [CrossRef]
130. Fan, Y.H.; Cheng, J.; Vasudevan, S.A.; Dou, J.; Zhang, H.; Patel, R.H.; Ma, I.T.; Rojas, Y.; Zhao, Y.; Yu, Y.; et al. USP7 inhibitor P22077 inhibits neuroblastoma growth via inducing p53-mediated apoptosis. *Cell Death Dis.* **2013**, *4*, e867. [CrossRef]
131. Cartel, M.; Mouchel, P.L.; Gotanegre, M.; David, L.; Bertoli, S.; Mansat-De Mas, V.; Besson, A.; Sarry, J.E.; Manenti, S.; Didier, C. Inhibition of ubiquitin-specific protease 7 sensitizes acute myeloid leukemia to chemotherapy. *Leukemia* **2021**, *35*, 417–432. [CrossRef] [PubMed]
132. Pozhidaeva, A.; Valles, G.; Wang, F.; Wu, J.; Sterner, D.E.; Nguyen, P.; Weinstock, J.; Kumar, K.G.S.; Kanyo, J.; Wright, D.; et al. USP7-Specific Inhibitors Target and Modify the Enzyme's Active Site via Distinct Chemical Mechanisms. *Cell Chem. Biol.* **2017**, *24*, 1501–1512. [CrossRef] [PubMed]
133. Chauhan, D.; Tian, Z.; Nicholson, B.; Kumar, K.G.; Zhou, B.; Carrasco, R.; McDermott, J.L.; Leach, C.A.; Fulcinniti, M.; Kodrasov, M.P.; et al. A small molecule inhibitor of ubiquitin-specific protease-7 induces apoptosis in multiple myeloma cells and overcomes bortezomib resistance. *Cancer Cell* **2012**, *22*, 345–358. [CrossRef]
134. An, T.; Gong, Y.; Li, X.; Kong, L.; Ma, P.; Gong, L.; Zhu, H.; Yu, C.; Liu, J.; Zhou, H.; et al. USP7 inhibitor P5091 inhibits Wnt signaling and colorectal tumor growth. *Biochem. Pharmacol.* **2017**, *131*, 29–39. [CrossRef]
135. Granieri, L.; Marocchi, F.; Melixetian, M.; Mohammadi, N.; Nicoli, P.; Cuomo, A.; Bonaldi, T.; Confalonieri, S.; Pisati, F.; Giardina, G.; et al. Targeting the USP7/RRM2 axis drives senescence and sensitizes melanoma cells to HDAC/LSD1 inhibitors. *Cell Rep.* **2022**, *40*, 111396. [CrossRef]
136. Weinstock, J.; Wu, J.; Cao, P.; Kingsbury, W.D.; McDermott, J.L.; Kodrasov, M.P.; McKelvey, D.M.; Suresh Kumar, K.G.; Goldenberg, S.J.; Mattern, M.R.; et al. Selective Dual Inhibitors of the Cancer-Related Deubiquitylating Proteases USP7 and USP47. *ACS Med. Chem. Lett.* **2012**, *3*, 789–792. [CrossRef] [PubMed]
137. Colombo, M.; Vallese, S.; Peretto, I.; Jacq, X.; Rain, J.C.; Colland, F.; Guedat, P. Synthesis and biological evaluation of 9-oxo-9H-indeno[1,2-b]pyrazine-2,3-dicarbonitrile analogues as potential inhibitors of deubiquitinating enzymes. *ChemMedChem* **2010**, *5*, 552–558. [CrossRef] [PubMed]
138. Clancy, A.; Heride, C.; Pinto-Fernandez, A.; Elcocks, H.; Kallinos, A.; Kayser-Bricker, K.J.; Wang, W.; Smith, V.; Davis, S.; Fessler, S.; et al. The deubiquitylase USP9X controls ribosomal stalling. *J. Cell Biol.* **2021**, *220*, e202004211. [CrossRef]
139. Yu, M.; Fang, Z.X.; Wang, W.W.; Zhang, Y.; Bu, Z.L.; Liu, M.; Xiao, X.H.; Zhang, Z.L.; Zhang, X.M.; Cao, Y.; et al. Wu-5, a novel USP10 inhibitor, enhances crenolanib-induced FLT3-ITD-positive AML cell death via inhibiting FLT3 and AMPK pathways. *Acta Pharmacol. Sin.* **2021**, *42*, 604–612. [CrossRef]

140. Yang, J.; Meng, C.; Weisberg, E.; Case, A.; Lamberto, I.; Magin, R.S.; Adamia, S.; Wang, J.; Gray, N.; Liu, S.; et al. Inhibition of the deubiquitinase USP10 induces degradation of SYK. *Br. J. Cancer* **2020**, *122*, 1175–1184. [CrossRef] [PubMed]
141. Liu, J.; Xia, H.; Kim, M.; Xu, L.; Li, Y.; Zhang, L.; Cai, Y.; Norberg, H.V.; Zhang, T.; Furuya, T.; et al. Beclin1 controls the levels of p53 by regulating the deubiquitination activity of USP10 and USP13. *Cell* **2011**, *147*, 223–234. [CrossRef] [PubMed]
142. Ndubaku, C.; Tsui, V. Inhibiting the deubiquitinating enzymes (DUBs). *J. Med. Chem.* **2015**, *58*, 1581–1595. [CrossRef] [PubMed]
143. Kunimasa, K.; Ikeda-Ishikawa, C.; Tani, Y.; Tsukahara, S.; Sakurai, J.; Okamoto, Y.; Koido, M.; Dan, S.; Tomida, A. Spautin-1 inhibits mitochondrial complex I and leads to suppression of the unfolded protein response and cell survival during glucose starvation. *Sci. Rep.* **2022**, *12*, 11533. [CrossRef]
144. Burkhart, R.A.; Peng, Y.; Norris, Z.A.; Tholey, R.M.; Talbott, V.A.; Liang, Q.; Ai, Y.; Miller, K.; Lal, S.; Cozzitorto, J.A.; et al. Mitoxantrone targets human ubiquitin-specific peptidase 11 (USP11) and is a potent inhibitor of pancreatic cancer cell survival. *Mol. Cancer Res.* **2013**, *11*, 901–911. [CrossRef]
145. McClurg, U.L.; Azizyan, M.; Dransfield, D.T.; Namdev, N.; Chit, N.; Nakjang, S.; Robson, C.N. The novel anti-androgen candidate galeterone targets deubiquitinating enzymes, USP12 and USP46, to control prostate cancer growth and survival. *Oncotarget* **2018**, *9*, 24992–25007. [CrossRef] [PubMed]
146. Lee, B.H.; Lee, M.J.; Park, S.; Oh, D.C.; Elsasser, S.; Chen, P.C.; Gartner, C.; Dimova, N.; Hanna, J.; Gygi, S.P.; et al. Enhancement of proteasome activity by a small-molecule inhibitor of USP14. *Nature* **2010**, *467*, 179–184. [CrossRef]
147. Xu, L.; Wang, J.; Yuan, X.; Yang, S.; Xu, X.; Li, K.; He, Y.; Wei, L.; Zhang, J.; Tian, Y. IU1 suppresses proliferation of cervical cancer cells through MDM2 degradation. *Int. J. Biol. Sci.* **2020**, *16*, 2951–2963. [CrossRef] [PubMed]
148. Boselli, M.; Lee, B.H.; Robert, J.; Prado, M.A.; Min, S.W.; Cheng, C.; Silva, M.C.; Seong, C.; Elsasser, S.; Hatle, K.M.; et al. An inhibitor of the proteasomal deubiquitinating enzyme USP14 induces tau elimination in cultured neurons. *J. Biol. Chem.* **2017**, *292*, 19209–19225. [CrossRef]
149. Wang, Y.; Jiang, Y.; Ding, S.; Li, J.; Song, N.; Ren, Y.; Hong, D.; Wu, C.; Li, B.; Wang, F.; et al. Small molecule inhibitors reveal allosteric regulation of USP14 via steric blockade. *Cell Res.* **2018**, *28*, 1186–1194. [CrossRef]
150. Li, P.; Zhu, X.; Qu, H.; Han, Z.; Yao, X.; Wei, Y.; Li, B.; Chen, H. Synergistic Effect of Ubiquitin-Specific Protease 14 and Poly(ADP-Ribose) Glycohydrolase Co-Inhibition in BRCA1-Mutant, Poly(ADP-Ribose) Polymerase Inhibitor-Resistant Triple-Negative Breast Cancer Cells. *Onco Targets Ther.* **2024**, *17*, 741–753. [CrossRef]
151. Chen, X.; Yang, Q.; Xiao, L.; Tang, D.; Dou, Q.P.; Liu, J. Metal-based proteasomal deubiquitinase inhibitors as potential anticancer agents. *Cancer Metastasis Rev.* **2017**, *36*, 655–668. [CrossRef]
152. Liu, N.; Li, X.; Huang, H.; Zhao, C.; Liao, S.; Yang, C.; Liu, S.; Song, W.; Lu, X.; Lan, X.; et al. Clinically used antirheumatic agent auranofin is a proteasomal deubiquitinase inhibitor and inhibits tumor growth. *Oncotarget* **2014**, *5*, 5453–5471. [CrossRef]
153. Liu, N.; Liu, C.; Li, X.; Liao, S.; Song, W.; Yang, C.; Zhao, C.; Huang, H.; Guan, L.; Zhang, P.; et al. A novel proteasome inhibitor suppresses tumor growth via targeting both 19S proteasome deubiquitinases and 20S proteolytic peptidases. *Sci. Rep.* **2014**, *4*, 5240. [CrossRef] [PubMed]
154. Zhao, C.; Chen, X.; Yang, C.; Zang, D.; Lan, X.; Liao, S.; Zhang, P.; Wu, J.; Li, X.; Liu, N.; et al. Repurposing an antidandruff agent to treating cancer: Zinc pyrithione inhibits tumor growth via targeting proteasome-associated deubiquitinases. *Oncotarget* **2017**, *8*, 13942–13956. [CrossRef] [PubMed]
155. Zhao, C.; Chen, X.; Zang, D.; Lan, X.; Liao, S.; Yang, C.; Zhang, P.; Wu, J.; Li, X.; Liu, N.; et al. Platinum-containing compound platinum pyrithione is stronger and safer than cisplatin in cancer therapy. *Biochem. Pharmacol.* **2016**, *116*, 22–38. [CrossRef]
156. Zhao, C.; Chen, X.; Zang, D.; Lan, X.; Liao, S.; Yang, C.; Zhang, P.; Wu, J.; Li, X.; Liu, N.; et al. A novel nickel complex works as a proteasomal deubiquitinase inhibitor for cancer therapy. *Oncogene* **2016**, *35*, 5916–5927. [CrossRef] [PubMed]
157. Wang, F.; Ning, S.; Yu, B.; Wang, Y. USP14: Structure, Function, and Target Inhibition. *Front. Pharmacol.* **2021**, *12*, 801328. [CrossRef]
158. Wang, X.; Stafford, W.; Mazurkiewicz, M.; Fryknas, M.; Brjnic, S.; Zhang, X.; Gullbo, J.; Larsson, R.; Arner, E.S.; D'Arcy, P.; et al. The 19S Deubiquitinase inhibitor b-AP15 is enriched in cells and elicits rapid commitment to cell death. *Mol. Pharmacol.* **2014**, *85*, 932–945. [CrossRef]
159. Ward, J.A.; Pinto-Fernandez, A.; Cornelissen, L.; Bonham, S.; Diaz-Saez, L.; Riant, O.; Huber, K.V.M.; Kessler, B.M.; Feron, O.; Tate, E.W. Re-Evaluating the Mechanism of Action of alpha, beta-Unsaturated Carbonyl DUB Inhibitors b-AP15 and VLX1570: A Paradigmatic Example of Unspecific Protein Cross-linking with Michael Acceptor Motif-Containing Drugs. *J. Med. Chem.* **2020**, *63*, 3756–3762. [CrossRef]
160. Cai, J.; Xia, X.; Liao, Y.; Liu, N.; Guo, Z.; Chen, J.; Yang, L.; Long, H.; Yang, Q.; Zhang, X.; et al. A novel deubiquitinase inhibitor b-AP15 triggers apoptosis in both androgen receptor-dependent and -independent prostate cancers. *Oncotarget* **2017**, *8*, 63232–63246. [CrossRef]

161. Gubat, J.; Selvaraju, K.; Sjostrand, L.; Kumar Singh, D.; Turkina, M.V.; Schmierer, B.; Sabatier, P.; Zubarev, R.A.; Linder, S.; D'Arcy, P. Comprehensive Target Screening and Cellular Profiling of the Cancer-Active Compound b-AP15 Indicate Abrogation of Protein Homeostasis and Organelle Dysfunction as the Primary Mechanism of Action. *Front. Oncol.* **2022**, *12*, 852980. [CrossRef] [PubMed]
162. Liu, K.; Gao, Q.; Jia, Y.; Wei, J.; Chaudhuri, S.M.; Wang, S.; Tang, A.; Mani, N.L.; Iyer, R.; Cheng, Y.; et al. Ubiquitin-specific peptidase 22 controls integrin-dependent cancer cell stemness and metastasis. *iScience* **2024**, *27*, 110592. [CrossRef]
163. Prokakis, E.; Dyas, A.; Grun, R.; Fritzsche, S.; Bedi, U.; Kazerouni, Z.B.; Kosinsky, R.L.; Johnsen, S.A.; Wegwitz, F. USP22 promotes HER2-driven mammary carcinoma aggressiveness by suppressing the unfolded protein response. *Oncogene* **2021**, *40*, 4004–4018. [CrossRef] [PubMed]
164. Montauti, E.; Weinberg, S.E.; Chu, P.; Chaudhuri, S.; Mani, N.L.; Iyer, R.; Zhou, Y.; Zhang, Y.; Liu, C.; Xin, C.; et al. A deubiquitination module essential for T(reg) fitness in the tumor microenvironment. *Sci. Adv.* **2022**, *8*, eabo4116. [CrossRef] [PubMed]
165. Prokakis, E.; Bamahmoud, H.; Jansari, S.; Fritsche, L.; Dietz, A.; Boshnakovska, A.; Rehling, P.; Johnsen, S.A.; Gallwas, J.; Wegwitz, F. USP22 supports the aggressive behavior of basal-like breast cancer by stimulating cellular respiration. *Cell Commun. Signal* **2024**, *22*, 120. [CrossRef]
166. Wang, S.A.; Wu, Y.C.; Yang, F.M.; Hsu, F.L.; Zhang, K.; Hung, J.J. NCI677397 targeting USP24-mediated induction of lipid peroxidation induces ferroptosis in drug-resistant cancer cells. *Mol. Oncol.* **2024**, *18*, 2255–2276. [CrossRef] [PubMed]
167. Young, M.J.; Wang, S.A.; Chen, Y.C.; Liu, C.Y.; Hsu, K.C.; Tang, S.W.; Tseng, Y.L.; Wang, Y.C.; Lin, S.M.; Hung, J.J. USP24-i-101 targeting of USP24 activates autophagy to inhibit drug resistance acquired during cancer therapy. *Cell Death Differ.* **2024**, *31*, 574–591. [CrossRef] [PubMed]
168. Wrigley, J.D.; Gavory, G.; Simpson, I.; Preston, M.; Plant, H.; Bradley, J.; Goeppert, A.U.; Rozycka, E.; Davies, G.; Walsh, J.; et al. Identification and Characterization of Dual Inhibitors of the USP25/28 Deubiquitinating Enzyme Subfamily. *ACS Chem. Biol.* **2017**, *12*, 3113–3125. [CrossRef] [PubMed]
169. Patzke, J.V.; Sauer, F.; Nair, R.K.; Endres, E.; Proschak, E.; Hernandez-Olmos, V.; Sotriffer, C.; Kisker, C. Structural basis for the bi-specificity of USP25 and USP28 inhibitors. *EMBO Rep.* **2024**, *25*, 2950–2973. [CrossRef] [PubMed]
170. Zhou, D.; Xu, Z.; Huang, Y.; Wang, H.; Zhu, X.; Zhang, W.; Song, W.; Gao, T.; Liu, T.; Wang, M.; et al. Structure-based discovery of potent USP28 inhibitors derived from Vismodegib. *Eur. J. Med. Chem.* **2023**, *254*, 115369. [CrossRef]
171. Aditya, S.; Rattan, A. Vismodegib: A smoothened inhibitor for the treatment of advanced basal cell carcinoma. *Indian. Dermatol. Online J.* **2013**, *4*, 365–368. [CrossRef]
172. Wang, H.; Meng, Q.; Ding, Y.; Xiong, M.; Zhu, M.; Yang, Y.; Su, H.; Gu, L.; Xu, Y.; Shi, L.; et al. USP28 and USP25 are downregulated by Vismodegib in vitro and in colorectal cancer cell lines. *FEBS J.* **2021**, *288*, 1325–1342. [CrossRef]
173. Kluge, A.F.; Lagu, B.R.; Maiti, P.; Jaleel, M.; Webb, M.; Malhotra, J.; Mallat, A.; Srinivas, P.A.; Thompson, J.E. Novel highly selective inhibitors of ubiquitin specific protease 30 (USP30) accelerate mitophagy. *Bioorg. Med. Chem. Lett.* **2018**, *28*, 2655–2659. [CrossRef]
174. Zhang, X.; Han, Y.; Liu, S.; Guo, B.; Xu, S.; He, Y.; Liu, L. MF-094 nanodelivery inhibits oral squamous cell carcinoma by targeting USP30. *Cell Mol. Biol. Lett.* **2022**, *27*, 107. [CrossRef]
175. Li, X.; Wang, T.; Tao, Y.; Wang, X.; Li, L.; Liu, J. MF-094, a potent and selective USP30 inhibitor, accelerates diabetic wound healing by inhibiting the NLRP3 inflammasome. *Exp. Cell Res.* **2022**, *410*, 112967. [CrossRef]
176. Wang, W.; Liu, H.; Wang, S.; Hao, X.; Li, L. A diterpenoid derivative 15-oxospiramilactone inhibits Wnt/beta-catenin signaling and colon cancer cell tumorigenesis. *Cell Res.* **2011**, *21*, 730–740. [CrossRef] [PubMed]
177. Yue, W.; Chen, Z.; Liu, H.; Yan, C.; Chen, M.; Feng, D.; Yan, C.; Wu, H.; Du, L.; Wang, Y.; et al. A small natural molecule promotes mitochondrial fusion through inhibition of the deubiquitinase USP30. *Cell Res.* **2014**, *24*, 482–496. [CrossRef]
178. Waters, C.S.; Angenent, S.B.; Altschuler, S.J.; Wu, L.F. A PINK1 input threshold arises from positive feedback in the PINK1/Parkin mitophagy decision circuit. *Cell Rep.* **2023**, *42*, 113260. [CrossRef]
179. Rusilowicz-Jones, E.V.; Barone, F.G.; Lopes, F.M.; Stephen, E.; Mortiboys, H.; Urbe, S.; Clague, M.J. Benchmarking a highly selective USP30 inhibitor for enhancement of mitophagy and pexophagy. *Life Sci. Alliance* **2022**, *5*, e202101287. [CrossRef]
180. Li, M.; Wang, X.; Qiu, Y.; Zhang, Y.; Pan, X.; Tang, N.; Chen, T.; Ruan, B.; Shao, S.; He, L.; et al. Safety reassessment of cinobufotalin injection: New findings into cardiotoxicity. *Toxicol. Res.* **2020**, *9*, 390–398. [CrossRef]
181. Kim, Y.J.; Lee, Y.; Shin, H.; Hwang, S.; Park, J.; Song, E.J. Ubiquitin-proteasome system as a target for anticancer treatment-an update. *Arch. Pharm. Res.* **2023**, *46*, 573–597. [CrossRef]
182. Gao, H.; Xi, Z.; Dai, J.; Xue, J.; Guan, X.; Zhao, L.; Chen, Z.; Xing, F. Drug resistance mechanisms and treatment strategies mediated by Ubiquitin-Specific Proteases (USPs) in cancers: New directions and therapeutic options. *Mol. Cancer* **2024**, *23*, 88. [CrossRef]

183. Gao, H.; Yin, J.; Ji, C.; Yu, X.; Xue, J.; Guan, X.; Zhang, S.; Liu, X.; Xing, F. Targeting ubiquitin specific proteases (USPs) in cancer immunotherapy: From basic research to preclinical application. *J. Exp. Clin. Cancer Res.* **2023**, *42*, 225. [CrossRef]
184. Zhu, X.; Wang, P.; Zhan, X.; Zhang, Y.; Sheng, J.; He, S.; Chen, Y.; Nie, D.; You, X.; Mai, H.; et al. USP1-regulated reciprocal differentiation of Th17 cells and Treg cells by deubiquitinating and stabilizing TAZ. *Cell Mol. Immunol.* **2023**, *20*, 252–263. [CrossRef] [PubMed]
185. van Loosdregt, J.; Fleskens, V.; Fu, J.; Brenkman, A.B.; Bekker, C.P.; Pals, C.E.; Meerding, J.; Berkers, C.R.; Barbi, J.; Grone, A.; et al. Stabilization of the transcription factor Foxp3 by the deubiquitinase USP7 increases Treg-cell-suppressive capacity. *Immunity* **2013**, *39*, 259–271. [CrossRef]
186. Wang, L.; Kumar, S.; Dahiya, S.; Wang, F.; Wu, J.; Newick, K.; Han, R.; Samanta, A.; Beier, U.H.; Akimova, T.; et al. Ubiquitin-specific Protease-7 Inhibition Impairs Tip60-dependent Foxp3+ T-regulatory Cell Function and Promotes Antitumor Immunity. *EBioMedicine* **2016**, *13*, 99–112. [CrossRef]
187. Wang, Z.; Kang, W.; Li, O.; Qi, F.; Wang, J.; You, Y.; He, P.; Suo, Z.; Zheng, Y.; Liu, H.M. Abrogation of USP7 is an alternative strategy to downregulate PD-L1 and sensitize gastric cancer cells to T cells killing. *Acta Pharm. Sin. B* **2021**, *11*, 694–707. [CrossRef]
188. Selvaraju, K.; Mazurkiewicz, M.; Wang, X.; Gullbo, J.; Linder, S.; D'Arcy, P. Inhibition of proteasome deubiquitinase activity: A strategy to overcome resistance to conventional proteasome inhibitors? *Drug Resist. Updat.* **2015**, *21–22*, 20–29. [CrossRef]
189. Li, J.; Yuan, S.; Norgard, R.J.; Yan, F.; Yamazoe, T.; Blanco, A.; Stanger, B.Z. Tumor Cell-Intrinsic USP22 Suppresses Antitumor Immunity in Pancreatic Cancer. *Cancer Immunol. Res.* **2020**, *8*, 282–291. [CrossRef]
190. Huang, X.; Zhang, Q.; Lou, Y.; Wang, J.; Zhao, X.; Wang, L.; Zhang, X.; Li, S.; Zhao, Y.; Chen, Q.; et al. USP22 Deubiquitinates CD274 to Suppress Anticancer Immunity. *Cancer Immunol. Res.* **2019**, *7*, 1580–1590. [CrossRef] [PubMed]
191. Cui, X.; Yu, H.; Yao, J.; Li, J.; Li, Z.; Jiang, Z. ncRNA-mediated overexpression of ubiquitin-specific proteinase 13 contributes to the progression of prostate cancer via modulating AR signaling, DNA damage repair and immune infiltration. *BMC Cancer* **2022**, *22*, 1350. [CrossRef] [PubMed]
192. Shi, D.; Wu, X.; Jian, Y.; Wang, J.; Huang, C.; Mo, S.; Li, Y.; Li, F.; Zhang, C.; Zhang, D.; et al. USP14 promotes tryptophan metabolism and immune suppression by stabilizing IDO1 in colorectal cancer. *Nat. Commun.* **2022**, *13*, 5644. [CrossRef] [PubMed]
193. Jaiswal, A.; Murakami, K.; Elia, A.; Shibahara, Y.; Done, S.J.; Wood, S.A.; Donato, N.J.; Ohashi, P.S.; Reedijk, M. Therapeutic inhibition of USP9x-mediated Notch signaling in triple-negative breast cancer. *Proc. Natl. Acad. Sci. USA* **2021**, *118*, e2101592118. [CrossRef]
194. Lim, K.S.; Li, H.; Roberts, E.A.; Gaudiano, E.F.; Clairmont, C.; Sambel, L.A.; Ponnienselvan, K.; Liu, J.C.; Yang, C.; Kozono, D.; et al. USP1 Is Required for Replication Fork Protection in BRCA1-Deficient Tumors. *Mol. Cell* **2018**, *72*, 925–941 e924. [CrossRef] [PubMed]
195. Rennie, M.L.; Gundogdu, M.; Arkinson, C.; Liness, S.; Frame, S.; Walden, H. Structural and Biochemical Insights into the Mechanism of Action of the Clinical USP1 Inhibitor, KSQ-4279. *J. Med. Chem.* **2024**, *67*, 15557–15568. [CrossRef] [PubMed]
196. Cadzow, L.; Brenneman, J.; Tobin, E.; Sullivan, P.; Nayak, S.; Ali, J.A.; Shenker, S.; Griffith, J.; McGuire, M.; Grasberger, P.; et al. The USP1 Inhibitor KSQ-4279 Overcomes PARP Inhibitor Resistance in Homologous Recombination-Deficient Tumors. *Cancer Res.* **2024**, *84*, 3419–3434. [CrossRef]
197. Li, H.; Liu, B.J.; Xu, J.; Song, S.S.; Ba, R.; Zhang, J.; Huan, X.J.; Wang, D.; Miao, Z.H.; Liu, T.; et al. Design, synthesis, and biological evaluation of pyrido[2,3-d]pyrimidin-7(8H)-one derivatives as potent USP1 inhibitors. *Eur. J. Med. Chem.* **2024**, *275*, 116568. [CrossRef]
198. Wang, X.; Mazurkiewicz, M.; Hillert, E.K.; Olofsson, M.H.; Pierrou, S.; Hillertz, P.; Gullbo, J.; Selvaraju, K.; Paulus, A.; Akhtar, S.; et al. The proteasome deubiquitinase inhibitor VLX1570 shows selectivity for ubiquitin-specific protease-14 and induces apoptosis of multiple myeloma cells. *Sci. Rep.* **2016**, *6*, 26979. [CrossRef]
199. Rowinsky, E.K.; Paner, A.; Berdeja, J.G.; Paba-Prada, C.; Venugopal, P.; Porkka, K.; Gullbo, J.; Linder, S.; Loskog, A.; Richardson, P.G.; et al. Phase 1 study of the protein deubiquitinase inhibitor VLX1570 in patients with relapsed and/or refractory multiple myeloma. *Investig. New Drugs* **2020**, *38*, 1448–1453. [CrossRef]
200. Simoneau, A.; Pratt, C.; Comer, G.; Wu, H.-J.; Choi, A.; Khendu, T.; Meier, S.; Yu, Y.; Liu, S.; Zhang, W.; et al. Abstract B054: TNG348, a selective USP1 inhibitor, shows strong preclinical combination activity with PARP inhibitors and other agents targeting DNA repair. *Mol. Cancer Ther.* **2023**, *22*, B054. [CrossRef]
201. Simoneau, A.; Wu, H.-J.; Bandi, M.; Lazarides, K.; Sun, S.; Liu, S.; Meier, S.; Choi, A.; Zhang, H.; Shen, B.; et al. Abstract 4968: Characterization of the clinical development candidate TNG348 as a potent and selective inhibitor of USP1 for the treatment of BRCA1/2mut cancers. *Cancer Res.* **2023**, *83*, 4968. [CrossRef]
202. Phase 1/2, Multi-Center, Open-Label Study to Evaluate the Safety, Tolerability, and Preliminary Antitumor Activity of TNG348 Single Agent and in Combination With a PARP Inhibitor in Patients With BRCA 1/2 Mutant or Other HRD+ Solid Tumors. 2024. Available online: https://www.dana-farber.org/clinical-trials/23-643 (accessed on 13 August 2024).

203. A Phase 1 Study of RO7623066 Alone and in Combination in Patients With Advanced Solid Tumors. 2022. Available online: https://www.cancer.gov/research/participate/clinical-trials-search/v?id=NCI-2022-01620 (accessed on 4 February 2025).
204. Yap, T.A.; Lakhani, N.J.; Patnaik, A.; Lee, E.K.; Gutierrez, M.; Moore, K.N.; Carneiro, B.A.; Hays, J.L.; Huang, M.; LoRusso, P.; et al. First-in-human phase I trial of the oral first-in-class ubiquitin specific peptidase 1 (USP1) inhibitor KSQ-4279 (KSQi), given as single agent (SA) and in combination with olaparib (OLA) or carboplatin (CARBO) in patients (pts) with advanced solid tumors, enriched for deleterious homologous recombination repair (HRR) mutations. *J. Clin. Oncol.* **2024**, *42*, 3005. [CrossRef]
205. Li, Y.; Wu, J.; Liu, J.; Qin, L.; Cai, X.; Qiao, J.; Wang, L.; Rao, S.; Ren, F.; Zhavoronkov, A. Abstract 502: ISM3091, a novel selective USP1 inhibitor as a targeted anticancer therapy. *Cancer Res.* **2023**, *83*, 502. [CrossRef]
206. An Open-Label, Multicenter Study to Investigate the Safety, Tolerability, Pharmacokinetics, and Preliminary Efficacy of XL309 (ISM3091) as Single-Agent and Combination Therapy in Patients With Advanced Solid Tumors. 2025. Available online: https://clinicaltrials.gov/study/NCT05932862 (accessed on 24 January 2025).
207. A Phase I First-in-human, Open-label Trial to Investigate the Safety, Pharmacokinetics and Antitumor Activity of SIM0501 as Monotherapy and in Combination in Participants With Advanced Solid Tumors. 2024. Available online: https://www.medifind.com/articles/clinical-trial/507724029 (accessed on 4 February 2025).
208. Yang, S.; Li, Z.; Zhou, L.; Zhou, F.; Qin, X.; Xue, L.; Xu, P.; Tang, J.; Wang, W.; Tang, R. Abstract LB271: Synergistic anti-tumor efficacy in olaparib-sensitive and -resistant models via simultaneously inhibition of USP1 and PARP. *Cancer Res.* **2024**, *84*, LB271. [CrossRef]
209. Xu, J.; Li, Y.; Wang, J.; Tang, P.; Wang, J.; Zhang, C.; Yan, P. Abstract 7146: HSK39775: A USP1 inhibitor for the treatment of cancers with homologous recombination deficiencies. *Cancer Res.* **2024**, *84*, 7146. [CrossRef]
210. A Phase I/II Study to Assess the Safety, Tolerability, PK/PD, and Preliminary Efficacy of HSK39775 Tablet Monotherapy in Participants With Advanced Solid Malignancies. 2024. Available online: https://adisinsight.springer.com/trials/700372061 (accessed on 22 March 2024).
211. VLX1570 and Low-Dose Dexamethasone in Relapsed or Relapsed and Refractory Multiple Myeloma: A Clinical and Correlative Phase 1/2 Study. 2018. Available online: https://clinicaltrials.gov/study/NCT02372240 (accessed on 11 May 2018).
212. Kharel, P.; Uprety, D.; Chandra, A.B.; Hu, Y.; Belur, A.A.; Dhakal, A. Bortezomib-Induced Pulmonary Toxicity: A Case Report and Review of Literature. *Case Rep. Med.* **2018**, *2018*, 2913124. [CrossRef]

biomolecules

MDPI

Review

Targeting the PARylation-Dependent Ubiquitination Signaling Pathway for Cancer Therapies

Daoyuan Huang [1], Jingchao Wang [1], Li Chen [1], Weiwei Jiang [1], Hiroyuki Inuzuka [1], David K. Simon [2] and Wenyi Wei [1,*]

[1] Department of Pathology, Beth Israel Deaconess Medical Center, Harvard Medical School, Boston, MA 02215, USA
[2] Department of Neurology, Beth Israel Deaconess Medical Center, Harvard Medical School, Boston, MA 02215, USA; dsimon1@bidmc.harvard.edu
* Correspondence: wwei2@bidmc.harvard.edu

Abstract: Poly(ADP-ribosyl)ation (PARylation) is a dynamic protein post-translational modification (PTM) mediated by ADP-ribosyltransferases (ARTs), which regulates a plethora of essential biological processes, such as DNA repair, gene expression, and signal transduction. Among these, PAR-dependent ubiquitination (PARdU) plays a pivotal role in tagging PARylated substrates for subsequent ubiquitination and degradation events through the coordinated action of enzymes, including the E3 ligase RNF146 and the ADP-ribosyltransferase tankyrase. Notably, this pathway has emerged as a key regulator of tumorigenesis, immune modulation, and cell death. This review elucidates the molecular mechanisms of the PARdU pathway, including the RNF146–tankyrase interaction, substrate specificity, and upstream regulatory pathways. It also highlights the biological functions of PARdU in DNA damage repair, signaling pathways, and metabolic regulation, with a focus on its therapeutic potential in cancer treatment. Strategies targeting PARdU, such as tankyrase and RNF146 inhibitors, synthetic lethality approaches, and immune checkpoint regulation, offer promising avenues for precision oncology. These developments underscore the potential of PARdU as a transformative therapeutic target in combating various types of human cancer.

Keywords: PARylation ubiquitination; RNF146; tankyrase

1. Mechanism of Poly(ADP-Ribosyl)ation

Protein ADP-ribosylation is a reversible protein post-translational modification (PTM) in which ADP-ribose (ADPr) units are covalently transferred from nicotinamide adenine dinucleotide (NAD) to target molecules, releasing nicotinamide (NAM) as a byproduct [1–3]. This modification includes both mono-ADP-ribosylation (MAR) and poly-ADP-ribosylation (PAR) events and involves the coordinated actions of "writers" (enzymes that catalyze the modification), "erasers" (enzymes that remove it), and "readers" (proteins or domains that recognize and mediate downstream signaling), along with essential cofactors [4,5] (Figure 1A,B).

ADP-ribosylation is mediated primarily by the ADP-ribosyltransferase (ART) family of enzymes, which catalyze the attachment of ADPr units to specific amino acid residues such as serine (Ser), threonine (Thr), lysine (Lys), arginine (Arg), glutamate (Glu), aspartate (Asp), and cysteine (Cys) within the protein substrates [6–9]. Mono-ADP-ribosylating (MARylating) enzymes attach a single ADPr unit, while poly-ADP-ribosylating (PARylating) enzymes synthesize and attach PAR polymers. In humans, the ART family includes

17 known enzymes [2,10], with PARP1, PARP2, and tankyrases (TNKS1/2) catalyzing PARylation, while most others function primarily as MAR transferases [3].

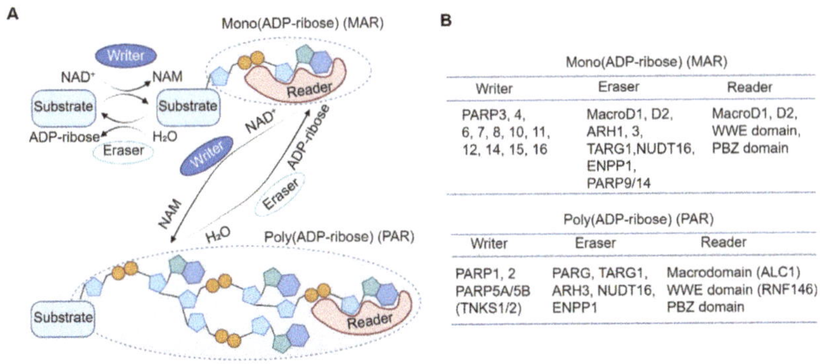

Figure 1. The protein ADP-ribosylation reaction and its working enzymes. (**A**) Protein ADP-ribosylation is dynamically regulated by writers (enzymes that add ADP-ribose) and erasers (enzymes that remove ADP-ribose) and recognized by readers (proteins or domains that bind ADP-ribose modifications). (**B**) Key writers, erasers, and readers in MARylation and PARylation.

Although proteins are the primary targets of ADP-ribosylation, recent studies have revealed its role in modifying DNA and RNA [11–14], thereby expanding its regulatory influence on critical biological processes such as DNA repair [6], gene expression [15], signal transduction [16,17], and energy metabolism [5,18]. This review focuses on the mechanisms, functions, and therapeutic potential of PAR-dependent ubiquitination.

2. Mechanism of PARylation-Dependent Ubiquitination

The understanding of PAR-dependent ubiquitination (PARdU) has significantly advanced in recent years. In 2009, researchers discovered that the tankyrase inhibitor XAV939 stabilizes Axin and suppresses Wnt signaling in part by blocking its degradation via the ubiquitin–proteasome pathway [19]. This process is largely mediated by tankyrase isoforms (TNKS1/2), which bind conserved domains in Axin for its PARylation, and the E3 ligase RNF146, which recognizes poly(ADP-ribose) (PAR) chains through its WWE domain to facilitate the subsequent ubiquitination and degradation of PARylated proteins [19–22].

Substrates of TNKS1/2, such as Axin1/2, rely on the conserved "tankyrase-binding motif" (TBM, RXXPDG) for PARdU [1,23,24]. Additional protein targets, including 3BP2 [25] and PTEN [16], have expanded the known substrate repertoire. Moreover, proteomic analyses have identified hundreds of potential PARdU substrates [20,26–29], emphasizing PARylation's pivotal role in ubiquitination and protein degradation and providing a deeper understanding of its regulatory mechanisms.

2.1. Insights into RNF146 and Tankyrase Interactions

RNF146 features an N-terminal WWE domain, which binds *iso*-ADP-ribose units within PAR chains, and a RING domain that mediates the ubiquitination process (Figure 2). Structural studies further reveal that PAR binding induces conformational changes in RNF146, activating its ubiquitin ligase activity [30]. This specificity distinguishes RNF146 from other E3 ubiquitin ligases and highlights its potential as a therapeutic target.

The C-terminal domain of RNF146 is critical for interacting with tankyrase (TNKS1/2). Five putative tankyrase-binding motifs (TBMs) have been identified within this domain, and at least four are essential for RNF146–tankyrase interactions [30,31] (Figure 2). The traditional TBM consensus sequence (RXXPDG) has been expanded to include broader variations such as RXX(A/G/P/V)XG or RXXX(A/G/P/V)XG [31], offering valuable insights into tankyrase-interacting proteins and potential PARdU substrates. This enhanced understanding provides a foundation for predicting and validating PARdU components in various cellular processes.

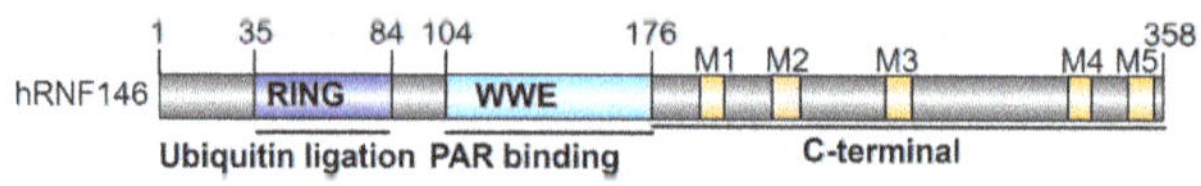

Figure 2. The protein domain features of human RNF146 E3 ubiquitin ligase. M1–M5 are putative TBMs.

2.2. Mechanism of RNF146-Mediated Ubiquitination of PARylated Substrates

Studies have revealed that RNF146 exists in cells as part of a protein complex with tankyrase through its C-terminal domain [30]. Initially, the RING domain of RNF146 remains in an inactive state (Figure 3A). In this complex, tankyrase is responsible for substrate selection and PARylation. Upon PARylation, the PAR groups on the substrate bind to RNF146, thereby triggering a conformational switch that activates its E3 ligase activity [30] (Figure 3B,C). During this process, the PARylated substrate remains tethered within the RNF146–tankyrase complex through its interaction with tankyrase, providing the molecular basis for substrate ubiquitination in cells.

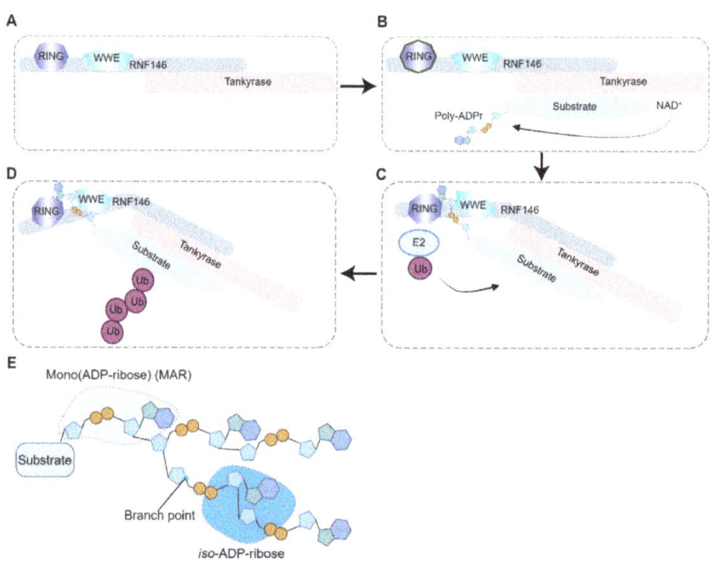

Figure 3. Molecular mechanism of RNF146 activation and substrate recognition in PAR-dependent ubiquitination. (**A**) RNF146 forms a protein complex with tankyrase via its C-terminal domain, while its RING domain remains in an inactive state, lacking E3 ubiquitin ligase activity. (**B**) Tankyrase catalyzes the poly(ADP-ribose) (PAR) modification of the downstream protein substrate. (**C**) The *iso*-ADP-ribose units within the PAR chain bind to the interface between the RING and WWE domains of RNF146, thereby inducing a conformational change in the RING domain that activates its E3 ubiquitin ligase activity. (**D**) The E3 ubiquitin ligase activity of the RNF146 RING domain is activated, enabling the transfer of ubiquitin to its substrate. (**E**) A schematic representation of the PAR chain on the substrate, with the *iso*-ADP-ribose unit highlighted in blue to indicate its critical role in RNF146 activation.

Iso-ADPr, the smallest internal structural unit of poly(ADP-ribose) (PAR), binds between the WWE and RING domains of RNF146 [30,32] (Figure 3D,E). This binding induces a conformational change in the RING domain, transitioning it from a catalytically inactive to an active state. In the absence of PAR, the RING domain cannot effectively bind to the E2 enzymes. To this end, the interaction between *iso*-ADPr and RNF146 enables the functional activation of the RING domain [30,32].

Since substrate PARylation and PARdU (PAR-dependent ubiquitination) are both catalyzed by enzymes within the same protein complex, substrate specificity in PARdU is likely governed primarily by substrate–tankyrase interactions. Maintaining the unliganded RNF146 in an inactive state may contribute to the stability of the RNF146–tankyrase complex, thereby regulating the steady-state activity of PARdU in cells.

3. Known Substrates of RNF146 and Their Biological Functions

The PAR-dependent ubiquitination (PARdU) mechanism, mediated by RNF146 and tankyrase, plays a pivotal role in various physiological and pathological processes [5,6]. This review comprehensively summarizes the known substrates of RNF146-mediated ubiquitination and the biological processes influenced by PARdU (Table 1). This mechanism involves the targeted ubiquitination of poly(ADP-ribosyl)ated substrates, playing a pivotal role in immune modulation, DNA repair, cell death regulation, neurodegeneration, and tumorigenesis. Through the precise regulation of substrate degradation, PARdU orchestrates a wide array of cellular functions and pathophysiological responses. Below, we provide a detailed overview of its important roles, organized into physiological and pathological contexts, with a specific focus on its regulatory mechanisms and potential therapeutic applications.

3.1. *Physiological Functions of PARdU*

In normal physiological contexts, PARdU plays an essential role in maintaining cellular homeostasis and regulating key signaling pathways. One prominent function is the regulation of epithelial lumen formation, where RNF146 facilitates the degradation of SH2BP5 and SH3BP5L, thereby inhibiting Rab11a activation [33]. This ensures proper morphogenesis and epithelial architecture. Additionally, PARdU stabilizes the Wnt/β-catenin signaling pathway through tankyrase-mediated poly(ADP-ribosyl)ation and RNF146-driven Axin1/2 degradation, a process critical for neurite outgrowth, synapse formation, and energy metabolism [18,34]. Another important role of PARdU is in cell cycle regulation, where it targets PARylated p21 for degradation [35], contributing to the progression of the cell cycle.

Proteomic studies have expanded our understanding of PARdU by identifying a diverse array of substrates. For instance, BLZF1 regulates gene transcription [20], while CASC3 disrupts spliceosome function [20,36]. Other notable substrates include OTUD5 [26], which influences de-ubiquitination processes [37], and PARP10 [26], a regulator of PARylation signaling [38]. Additionally, SARDH modulates sarcosine demethylation [26], showcasing PARdU's impact on metabolic pathways. Further research has identified substrates such as Disc1, Striatin, Fat4, BCR, MERIT40, and RAD54 [25], emphasizing the complexity and breadth of PARdU-mediated cellular regulation.

3.2. *Pathological Functions of PARdU*

3.2.1. Tumorigenesis and Immune Regulation

In pathological conditions, PARdU plays a significant role in tumor development and immune modulation. Furthermore, PARdU regulates tumorigenesis by targeting Axin1/2 for degradation, influencing the Wnt/β-catenin signaling pathway [19,20,39]. Additionally, PARdU-mediated degradation of AMOT family proteins regulates the YAP/TAZ pathway [17,40], and PTEN ubiquitination drives the activation of the PTEN-AKT signaling cascade [16], which accelerates tumor progression. It suppresses innate immunity by degrading MAVS [41] while also promoting tumor immune evasion through PARP1 degradation, which enhances STAT3 activity and upregulates PD-L1 expression [42]. In autoinflammatory diseases, tankyrase negatively regulates the homeostasis of 3BP2, also termed SH3 domain-binding protein 2, which functions as a signaling adaptor protein, through RNF146-mediated ubiquitination [25,27]. Mutations in 3BP2 disrupt this regulation [27,43], leading to Cherubism, an autosomal dominant disorder associated with craniofacial abnormalities [44].

3.2.2. Parkinson's Disease and PARthanatos

Neurodegenerative diseases, particularly Parkinson's disease (PD), are another major area where PARdU exhibits regulatory functions [45]. Poly(ADP-ribose)-dependent cell death (PARthanatos) is characterized by the excessive synthesis of poly(ADP-ribose) (PAR) by PARP1 and is closely associated with neurodegenerative diseases and acute injuries such as stroke, MPTP-induced mitochondrial dysfunction, and NMDA-induced neuronal excitotoxicity [46–48]. Recently, PARthanatos has gained attention as a key focus in the study of progressive neurodegenerative diseases including Parkinson's disease (PD) [48]. Notably, studies in PD mouse models have shown that PARthanatos is likely a primary driver of selective dopaminergic neuron loss during their gradual degeneration [49]. In PD, PAR is primarily synthesized by PARP1, although TNKS1/2-derived PAR may also contribute to neuronal PARthanatos. Furthermore, as we mentioned before, PAR synthesized by PARP1 and TNKS1/2 can mediate substrate ubiquitination through PAR-dependent ubiquitination (PARdU). This suggests that PARdU signaling could be a potential regulator of dopaminergic neuron loss in PD. To this end, studies have revealed that Crocetin, through interaction with the PD-related protein HK-I, inhibits RNF146-mediated degradation of PARylated HK-I, thereby preventing mitochondrial dysfunction and DNA damage during the later stages of PARthanatos and helping cells resist irreversible death [50]. Additionally, in SH-SY5Y cells, chlorogenic acid activates the Akt1-CREB-RNF146 signaling pathway, promoting RNF146 transcription [51]. RNF146, by binding PAR rather than relying on its E3 ubiquitin ligase activity, suppresses PARP1 activity, enhancing cell survival against 6-OHDA and α-synuclein aggregation [51]. In mice, chlorogenic acid similarly activates the Akt1-CREB-RNF146 signaling pathway in the brain, providing RNF146-dependent neuroprotection [51]. However, in post-mortem brain samples of PD patients, this pathway was found to be dysregulated [51]. These findings indicate that PARdU signaling plays an important regulatory role in PD pathogenesis. However, further investigation is needed to further elucidate its precise mechanisms and to develop PARdU-dependent therapeutic strategies for PD.

3.2.3. DNA Damage Repair and Cell Death

PARdU also plays a pivotal role in DNA damage repair by targeting key proteins for degradation, including 53BP1 [52], CtIP [53], and BRD7 [54], which are essential for effective DNA damage responses and chemotherapy efficacy. In addition, it regulates cell death pathways, such as necroptosis via RIPK1 ubiquitination [55,56], oxidant-induced cell death through PARP1 degradation [57], and PARthanatos by targeting PARylated HK-I [50]. These functions underline PARdU's significance in maintaining cellular integrity under stress conditions.

3.2.4. Other Cellular Functions

Other notable functions include the regulation of cartilage matrix synthesis via SOX9 PARylation [58], which impacts osteoarthritis progression, and the degradation of p65 [59], which supports the efficacy of PARP inhibitors in BRCA wild-type cancer cells. Finally, PARdU-mediated degradation of PARP1 alleviates macrophage inflammation [60].

The majority of reported PARdU substrates undergo poly(ADP-ribosyl)ation mediated by tankyrase (TNKS1/2). However, the poly(ADP-ribosyl)ation of substrates such as p65, BRD7, PARP1, and CtIP is catalyzed by PARP1, suggesting that among the known PARylation writers (TNKS1/2, PARP1/2), only PARP2 has not yet been exactly reported to assemble the PARdU machinery to facilitate the PARdU process. However, the substrate-specific recognition mechanisms of TNKS1/2 and PARP1 remain to be elucidated and warrant further investigation.

By systematically analyzing PARdU substrates and their associated functions, we will gain a deeper understanding of their roles in physiological and pathological contexts, thereby providing a robust foundation for developing PARdU-based therapeutic strategies (Figure 4).

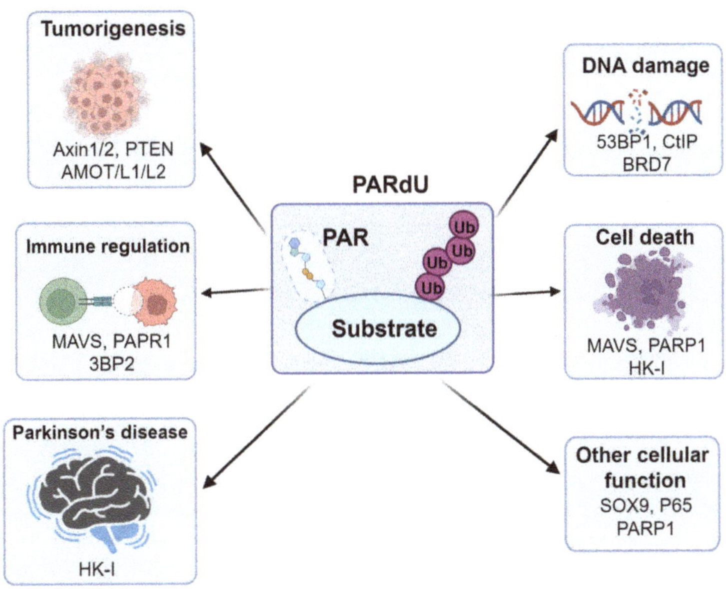

Figure 4. A schematic summary of PARdU substrates involved in various pathological signaling pathways.

Table 1. Summary of known PARdU substrates and their functions.

PARylation Enzyme	Substrate	Regulation	Biological Pathway	Reference
Tankyrase	SH3BP5	Degradation	Regulates epithelial lumen formation	[33]
	SH3BP5L	Degradation	Regulates epithelial lumen formation	[33]
	3BP2	Degradation	Cherubism disease	[25,27,43]
	AXIN1/2	Degradation	Regulates neurite outgrowth	[34]
	AXIN1/2	Degradation	Regulates bone dynamics and energy metabolism	[18]

Table 1. *Cont.*

PARylation Enzyme	Substrate	Regulation	Biological Pathway	Reference
Tankyrase	AXIN1/2	Degradation	Wnt pathway	[19,20,39]
	RNF146	Degradation	E3 ligase	[20,27]
	p21	Degradation	Cell cycle	[35]
	LKB1	Block activation	Inhibits LKB1-AMPK signaling	[61]
	PTEN	Degradation	Promotes tumor growth	[16]
	FADD	Degradation	Inhibits TNF-induced cell death	[55]
	OTUD5	Degradation	De-ubiquitination	[26]
	PARP10	Degradation	PARylation	[26]
	SARDH	Degradation	Sarcosine demethylation	[26]
	AMOT/L1/L2	Degradation	Regulates YAP oncogenic activity	[17]
	Motins	Degradation	Regulates YAP/TAZ pathway	[40]
	MAVS	Degradation	Inhibits innate antiviral response	[41]
	53BP1	Degradation	DNA damage	[52]
	BLZF1	Degradation	Transcription factor	[20]
	SOX9	Degradation	Regulates osteoarthritic cartilage	[58]
	CASC3	Degradation	Spliceosome	[20,26]
TNKS1	RIPK1	Degradation	Block necroptosis	[56]
PARP1	p65	Degradation	Directs to combine bortezomib with niraparib for thyroid cancer	[59]
	BRD7	Degradation	DNA-damaging chemotherapeutic	[54]
	PARP1	Degradation	Regulates PD-L1 expression	[42]
	PAPR1	Degradation	Inhibits oxidant-induced cell death	[57]
	CtIP	Degradation	Homologous recombination repair	[53]
Undefined	PARP1	Degradation	Ameliorates macrophage inflammation	[60]
	HK-I	Degradation	Regulates PARthanatos	[50]

4. Targeting the PARdU Pathway for Cancer Therapy

Targeting the PARdU pathway represents a promising approach for cancer therapy, particularly in malignancies driven by dysregulated components such as TNKS1/2 and RNF146. Tankyrase inhibitors (e.g., XAV939, IWR-1, G007-LK) have shown potential by stabilizing substrates like Axin1/2 and suppressing hyperactive Wnt/β-catenin signaling [19,62], frequently observed in colorectal cancer and hepatocellular carcinoma [63–66]. However, the diverse substrates of tankyrase present challenges, necessitating the development of substrate-selective inhibitors to minimize off-target effects. Alternatively, directly targeting RNF146, the E3 ligase central to the PARdU pathway, offers another strategy. This can be achieved by inhibiting its WWE domain, which binds PAR, or its RING domain, which mediates ubiquitination. Although selective RNF146 inhibitors are currently lacking, efforts should focus on developing small molecules or peptides that disrupt WWE-PAR interactions or inhibit RING activity [67].

4.1. Regulating RNF146 to Enhance Therapeutic Strategies

Understanding the upstream regulatory signals of RNF146 is crucial for effectively targeting the PARdU pathway and identifying novel strategies for cancer therapy. Here, we summarize the currently known mechanisms regulating RNF146 expression and activity (Table 2).

In HCC, *RNF146* is identified as a novel HIF1α/2α target gene that is transcriptionally activated under hypoxic conditions. It promotes PTEN ubiquitination and degradation, thereby activating the AKT/mTOR signaling pathway, which drives HCC proliferation, clonogenicity, and glycolysis. Due to its critical role in HCC progression, RNF146 is considered a promising therapeutic target for anti-HCC strategies [68]. Additionally, Akt1, activated by chlorogenic acid, promotes CREB-dependent transcriptional activation of RNF146, which inhibits PARP1 by sequestering PAR rather than through its E3 ligase activity. This RNF146-mediated mechanism protects dopaminergic neurons from 6-OHDA toxicity and α-synuclein aggregation, highlighting its critical neuroprotective role in Parkinson's disease [51]. Moreover, RNF146 is upregulated in the prefrontal cortex of valproic acid (VPA)-exposed mice, leading to dysregulation of the Wnt/β-catenin signaling pathway and impaired social behaviors. Overexpression of RNF146 in the prefrontal cortex enhances excitatory synaptic transmission, contributing to the social deficits observed in the VPA-induced ASD model [69]. However, the mechanism by which VPA exposure induces the upregulation of RNF146 protein remains unknown. Neuroprotectin D1 (NPD1) upregulates RNF146 (Iduna) expression, enhancing its poly(ADP-ribose) (PAR)-dependent activity to facilitate DNA repair and protect against cell death under oxidative stress. Systemic administration of DHA, the precursor of NPD1, increases Iduna levels in astrocytes and neurons, providing significant neuroprotection following ischemic stroke [70].

Table 2. Summary of the upstream regulators of RNF146.

Regulator	Positive	Negative	Function	Reference
HIF1α/2α	+	−	Promotes *RNF146* transcription	[68]
Chlorogenic acid	+	−	Promotes *RNF146* transcription	[51]
Valproic acid exposed	+	−	Upregulates RNF146 protein	[69]
NPD1	+	−	Upregulates RNF146 protein	[70]
Sumo3	+	−	Accelerates Axin degradation	[71]
miR-79/306	−	+	Downregulates RNF146	[72]
RANKL	−	+	Suppresses *RNF146* transcription	[73]

Post-translational modifications also play a critical role in regulating RNF146 E3 ligase activity. To this end, RNF146 undergoes SUMOylation primarily at lysine residues K19, K61, K174, and K175, with UBC9/PIAS3/MMS21 mediating SUMO3 conjugation and SENP1/2/6 facilitating its deconjugation. SUMOylation promotes RNF146 nuclear localization and enhances its interaction with Axin, accelerating Axin ubiquitination and degradation, thereby activating β-catenin signaling and driving HCC progression. Inhibiting RNF146 SUMOylation suppresses HCC progression, highlighting its SUMOylation as a potential therapeutic target [71]. Conversely, certain pathways negatively regulate RNF146. RNF146 is regulated by tumor-suppressive miRNAs miR-306 and miR-79, which target RNF146 mRNA, reducing its expression. This downregulation inhibits RNF146-mediated degradation of tankyrase, leading to hyperactivation of JNK signaling through a non-canonical pathway. While JNK activity is critical for tumor growth, excessive activation induced by miR-306 and miR-79 drives tumor elimination, highlighting RNF146 as a key modulator in JNK signaling and a potential target for miRNA-based cancer therapies [72]. Moreover, *RNF146* transcription is suppressed by RANKL through an NF-κB-related inhibitory element in its promoter, stabilizing its substrates 3BP2 and AXIN1. This leads to SRC activation and β-catenin downregulation, essential for osteoclastogenesis. RNF146 acts as a regulatory switch to control osteoclast differentiation and cytokine production, implicating it as a potential target in chronic inflammatory diseases [73]. The regulation of protein homeostasis through ubiquitination and de-ubiquitination presents a promising therapeutic avenue for cancer treatment [74]. RNF146 mediates PARdU-dependent degradation of key tumor suppressors such as AXIN1/2 and PTEN, thereby promoting tumorigenesis. Hence, investigating the regulatory mechanisms of RNF146 protein homeostasis, particularly its deubiquitinases, may offer new strategies for therapeutic intervention. Targeting RNF146-specific deubiquitinase inhibitors to lower RNF146 protein levels could be an effective treatment for HCC and other Wnt-dependent cancers.

These insights suggest opportunities for therapeutic interventions, including the use of HIF1α/2α inhibitors [75,76] and CREB inhibitors [77], the development of SUMOylation mimetics and *miRNA mimics*, and the identification of potential RNF146-specific deubiquitinase (DUB) inhibitors to suppress RNF146 pathways, thereby improving therapeutic outcomes (Figure 5).

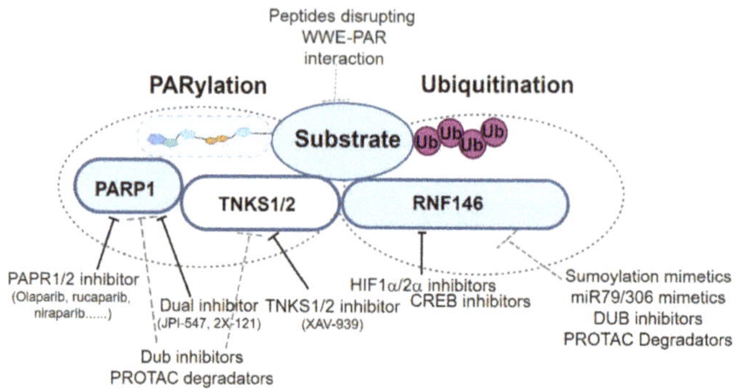

Figure 5. Strategies for targeting the PARdU signaling pathway for cancer treatments.

4.2. Restoring Tumor Suppressors and Enhancing Combination Therapies

Restoring the stability of tumor suppressors degraded by PARdU, such as PTEN, AMOT family proteins, and Motins, presents a promising strategy to bolster cancer treatments. This can be achieved through deubiquitinase (DUB) activators, which counteract protein ubiquitination induced by RNF146 activity, or proteasome inhibitors, which block substrate degradation. These approaches are particularly relevant in cancers with PTEN-AKT pathway dysregulation or YAP/TAZ activation.

Since PARP1 and tankyrase are key writers of the PARdU pathway, their inhibitors have been a focus of therapeutic development. A comprehensive summary of PARP1 and tankyrase inhibitors [78], including those approved for clinical use or undergoing clinical trials, is provided in Table 3. PARP1/2 inhibitors function primarily through two mechanisms: enzymatic inhibition and PARP1 trapping [79,80]. By blocking PARP1 activity, they prevent single-strand break (SSB) repair, while PARP1 trapping, achieved by mimicking NAD^+ and competitively binding to PARP1's catalytic domain, enhances its affinity for DNA damage sites [81,82]. This trapping obstructs DNA repair protein recruitment and results in replication-associated double-strand breaks (DSBs). In homologous recombination-deficient (HRD) tumors, such as BRCA1/2-mutant cancers, this synthetic lethality induces tumor cell death while sparing normal cells. Dual-target PARP-TNKS inhibitors, like 2X-121/JPI-547, combine PARP1/2 inhibition with tankyrase (PARP5a/b) suppression, disrupting DNA repair and Wnt/β-catenin signaling [83,84]. While this dual approach enhances anti-cancer efficacy, it also poses a higher risk of toxicity due to tankyrase's diverse cellular roles. These inhibitors represent promising strategies for targeting HRD tumors and pathways critical for tumor survival. These inhibitors have demonstrated efficacy in targeting key pathways regulated by PARdU signaling. As writers of the PARdU pathway, tankyrases regulate critical processes such as Wnt/β-catenin signaling [19,20,39] and adherens junction (AJ) assembly [85], both of which play central roles in tumorigenesis. Tankyrase inhibitors, such as XAV939, RK-582, and LZZ-02 (binding the nicotinamide subsite, NS) [86–88] and G007-LK and K-756 (targeting the adenosine subsite, AS) [89,90], have shown therapeutic potential. For instance, these inhibitors stabilize PARylated substrates like Axin1/2, suppressing hyperactive Wnt signaling, a hallmark of APC-mutant cancers. Furthermore, tankyrase inhibition disrupts AJ assembly by preventing F-actin anchoring to cadherins, a process associated with cancer cell adhesion and metastasis. Combining tankyrase inhibitors with other targeted therapies, such as PI3K inhibitors (e.g., BKM120) and EGFR inhibitors (e.g., erlotinib), has been proposed as a promising approach for treating APC-mutant colorectal cancers [91,92]. These strategies can restore tumor suppressor stability, inhibit oncogenic signaling, and disrupt key mechanisms of cancer progression.

Table 3. Overview of PARP1/2 and tankyrase inhibitors approved for clinical application and undergoing clinical trials [78].

Inhibitor	Indication	Targets
Olaparib	Recurrent ovarian cancer, BRCA mutation	PARP1/2
Rucaparib	BRCA-mutated ovarian cancer Recurrent or relapsed ovarian cancer	PARP1/2
Niraparib	Ovarian cancer without germ-line or somatic mutation	PARP1/2
Talazoparib	Germline BRCA mutations, EGF2-negative, locally advanced or metastatic breast cancer	PARP1/2
Fluzoparib	Platinum-sensitive recurrent ovarian cancer Fallopian tube cancer Primary peritoneal cancer with germline BRCA mutation	PARP1/2
Pamiparib	Germline BRCA mutation-associated Recurrent advanced ovarian Fallopian tube or primary peritoneal cancer	PARP1/2
CVL218	Advanced solid tumor	PARP1/2
RP12146	Solid tumor Extensive-stage small-cell lung cancer Locally advanced breast cancer	PARP1/2
IDX-1197	Solid tumors; HR repair gene Mutation or deficiency gastric cancer	PARP1/2
Senaparib	Solid tumor Advanced solid tumors	PARP1/2
AMXI-5001	Advanced malignant neoplasm Breast cancer Ovarian cancer	PARP1/2
AZD5305	Breast cancer Ovarian cancer Pancreatic cancer	PARP1
AZD9574	Advanced solid malignancies	PARP1
JPI-547	Ovarian cancer pancreatic ducal Adenocarcinoma	PARP1/2 TNKS1/2
2X-121	Metastatic breast cancer	PARP1/2 TNKS1/2

Tankyrase inhibition, coupled with emerging insights into PARdU signaling, represents a cornerstone of precision oncology, offering transformative potential in cancer therapy.

4.3. Targeting the Tankyrase-Binding Motif (TBM) and the PARdU Pathway

Tankyrase-binding motifs (TBMs) in PARdU substrates present another therapeutic target. Small molecules or mimetics that competitively inhibit TBM binding to tankyrase could prevent substrate PARylation and degradation. Furthermore, combining PARdU inhibitors with targeted therapies, such as Wnt pathway inhibitors [93], PI3K/AKT inhibitors [93,94], or immunotherapies [95], may improve efficacy and overcome resistance.

4.4. Exploiting PARdU in DNA Damage Repair and Cell Death Regulation

The PARdU pathway is critical in DNA damage repair, with RNF146 mediating the degradation of proteins like 53BP1 and BRD7, which are key to DNA damage response pathways. Modulating this process could sensitize cancer cells to DNA-damaging agents, enhancing the efficacy of chemotherapies or radiotherapy [29,54,96,97]. Similarly, PARdU regulates cell death pathways by targeting proteins such as RIPK1 (necroptosis), PARP1 (oxidant-induced cell death), and HK-I (PARthanatos). Hence, inhibiting PARdU could amplify pro-apoptotic signals or disrupt pro-survival mechanisms in cancer cells [55,98].

4.5. Future Directions and Drug Development

Given the complexity of PARdU signaling and potential resistance to monotherapies, targeting the PARdU pathway offers a multifaceted strategy to mitigate such resistance by intersecting multiple signaling networks. This could be achieved through the following possible approaches: developing proteolysis-targeting chimeras (PROTACs) for selective degradation of PARdU components and their positive regulators (Figure 5); conducting proteomics and CRISPR screens to identify new PARdU substrates and their interactions; designing *iso*-ADPr mimetics to disrupt RNF146-substrate interactions; and combining PARdU inhibitors with immune checkpoint blockade therapies to modulate PD-L1 degradation and enhance anti-tumor immunity.

By balancing specificity, efficacy, and safety, PARdU modulation has the potential to revolutionize cancer therapy. Integrating pathway-specific inhibitors, synthetic lethality strategies, and precision medicine approaches will be essential to fully harness the therapeutic potential of this pathway.

5. Conclusions: The Complexity of the PARdU Signaling Pathway

The poly(ADP-ribose)-dependent ubiquitination (PARdU) mechanism highlights a multifaceted regulatory system where poly(ADP-ribose) (PAR) synthesis and interaction with specific E3 ubiquitin ligases, such as RNF146, play critical roles. However, the complexity of this system raises several open questions and avenues for further exploration. First, while RNF146's activation through its WWE domain binding to iso-ADPr has been well documented [20,32,99,100], the potential involvement of other PARP enzymes beyond PARP5a/b in synthesizing PAR modifications for E3 ligase activation remains less clear. For instance, PARP1 and PARP2 are suggested to contribute to PARdU, as evidenced by their interactions with RNF146 [2,101,102], while the diverse subcellular localizations of different PARPs may add specificity to the substrates targeted for degradation [103].

Furthermore, the possibility of other E3 ubiquitin ligases with WWE domains, such as HUWE1, TRIP12, or DTX1-4 [104], participating in PARdU suggests a broader spectrum of mechanisms beyond RNF146 [32]. These ligases, with their unique catalytic mechanisms (RING or HECT), may exhibit distinct modes of activation and substrate specificity. Whether PAR binding is essential for their activation or simply facilitates proximity to substrates remains an open question, as does the role of non-WWE domain-containing ligases in PARdU.

Another layer of complexity lies in the biochemical properties of PAR itself. PAR may function as a direct scaffold for the E3 ligase–substrate interaction, as seen in RNF146's PARP1-mediated ubiquitination [100], or act indirectly by stabilizing E3–substrate complexes. The observation that free PAR polymers can activate RNF146 and promote auto-ubiquitination introduces the intriguing possibility that free PAR molecules, rather than PARylated substrates, may serve as key regulators in specific contexts [100].

Finally, emerging evidence suggests that PARdU is not a monolithic process but instead a dynamic and context-dependent mechanism. For example, PARP-mediated mono-ADP-ribosylation (MARylation) could serve as a precursor for PAR synthesis, with subsequent elongation catalyzed by PARP5a/b or other PARPs [5]. This sequential process raises questions about the interplay between MARylation and PARylation in controlling E3 ligase activity.

In summary, the complexity of PARdU reflects its integration of diverse molecular players and regulatory mechanisms, with spatial, temporal, and substrate-specific factors adding layers of control. Future studies should aim to unravel the interplay between different PARPs, E3 ligases, and free PAR molecules, which could provide deeper insights into PARdU's physiological role in cellular homeostasis and disease and guide the development of novel therapeutic strategies targeting this intricate system.

Funding: This work was supported by an NIH grant (NS133187 to D.K.S. and W.W.). We apologize that due to space limitations, not all related studies are included in this review.

Institutional Review Board Statement: Not applicable.

Informed Consent Statement: Not applicable.

Data Availability Statement: No new data were created or analyzed in this study.

Conflicts of Interest: W.W. is a co-founder and consultant for ReKindle Therapeutics. The other authors declare no competing financial interests.

References

1. Barkauskaite, E.; Jankevicius, G.; Ahel, I. Structures and Mechanisms of Enzymes Employed in the Synthesis and Degradation of PARP-Dependent Protein ADP-Ribosylation. *Mol. Cell* **2015**, *58*, 935–946. [CrossRef]
2. Vyas, S.; Matic, I.; Uchima, L.; Rood, J.; Zaja, R.; Hay, R.T.; Ahel, I.; Chang, P. Family-wide analysis of poly(ADP-ribose) polymerase activity. *Nat. Commun.* **2014**, *5*, 4426. [CrossRef]
3. Gibson, B.A.; Kraus, W.L. New insights into the molecular and cellular functions of poly(ADP-ribose) and PARPs. *Nat. Reviews. Mol. Cell Biol.* **2012**, *13*, 411–424. [CrossRef] [PubMed]
4. Gupte, R.; Liu, Z.; Kraus, W.L. PARPs and ADP-ribosylation: Recent advances linking molecular functions to biological outcomes. *Genes Dev.* **2017**, *31*, 101–126. [CrossRef] [PubMed]
5. Suskiewicz, M.J.; Prokhorova, E.; Rack, J.G.M.; Ahel, I. ADP-ribosylation from molecular mechanisms to therapeutic implications. *Cell* **2023**, *186*, 4475–4495. [CrossRef]
6. Li, Z.; Luo, A.; Xie, B. The Complex Network of ADP-Ribosylation and DNA Repair: Emerging Insights and Implications for Cancer Therapy. *Int. J. Mol. Sci.* **2023**, *24*, 15028. [CrossRef] [PubMed]
7. Lüscher, B.; Ahel, I.; Altmeyer, M.; Ashworth, A.; Bai, P.; Chang, P.; Cohen, M.; Corda, D.; Dantzer, F.; Daugherty, M.D.; et al. ADP-ribosyltransferases, an update on function and nomenclature. *FEBS J.* **2022**, *289*, 7399–7410. [CrossRef] [PubMed]
8. Palazzo, L.; Leidecker, O.; Prokhorova, E.; Dauben, H.; Matic, I.; Ahel, I. Serine is the major residue for ADP-ribosylation upon DNA damage. *eLife* **2018**, *7*, e34334. [CrossRef] [PubMed]
9. Yan, F.; Huang, C.; Wang, X.; Tan, J.; Cheng, S.; Wan, M.; Wang, Z.; Wang, S.; Luo, S.; Li, A.; et al. Threonine ADP-Ribosylation of Ubiquitin by a Bacterial Effector Family Blocks Host Ubiquitination. *Mol. Cell* **2020**, *78*, 641–652.e9. [CrossRef]
10. Hottiger, M.O.; Hassa, P.O.; Lüscher, B.; Schüler, H.; Koch-Nolte, F. Toward a unified nomenclature for mammalian ADP-ribosyltransferases. *Trends Biochem. Sci.* **2010**, *35*, 208–219. [CrossRef]
11. Weixler, L.; Schäringer, K.; Momoh, J.; Lüscher, B.; Feijs, K.L.H.; Žaja, R. ADP-ribosylation of RNA and DNA: From in vitro characterization to in vivo function. *Nucleic Acids Res.* **2021**, *49*, 3634–3650. [CrossRef] [PubMed]
12. Wondisford, A.R.; Lee, J.; Lu, R.; Schuller, M.; Groslambert, J.; Bhargava, R.; Schamus-Haynes, S.; Cespedes, L.C.; Opresko, P.L.; Pickett, H.A.; et al. Deregulated DNA ADP-ribosylation impairs telomere replication. *Nat. Struct. Mol. Biol.* **2024**, *31*, 791–800. [CrossRef] [PubMed]

13. Takamura-Enya, T.; Watanabe, M.; Totsuka, Y.; Kanazawa, T.; Matsushima-Hibiya, Y.; Koyama, K.; Sugimura, T.; Wakabayashi, K. Mono(ADP-ribosyl)ation of 2′-deoxyguanosine residue in DNA by an apoptosis-inducing protein, pierisin-1, from cabbage butterfly. *Proc. Natl. Acad. Sci. USA* **2001**, *98*, 12414–12419. [CrossRef] [PubMed]
14. Zarkovic, G.; Belousova, E.A.; Talhaoui, I.; Saint-Pierre, C.; Kutuzov, M.M.; Matkarimov, B.T.; Biard, D.; Gasparutto, D.; Lavrik, O.I.; Ishchenko, A.A. Characterization of DNA ADP-ribosyltransferase activities of PARP2 and PARP3. New insights into DNA ADP-ribosylation. *Nucleic Acids Res.* **2018**, *46*, 2417–2431. [CrossRef] [PubMed]
15. Kim, S.; Song, G.; Lee, T.; Kim, M.; Kim, J.; Kwon, H.; Kim, J.; Jeong, W.; Lee, U.; Na, C.; et al. PARsylated transcription factor EB (TFEB) regulates the expression of a subset of Wnt target genes by forming a complex with β-catenin-TCF/LEF1. *Cell Death Differ.* **2021**, *28*, 2555–2570. [CrossRef] [PubMed]
16. Li, N.; Zhang, Y.; Han, X.; Liang, K.; Wang, J.; Feng, L.; Wang, W.; Songyang, Z.; Lin, C.; Yang, L.; et al. Poly-ADP ribosylation of PTEN by tankyrases promotes PTEN degradation and tumor growth. *Genes Dev.* **2015**, *29*, 157–170. [CrossRef] [PubMed]
17. Wang, W.; Li, N.; Li, X.; Tran, M.K.; Han, X.; Chen, J. Tankyrase Inhibitors Target YAP by Stabilizing Angiomotin Family Proteins. *Cell Rep.* **2015**, *13*, 524–532. [CrossRef] [PubMed]
18. Matsumoto, Y.; La Rose, J.; Lim, M.; Adissu, H.A.; Law, N.; Mao, X.; Cong, F.; Mera, P.; Karsenty, G.; Goltzman, D.; et al. Ubiquitin ligase RNF146 coordinates bone dynamics and energy metabolism. *J. Clin. Investig.* **2017**, *127*, 2612–2625. [CrossRef] [PubMed]
19. Huang, S.M.; Mishina, Y.M.; Liu, S.; Cheung, A.; Stegmeier, F.; Michaud, G.A.; Charlat, O.; Wiellette, E.; Zhang, Y.; Wiessner, S.; et al. Tankyrase inhibition stabilizes axin and antagonizes Wnt signalling. *Nature* **2009**, *461*, 614–620. [CrossRef]
20. Zhang, Y.; Liu, S.; Mickanin, C.; Feng, Y.; Charlat, O.; Michaud, G.A.; Schirle, M.; Shi, X.; Hild, M.; Bauer, A.; et al. RNF146 is a poly(ADP-ribose)-directed E3 ligase that regulates axin degradation and Wnt signalling. *Nat. Cell Biol.* **2011**, *13*, 623–629. [CrossRef]
21. Callow, M.G.; Tran, H.; Phu, L.; Lau, T.; Lee, J.; Sandoval, W.N.; Liu, P.S.; Bheddah, S.; Tao, J.; Lill, J.R.; et al. Ubiquitin ligase RNF146 regulates tankyrase and Axin to promote Wnt signaling. *PLoS ONE* **2011**, *6*, e22595. [CrossRef] [PubMed]
22. Zhou, Z.D.; Chan, C.H.; Xiao, Z.C.; Tan, E.K. Ring finger protein 146/Iduna is a poly(ADP-ribose) polymer binding and PARsylation dependent E3 ubiquitin ligase. *Cell Adhes. Migr.* **2011**, *5*, 463–471. [CrossRef] [PubMed]
23. Chi, N.W.; Lodish, H.F. Tankyrase is a golgi-associated mitogen-activated protein kinase substrate that interacts with IRAP in GLUT4 vesicles. *J. Biol. Chem.* **2000**, *275*, 38437–38444. [CrossRef] [PubMed]
24. Sbodio, J.I.; Chi, N.W. Identification of a tankyrase-binding motif shared by IRAP, TAB182, and human TRF1 but not mouse TRF1. NuMA contains this RXXPDG motif and is a novel tankyrase partner. *J. Biol. Chem.* **2002**, *277*, 31887–31892. [CrossRef]
25. Guettler, S.; LaRose, J.; Petsalaki, E.; Gish, G.; Scotter, A.; Pawson, T.; Rottapel, R.; Sicheri, F. Structural basis and sequence rules for substrate recognition by Tankyrase explain the basis for cherubism disease. *Cell* **2011**, *147*, 1340–1354. [CrossRef] [PubMed]
26. Nie, L.; Wang, C.; Li, N.; Feng, X.; Lee, N.; Su, D.; Tang, M.; Yao, F.; Chen, J. Proteome-wide Analysis Reveals Substrates of E3 Ligase RNF146 Targeted for Degradation. *Mol. Cell. Proteom. MCP* **2020**, *19*, 2015–2030. [CrossRef] [PubMed]
27. Levaot, N.; Voytyuk, O.; Dimitriou, I.; Sircoulomb, F.; Chandrakumar, A.; Deckert, M.; Krzyzanowski, P.M.; Scotter, A.; Gu, S.; Janmohamed, S.; et al. Loss of Tankyrase-mediated destruction of 3BP2 is the underlying pathogenic mechanism of cherubism. *Cell* **2011**, *147*, 1324–1339. [CrossRef]
28. Li, X.; Han, H.; Zhou, M.T.; Yang, B.; Ta, A.P.; Li, N.; Chen, J.; Wang, W. Proteomic Analysis of the Human Tankyrase Protein Interaction Network Reveals Its Role in Pexophagy. *Cell Rep.* **2017**, *20*, 737–749. [CrossRef]
29. Bhardwaj, A.; Yang, Y.; Ueberheide, B.; Smith, S. Whole proteome analysis of human tankyrase knockout cells reveals targets of tankyrase-mediated degradation. *Nat. Commun.* **2017**, *8*, 2214. [CrossRef] [PubMed]
30. DaRosa, P.A.; Wang, Z.; Jiang, X.; Pruneda, J.N.; Cong, F.; Klevit, R.E.; Xu, W. Allosteric activation of the RNF146 ubiquitin ligase by a poly(ADP-ribosyl)ation signal. *Nature* **2015**, *517*, 223–226. [CrossRef] [PubMed]
31. DaRosa, P.A.; Klevit, R.E.; Xu, W. Structural basis for tankyrase-RNF146 interaction reveals noncanonical tankyrase-binding motifs. *Protein Sci. A Publ. Protein Soc.* **2018**, *27*, 1057–1067. [CrossRef] [PubMed]
32. Wang, Z.; Michaud, G.A.; Cheng, Z.; Zhang, Y.; Hinds, T.R.; Fan, E.; Cong, F.; Xu, W. Recognition of the iso-ADP-ribose moiety in poly(ADP-ribose) by WWE domains suggests a general mechanism for poly(ADP-ribosyl)ation-dependent ubiquitination. *Genes Dev.* **2012**, *26*, 235–240. [CrossRef] [PubMed]
33. Chandrakumar, A.A.; Coyaud, É.; Marshall, C.B.; Ikura, M.; Raught, B.; Rottapel, R. Tankyrase regulates epithelial lumen formation via suppression of Rab11 GEFs. *J. Cell Biol.* **2021**, *220*, e20208037. [CrossRef]
34. Mashimo, M.; Kita, M.; Uno, A.; Nii, M.; Ishihara, M.; Honda, T.; Gotoh-Kinoshita, Y.; Nomura, A.; Nakamura, H.; Murayama, T.; et al. Tankyrase Regulates Neurite Outgrowth through Poly(ADP-ribosyl)ation-Dependent Activation of β-Catenin Signaling. *Int. J. Mol. Sci.* **2022**, *23*, 2834. [CrossRef]
35. Jung, M.; Kim, W.; Cho, J.W.; Yang, W.H.; Chung, I.K. Poly-ADP ribosylation of p21 by tankyrases promotes p21 degradation and regulates cell cycle progression. *Biochem. J.* **2022**, *479*, 2379–2394. [CrossRef] [PubMed]

36. Gerbracht, J.V.; Boehm, V.; Britto-Borges, T.; Kallabis, S.; Wiederstein, J.L.; Ciriello, S.; Aschemeier, D.U.; Krüger, M.; Frese, C.K.; Altmüller, J.; et al. CASC3 promotes transcriptome-wide activation of nonsense-mediated decay by the exon junction complex. *Nucleic Acids Res.* **2020**, *48*, 8626–8644. [CrossRef] [PubMed]
37. Zhao, Y.; Fan, S.; Zhu, H.; Zhao, Q.; Fang, Z.; Xu, D.; Lin, W.; Lin, L.; Hu, X.; Wu, G.; et al. Podocyte OTUD5 alleviates diabetic kidney disease through deubiquitinating TAK1 and reducing podocyte inflammation and injury. *Nat. Commun.* **2024**, *15*, 5441. [CrossRef] [PubMed]
38. Weixler, L.; Feijs, K.L.H.; Zaja, R. ADP-ribosylation of RNA in mammalian cells is mediated by TRPT1 and multiple PARPs. *Nucleic Acids Res.* **2022**, *50*, 9426–9441. [CrossRef]
39. Shen, J.; Yu, Z.; Li, N. The E3 ubiquitin ligase RNF146 promotes colorectal cancer by activating the Wnt/β-catenin pathway via ubiquitination of Axin1. *Biochem. Biophys. Res. Commun.* **2018**, *503*, 991–997. [CrossRef] [PubMed]
40. Wang, Y.; Zhu, Y.; Gu, Y.; Ma, M.; Wang, Y.; Qi, S.; Zeng, Y.; Zhu, R.; Wang, X.; Yu, P.; et al. Stabilization of Motin family proteins in NF2-deficient cells prevents full activation of YAP/TAZ and rapid tumorigenesis. *Cell Rep.* **2021**, *36*, 109596. [CrossRef] [PubMed]
41. Xu, Y.R.; Shi, M.L.; Zhang, Y.; Kong, N.; Wang, C.; Xiao, Y.F.; Du, S.S.; Zhu, Q.Y.; Lei, C.Q. Tankyrases inhibit innate antiviral response by PARylating VISA/MAVS and priming it for RNF146-mediated ubiquitination and degradation. *Proc. Natl. Acad. Sci. USA* **2022**, *119*, e2122805119. [CrossRef]
42. Zhou, B.; Yan, J.; Guo, L.; Zhang, B.; Liu, S.; Yu, M.; Chen, Z.; Zhang, K.; Zhang, W.; Li, X.; et al. Hepatoma cell-intrinsic TLR9 activation induces immune escape through PD-L1 upregulation in hepatocellular carcinoma. *Theranostics* **2020**, *10*, 6530–6543. [CrossRef] [PubMed]
43. Ueki, Y.; Tiziani, V.; Santanna, C.; Fukai, N.; Maulik, C.; Garfinkle, J.; Ninomiya, C.; doAmaral, C.; Peters, H.; Habal, M.; et al. Mutations in the gene encoding c-Abl-binding protein SH3BP2 cause cherubism. *Nat. Genet.* **2001**, *28*, 125–126. [CrossRef] [PubMed]
44. Verma, R.; Turbin, R.E.; Langer, P.D. Cherubism. *Ophthalmology* **2024**. [CrossRef]
45. Lee, Y.; Kang, H.C.; Lee, B.D.; Lee, Y.I.; Kim, Y.P.; Shin, J.H. Poly (ADP-ribose) in the pathogenesis of Parkinson's disease. *BMB Rep.* **2014**, *47*, 424–432. [CrossRef] [PubMed]
46. Virág, L.; Robaszkiewicz, A.; Rodriguez-Vargas, J.M.; Oliver, F.J. Poly(ADP-ribose) signaling in cell death. *Mol. Asp. Med.* **2013**, *34*, 1153–1167. [CrossRef] [PubMed]
47. Xu, X.; Sun, B.; Zhao, C. Poly (ADP-Ribose) polymerase 1 and parthanatos in neurological diseases: From pathogenesis to therapeutic opportunities. *Neurobiol. Dis.* **2023**, *187*, 106314. [CrossRef] [PubMed]
48. Huang, P.; Chen, G.; Jin, W.; Mao, K.; Wan, H.; He, Y. Molecular Mechanisms of Parthanatos and Its Role in Diverse Diseases. *Int. J. Mol. Sci.* **2022**, *23*, 7292. [CrossRef]
49. Lee, Y.; Karuppagounder, S.S.; Shin, J.H.; Lee, Y.I.; Ko, H.S.; Swing, D.; Jiang, H.; Kang, S.U.; Lee, B.D.; Kang, H.C.; et al. Parthanatos mediates AIMP2-activated age-dependent dopaminergic neuronal loss. *Nat. Neurosci.* **2013**, *16*, 1392–1400. [CrossRef]
50. Wu, H.; Li, Y.; Zhang, Q.; Wang, H.; Xiu, W.; Xu, P.; Deng, Y.; Huang, W.; Wang, D.O. Crocetin antagonizes parthanatos in ischemic stroke via inhibiting NOX2 and preserving mitochondrial hexokinase-I. *Cell Death Dis.* **2023**, *14*, 50. [CrossRef] [PubMed]
51. Kim, H.; Park, J.; Kang, H.; Yun, S.P.; Lee, Y.S.; Lee, Y.I.; Lee, Y. Activation of the Akt1-CREB pathway promotes RNF146 expression to inhibit PARP1-mediated neuronal death. *Sci. Signal.* **2020**, *13*, eaax7119. [CrossRef]
52. Zhang, F.; Lou, L.; Peng, B.; Song, X.; Reizes, O.; Almasan, A.; Gong, Z. Nudix Hydrolase NUDT16 Regulates 53BP1 Protein by Reversing 53BP1 ADP-Ribosylation. *Cancer Res.* **2020**, *80*, 999–1010. [CrossRef] [PubMed]
53. Zhang, Z.; Samsa, W.E.; Gong, Z. NUDT16 regulates CtIP PARylation to dictate homologous recombination repair. *Nucleic Acids Res.* **2024**, *52*, 3761–3777. [CrossRef]
54. Hu, K.; Wu, W.; Li, Y.; Lin, L.; Chen, D.; Yan, H.; Xiao, X.; Chen, H.; Chen, Z.; Zhang, Y.; et al. Poly(ADP-ribosyl)ation of BRD7 by PARP1 confers resistance to DNA-damaging chemotherapeutic agents. *EMBO Rep.* **2019**, *20*, e46166. [CrossRef]
55. Liu, L.; Sandow, J.J.; Leslie Pedrioli, D.M.; Samson, A.L.; Silke, N.; Kratina, T.; Ambrose, R.L.; Doerflinger, M.; Hu, Z.; Morrish, E.; et al. Tankyrase-mediated ADP-ribosylation is a regulator of TNF-induced death. *Sci. Adv.* **2022**, *8*, eabh2332. [CrossRef]
56. Hou, S.; Zhang, J.; Jiang, X.; Yang, Y.; Shan, B.; Zhang, M.; Liu, C.; Yuan, J.; Xu, D. PARP5A and RNF146 phase separation restrains RIPK1-dependent necroptosis. *Mol. Cell* **2024**, *84*, 938–954.e8. [CrossRef]
57. Gerö, D.; Szoleczky, P.; Chatzianastasiou, A.; Papapetropoulos, A.; Szabo, C. Modulation of poly(ADP-ribose) polymerase-1 (PARP-1)-mediated oxidative cell injury by ring finger protein 146 (RNF146) in cardiac myocytes. *Mol. Med. (Camb. Mass.)* **2014**, *20*, 313–328. [CrossRef] [PubMed]
58. Kim, S.; Han, S.; Kim, Y.; Kim, H.S.; Gu, Y.R.; Kang, D.; Cho, Y.; Kim, H.; Lee, J.; Seo, Y.; et al. Tankyrase inhibition preserves osteoarthritic cartilage by coordinating cartilage matrix anabolism via effects on SOX9 PARylation. *Nat. Commun.* **2019**, *10*, 4898. [CrossRef]

59. Hou, X.; Tian, M.; Ning, J.; Wang, Z.; Guo, F.; Zhang, W.; Hu, L.; Wei, S.; Hu, C.; Yun, X.; et al. PARP inhibitor shuts down the global translation of thyroid cancer through promoting Pol II binding to DIMT1 pause. *Int. J. Biol. Sci.* **2023**, *19*, 3970–3986. [CrossRef] [PubMed]
60. Yang, L.; Du, M.; Liu, K.; Wang, P.; Zhu, J.; Li, F.; Wang, Z.; Huang, K.; Liang, M. Pimpinellin ameliorates macrophage inflammation by promoting RNF146-mediated PARP1 ubiquitination. *Phytother. Res. PTR* **2024**, *38*, 1783–1798. [CrossRef]
61. Li, N.; Wang, Y.; Neri, S.; Zhen, Y.; Fong, L.W.R.; Qiao, Y.; Li, X.; Chen, Z.; Stephan, C.; Deng, W.; et al. Tankyrase disrupts metabolic homeostasis and promotes tumorigenesis by inhibiting LKB1-AMPK signalling. *Nat. Commun.* **2019**, *10*, 4363. [CrossRef] [PubMed]
62. Voronkov, A.; Holsworth, D.D.; Waaler, J.; Wilson, S.R.; Ekblad, B.; Perdreau-Dahl, H.; Dinh, H.; Drewes, G.; Hopf, C.; Morth, J.P.; et al. Structural basis and SAR for G007-LK, a lead stage 1,2,4-triazole based specific tankyrase 1/2 inhibitor. *J. Med. Chem.* **2013**, *56*, 3012–3023. [CrossRef] [PubMed]
63. Jang, M.K.; Mashima, T.; Seimiya, H. Tankyrase Inhibitors Target Colorectal Cancer Stem Cells via AXIN-Dependent Downregulation of c-KIT Tyrosine Kinase. *Mol. Cancer Ther.* **2020**, *19*, 765–776. [CrossRef] [PubMed]
64. Wang, W.; Liu, P.; Lavrijsen, M.; Li, S.; Zhang, R.; Li, S.; van de Geer, W.S.; van de Werken, H.J.G.; Peppelenbosch, M.P.; Smits, R. Evaluation of AXIN1 and AXIN2 as targets of tankyrase inhibition in hepatocellular carcinoma cell lines. *Sci. Rep.* **2021**, *11*, 7470. [CrossRef] [PubMed]
65. Tai, D.; Wells, K.; Arcaroli, J.; Vanderbilt, C.; Aisner, D.L.; Messersmith, W.A.; Lieu, C.H. Targeting the WNT Signaling Pathway in Cancer Therapeutics. *Oncol.* **2015**, *20*, 1189–1198. [CrossRef] [PubMed]
66. Xu, C.; Xu, Z.; Zhang, Y.; Evert, M.; Calvisi, D.F.; Chen, X. β-Catenin signaling in hepatocellular carcinoma. *J. Clin. Investig.* **2022**, *132*, e154515. [CrossRef]
67. Peng, K.; Anmangandla, A.; Jana, S.; Jin, Y.; Lin, H. Iso-ADP-Ribose Fluorescence Polarization Probe for the Screening of RNF146 WWE Domain Inhibitors. *ACS Chem. Biol.* **2024**, *19*, 300–307. [CrossRef] [PubMed]
68. Shen, G.; Wang, H.; Zhu, N.; Lu, Q.; Liu, J.; Xu, Q.; Huang, D. HIF-1/2α-Activated RNF146 Enhances the Proliferation and Glycolysis of Hepatocellular Carcinoma Cells via the PTEN/AKT/mTOR Pathway. *Front. Cell Dev. Biol.* **2022**, *10*, 893888. [CrossRef] [PubMed]
69. Park, G.; Jang, W.E.; Kim, S.; Gonzales, E.L.; Ji, J.; Choi, S.; Kim, Y.; Park, J.H.; Mohammad, H.B.; Bang, G.; et al. Dysregulation of the Wnt/β-catenin signaling pathway via Rnf146 upregulation in a VPA-induced mouse model of autism spectrum disorder. *Exp. Mol. Med.* **2023**, *55*, 1783–1794. [CrossRef]
70. Belayev, L.; Mukherjee, P.K.; Balaszczuk, V.; Calandria, J.M.; Obenaus, A.; Khoutorova, L.; Hong, S.H.; Bazan, N.G. Neuroprotectin D1 upregulates Iduna expression and provides protection in cellular uncompensated oxidative stress and in experimental ischemic stroke. *Cell Death Differ.* **2017**, *24*, 1091–1099. [CrossRef]
71. Li, W.; Han, Q.; Zhu, Y.; Zhou, Y.; Zhang, J.; Wu, W.; Li, Y.; Liu, L.; Qiu, Y.; Hu, K.; et al. SUMOylation of RNF146 results in Axin degradation and activation of Wnt/β-catenin signaling to promote the progression of hepatocellular carcinoma. *Oncogene* **2023**, *42*, 1728–1740. [CrossRef] [PubMed]
72. Wang, Z.; Xia, X.; Li, J.; Igaki, T. Tumor elimination by clustered microRNAs miR-306 and miR-79 via noncanonical activation of JNK signaling. *eLife* **2022**, *11*, e77340. [CrossRef] [PubMed]
73. Matsumoto, Y.; Larose, J.; Kent, O.A.; Lim, M.; Changoor, A.; Zhang, L.; Storozhuk, Y.; Mao, X.; Grynpas, M.D.; Cong, F.; et al. RANKL coordinates multiple osteoclastogenic pathways by regulating expression of ubiquitin ligase RNF146. *J. Clin. Investig.* **2017**, *127*, 1303–1315. [CrossRef] [PubMed]
74. Liu, F.; Chen, J.; Li, K.; Li, H.; Zhu, Y.; Zhai, Y.; Lu, B.; Fan, Y.; Liu, Z.; Chen, X.; et al. Ubiquitination and deubiquitination in cancer: From mechanisms to novel therapeutic approaches. *Mol. Cancer* **2024**, *23*, 148. [CrossRef] [PubMed]
75. Zhang, Y.; Wang, S.; Zhang, J.; Liu, C.; Li, X.; Guo, W.; Duan, Y.; Chen, X.; Zong, S.; Zheng, J.; et al. Elucidating minimal residual disease of paediatric B-cell acute lymphoblastic leukaemia by single-cell analysis. *Nat. Cell Biol.* **2022**, *24*, 242–252. [CrossRef]
76. Ilegems, E.; Bryzgalova, G.; Correia, J.; Yesildag, B.; Berra, E.; Ruas, J.L.; Pereira, T.S.; Berggren, P.-O. HIF-1α inhibitor PX-478 preserves pancreatic β cell function in diabetes. *Sci. Transl. Med.* **2022**, *14*, eaba9112. [CrossRef]
77. Sapio, L.; Salzillo, A.; Ragone, A.; Illiano, M.; Spina, A.; Naviglio, S. Targeting CREB in Cancer Therapy: A Key Candidate or One of Many? An Update. *Cancers* **2020**, *12*, 3166. [CrossRef] [PubMed]
78. Duan, M.; Gao, J.; Li, J.; Huang, X.; Ren, Y.; Li, Y.; Liao, M.; Zhang, Y. Targeting selective inhibitors of PARPs in drug discovery and development. *Med. Chem. Res.* **2024**, *33*, 1734–1756. [CrossRef]
79. Murai, J.; Huang, S.Y.; Das, B.B.; Renaud, A.; Zhang, Y.; Doroshow, J.H.; Ji, J.; Takeda, S.; Pommier, Y. Trapping of PARP1 and PARP2 by Clinical PARP Inhibitors. *Cancer Res.* **2012**, *72*, 5588–5599. [CrossRef]

80. Murai, J.; Zhang, Y.; Morris, J.; Ji, J.; Takeda, S.; Doroshow, J.H.; Pommier, Y. Rationale for poly(ADP-ribose) polymerase (PARP) inhibitors in combination therapy with camptothecins or temozolomide based on PARP trapping versus catalytic inhibition. *J. Pharmacol. Exp. Ther.* **2014**, *349*, 408–416. [CrossRef] [PubMed]
81. D'Andrea, A.D. Mechanisms of PARP inhibitor sensitivity and resistance. *DNA Repair* **2018**, *71*, 172–176. [CrossRef] [PubMed]
82. Mateo, J.; Lord, C.J.; Serra, V.; Tutt, A.; Balmaña, J.; Castroviejo-Bermejo, M.; Cruz, C.; Oaknin, A.; Kaye, S.B.; de Bono, J.S. A decade of clinical development of PARP inhibitors in perspective. *Ann. Oncol. Off. J. Eur. Soc. Med. Oncol.* **2019**, *30*, 1437–1447. [CrossRef] [PubMed]
83. Hu, X.; Zhang, J.; Zhang, Y.; Jiao, F.; Wang, J.; Chen, H.; Ouyang, L.; Wang, Y. Dual-target inhibitors of poly (ADP-ribose) polymerase-1 for cancer therapy: Advances, challenges, and opportunities. *Eur. J. Med. Chem.* **2022**, *230*, 114094. [CrossRef] [PubMed]
84. Im, S.-A.; Lee, S.; Lee, K.W.; Lee, Y.; Sohn, J.; Kim, J.H.; Im, Y.-H.; Park, K.H.; Oh, D.-Y.; Kim, M.H.; et al. A phase I dose-escalation and expansion study of JPI-547, a dual inhibitor of PARP/tankyrase in patients with advanced solid tumors. *J. Clin. Oncol.* **2021**, *39*, 3113. [CrossRef]
85. Shultz, M.D.; Cheung, A.K.; Kirby, C.A.; Firestone, B.; Fan, J.; Chen, C.H.; Chen, Z.; Chin, D.N.; Dipietro, L.; Fazal, A.; et al. Identification of NVP-TNKS656: The use of structure-efficiency relationships to generate a highly potent, selective, and orally active tankyrase inhibitor. *J. Med. Chem.* **2013**, *56*, 6495–6511. [CrossRef] [PubMed]
86. Ferri, M.; Liscio, P.; Carotti, A.; Asciutti, S.; Sardella, R.; Macchiarulo, A.; Camaioni, E. Targeting Wnt-driven cancers: Discovery of novel tankyrase inhibitors. *Eur. J. Med. Chem.* **2017**, *142*, 506–522. [CrossRef] [PubMed]
87. Ryu, H.; Nam, K.Y.; Kim, H.J.; Song, J.Y.; Hwang, S.G.; Kim, J.S.; Kim, J.; Ahn, J. Discovery of a Novel Triazolopyridine Derivative as a Tankyrase Inhibitor. *Int. J. Mol. Sci.* **2021**, *22*, 7330. [CrossRef]
88. Shirai, F.; Tsumura, T.; Yashiroda, Y.; Yuki, H.; Niwa, H.; Sato, S.; Chikada, T.; Koda, Y.; Washizuka, K.; Yoshimoto, N.; et al. Discovery of Novel Spiroindoline Derivatives as Selective Tankyrase Inhibitors. *J. Med. Chem.* **2019**, *62*, 3407–3427. [CrossRef]
89. Li, B.; Liang, J.; Lu, F.; Zeng, G.; Zhang, J.; Ma, Y.; Liu, P.; Wang, Q.; Zhou, Q.; Chen, L. Discovery of Novel Inhibitor for WNT/β-Catenin Pathway by Tankyrase 1/2 Structure-Based Virtual Screening. *Molecules* **2020**, *25*, 1680. [CrossRef]
90. Bregman, H.; Chakka, N.; Guzman-Perez, A.; Gunaydin, H.; Gu, Y.; Huang, X.; Berry, V.; Liu, J.; Teffera, Y.; Huang, L.; et al. Discovery of novel, induced-pocket binding oxazolidinones as potent, selective, and orally bioavailable tankyrase inhibitors. *J. Med. Chem.* **2013**, *56*, 4320–4342. [CrossRef]
91. Solberg, N.T.; Waaler, J.; Lund, K.; Mygland, L.; Olsen, P.A.; Krauss, S. TANKYRASE Inhibition Enhances the Antiproliferative Effect of PI3K and EGFR Inhibition, Mutually Affecting β-CATENIN and AKT Signaling in Colorectal Cancer. *Mol. Cancer Res. MCR* **2018**, *16*, 543–553. [CrossRef]
92. Zamudio-Martinez, E.; Herrera-Campos, A.B.; Muñoz, A.; Rodríguez-Vargas, J.M.; Oliver, F.J. Tankyrases as modulators of pro-tumoral functions: Molecular insights and therapeutic opportunities. *J. Exp. Clin. Cancer Res. CR* **2021**, *40*, 144. [CrossRef]
93. Arqués, O.; Chicote, I.; Puig, I.; Tenbaum, S.P.; Argilés, G.; Dienstmann, R.; Fernández, N.; Caratù, G.; Matito, J.; Silberschmidt, D.; et al. Tankyrase Inhibition Blocks Wnt/β-Catenin Pathway and Reverts Resistance to PI3K and AKT Inhibitors in the Treatment of Colorectal Cancer. *Clin. Cancer Res. Off. J. Am. Assoc. Cancer Res.* **2016**, *22*, 644–656. [CrossRef] [PubMed]
94. Pećina-Šlaus, N.; Aničić, S.; Bukovac, A.; Kafka, A. Wnt Signaling Inhibitors and Their Promising Role in Tumor Treatment. *Int. J. Mol. Sci.* **2023**, *24*, 6733. [CrossRef] [PubMed]
95. Kinosada, H.; Okada-Iwasaki, R.; Kunieda, K.; Suzuki-Imaizumi, M.; Takahashi, Y.; Miyagi, H.; Suzuki, M.; Motosawa, K.; Watanabe, M.; Mie, M.; et al. The dual pocket binding novel tankyrase inhibitor K-476 enhances the efficacy of immune checkpoint inhibitor by attracting CD8(+) T cells to tumors. *Am. J. Cancer Res.* **2021**, *11*, 264–276. [PubMed]
96. Roy, S.; Roy, S.; Kar, M.; Chakraborty, A.; Kumar, A.; Delogu, F.; Asthana, S.; Hande, M.P.; Banerjee, B. Combined treatment with cisplatin and the tankyrase inhibitor XAV-939 increases cytotoxicity, abrogates cancer-stem-like cell phenotype and increases chemosensitivity of head-and-neck squamous-cell carcinoma cells. *Mutat. Res. /Genet. Toxicol. Environ. Mutagen.* **2019**, *846*, 503084. [CrossRef] [PubMed]
97. Okamoto, K.; Ohishi, T.; Kuroiwa, M.; Iemura, S.I.; Natsume, T.; Seimiya, H. MERIT40-dependent recruitment of tankyrase to damaged DNA and its implication for cell sensitivity to DNA-damaging anticancer drugs. *Oncotarget* **2018**, *9*, 35844–35855. [CrossRef]
98. Tian, X.-H.; Hou, W.-J.; Fang, Y.; Fan, J.; Tong, H.; Bai, S.-L.; Chen, Q.; Xu, H.; Li, Y. XAV939, a tankyrase 1 inhibitior, promotes cell apoptosis in neuroblastoma cell lines by inhibiting Wnt/β-catenin signaling pathway. *J. Exp. Clin. Cancer Res.* **2013**, *32*, 100. [CrossRef]
99. Pollock, K.; Ranes, M.; Collins, I.; Guettler, S. Identifying and Validating Tankyrase Binders and Substrates: A Candidate Approach. *Methods Mol. Biol. (Clifton N.J.)* **2017**, *1608*, 445–473. [CrossRef]

100. Kang, H.C.; Lee, Y.I.; Shin, J.H.; Andrabi, S.A.; Chi, Z.; Gagné, J.P.; Lee, Y.; Ko, H.S.; Lee, B.D.; Poirier, G.G.; et al. Iduna is a poly(ADP-ribose) (PAR)-dependent E3 ubiquitin ligase that regulates DNA damage. *Proc. Natl. Acad. Sci. USA* **2011**, *108*, 14103–14108. [CrossRef] [PubMed]
101. Rolli, V.; O'Farrell, M.; Menissier-de Murcia, J.; de Murcia, G. Random mutagenesis of the poly(ADP-ribose) polymerase catalytic domain reveals amino acids involved in polymer branching. *Biochemistry* **1997**, *36*, 12147–12154. [CrossRef]
102. Chen, Q.; Kassab, M.A.; Dantzer, F.; Yu, X. PARP2 mediates branched poly ADP-ribosylation in response to DNA damage. *Nat. Commun.* **2018**, *9*, 3233. [CrossRef] [PubMed]
103. Vyas, S.; Chesarone-Cataldo, M.; Todorova, T.; Huang, Y.H.; Chang, P. A systematic analysis of the PARP protein family identifies new functions critical for cell physiology. *Nat. Commun.* **2013**, *4*, 2240. [CrossRef] [PubMed]
104. Aravind, L. The WWE domain: A common interaction module in protein ubiquitination and ADP ribosylation. *Trends Biochem. Sci.* **2001**, *26*, 273–275. [CrossRef] [PubMed]

biomolecules

MDPI

Review

Epigenetic Properties of Compounds Contained in Functional Foods Against Cancer

Giulia Casari [1], Brenda Romaldi [1], Andrea Scirè [2], Cristina Minnelli [2], Daniela Marzioni [3], Gianna Ferretti [1] and Tatiana Armeni [1,*]

[1] Department of Clinical and Specialist Sciences (DISCO), Università Politecnica delle Marche, 60131 Ancona, Italy; g.casari@staff.univpm.it (G.C.); b.romaldi@pm.univpm.it (B.R.); g.ferretti@staff.univpm.it (G.F.)

[2] Department of Life and Environmental Sciences (DISVA), Università Politecnica delle Marche, 60131 Ancona, Italy; a.a.scire@staff.univpm.it (A.S.); c.minnelli@staff.univpm.it (C.M.)

[3] Department of Experimental and Clinical Medicine, Università Politecnica delle Marche, 60131 Ancona, Italy; d.marzioni@staff.univpm.it

* Correspondence: t.armeni@staff.univpm.it

Abstract: Epigenetics encompasses reversible and heritable genomic changes in histones, DNA expression, and non-coding RNAs that occur without modifying the nucleotide DNA sequence. These changes play a critical role in modulating cell function in both healthy and pathological conditions. Dysregulated epigenetic mechanisms are implicated in various diseases, including cardiovascular disorders, neurodegenerative diseases, obesity, and mainly cancer. Therefore, to develop innovative therapeutic strategies, research for compounds able to modulate the complex epigenetic landscape of cancer is rapidly surging. Dietary phytochemicals, mostly flavonoids but also tetraterpenoids, organosulfur compounds, and isothiocyanates, represent biologically active molecules found in vegetables, fruits, medicinal plants, and beverages. These natural organic compounds exhibit epigenetic modulatory properties by influencing the activity of epigenetics key enzymes, such as DNA methyltransferases, histone acetyltransferases and deacetylases, and histone methyltransferases and demethylases. Due to the reversibility of the modifications that they induce, their minimal adverse effects, and their potent epigenetic regulatory activity, dietary phytochemicals hold significant promise as antitumor agents and warrant further investigation. This review aims to consolidate current data on the diverse epigenetic effects of the six major flavonoid subclasses, as well as other natural compounds, in the context of cancer. The goal is to identify new therapeutic epigenetic targets for drug development, whether as stand-alone treatments or in combination with conventional antitumor approaches.

Keywords: epigenetic; cancer; phytochemicals; flavonoids; DNA-methylation; histone-modifications; non-coding RNAs

1. Introduction

Epigenetics investigates reversible yet heritable genomic modifications at both the DNA and histone protein levels, regulating gene expression and phenotypic outcomes without modifying the nucleotide sequence itself. These epigenetic changes are intricately linked to chromatin condensation states, which are regulated by histone post-translational modifications and gene methylation patterns that control gene silencing [1]. Epigenetics plays a crucial role in modulating cellular functions in both healthy conditions (including gene expression of each tissue, X-chromosome inactivation, and non-coding DNA modulation) and pathological conditions (such as cancer, Alzheimer's disease, and neurological diseases) [2]. In both health and disease, gene expression regulation is mediated mainly by DNA methylation processes, covalent histone modifications, and noncoding RNA (ncRNA) activity, all of which have a marked impact on cellular homeostasis [3].

Cancer, as well as other pathologies, represents a complex condition whose etiology, development, and progression are strictly related to both genetic alterations and microenvironment-connected epigenetic factors. Epigenetic mechanisms can directly modulate the expression of proto-oncogenes and tumor suppressor genes, frequently occurring in the early steps of tumorigenesis, often preceding genetic mutations. These features make epigenetic targets highly promising for cancer prevention and therapy [4].

Environmental stimuli, including the intake of dietary phytochemicals, can exert a strong influence on regulating these epigenetic processes [2]. Dietaries are bioactive plant-derived molecules that exhibit several biological effects, such as anticancer, antioxidant, and immunomodulatory features, also thanks to their epigenetic modulation mechanisms. These compounds exert their effects via complementary and overlapping mechanisms, such as the activation of detoxification enzymes and inhibition of nitrosamine formation. Specifically, they target the activity of key epigenetic regulators, including DNA methyltransferases (DNMTs) and histone deacetylases (HDACs) [5,6].

Flavonoids, the most abundant class of natural polyphenolic compounds consumed daily, are particularly effective at modulating several epigenetic processes implicated in cancer and other diseases. Flavonoids are categorized into six subclasses: flavan-3-ols, flavonols, flavones, flavanones, isoflavones, and anthocyanidins, whose roles as epigenetic modulators are extensively reviewed here [7]. Together with flavonoids, other dietary phytochemicals present in vegetables, fruits, beverages, and spices also act as potent epigenetic modulators by inhibiting DNMTs, modifying histones, and modulating micro-RNA (miRNA) expression. Examples include curcumin, lycopene, organosulfur compounds, phenethyl isothiocyanate, and resveratrol, which are highlighted in this review.

Given their properties, these dietary compounds hold potential as epigenetic modulators for use in cancer prevention and therapy, either as standalone treatments or in combination with conventional pharmacological approaches [8]. Starting from these data from the literature, recent research progress concerning epigenetic properties of compounds contained in functional foods, including all types of flavonoids but also other phytochemicals like resveratrol, lycopene, and curcumin, is reviewed here. The focus is on their potential for tumor treatment and prevention, aiming to promote the development of innovative strategies and mitigate the numerous adverse effects associated with traditional approaches.

2. Epigenetic Mechanisms

Among the several epigenetic mechanisms, DNA methylation, post-translational histone modifications (e.g., phosphorylation, deacetylation, methylation), and expression variation in ncRNAs (e.g., siRNA, miRNA, lncRNA, cirRNA) represent the most epigenetic modulators on which phytochemicals could exert their potential [9] (Figure 1). These reversible epigenetic alterations play a critical role in the onset, development, and progression of tumors. Indeed, abnormal promoter hypermethylation, aberrant histone acetylation, and miRNA expression dysregulation can promote improper gene silencing within each step of tumor progression [10].

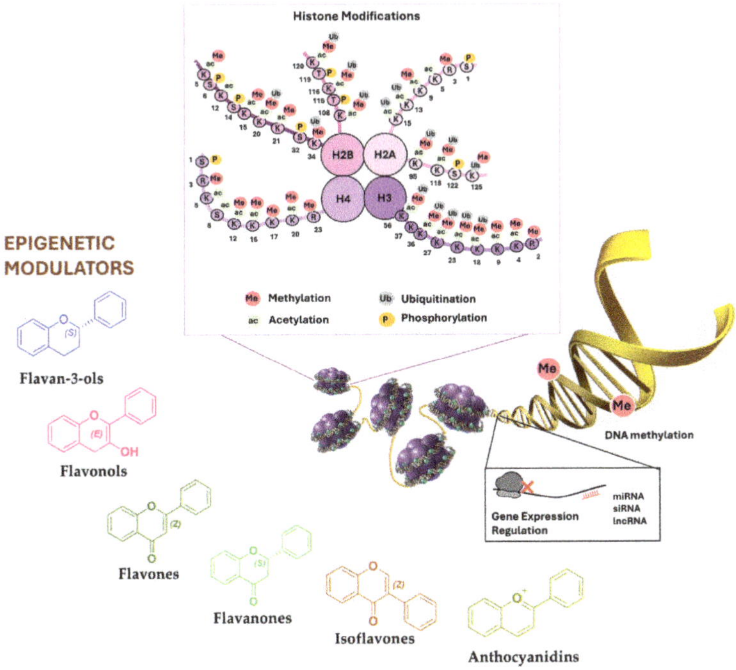

Figure 1. Summary of key epigenetic mechanisms affecting the development and progression of tumors and the chemical scaffold of the main flavonoids acting as epigenetic modulators. Epigenetic mechanisms, including DNA methylation and histone modifications like methylation, acetylation, phosphorylation, and ubiquitination, regulate the transcriptional activity of genes. Natural epigenetic modulators, such as flavan-3-ols, flavonols, flavones, flavanones, isoflavones, and anthocyanidins, which influence these epigenetic modifications, are highlighted. The image also includes epigenetic modifications mediated by ncRNAs, such as miRNAs, siRNAs, cirRNA, and lncRNAs, suggesting that phytochemicals may exert a modulating role through these epigenetic mechanisms.

2.1. DNA Methylation

In DNA methylation, a methyl group (-CH_3) is transferred 5′ to a cytosine base, particularly when it precedes a guanine in a CpG dinucleotide, to form 5-methyl cytosine. This occurs most frequently in CpG islands, regions of DNA with a high frequency of CpG sites, which are often located in gene promoter regions and enhancers. CpG islands are typically characterized by a cytosine and guanine content greater than 50% and a sequence length greater than 200 base pairs. Methylation of these regions often leads to gene silencing, playing a vital role in regulating gene expression during development and in disease processes such as cancer [11]. Although CpG methylation is the most well-studied in mammals, DNA methylation can also occur at other bases, such as adenine, to form N6-methyladenine. DNA methylation is mediated by a group of enzymes called DNA methyltransferases (DNMTs), which includes the isoenzymes DNMT1, DNMT3a, and DNMT3b [12]. DNMT1 represents the main enzyme responsible for the preservation of DNA methylation sites through the maintenance of existing DNA methylation patterns during DNA replication. This enzyme is essential for the inheritance of epigenetic marks, ensuring that methylation patterns are copied to daughter cells after cell division. DNMT3a and DNMT3b, deeply expressed in embryogenesis and slightly expressed in adults, are mainly involved in de novo methylation as they can add methyl groups to previously unmethylated DNA sequences. These enzymes play a key role in early embryogenesis, where widespread DNA methylation is essential for normal development [13]. Moreover, there is also DNMT-3L, which facilitates retrotransposon methylation through the interaction between DNMT3a and DNMT3b [14]. Aberrant DNA methylation patterns are a hallmark of cancer. Indeed, in tumor cells, the CpG methylation pattern significantly differs from that of healthy cells and is often deregulated. The promoters of oncogene are often characterized by hypomethylation in CpG islands, resulting in their overexpression and uncontrolled cell proliferation. Instead, promoters of tumor suppressor genes frequently are hypermethylated, leading to gene silencing. This prevents these genes from performing their normal role in halting tumor cell growth, thus promoting the survival and expansion of cancerous cells [15]. Moreover, aberrant DNA methylation affects genes implicated in apoptosis, migration, and DNA repair, all of which contribute to tumor progression and metastasis [16,17]. The reversible nature of DNA methylation makes it an attractive target for cancer therapy. Therefore, compounds able to inhibit DNMTs can induce the re-expression of silenced genes, such as tumor suppressor genes, thus playing an anticancer role. Indeed, several DNMT inhibitors (DNMTi) are currently being investigated within clinical trials or are already used in clinical practice against cancer, given their ability to reverse aberrant methylation patterns. By inhibiting DNMT activity, these compounds can reactivate silenced tumor suppressor genes, restoring their function and thereby inhibiting tumor growth.

However, DNMT inhibitors are often associated with toxicity, which creates a major limitation to their use in clinical settings. The Food and Drug Administration (FDA) has approved two DNMT inhibitors, Azacitidine (Vidaza) and Decitabine (Dacogen) [18], for the treatment of certain types of cancer, particularly myelodysplastic syndromes and acute myeloid leukemia. Both drugs incorporate into DNA and inhibit DNMT activity, leading to the passive demethylation of DNA during cell division. However, their effectiveness can be compromised by side effects, such as myelosuppression, resulting in neutropenia, thrombocytopenia, and other toxicities. Among the most frequent nonhematological effects are gastrointestinal toxicities, including diarrhea, vomiting, nausea, and constipation, which are usually moderate and transient and easily treated with antiemetics and antidiarrheal drugs [18,19]. In addition to synthetic DNMT inhibitors, several natural compounds, particularly flavonoids, and other plant-derived molecules, have been shown to be promising as DNMT inhibitors. These compounds may modulate DNMT activity either directly or indirectly, offering a potentially less toxic alternative to synthetic inhibitors. Further research is needed to evaluate their therapeutic potential and safety in humans [20,21].

2.2. Histone Modifications

Histones are positively charged proteins involved in DNA condensation processes into chromatin within the nucleus. Histones have tails, particularly at their N-terminus, that are subject to various post-translational modifications (PTMs) (Figure 1). These PTMs include both de/acetylation and de/methylation of the histone N-terminal, as well as phosphorylation. They directly influence chromatin structure and, thus, the accessibility of DNA to transcription factors, thereby affecting gene expression [7,22]. Histone acetylation and deacetylation are among the most studied histone modifications, playing a pivotal role in gene expression regulation.

Histone deacetylation is regulated by HDACs, divided into zinc/iron-dependent HDACs and NAD^+-dependent ones [23]. Histone acetylation, mediated by histone acetyltransferases (HATs), consists of adding an acetyl group to lysine residues at the histone N-terminal. This PTM neutralizes the positive charge of lysine, reducing the affinity between histones and the negatively charged DNA. This results in a relaxed chromatin structure and allows for transcriptional activity [24]. Instead, the histone deacetylation of *e*-amino groups of lysine residues, mediated by HDACs, inhibits gene expression by compacting chromatin structure [25]. HDACs are classified into two main categories: Zinc/iron-dependent HDACs, which require metal ions for their activity, and NAD+-dependent HDACs, which utilize the NAD+ molecule for their enzymatic function [23].

The pattern of acetylation and deacetylation is markedly different between healthy and tumor cells [26]. Tumor cells frequently exhibit aberrant deacetylation of histones, resulting in the silencing of tumor suppressor genes and promoting malignant growth. Consequently, HDAC inhibitors (HDACi) have emerged as promising agents in cancer therapy because they can restore the expression of these silenced genes. Several HDACi have shown antitumor activity in both preclinical studies and clinical trials (reviewed in [27]), including Tucidinostat for breast cancer and Entinostat for advanced epithelial ovarian cancers. Although distinct HDACi have shown promising results, they display low specificity to the different HDAC isoforms, thus leading to several adverse effects such as cardiotoxicities, as well as hematologic and gastrointestinal alterations. This highlights the need to improve the therapeutic efficacy of these drugs through the development of more selective HDACi. In this regard, natural compounds like flavonoids have been investigated for their HDAC inhibitory activity, offering a potential alternative with fewer side effects [20,28].

Histone de/methylation mostly occurs on arginine and lysine residues within the tails of histone proteins by histone methyltransferases (HMTs) and histone demethylases (HDMs) and plays a key role in modulating transcriptional activation or gene silencing in cancer [29]. HMTs add methyl groups to histones, influencing gene expression. Unlike acetylation, methylation does not affect the charge of histones but instead creates a binding platform for other regulatory proteins. HDMs remove methyl groups, reversing the effects of methylation and dynamically regulating gene expression. The functional outcome of histone methylation depends on the methylated residue and its degree of methylation (mono-, di-, or tri-methylation). For instance, H3K4 methylation is associated with active transcription, while H3K9 and H3K27 methylation are linked to gene repression and heterochromatin formation [29]. Dysregulation of histone methylation is common in cancer. For example, overexpression of EZH2, a methyltransferase that methylates H3K27, leads to the silencing of tumor suppressor genes and is implicated in the progression of several cancers. Targeting the enzymes that mediate histone methylation and demethylation has become an area of intense research in cancer therapy.

Finally, histone phosphorylation represents another epigenetic modification carried out by cell-cycle-related kinases. Histone phosphorylation is typically associated with chromatin remodeling during processes such as DNA damage repair, cell division, and apoptosis. This modification is mediated by various kinases and occurs primarily on serine, threonine, and tyrosine residues of histones. Phosphorylation can alter the interaction between histones and DNA, facilitating access to repair proteins and other factors [30]. During mitosis, phosphorylation of H3S10 is crucial for chromatin condensation.

2.3. Non-Coding RNAs

An important regulation of gene expression at the post-transcriptional level is mediated by non-coding RNAs (ncRNAs). NcRNAs are a group of transcripts that are not translated into proteins but exert specific functions at the RNA level. Among the regulatory ncRNAs, there are miRNAs, small interfering RNAs (siRNAs), long non-coding RNAs (lncRNAs), and circular RNA (cirRNA) with crucial biological functions. MiR-

NAs are small ncRNAs of 18–25 nucleotides, able to influence both mRNA stability and translation-inducing RNA-silencing through the interaction with the mRNA 3′ region. Different genes can be modulated by the same miRNAs, and several miRNAs can regulate the transcription profiles of the same gene, creating a complex regulatory network that fine-tunes gene expression [31,32]. MiRNAs play a crucial role in various physiological processes, including cell differentiation, proliferation, apoptosis, and development. Similarly, cirRNAs have emerged as important regulators of gene expression. By sequestering miRNAs, cirRNAs prevent them from binding their mRNA targets, influencing mRNA stability and translation [33]. However, their dysregulation is a characteristic feature of several diseases, including cancer. Different miRNAs, called onco-miRNAs, play a crucial role in cancer growth and progression by influencing the transcription of both oncogenes and tumor suppressor genes because their misexpression has been related to tumorigenesis processes [34]. A notable example is miR-21, one of the most studied onco-miRNAs, which is frequently overexpressed in various cancers, including breast, lung, and colon cancers. MiR-21 promotes tumor growth by targeting tumor suppressors such as *PTEN* and *PDCD4* [35].

In addition to miRNAs, lncRNAs are polyadenylated RNAs that last longer than 200 nucleotides and are able to interact with DNA, RNA, and proteins. LncRNAs participate in gene regulation at multiple levels, including chromatin remodeling, transcriptional activation/repression, splicing regulation, and mRNA stability [36]. LncRNAs are known to function as scaffolds, decoys, or guides in the regulation of gene expression. Scaffold lncRNAs serve as platforms for the assembly of protein complexes, facilitating interactions between proteins or between proteins and chromatin. Decoy lncRNAs bind and sequester transcription factors or other regulatory proteins, preventing them from interacting with their target genes. Guide lncRNAs direct chromatin-modifying complexes to specific genomic loci to influence transcription [37].

In the context of cancer, lncRNAs can act as tumor suppressors or oncogenes by regulating a wide range of pathways underlying cancer [38]. For example, HOTAIR is a well-characterized oncogenic lncRNA that is overexpressed in various cancers, including breast cancer, where it promotes metastasis by altering the epigenetic landscape through interaction with PRC2 (Polycomb Repressive Complex 2) [39]. Conversely, lncRNAs like MEG3 function as tumor suppressors by enhancing p53 activity and promoting cell cycle arrest [40].

The misregulation of ncRNAs is a common feature of cancer. CirRNAs are also strongly implicated in tumor biology. Dysregulated expression of cirRNAs may contribute to oncogenesis by acting as competitive endogenous RNAs (ceRNAs) for onco-miRNAs or by modulating signaling pathways for tumor growth and metastasis [33]. Recent studies have shown that cirRNAs play a role in the regulation of epigenetic enzymes such as HDACs and DNMTs. Behaving as competing endogenous RNAs (ceRNAs), cirRNAs bind and sequester miRNAs, thus indirectly influencing the expression and activity of these epigenetic enzymes. This regulatory mechanism is crucial in diseases such as cancer and neurodegenerative disorders, where changes in HDAC and DNMT activity lead to altered gene expression and contribute to disease progression [41]. CirRNAs can influence signaling pathways involved in cell proliferation, apoptosis, and metastasis. In particular, they regulate the expression of oncogenes and tumor suppressors through interaction with specific miRNAs, and their alteration may contribute to gene dysregulation and tumorigenesis [42]. Both miRNAs and lncRNAs are involved in the initiation, progression, and metastasis of tumors by modulating the expression of oncogenes and tumor suppressor genes [34,43]. Anyway, ncRNAs are promising targets for therapeutic intervention. Therapies targeting cirRNAs are also being explored, focusing on either restoring normal cirRNA expression or inhibiting oncogenic cirRNAs that contribute to tumor progression [44]. MiRNA mimic drugs and anti-miRNAs (antagomirs) are being developed to restore normal miRNA function in cancer cells [45]. Similarly, lncRNA-based therapies are being investigated, including approaches to silence oncogenic lncRNAs or restore the function of tumor-suppressive lncRNAs [46].

3. Phytochemicals Targeting Enzymes Involved in Epigenetic Regulation

Among phytochemicals, flavonoids are natural polyphenolic compounds characterized by the flavan backbone, which consists of two phenolic rings linked by an oxygen-containing heterocycle (Figure 2). These plant-derived metabolites, although not produced within the human body, can be consumed daily by eating different fruits, vegetables, nuts, cereal, and beverages like coffee and tea [47,48]. Flavonoids are responsible for several health benefits for mammals in the nutraceutical, medical, and pharmaceutical fields. Indeed, they have pro-apoptotic, antibacterial, antiviral, and analgesic properties through a wide range of molecular processes, like epigenetic mechanisms, cell cycle modulation, and regulation of detoxification enzyme activities [48,49]. Furthermore, flavonoids show anticancer potential by both inhibiting oncogene expression and promoting the expression of tumor suppressor genes [50,51]. This modulation also occurs at the epigenetic level, inhibiting the activities of several enzymes involved in epigenetic regulation, like HATs, HDACs, and DNMTs, and modulating the expression of ncRNAs [4]. A study also explored the correlation between cirRNAs, polyphenols, and cancer, showing that polyphenols, known for their anti-cancer properties, can influence cirRNA expression. Polyphenols such as resveratrol and curcumin have been observed to modulate certain cirRNAs, with significant effects on cancer cell behavior, including the regulation of pathways associated with cell cycle control and apoptosis. This interaction is particularly relevant for cancer therapies, where modulation of cirRNAs could improve treatment efficacy or overcome drug resistance [52,53].

Flavonoids, thanks to their lower toxicity against healthy and their few side effects compared to the traditional treatments, alone or combined with other drugs, might be an attractive strategy to treat cancer conditions, even if additional studies of safety and efficacy are required [7,20,54]. The potential of flavonoids as anti-tumor agents has been extensively studied, as demonstrated by the increasing number of publications related to the synthesis of flavonoid derivatives with improved biological activities. These modifications are aimed at enhancing their interaction with specific tumor targets such as EGFR [55] or increasing their selective toxicity towards tumor cells as opposed to healthy ones [56]. Additionally, researchers have focused on optimizing the structure of flavonoid compounds to improve their ability to selectively inhibit the activity of enzymes involved in epigenetic regulation, such as HDAC [57] and DNMT inhibitors [58].

According to their chemical structures, flavonoids can be divided into the following six groups: flavan-3-ols, flavonols, flavones, flavanones, isoflavones, and anthocyanidins (Figure 2) [59].

Flavan-3-ols (also known as catechins) are characterized by the absence of a double bond between the C2 and C3 carbons, which differentiates them from other flavonoids. Key compounds in this subclass, such as catechins and epigallocatechin gallate (EGCG), are commonly found in green tea and cocoa. Flavan-3-ols have several properties, including anticancer potential, particularly EGCG from green tea. EGCG induces apoptosis, inhibits tumor proliferation, reduces inflammation, and epigenetically modulates DNMTs, HATs, and miRNA, reactivating tumor suppressor genes [60–62]. Other flavan-3-ols, such as EGC and ECG, also show similar effects, making them promising cancer therapeutic agents [20,63,64].

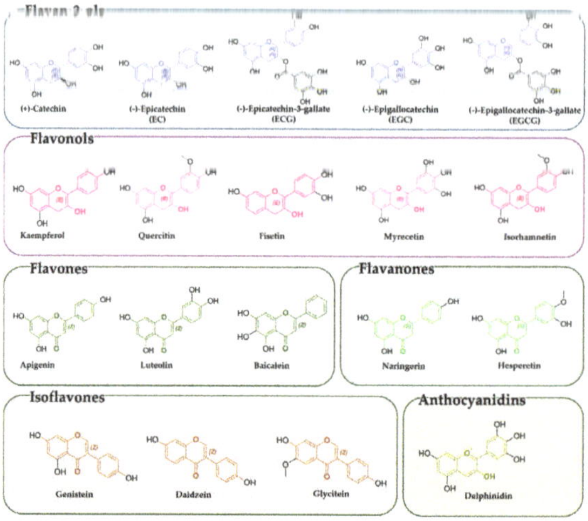

Figure 2. 2D representations of various classes of flavonoids, with the characteristic central core highlighted in different colors.

Another important subclass is the flavonols, which are structurally similar to flavones but differ by having a hydroxyl group attached to the C3 position. Quercetin and kaempferol, for example, are found in foods like onions, apples, and green tea. These compounds exhibit antioxidant, anti-inflammatory, and anticancer properties, acting through epigenetic regulation, including inhibition of DNMTs, HDACs, and HMTs and modulation of tumor suppressor genes and miRNAs [65,66].

The flavones subclass, characterized by a double bond between the C2 and C3 positions in the central ring, have anticancer pharmacological properties due to their ability to epigenetically regulate gene expression. Apigenin, for example, inhibits the activity of HDACs and DNMTs and promotes cell cycle arrest and apoptosis in cancer cells [67,68]. Luteolin has a similar role by modulating miRNA expression and inhibiting metastasis in breast cancer [4,69].

Flavanones subclass, which differs from flavones by having a saturated bond between C2 and C3, are particularly stable and primarily found in citrus fruits. Compounds such as naringenin and hesperetin inhibit the activity of HDACs and DNMTs, promoting apoptosis and antioxidant activity [70]. Also, hesperidin exhibits neuroprotective, anti-inflammatory, and cardioprotective effects, making them essential for overall health.

Isoflavones are noteworthy due to their structural similarity to human estrogens. Isoflavones, found in legumes such as soya, have phytoestrogenic and anticancer properties. They work by inhibiting DNMT enzyme activity, activating HATs, and modifying DNA methylation, with beneficial effects against prostate and breast cancer. Genistein and daidzein act as phytoestrogens, influencing hormonal health, preventing osteoporosis, and reducing the risk of certain chronic diseases. Genistein, in particular, also regulates miRNA expression and increases estrogen receptor α expression [71–74].

Anthocyanins are responsible for the red, purple, and blue colors in many fruits and vegetables. Compounds like cyanidin and malvidin, found in berries, grapes, and eggplants, are recognized for their antioxidant and anti-inflammatory properties. Anthocyanins act by activating the Nrf2-ARE pathway and promote apoptosis in cancer cells by regulating methylation and HDAC activity [75,76].

3.1. Flavan-3-ols

Flavan-3-ols are a subclass of flavonoids with two chiral carbons in the central ring. They include catechin and epicatechin (EC), which are the simplest monomers, and epicatechin-3-gallate (ECG), epigallocatechin (EGC), and epigallocatechin-3-gallate (EGCG), which are formed by esterification at position 3 of the central ring. High levels of flavan-3-ols are present in tea, wine, apples, grapes, and cacao. In particular, EC, ECG, EGC, and EGCG are the major polyphenols in green tea, whereas catechins and thearubigens are the major ones in black tea [77].

Among these flavan-3-ols, EGCG stands out, representing more than 50% of total flavan-3-ols in green tea. EGCG promotes apoptosis in cancer cells through various mechanisms, including the inhibition of the Epidermal Growth Factor receptor (EGFR) [78], the activation of caspase (specifically, caspase 9, 8, and 3), and the upregulation of pro-apoptotic proteins such as Bax. Furthermore, it induces cell cycle arrest by enhancing the expression of cell cycle regulator proteins, like p21 and p27, thereby inhibiting tumor cell proliferation while preserving healthy cells. It also reduces oxidative stress, angiogenesis, and inflammation and exhibits antiviral properties (Table 1) [20]. These multifaceted actions help to prevent the formation of new blood vessels, which are crucial for tumor growth.

At the epigenetic level, it has been demonstrated that EGCG inhibits the activity of DNMTs (particularly DNMT1) in several tumor cell lines, including breast, colon, prostate, and esophageal tumor cells, thereby promoting the reactivation of antioxidant enzymes. Breast cancer cells (but also lung, oral cavity, liver, and thyroid tumor cells) treated with EGCG showed a reduced level of telomerase, a key enzyme involved in cellular immortality, due to the decreased methylation of *human telomerase reverse transcriptase (hTERT)* promoter, thus leading to a cell growth inhibition and an enhancement in the sensitivity to chemotherapy [61]. In both estrogen receptor (ER)-positive (MCF-7) and ER-negative (MDA-MB-231) breast cancer cells, EGCG also increases the binding of hTERT repressor E2F-1 to its promoter [79]. Moreover, by inhibiting DNMT activity, EGCG demethylates *signal peptide CUB-EGF-domain-containing protein 2 (SCUBE2)* promoter, revoking the epithelial-mesenchymal transition and thus easing tumor progression [80]. Along with the modulation of DNMT activity, EGCG also regulates HAT activity, thus playing a crucial role in chromatin reorganization. This regulation can lead to the re-expression of various tumor suppressor genes that are often silenced in cancer cells [60]. "In vivo" studies also demonstrated that EGCG could modulate miRNAs in cancers such as hepatocellular carcinoma and gastric cancer, targeting critical genes like *c-Kit*, *Bcl2*, *E2F*, and *RAS*. The dysregulation of these genes is frequently associated with cancer progression, and EGCG's ability to modulate their expression through miRNAs underscores its potential as a therapeutic agent [62]. Regulation of DNMTs, HATs, and miRNAs alters the expression of different tumor suppressor genes. Taken together, data obtained from both in vitro and in vivo studies suggest that EGCG has proapoptotic, antiangiogenic, and, therefore, anti-cancer potential through its action as a broad epigenetic modulator.

Other flavan-3-ols found in green and black tea, including gallocatechin, EC, ECG, EGC, and thearubigens, showed similar effects compared to EGCG in promoting tumor cell apoptosis, inhibiting growth and reactivating tumor suppressor genes in different cancer cell lines (Table 1) [7]. For instance, EGC showed the ability to inhibit prostate cancer cell growth and induce apoptosis by regulating several signaling pathways [7,20].

Table 1. Epigenetic effects of flavan-3-ols and their role in pathological conditions.

Compounds FLAVAN-3-OLS	Epigenetic Effects	High Content Food Quantity mg/100 g [81]	Role in Pathological Conditions
(−)-Epigallocatechin 3-gallate (EGCG) SubClass: Flavan-3-ols Class: Flavonoids Family: Polyphenols	Inhibition of DNMT activity (mainly DNMT1) in breast, colon, prostate, and esophageal cancer cells [61] Inhibition of *hTERT* promoter methylation in breast, lung, oral cavity, liver, and thyroid tumor cells [61] Demethylation of *SCUBE2* promoter, inhibiting the epithelial-mesenchymal transition [80] Modulation of HAT activity [60] Regulation of miRNA expression in hepatocellular carcinoma and gastric cancer in vivo, targeting *c-Kit*, *Bcl2*, *E2F*, and *RAS* [62]	Carob Flour (109.40), Green tea (70.20), White tea (42.45), Oolong tea (34.48), Black tea (9.36), Nuts pecans (2.30), Fuji Apple with skin (1.93), Hazelnut (1.06), Cranberry (0.97), Blackberry (0.68), Raspberry (0.54), Plum (0.48), Pistachio nuts (0.40), Peach (*Prunus persica*) (0.30), Granny Smith Apple, with skin (0.24), Golden Delicious Apple, with skin (0.19), Pear (0.19), Avocado (0.15), Red Delicious Apple, with skin (0.13), Gala Apple, with skin (0.11), Strawberry (0.11)	Reduction in oxidative stress, angiogenesis, and inflammation in tumors [82] Cancer prevention [83,84] Induction of apoptosis by activating caspase 9, 8, and 3 and upregulating Bax protein in cancer cells [85] Induction of cell cycle arrest by enhancing p21 and p27 expression in tumor cells [85]
(−)-Epicatechin (EC), Epicatechin-3-gallate (ECG) Epigallocatechin (EGC) SubClass: Flavan-3-ols Class: Flavonoids Family: Polyphenols	Inhibition of DNMT activity [64] Regulation of HAT activity [20] Modulation of miRNA expression in cancer [63]	**(−)-Epicatechin:** Cocoa powder (158.3) Baking chocolate (141.83), Cacao seed (99.18), Grape seed (93.31), Dark chocolate (84.80), Soybean, mature seed (37.41), Broad beans, immature seed (*Vicia faba*) (28.96), Apple, skin only (28.73), Blueberry (25.66), Green tea, Quingmao (20.80), Sour cherry juice (12.97), Milk chocolate (10.88), Alcoholic beverage, red wine (10.66–3.79), Apple (9.83–4.09), Grape, black (*Vitis vinifera*) (8.68), Cherry (*Prunus avium*) (5.00), Apricot (*Prunus armeniaca*) (4.74), Apple juice (4.70), Blackberry (4.66), Cranberry (4.37), Peache, white, (4.09), Pear, raw (*Pyrus communis*) (3.76), Raspberry (*Rubus* spp.) (3.52), Plum (3.20), Nectarine (*Prunus persica*) (3.06–2.34), Buckwheat flour, whole-groat (3.02), Cranberry bush berries (*Viburnum opulus*) (2.69), Oolong tea (2.54), Vinegar, wine, red (2.20),	Antioxidative, anti-inflammatory, and anticancer effects [20]

Table 1. *Cont.*

Compounds FLAVAN-3-OLS	Epigenetic Effects	High Content Food Quantity mg/100 g [81]	Role in Pathological Conditions
(−)-Epicatechin (EC), Epicatechin-3-gallate (ECG) Epigallocatechin (EGC) SubClass: Flavan-3-ols Class: Flavonoids Family: Polyphenols	Inhibition of DNMT activity [64] Regulation of HAT activity [20] Modulation of miRNA expression in cancer [63]	Black tea (2.13), Arctic bramble berries (1.80), Grape, white (1.70), Strawberry tree fruit (*Arbutus unedo*) (1.56), Pistachio nuts (*Pistacia vera*) (0.83), Nut pecans (*Carya illinoinensis*) (0.82), Vinegar, cider (0.82), Cloudberry (0.80), Kiwifruit, gold (*Actinidia chinensis*) (0.64), Nut almonds (*Prunus dulcis*) (0.60), Alcoholic beverage, white wine (0.55), Medlar (*Mespilus germanica*) (0.53), Rhubarb (*Rheum rhabarbarum*) (0.51), Fig (*Ficus carica*) (0.50), Currants, European black (*Ribes nigrum*) (0.47) **Epicatechin-3-gallate and Epigallocatechin:** Cocoa powder (196.43), Cacao Bean (99.18), Carob Flour (30.06), Green tea (8.33), White tea (8.35), Oolong tea (2.54), Black tea (2.11), Nut almond (0.60), Nut pecan (0.82)	Antioxidative, anti-inflammatory, and anticancer effects [20]
(+)-Catechin, (+)-Gallocatechin SubClass: Flavan-3-ols Class: Flavonoids Family: Polyphenols	Inhibition of DNMT activity [64] Regulation of HAT activity [20] Modulation of miRNA expression in cancer [63]	**(+)-Catechin:** Blueberry (98.47), Cacao Bean (88.45), Grape seed (74.63), Tea green, Quingmao (67.60), Cacao powder (64.82), Carob Fluor (50.75), Blackberry (37.06), Cranberry (29.04), Chocolate, dark (24.20), Cocoa powder (21.51), Plum Black Diamond(17.22), Broad bean (*Vicia faba*) (14.29), Peache white (12.25), Grape Black (10.14), Nectarine (*Prunus persica*) (7.58), Apple skin only (7.40), Nut Pecan (7.24), Alcoholic beverage, wine red (from 6.21 to 7.70) Strawberry (6.70), Chard, red leaf (*Beta vulgaris*) (6.70), Apple (6.67–0.75) Banana (6.10), Blueberry (5.29), Bean, pinto (*Phaseolus vulgaris*) (5.07). Cider (4.85), Green tea (4.47), Cherry (4.36), Milk chocolate (4.16), Grape, white (3.73), Apricot (3.67),	Antioxidative, anti-inflammatory, and anticancer effects [20]

Table 1. *Cont.*

Compounds FLAVAN-3-OLS	Epigenetic Effects	High Content Food Quantity mg/100 g [81]	Role in Pathological Conditions
(+)-Catechin, (+)-Gallocatechin SubClass: Flavan-3-ols Class: Flavonoids Family: Polyphenols	Inhibition of DNMT activity [64] Regulation of HAT activity [20] Modulation of miRNA expression in cancer [63]	Vinegar, wine (3.60), Pistachio nuts, (3.57), Jujube (*Ziziphus jujuba*) (3.21), Strawberry (3.11), Juice sour cherry (3.18), Plum (2.89), Barley, hulled (*Hordeum vulgare* L.) (2.39), Arctic bramble berry (2.30), Rhubarb (*Rheum rhabarbarum*) (2.17), Mango (1.72), Gooseberry (1.67), Raspberry (1.31), Currants, red (1.27), Apple juice (1.25), Black tea (1.51), Nut, hazelnut or peanut (*Corylus* spp.) (1.19), Quinces (0.75) **(+)-Gallocatechin:** Cacao Bean (8262.00), Broad bean, immature seeds (*Vicia faba*) (4.15), Strawberry tree fruit (*Arbutus*) (1.60), Green Tea (1.54), Currants, red (1.28), Black tea (1.25), Pomegranate, raw (0.17), Persimmon (0.17), Star apple (0.53)	Antioxidative, anti-inflammatory, and anticancer effects [20]
Thearubigins SubClass: Flavan-3-ols Class: Flavonoids Family: Polyphenols	Inhibition of DNMT activity [86]	Black tea (81.30)	Antioxidative, anti-inflammatory, and anticancer effects [87,88]

DNMT: DNA methyltransferase; HAT: histone acetyltransferase; miRNA: microRNA.

3.2. Flavonols

Flavonols, the most common flavonoids present in food, are characterized by the presence of a 3-hydroxyflavone backbone. The different positions of the phenolic -OH groups give rise to their diversity. Kaempferol, quercetin, fisetin, and myricetin are the main plant-derived flavonols, which can be found in different vegetables and fruits like strawberries, apples, asparagus, and onions but also in wine [89].

Kaempferol, found in *Zingiberaceae* vegetables and fruits, including strawberries, hop, tomatoes, and grapefruit, shows several physiological functions, such as inhibition of fat formation, nervous system, and heart protection, antioxidant, antiallergic, and anticancer effects (Table 2) [87]. Due to its capability to regulate gene expression through epigenetic modification, kaempferol has shown promising results as an anticancer agent. Specifically, it showed the ability to inhibit the activity of different human HDAC enzymes and promote hyperacetylation of histone H3 in hepatoma and colon tumor cell lines [90]. Moreover, in colorectal cancer cell lines, kaempferol can bind DNMT1 and downregulate the methylation of tumor suppressor gene *DACT2*, thus inhibiting the Wnt/β-catenin pathway [90]. In gastric tumor cells, kaempferol promotes autophagic cell death, whereas, in lung tumors, it inhibits cancer cell proliferation and induces cell apoptosis by regulating miR-340 expression [65].

Quercetin, the predominant flavonol present in buckwheat and citrus fruits, has antioxidant, anticancer, anti-inflammatory, antiviral, antibacterial, neuroprotective, and hypolipidemic impacts (Table 2). It reduces the activity of DNMTs, HMTs, and HDACs in cervical tumor cells, along with the global concentrations of DNA methylation, whereas it activates HATs, thus restoring the expression of different tumor suppressor genes [66]. In pancreatic tumor cells, it inhibits cell proliferation by promoting let-7c and, in turn, Numb1 and Notch, and induces apoptosis. It has also been related to anti-inflammatory properties and angiogenesis inhibition [91]. In triple-negative breast cancer, quercetin can control the beta-catenin signal, leading to the reversion of epithelial-mesenchymal transition and the increase in *BRCA1* expression [92]. Combined with curcumin, quercetin further increases *BRCA1* levels by enhancing the histone H3K9 acetylation of the *BRCA1* promoter [93]. Moreover, quercetin regulates different miRNAs, like miR-145 and miR-146a, and regulatory axes including miR-22/WNT1/β-catenin, miR-197/IGFBP5, TP53/miR-15/miR-16, miR-16/HOXA10 and p53/miR-34a/SIRT1 in several tumor cell lines [7]. Inhibition of both proliferation and invasion of breast cancer cells is mediated by quercetin through the upregulation in miR-146a expression and the subsequent activation of caspase-3 and Bax, thus resulting in mitochondrial-dependent apoptotic pathway triggering [94].

Fisetin is a flavonol present in onions, apples, strawberries, tea, and wine. It can interfere with cancer cell growth and cell cycle progression and promote PARP cleavage and apoptosis through DNMT inhibition (Table 2) [95]. Furthermore, fisetin regulates the Bcl-2 family protein expression and inhibits signaling pathways in which p38 MAPK, ERK1/2, or NF-kB are involved [96]. Fisetin, as well as quercetin, activates sirtuins (SIRTs) [97].

Myricetin is contained in berries, grapes, tea, and red wine. It plays a crucial role in the prevention of cardiovascular disorders, has anticancer ability, and reduces blood lipid levels, blood pressure, and diabetes and bacteriostasis complications (Table 2) [4]. It is considered the most potent flavonol, and it inhibits DNMT activity in a concentration-dependent manner [20]. Furthermore, myricetin indirectly modulates deacetylation, activating HDAC SIRT1, such that it can induce *HIF-1α* expression and suppress *cMyc* and *β-catenin* expression [98].

Table 2. Epigenetic effects of flavonols and their role in pathological conditions. Content in food [81].

Compounds FLAVONOLS	Epigenetic Effects	High Content Food Quantity mg/100 g	Role in PATHOLOGICAL Conditions
Kaempferol SubClass: Flavonols Class: Flavonoids Family: Polyphenols	Inhibitory activity towards HDAC enzymes in human hepatoma and colon cancer cell lines [65] Promotion of H3 histone hyperacetylation in hepatoma and colon tumor cells [65] Binding to DNMT1 in colorectal cancer cell lines, thus downregulating *DACT2* methylation and inhibiting Wnt/β-catenin pathway [90] Regulation of miR-340 expression in lung tumor [7,66]	Caper (259.19), Caper canned (131.34), Kale (46.80), Arugula (34.89), Mustard green (38.30), Ginger (33.60), Watercress (23.03), Radish (21.85), Chia seed (12.30), Chives (10.0), Chard (9.20), Collards (8.74), Broccoli (7.84), Lovage (*Levisticum officinale*) (7.0), Fennel leaves (6.50), Dried Goji berries (6.20), Cherry powder (5.14), Thistle (3.80), Chicory green (2.45), Corn poppy (*Papaver rhoeas*) (2.30), Blueberry (1.66), Black Tea (1.41), Asparagus (1.39), Bee Pollen granules (1.12), Acerola (1.05), Green Tea (1.00), Ribes Nigrum (0.71), Red onion (0.70), Elderberry (0.51), Strawberry (0.49), Carob Flour (0.44), Grapefruit (0.40), Lingonberry (0.38), Blackberry (0.27), Apple (0.14), Cranberry (0.12)	Inhibition of fat formation, nervous system protection, and heart protection [87] Antioxidant, antiallergic and anticancer effects [87] Promotion of autophagic cell death in gastric tumor cells; inhibition of cancer cell proliferation and induction of cell apoptosis in lung tumor [7,66]
Quercetin SubClass: Flavonols Class: Flavonoids Family: Polyphenols Current dosage: 500 mg quercetin/day, 12 weeks Mean intake by fruits or vegetables: 5–500 mg/day IC_{50}: 1.6 μmol/L	Reduction in the activity of DNMTs, HMTs, and HDACs, and activation of HATs in cervical tumor cells [66] Regulation of different miRNAs (like miR-145 and miR-146) and regulatory axes (such as miR-22/WNT1/β-catenin and p53/miR-34a/SIRT1) in several tumor cell lines [7]	Caper (233.84), Caper canned (172.55), Levisticum (170.0), Juice concentrate Elderberry (108.16), Radish (70.37), Arugula (66.19), Corianders leaves (52.90), Peppers yellow wax (50.63), Fennel leaves (48.80), Juniper berries ripe (46.16), Red onion ((39.21), Watercress (29.99), Elderberries (26.77), Corn poppy (*Papaver rhoeas*) (26.30), Carob flour (38.78), Onion cooked boiled drained (24.36), Kale (22.58), Been pollen (20.95), Apple skin (19.36), Chia seed (18.20), Cherries powder (17.44), Thistle (16.50),	Antioxidant, anticancer, anti-inflammatory, antiviral, antibacterial, neuroprotective and hypolipidemic impact [66] Inhibition of cell proliferation by promoting let-7c, Numb1, and Notch, and induction of cell apoptosis in pancreatic tumor cells [91] Regulation of β-catenin signal and promotion of *BRCA1* expression in triple-negative breast cancer [92]

Table 2. *Cont.*

Compounds FLAVONOLS	Epigenetic Effects	High Content Food Quantity mg/100 g	Role in PATHOLOGICAL Conditions
Quercetin SubClass: Flavonols Class: Flavonoids Family: Polyphenols Current dosage: 500 mg quercetin/day, 12 weeks Mean intake by fruits or vegetables: 5–500 mg/day IC_{50}: 1.6 µmol/L	Reduction in the activity of DNMTs, HMTs, and HDACs, and activation of HATs in cervical tumor cells [66] Regulation of different miRNAs (like miR-145 and miR-146) and regulatory axes (such as miR-22/WNT1/β-catenin and p53/miR-34a/SIRT1) in several tumor cell lines [7]	Arugula (15.16), Asparagu cook boiled-drained (15.16), Cranberries (14.84), Asparagus (13.98), Goji berry dried (13.60), Lingonberries (13.30), Plums black Diamond (12.45), Lovage Mustard green (8.80), Blueberries (7.67), Chard (7.50), Chicory green (6.49), Chives (4.77), Acerola (4.74), Ribes Nigrum (4.45), Apple (4.01), Blackberries (3.58), Broccoli (3.26), Collards (2.57)	Antioxidant, anticancer, anti-inflammatory, antiviral, antibacterial, neuroprotective and hypolipidemic impact [66] Inhibition of cell proliferation by promoting let-7c, Numb1, and Notch, and induction of cell apoptosis in pancreatic tumor cells [91] Regulation of β-catenin signal and promotion of *BRCA1* expression in triple-negative breast cancer [92]
Fisetin SubClass: Flavonols Class: Flavonoids Family: Polyphenols IC_{50}: 3.5 µmol/L	Inhibition of DNMT activity [95] Activation of SIRTs [97]	Strawberry (16.0), Apple (2.69), Persimmons (1.05), Onion (0.48), Grapes (0.39), Kiwi (0.2)	Interference with cancer cell growth, cell cycle progression, and promotion of cell apoptosis [95] Regulation of the Bcl-2 family protein expression and inhibition of signaling pathways in which p38 MAPK, ERK 1/2, or NF-kB are involved [96]
Myricetin SubClass: Flavonols Class: Flavonoids Family: Polyphenols IC_{50}: 1.2 µmol/L	Inhibition of DNMT activity in a concentration-dependent manner [20] Indirect modulation of deacetylation to induce *HIF-1α* expression and suppress *cMyc* and *β-catenin* expression [98]	Juice concentrate black currant (20.85), Fennel leaves (19.80), Goji berry dried (11.40), Arugula (7.92), Carob flour (6.73), Cranberries (6.63), Thistle (3.60), Ribes Nigrum (6.18), Been pollen (3.34), Swiss Chard (2.20), Red onion (2.16), Blueberries (1.30), Corn poppy (*Papaver rhoeas*) (1.10), Tea green (1.00), Blackberry (0.67), Black tea (0.45), Strawberry (0.35), Broccoli (0.06), Apple (0.01), Plums Black Diamond (0.01)	Prevention of cardiovascular disorders, anticancer ability, and reduction in blood lipid levels, blood pressure, and both diabetes and bacteriostasis complications [4]

HDAC: histone deacetylase; DNMT: DNA methyltransferase; HMT: histone methyltransferase; HAT: histone acetyltransferase; miRNA: microRNA; SIRT: sirtuin.

Isorhamnetin is another type of flavonol, specifically an O-methylated flavonol, predominantly found in the fruits and leaves of several plants used in traditional herbal medicine. This compound has garnered attention for its potential benefits in managing diabetes and for its protective effects on the kidneys, which may involve the modulation of epigenetic regulators [99]. However, specific evidence linking isorhamnetin directly to the modulation of epigenetic genes and its direct implications for cancer treatment or prevention may not be well established.

3.3. Flavones and Flavanones

Flavones, characterized by the backbone of 2-phenylchromen-4-one, are present in herbs like parsley, celery, thyme, oregano, and cereal species. Among the most common flavones are apigenin, luteolin, baicalein, and tangerine [100].

It has been reported that apigenin, contained in chamomile, parsley, and celery, modifies chromatin architecture, leading to inhibition of class I HDACs activity in human prostate tumor cells, causing cell cycle arrest and apoptosis, ultimately reducing cell growth (Table 3). Apigenin blocks the proliferation and growth of breast cancer cells through *p21* transcription associated with the acetylation of histone H3 [68,101]. Furthermore, apigenin, as well as luteolin, inhibits the activity of 5-cytosine DNMTs, including DNMT1, -3a, and -3b, in skin epidermal cells, along with the silencing of *nuclear factor erythroid 2-related factor 2* (*Nrf2*), inducing antioxidant and anticancer mechanisms [67]. Lastly, it induces the expression of both miR-16 and miR215-5p, exerting an anticancer role in glioma and colon cancer [7].

Luteolin, commonly found in peppers, peppermint, rosemary, carrots, and thyme, has several pharmacological properties, and it is used against diabetes and Alzheimer's disorder (Table 3) [4]. It epigenetically regulates the expression of MMP9 and β-catenin, inhibiting proliferation and metastasis in breast tumors and inverting the epithelial-to-mesenchymal transition, respectively [102]. Luteolin also reduces the expression of DNMT1 and HDACs in a dose-dependent manner and downregulates calpain and UHRF1 in colorectal tumor cells, thus promoting apoptosis [69]. In prostate cancer cells, luteolin modifies the acetylation pattern of the gene promoter histone and inhibits the expression of 22 key genes involved in the cell cycle pathway [103]. Moreover, it regulates the expression of several miRNAs in different tumor cell lines [4].

Flavanones, unlike flavones, contain a saturated double bond between positions 2 and 3 (Figure 2). These compounds can be mainly found in the peel of citrus fruits and include a variety of phytochemicals, such as hesperetin, naringenin, naringin, and eriodictyol [104].

Hesperetin is a citrus flavanone that exerts its epigenetic regulation, diminishing the methylation of histone H3K79 to reduce metastasis recurrence (Table 3) [105].

Naringenin is a common citrus flavonoid and represents the aglycone form of naringin. Among their pharmacological properties, they have antidiabetic, liver- and heart-protective, antiviral, antioxidant, and antitumor effects. It is also used in sepsis treatment (Table 3) [106]. In neuroblastoma cancer cells, it promotes apoptosis, preserving the neighboring healthy cells due to its inhibition of HDAC activity. Moreover, both naringenin and hesperetin inhibit the activity of DNMTs [70]. Lastly, in human colon adenocarcinoma, naringenin can reduce the expression of several miRNAs, including miR-17-3p and miR-25-5p, involved in anti-inflammatory and antioxidant mechanisms, resulting in the upregulation of both glutathione peroxidase (GPX) 2 and manganese superoxide dismutase (SOD) [107].

Table 3. Epigenetic effects of flavones and flavanones, and their role in pathological conditions.

Compounds FLAVONES and FLAVANONES	Epigenetic Effects	High Content Food Quantity mg/100 g	Role in Pathological Conditions
Apigenin SubClass: Flavones Class: Flavonoids Family: Polyphenols Dosage: 500–1000 mg	Inhibition of class I HDAC activity in prostate tumor cells [20,68] Acetylation of H3 histone in breast tumor cells [101] Inhibition of the activity of 5-cytosine DNMTs and silencing of *NRF2* in skin epidermal cells [67] Induction of the expression of miR-16 and miR215-5p in glioma and colon cancer [7]	Parsley dried (4503.50), Celery seed (78.65), Vine Spinach (62.20), Kumquat (*Citrus japonica*) (21.87), Celery heart green (19.1), Red onion (0.24)	Inhibition of cell growth and promotion of cell cycle arrest and apoptosis in human prostate cancer cells [108] Blocking of the proliferation and growth of breast tumor cells through *p21* transcription [101] Antioxidant and anticancer effects [67]
Luteolin SubClass: Flavones Class: Flavonoids Family: Polyphenols	Inhibition of DNMT1 and HDAC enzymes in a dose-dependent manner in colorectal tumor cells [69] Modification of the acetylation pattern of the gene promoter histone in prostate cancer cells [103] Regulation of the expression of different miRNAs in several tumor cell lines [4]	Juniper berries ripe (*Juniperus communis* L.) (69.05), Parsley dried (19.75), Red onion (0.16)	Use against diabetes and Alzheimer's disorder [4] Inhibition of proliferation and metastasis in breast tumors by regulating MMP9 expression and reversal of the epithelial-to-mesenchymal transition through the regulation of β-catenin expression [102] Promotion of apoptosis in colorectal tumor cells [69]
Hesperetin SubClass: Flavanones Class: Flavonoids Family: Polyphenols	Reduction in methylation of the histone H3K79 in gastric cancer cells [105] Inhibition of DNMT activity [70]	Grapefruit (1.50), Citrus paradise (1.50)	Anticancer effects by reducing metastasis recurrence [20]
Naringenin SubClass: Flavanones Class: Flavonoids Family: Polyphenols Dose: 500–1000 mg/day	Inhibition of DNMT and HDAC activity [70] Reduction in the expression of several miRNAs involved in anti-inflammatory and antioxidant mechanisms (including miR-17-3p and miR-25-5p) in human colon adenocarcinoma to upregulate GPX and SOD expression [107])	Grapefruit (53.0), Citrus paradise (53.0)	Antidiabetic, liver and heart-protective, antiviral, antioxidant, and antitumor effects; use in sepsis treatment [106] Promotion of apoptosis in neuroblastoma cancer cells [70]

HDAC: histone deacetylase; DNMT: DNA methyltransferase; miRNA: microRNA.

3.4. Isoflavones and Anthocyanidins

Isoflavones are characterized by a 3-phenylchromen-4-one skeleton in which the hydrogen at position 3 is replaced by phenyl (Figure 2). They are mainly contained in legumes, soy foods, soybeans, and fava beans. Genistein, daidzein, and glycitein are the main known isoflavonoids [109]. In particular, genistein and daidzein show a chemical and conformational structure similar to 17β-estradiol, thus allowing their interaction with estrogen receptors and their consequent phytoestrogenic property [110]. In addition to their phytoestrogenic role, isoflavones can also be employed against cardiovascular disorders, tumors, or other diseases associated with hormone dysfunction [4]. Indeed, several studies conducted in Asian countries have demonstrated that isoflavone consumption diminishes breast tumor risk, whereas no association has been detected within Western countries [111,112]. Considering that isoflavone intake is greater in Japan, Korea, and China compared to other countries, this discrepancy could be related to the ingested dose, which represents a pivotal factor able to influence their effect on tumor prevention and mortality [113].

Genistein, mainly present in soy protein and soybeans, is the flavonoid with the strongest antitumor and anti-proliferative effect against several tumors both in vitro and in vivo, thanks to its chemopreventive properties (Table 4) [114]. In prostate cell lines, genistein reduces DNA methylation at different gene promoters (such as *GSTP1* and *BRCA1*), along with a genome-wide modulation of DNA methylation patterns [115]. It promotes the inactivation of DNMT1, DNMT3a, and DNMT3b, leading to the increased expression of several tumor suppressor genes [116]. Its potential is not only related to DNMT enzyme inactivation but also to post-translational histone modifications, including HAT activation, histone demethylation, and SIRT inhibition [20]. In breast cancer, genistein enhances the ERα expression, and its combination with trichostatin A, a known HDACi, further increases ERα expression by promoting histone acetylation and inhibiting DNMT1 expression [117]. In ER-negative breast tumors, genistein reduces *BRCA1* methylation, thus resulting in *BRCA1* activation, which is associated with aryl hydrocarbon receptor (AhR) action inhibition [118]. Furthermore, it modulates the expression of several miRNAs in different cancers, including the activation of the miR-34a/RTCB axis in both head and neck tumors, which results in reactive oxygen species (ROS)-mediated apoptosis and reduction in epithelial-mesenchymal transition, counteracting tumor growth and cell cancer proliferation [119].

Daidzein, like genistein, can inhibit DNMT enzymes and activate HAT activity (Table 4) [67]. Specifically, daidzein inhibits the methylation of tumor suppressor gene promoters and genes involved in p53 and NF-kB pathways within human prostate tumor cells [4]. In breast tumors, combined with genistein, daidzein decreases the methylation of both *BRCA1* and *BRCA2* promoters, thus enhancing the BRCA1 and BRCA2 protein levels, and inhibits the MeCP2 expression associated with the entire genome methylation status [118].

Anthocyanidins, based on the flavylium cation, are water-soluble molecules responsible for the red and blue colors of plants. They are commonly present in grapes, cherries, and berries. They are flavonoid aglycones, and their glycosides are known as anthocyanins. Anthocyanidins exert beneficial effects, like anticancer, antioxidative, and vision-protective functions, and they are also used in the prevention and treatment of both type-2 diabetes and obesity. Delphinidin and cyanidin are the most common anthocyanidins contained in vegetables and fruits like blueberry, but there are also pelargonidin and malvidin [4,120].

Delphinidin is responsible for epigenetic regulation in different cancers, activating the Nrf2-ARE pathway due to the demethylation of several CpG sites within the *Nrf2* promoter (Table 4) [76]. In prostate tumor cells, it induces an enhancement in caspase-3, -7, and -8 expression, together with a boosted HDAC (mainly class I HDACs) activity. Moreover, delphinidin upregulates different pro-apoptotic genes, whereas downregulates anti-apoptotic ones [75].

Table 4. Epigenetic effects of isoflavones and anthocyanidins, and their role in pathological conditions.

Compounds ISOFLAVONES and ANTHOCYANIDINS	Epigenetic Effects	High Content Food Quantity mg/100 g	Role in Pathological Conditions
Genistein SubClass: Isoflavones Class: Flavonoids Family: Polyphenols (phytoestrogen group)	Activation of HAT activity and histone demethylation [73,74] Inhibition of DNMT enzymes [72] Modulation of several miRNAs in cancer [119]	Soy flour (89.42), Instant beverage Soy powder (62.18), Soy protein drink (42.91), Soy milk (42.85), Soybean (39.57), Natto (37.66), Tempeh (36.15), Miso (23.24), Green soybeans (22.57), Soy fiber (21.68) Soybean mature seeds (18.77), Tofu (16.01) Soy Yogurt (16.59)	Antitumor and antiproliferative effects [114] Suppressor of oncogene antiestrogenic activity [121] Induction of ROS-mediated apoptosis and reduction in epithelial-mesenchymal transition in both head and neck tumors, thus inhibiting tumor growth and proliferation [119]
Daidzein SubClass: Isoflavones Class: Flavonoids Family: Polyphenol (phytoestrogen group)	Activation of HAT activity [122] Inhibition of DNMT enzymes [115]	Soy flour (67.69), Soy Milk, dried (40.85), Instant beverage soy powder (40.07), Natto (33.22), Soy protein drink (27.98), Tempeh (22.66), Soybeans (21.75), Green soybeans (20.34), Soy fiber (18.80), Miso (16.43), Tofu (15.59), Soy Yogurt (13.77), Soybean mature seeds (12.86)	Antiestrogenic activity [123]
Delphinidin Subclass: Anthocyanidins Class: Flavonoids Family: Polyphenols	Inhibition of DNMT enzymes, which causes the activation of the Nrf2-ARE pathway [76] Increase in HDAC (in particular class I HDACs) activity [75]	Bilberry (97.59), Black currants (89.62), Eggplant (85.69), Blueberries (35.43), Black beans (18.50), Jambul (17.73), Red currants (9.32), Cranberries (7.67), Bananas (7.39), Pecans nuts (7.28), Jostaberry (6.61), Red onion (4.28), Red grapes (2.27)	Anticancer, antioxidative, and vision-protective functions; use in the prevention and treatment of type-2 diabetes and obesity [120] Upregulation of different pro-apoptotic genes and downregulation of anti-apoptotic ones [75]

HAT: histone acetyltransferase; DNMT: DNA methyltransferase; miRNA: microRNA; ROS: oxygen reactive species; HDAC: histone deacetylase.

3.5. Other Compounds

In addition to flavonoids, there are several other phytochemicals present in fruits, vegetables, spices, and beverages, able to exert epigenetic modifications, thus resulting in different effects, like anti-inflammatory, anticancer, and/or anti-microbial impact. Along with flavonoids, these phytochemicals might have the potential to be used as epigenetic drug targets against cancer and other pathological diseases, thanks to the reversibility of the induced epigenetic alterations [20,124].

Among natural compounds, curcumin, found in the plants of the *Curcumae longae* species, is a diferuloylmethane responsible for several beneficial effects on human health, like anticancer and anti-inflammatory properties. It is able to regulate a wide range of signaling pathways associated with proliferation, migration, apoptosis, inflammation, and metastasis (Table 5) [125]. Epigenetically, curcumin inhibits the activity of DNMT1 but also regulates both HAT and HDAC activities. Indeed, according to the different types of tumors, curcumin exerts diversified epigenetic effects [126]. It has been demonstrated to inhibit the HAT activity of p300/CBP, blocking the acetylation of both histones and non-histone proteins, including p53. Moreover, curcumin downregulates the activity of NF-kB and Notch1 in Raji cells through the inhibition of p300/CBP, HDAC1, and HDAC3, subsequently repressing cell proliferation [124]. Lastly, anticancer properties associated with curcumin are also related to its capacity to regulate both miRNA and lncRNA expression in tumor cells. Human pancreatic carcinoma cell lines treated with curcumin showed a changed expression of 29 miRNAs [127]. Instead, curcumin downregulates Bcl-2 expression through the induction of miR-15a and miR-16 in breast cancer, it enhances the expression of miR-34a in gastric cancer [128], and it inhibits the expression of miR-21 directly associated with tumor cell invasion and metastasis in human colon cancer cells [129]. However, curcumin is characterized by low bioavailability and poor pharmacokinetics. Therefore, the study and development of innovative curcumin analogs, such as refined preparations, to achieve better bioavailability and pharmacokinetics are still ongoing [130].

Folic acid, contained in beans, green vegetables, grains, cereals, and pasta, is included in the folate family and belongs to the methyl-metabolism pathway (Table 5). In contrast to naturally occurring folate (vitamin B9), folic acid is a synthetic metabolite that is ingested as a dietary supplement or in food. In order to obtain the biologically active 5-methyltetrahydrofolate, folic acid needs to be reduced to dihydrofolate through the action of dihydrofolate reductase enzyme within the liver and subsequently to tetrahydrofolate [131]. Hepatic DNA methylation status can be modified by a deficit of folic acid, which might be related to liver (but also breast, lung, brain, and colorectal) cancer [124].

Indole-3-carbinol (I3C) is a product of glucosinolate that is commonly contained in vegetables like cauliflower, cabbage, broccoli, and radish (Table 5). I3C can also be transformed into diindolylmethane (DIM). I3C and DIM can regulate different nuclear receptor-mediated signaling and kinases, thus promoting apoptotic mechanisms in several tumor cell lines [132]. Furthermore, DIM provokes proteasomal degradation of class I HDAC enzymes, leading to regulation of p21 and p27 expression and consequent inhibition of tumor growth [133]. DIM also regulates the expression of miRNAs, including miR-let-7e, miR-let-7b, miR-200b, and miR-200c [124].

Among the isothiocyanates, phenethyl isothiocyanate (PEITC) is found in cruciferous vegetables, and its main feature is its chemopreventive properties (Table 5). Indeed, PEITC can inhibit the proliferation and growth of different tumors and promote apoptotic mechanisms in several cancer cells [134]. In human prostate tumor cells, it has been demonstrated that PEITC inhibits HDAC activity through the enhanced de-methylation of the Ras-association domain family 1 isoform A (RASSF1A), whereas it induces a targeted histone acetylation and methylation [135]. Instead, in breast cancer cells, PEITC can target breast cancer stem cells through the epigenetic reactivation of *cadherin 1*, a tumor suppressor gene, by inhibiting DNMT and HDAC activity, thus demethylating its promoter site [136]. Moreover, PEITC regulates the expression of several miRNAs, like miR-125b, miR-26a, miR-192, miR-99, and miR-123, thus affecting NF-kB, Ras and TGF-β activation, and cell growth and apoptotic mechanisms [137].

Lycopene belongs to the tetra-terpenoid class and is commonly contained in ripe tomatoes, tomato products, and watermelons (Table 5). Indeed, lycopene is the pigment mainly responsible for their deep-red color. Its main ability is to reduce oxidative stress conditions thanks to its antioxidant properties. In human prostate, lung, liver, and breast cancer, lycopene inhibits tumor proliferation and growth through multiple molecular pathways [138,139]. Moreover, it protects against ultraviolet-related tumorigenesis by inhibiting inflammatory mechanisms and avoiding DNA structure injury [140]. Lastly, lycopene shows a demethylating impact on several promoters [124].

Resveratrol is naturally present in different plants, such as mulberries, peanuts, blueberries, and grapes, as well as in red wine (Table 5). It has anti-inflammatory and anticancer properties, influencing tumor cell proliferation, growth, invasion, and apoptosis, along with anti-aging benefits by modulating oxidative stress mechanisms. Its antioxidant properties are strictly related to free hydroxyl groups, which can donate hydrogen atoms to prevent cellular lipid peroxidation, in addition to resveratrol's ability to promote antioxidative enzymes like catalase (CAT) and SOD [141]. It has been demonstrated that approximately 20 proteins, including Nrf2 and SIRT1, SIRT2 and SIRT3, could interact with resveratrol. Among different polyphenols, resveratrol is considered the strongest inducer of HDAC activity, thus influencing *SIRT* gene expression. Thanks to SIRT activation, resveratrol can positively regulate *AMPK*, which plays a crucial role in both aging mechanisms and energy metabolism and inhibits *NF-kB* expression, resulting in both anticancer and anti-inflammatory effects [142]. It has been reported that resveratrol promotes apoptotic mechanisms in colon, breast, and prostate tumor cell lines, as well as leukemia. In human breast tumors, this phytochemical enhances the expression of the *BRCA1* gene through H3 acetylation and AhR signal regulation, and in prostate cancer, it has been previously reported that resveratrol promotes tumor cell apoptosis by the deacetylation of *FOXO*. In breast cancer cells, resveratrol also increases the *ATP2A3* expression through the promotion of the H3 lysine 27 acetylation into the *ATP2A3* promoter, whereas it reduces methyl-DNA binding protein expression, such as MBD2 and MeCP2 [143]. Furthermore, in colon cancer cells, a reduction in the expression of different oncogenic miRNAs has been detected, which are able to regulate the RNaseIII Dicer1 [124]. It has been shown that after treatment with resveratrol, 22 miRNAs, like the tumor suppressor miR-663, have been upregulated, whereas 26 miRNAs, including miR-21 and miR-25, have been downregulated. Hence, resveratrol boots the TGFβ pathway but inhibits the transcriptional activity of its effector proteins called SMADs [144].

Lastly, sulforaphane is contained in cruciferous vegetables like cabbage, kale, and broccoli (Table 5). It is considered a chemopreventive and chemotherapeutic agent thanks to its ability to reduce oxidative stress levels, inhibit cell proliferation, and promote cell apoptosis in several tumor cell lines, including cervical, pancreatic, liver, lung, and ovarian cancers [145]. It has been reported that sulforaphane inhibits HDAC activity and regulates histone methylation through the promotion of the HDM named RBP2. It also diminishes histone H1 phosphorylation by increasing protein phosphatase 1β and 2A [146,147]. In human prostate cancer cells, HDAC inhibition, after sulforaphane treatment, is followed by a rise of global histone acetylation and consequent binding on *p21* and *Bax* gene promoters, leading to apoptotic processes. Together with *p21* and *Bax*, *FOXO* transcription factors can mediate sulforaphane-associated apoptosis. Sulforaphane also protects cells from ultraviolet-associated tumorigenesis [124,148]. Moreover, sulforaphane reduces the activity of DNMTs, mainly DNMT1, -3a, and -3b, in cervical, prostate, and breast tumor cell lines, restoring the expression of several silenced genes, like *cyclin D2* and *PTEN*, by demethylating their promoters [149,150]. Sulforaphane enhances the expression of several cytoprotective genes, such as *CAT*, *glutathione S-transferase (GST)*, and *SOD*, through the epigenetic reactivation of the Nrf2 pathway, which is involved in oxidative stress response and anticancer mechanisms [147]. Combined with EGCG, sulforaphane reduces breast tumor cell proliferation through the inhibition of both DNMT1 and HDAC1 activity, which increases *ERα* expression compared to single treatments [151]. Therefore, sulforaphane and EGCG intake diminishes the onset of ER-negative breast tumors and stimulates cancer sensitivity to tamoxifen [152]. Furthermore, in breast cancer cells, its combination with genistein induces HDAC activity inhibition and reduces hTERT protein levels than a single treatment [153]. Lastly, different miRNAs, including miR-21, miR200c, and miR-616-5p, have been demonstrated to be regulated by sulforaphane in several human tumors [147].

Table 5. Epigenetic effects of other phytochemicals and their role in pathological conditions.

OTHER PHYTOCHEMICALS	Epigenetic Effects	High Content Food	Role in Pathological Conditions
Curcumin Dosage: Curcuma longa powered 500 mg/day. Dosage Indian cook: 1–2 g/day	Regulation of both HAT and HDAC activities [126] Inhibition of DNMT activity [154] Inhibition of oncogenic miRNAs and promotion of tumor suppressor miRNAs [127]	*Curcuma longa* powder contains 2% curcumin; 1 g Curcuma contains 20 mg curcumin; 1 g Curry contains 2.9 mg curcumin	Modulation of intracellular pathways implicated in inflammation, proliferation, invasion, survival, and apoptosis in cancer [125] Cancer prevention and suppressor [127,129] Toxic effects at high doses in normal cells like human dermal fibroblasts, resulting in a pronounced arrest of cell cycle progression and higher levels of cell death [155]
Folic Acid Family: Folate The recommended daily intake of folate for adults is about 400 µg.	A key element in the methyl-metabolism pathway [156] Regulation of the hepatic DNA methylation status [157]	Breakfast cereals Fortified (100–400 µg per 100 g), Peanuts (240 µg per 100 g), Black-eyed Peas (210 µg per 100 g), Fortified pasta (100–200 µg per 100 g), Spinach (194 µg per 100 g), Lentils (181 µg per 100 g), Chickpeas (*Cicer arietinum*) (172 µg per 100 g), Asparagus (149 µg per 100 g) Brussels Sprouts (140 µg per 100 g), Beetroot (*Beta vulgaris subsp. vulgaris*) (109 µg per 100 g), Lettuce (136 µg per 100 g), Broccoli (108 µg per 100 g), Almonds (50 µg per 100 g), Grains (20–40 µg per 100 g), Oranges (30 µg per 100 g), Eggs (22–47 µg per 100 g.), Bananas (20 µg per 100 g), Pasta (not fortified) (10–20 µg per 100 g)	Deficiency can modify hepatic DNA methylation patterns and induce liver (but also breast, lung, brain, and colorectal) cancer [158]. Antioxidant and pro-oxidant effects [159]

Table 5. *Cont.*

OTHER PHYTOCHEMICALS	Epigenetic Effects	High Content Food	Role in Pathological Conditions
Indole-3-carbinol (I3C) and diindolylmethane (DIM)	Proteasomal degradation of class I HDAC enzymes [133] Modulation of the expression of several miRNAs in cancer [160]	*Cruciferae family* (*Brassicaceae*) as Broccoli (15–25 mg/100 g), Cabbage (10–15 mg/100 g), Cauliflower (20–40 mg/100 g), Brussels Sprouts (40–50 mg/100 g), Mustard (10–15 mg/100 g), radish (all of *Brassica genus*) (1–4 mg/100 g)	Attenuation of symptoms of cigarette smoke [161] Anticancer effects, including promotion of apoptotic mechanisms in several tumor cell lines [132]
Phenethyl isothiocyanate (PEITC) Family: Isothiocyanates	Inhibition of HDAC and DNMT activity [136] Induction of targeted histone acetylation and methylation in human prostate tumor cells [135] Regulation of the expression of several miRNAs in cancer [137]	*Cruciferae family* as Watercress, Cauliflower, Cabbage, Cress, Bok Choy (*Brassica rapa subsp. Chinensis*) (Chinese cabbage) (10–150 mg per 100 g); Radishes (*Raphanus sativus*) (1–4 mg per 100 g); Arugula (1–2 mg per 100 g).	Inhibition of carcinogenic processes and regulation of stress response and inflammation [134]
Lycopene Family: Tetraterpenoid	Inhibition of DNMTs [139]	Tomato (2.5–4.5 mg per 100 g) Watermelon (4–7 mg per 100 g) Pink Grapefruit (1–3 mg per 100 g), Pink Guava (5–6 mg per 100 g), Papaya (2–3 mg per 100 g)	Antioxidant capacity, DNA protection from oxidative damage, reduction in tumor growth and proliferation in breast, prostate, liver and lung cancer [138] Protective agent against ultraviolet-related tumorigenesis [140]
Resveratrol Family: Polyphenols	Promotion of HDAC activity [162] Resveratrol targets: SIRT1, -2 and -3 [162] H3 acetylation in breast cancer, enhancing the expression of BRCA1 [144] FOXO deacetylation in prostate cancer [163] Reduction in the expression of different oncogenic miRNAs in colon cancer cells [144]	Red grape skin (0.5–1.5 mg per 100 g), Red wine (0.2–5.8 mg per 100 mL), Blueberry, Mulberry, Cranberry (0.02–0.06 mg per 100 g), and Peanut (0.1–0.3 mg per 100 g)	Anti-inflammatory, anti-aging, and anticancer properties, which influence tumor cell proliferation, growth, invasion, and apoptosis [141] Promotion of apoptotic mechanisms in colon, breast, prostate tumor cells and leukemia [124] Anti-aging potential and suppression of mouse skin cancer in combination with tea polyphenols [162,164]

Table 5. *Cont.*

OTHER PHYTOCHEMICALS	Epigenetic Effects	High Content Food	Role in Pathological Conditions
Sulforaphane Family: Isothiocyanate	Inhibition of HDAC activity, regulation of histone methylation, and reduction in H1 histone phosphorylation [146] Inhibition of the activity of DNMTs, mainly DNMT1, -3a, and -3b, in cervical, prostate, and breast tumor cell lines [149,150] Regulation of the expression of several miRNAs in different human tumors [137,147]	Broccoli Sprouts (around 30–100 mg per 100 g), *Cruciferae* as Cabbage, Brussels Sprouts, Broccoli, Kale (*Brassica oleracea var. Sabellica*) (0.1–2.5 mg per 100 g)	Reduction in oxidative stress levels, inhibition of cell proliferation, and induction of cell apoptosis in cervical, pancreatic, liver, lung, and ovarian cancers [145] Promotion of the expression of several cytoprotective genes (such as catalase, glutathione S-transferase, and superoxide dismutase) through the epigenetic reactivation of the Nrf2 pathway [147] Cell protection from ultraviolet-associated tumorigenesis [165]

HAT: histone acetyltransferase; HDAC: histone deacetylase; DNMT: DNA methyltransferase; miRNA: microRNA; SIRT: sirtuin.

4. Bioavailability of Bioactive Molecules

As reviewed, the biological properties of phytonutrients and their protective roles in molecular mechanisms of some human diseases and cancer have been demonstrated using different experimental models. The low bioavailability and bioaccessibility restrict the application and represent the major limitation [166]. The term bioavailability refers to the quantity of bioactive molecules that pass through the digestive tract, are absorbed, and reach the peripheral tissues in the intact or metabolized form to perform their bioactivity or to be stored. Bioaccessibility is defined as the proportion of a compound consumed in a meal that is released from the food matrix during digestion, in which the luminal content is accessible for absorption in the small intestine and transformed by the microbiota. Bioactivity represents the activity of the absorbed compounds or their metabolites at the cellular level, resulting in biological effects on the body. The evaluation of the bioavailability of polyphenols and other bioactive compounds has recently been gaining increasing interest in developing nutraceutical products. Molecular structure, water solubility, and composition of food matrix modulate bioaccessibility and digestibility. In addition, membrane transporters and metabolizing enzymes modulate the bioefficacy of phytonutrients. Even technological processes and/or cooking and heat treatments modulate bioavailability, and the mean plasma levels of several phytonutrients after absorption are low. It has also to be taken into account that individual responses could affect phytonutrient's bioavailability and bioactivity. It is also well known that polyphenols and other bioactive molecules in the large intestine are biotransformed by gut microbiota. Recent advances have been made to improve the bioavailability of some molecules. As of today, new delivery strategies have been studied, including lipid carriers, nano-emulsions, molecular enhancers, and encapsulation systems [167,168]. Various delivery systems were developed for carotenoids [169] and polyphenols [170]. However, further studies are necessary. The health benefits of dietary phytonutrients must be demonstrated in humans at appropriate doses, and well-controlled trials must be planned.

5. Phytochemicals and Traditional Anticancer Therapies: A Comparative Analysis Including Clinical Studies

The use of phytochemicals to fight cancer represents an innovative and expanding strategy. The goal is to overcome some of the limitations of traditional therapies, such as the low accessibility to tumor burden, the high dosage requirement, and their non-selective action [171].

Conventional anticancer strategies include chemotherapy, radiotherapy, and immunotherapy, but epigenetic drugs (epidrugs) are also increasingly being used. In particular, epidrugs exploit the reversible nature of epigenetic modifications, thus resetting the tumor epigenome. To date, the United States FDA (USFDA) has approved both DNMTi and HDACi [172]. Among DNMTi, 5-azacitidine (Vidaza, Azacitidine) and 5-aza-2′-deoxycytidine (decitabine (DAC), Dacogen) have been approved for myelodysplastic syndromes and acute myeloid leukemia treatment by reactivating silenced cancer suppressor genes [173]. Besides DNMTi, HDACi also shows promising anticancer properties through the reactivation of silenced cancer suppressor genes, thus promoting apoptotic mechanisms, inhibiting cell cycle arrest and DNA repair, and interfering with angiogenesis. At present, four HDACi have been approved by the USFDA: Vorinostat (SAHA) for the cutaneous T cell lymphoma treatment, Belinostat against peripheral T cell lymphomas, Romidepsin for both these diseases and Panobinostat for drug-resistant multiple myeloma treatment [174,175]. Instead, the only HDACi that was clinically approved in Europe is Panobinostat. Furthermore, the efficacy of DNMTi and HDACi can be improved by combining them or using each with other therapies, especially for solid tumors, as demonstrated by several trials. Indeed, epidrugs may be useful in sensitizing to traditional chemotherapy or immunotherapy [176]. Several preclinical analyses have demonstrated the anticancer potential of HDACi as both monotherapy and combinatory strategies in different solid tumors [177]. However, the subsequent clinical studies display positive results, with some exceptions due to the imperfections associated with preclinical research and the limitations of epigenetic strategies [27]. Indeed, the action of epidrugs is not locus-specific and induces large-scale modifications in the epigenome, also activating the expression of oncogenes and different off-target and adverse effects. Moreover, they display low permeability and solubility, making the investigation of alternative and innovative approaches essential [178]. Hence, there is a critical need to investigate the current state of clinical studies concerning HDACi efficacy and safety better and discover new potential strategies.

Among emerging anticancer approaches, evidence of the efficacy of flavonoids and other functional foods as innovative and alternative compounds with DNMTi and HDACi activity is constantly growing. Indeed, they display activity against several modifications in the intricate epigenetic network, from noncoding RNA to histone modifications [4]. After promising in vitro and in vivo results of flavonoid's impact on epigenetics, several clinical trials have been conducted on flavonoids used alone or combined with other drugs [171]. The clinical efficacy of catechin, EGCG, and genistein has been extensively evaluated in several solid cancers, including lung, colorectal, breast, and prostate tumors, with encouraging results. In particular, green tea catechin showed anticancer potential against prostate cancer (Phase I) [171], colon cancer (Phase II) [179], breast cancer (Phase II) [180], bladder cancer (Phase II) [181], lung cancer (Phase II) [182], cervical tumor (Phase II) [183] and skin cancer [184]. EGCG was found effective for breast (Phase II) [185], lung (Phase II) [186], and colorectal tumor (Phase I) [187] treatment. Also, genistein against colon and rectal (Phase I/II) [188], breast (Phase II) and prostate cancer (Phase II), leukemia and lymphoma (Phase I/II), and quercetin for prostate (Phase I), and squamous cell cancer (Phase II) treatment have been clinically tested. Some of these clinical trials are still ongoing [171,189]. Promising results have also been obtained using daidzein for prostate cancer treatment (Phase II) [189]. Moreover, in order to overcome the poor solubility, quick metabolism, and low absorption within the gastrointestinal system, the nano-formulation of flavonoids is being investigated. Nanocarriers of catechins, fisetin, quercetin, and apigenin showed promising in vitro and in vivo results, and one clinical trial using nano-formulated luteolin has been launched for the treatment of tongue neoplasm [171,190]. However, even though clinical analysis performed on flavonoids showed encouraging results, the existing evidence for their clinical employment is still insufficient. Furthermore, the clinical trials performed are restricted by several limitations, such as the low number of participants. Therefore, additional clinical studies may be needed to further confirm the potential of functional foods for cancer treatment and develop innovative approaches (including combined administration and nano-drug carrier) to overcome bioavailability concerns.

6. Conclusions

The existing literature concerning dietary phytochemicals has demonstrated that compounds contained in functional foods have multifaceted epigenetic properties by modulating several epigenetic modifiers, such as HMTs, HATs, HDACs, DNMTs, and ncRNAs. Within this review, different flavonoids were investigated along with their epigenetic regulation and therapeutic effects, mainly against cancer. In addition to flavonoids, other phytochemicals characterized by potent anti-inflammatory, antioxidative, and anticancer properties were described. Thanks to the reversibility of the induced epigenetic modifications, phytochemicals show enormous potential in cancer prevention and treatment. Evaluation of the safety and efficacy of several natural compounds in preclinical and clinical trials is currently ongoing; the next crucial step would be to identify the most suitable doses of these natural compounds with the aim of obtaining favorable effects on humans. However, their potential is partially hindered by some limitations, like poor bioavailability, low water solubility, insufficient therapeutic index, and side effects on the liver. To overcome these restrictions, it is necessary to develop innovative strategies, such as nanomaterial encapsulation technology. Furthermore, combining plant-derived compounds with traditional chemo and/or radiotherapy may lead to better results and/or reduction in adverse effects. It is mandatory to continue deepening the global patterns of epigenetic modifications by phytochemicals with the aim of identifying new targets and attractive agents in fighting cancer.

Funding: This research received no external funding.

Institutional Review Board Statement: Not applicable.

Informed Consent Statement: Not applicable.

Data Availability Statement: No new data were created or analyzed in this study.

Conflicts of Interest: The authors declare no conflict of interest.

References

1. Hughes, V. Epigenetics: The sins of the father. *Nature* **2014**, *507*, 22–24. [CrossRef]
2. Shannar, A.; Sarwar, M.S.; Kong, A.T. A New Frontier in Studying Dietary Phytochemicals in Cancer and in Health: Metabolic and Epigenetic Reprogramming. *Prev. Nutr. Food Sci.* **2022**, *27*, 335–346. [CrossRef]
3. Zhang, L.; Lu, Q.; Chang, C. Epigenetics in Health and Disease. *Adv. Exp. Med. Biol.* **2020**, *1253*, 3–55. [PubMed]
4. Jiang, W.; Xia, T.; Liu, C.; Li, J.; Zhang, W.; Sun, C. Remodeling the Epigenetic Landscape of Cancer-Application Potential of Flavonoids in the Prevention and Treatment of Cancer. *Front. Oncol.* **2021**, *11*, 705903. [CrossRef]
5. Russo, G.L.; Vastolo, V.; Ciccarelli, M.; Albano, L.; Macchia, P.E.; Ungaro, P. Dietary polyphenols and chromatin remodeling. *Crit. Rev. Food Sci. Nutr.* **2017**, *57*, 2589–2599. [CrossRef] [PubMed]
6. Suter, M.A.; Aagaard-Tillery, K.M. Environmental influences on epigenetic profiles. *Semin. Reprod. Med.* **2009**, *27*, 380–390. [CrossRef]
7. Fatima, N.; Baqri, S.S.R.; Bhattacharya, A.; Koney, N.K.; Husain, K.; Abbas, A.; Ansari, R.A. Role of Flavonoids as Epigenetic Modulators in Cancer Prevention and Therapy. *Front. Genet.* **2021**, *12*, 758733. [CrossRef] [PubMed]
8. Shankar, E.; Kanwal, R.; Candamo, M.; Gupta, S. Dietary phytochemicals as epigenetic modifiers in cancer: Promise and challenges. *Semin. Cancer Biol.* **2016**, *40–41*, 82–99. [CrossRef]
9. Choudhuri, S. From Waddington's epigenetic landscape to small noncoding RNA: Some important milestones in the history of epigenetics research. *Toxicol. Mech. Methods* **2011**, *21*, 252–274. [CrossRef]
10. Shoaib, S.; Ansari, M.A.; Ghazwani, M.; Hani, U.; Jamous, Y.F.; Alali, Z.; Wahab, S.; Ahmad, W.; Weir, S.A.; Alomary, M.N.; et al. Prospective Epigenetic Actions of Organo-Sulfur Compounds against Cancer: Perspectives and Molecular Mechanisms. *Cancers* **2023**, *15*, 697. [CrossRef] [PubMed]
11. Bird, A. DNA methylation patterns and epigenetic memory. *Genes. Dev.* **2002**, *16*, 6–21. [CrossRef] [PubMed]
12. Shyh-Chang, N.; Locasale, J.W.; Lyssiotis, C.A.; Zheng, Y.; Teo, R.Y.; Ratanasirintrawoot, S.; Zhang, J.; Onder, T.; Unternaehrer, J.J.; Zhu, H.; et al. Influence of threonine metabolism on S-adenosylmethionine and histone methylation. *Science* **2013**, *339*, 222–226. [CrossRef] [PubMed]
13. Okano, M.; Bell, D.W.; Haber, D.A.; Li, E. DNA methyltransferases Dnmt3a and Dnmt3b are essential for de novo methylation and mammalian development. *Cell* **1999**, *99*, 247–257. [CrossRef]
14. Denis, H.; Ndlovu, M.N.; Fuks, F. Regulation of mammalian DNA methyltransferases: A route to new mechanisms. *EMBO Rep.* **2011**, *12*, 647–656. [CrossRef]
15. Baylin, S.B.; Jones, P.A. A decade of exploring the cancer epigenome—Biological and translational implications. *Nat. Rev. Cancer* **2011**, *11*, 726–734. [CrossRef]

16. Cheung, H.H.; Lee, T.L.; Rennert, O.M.; Chan, W.Y. DNA methylation of cancer genome. *Birth Defects Res. Part C Embryo Today* **2009**, *87*, 335–350. [CrossRef]
17. Jones, P.A.; Baylin, S.B. The epigenomics of cancer. *Cell* **2007**, *128*, 683–692. [CrossRef] [PubMed]
18. Issa, J.P.; Kantarjian, H.M. Targeting DNA methylation. *Clin. Cancer Res.* **2009**, *15*, 3938–3946. [CrossRef]
19. Derissen, E.J.; Beijnen, J.H.; Schellens, J.H. Concise drug review: Azacitidine and decitabine. *Oncologist* **2013**, *18*, 619–624. [CrossRef] [PubMed]
20. Busch, C.; Burkard, M.; Leischner, C.; Lauer, U.M.; Frank, J.; Venturelli, S. Epigenetic activities of flavonoids in the prevention and treatment of cancer. *Clin. Epigenet.* **2015**, *7*, 64. [CrossRef] [PubMed]
21. Vanden Berghe, W. Epigenetic impact of dietary polyphenols in cancer chemoprevention: Lifelong remodeling of our epigenomes. *Pharmacol. Res.* **2012**, *65*, 565–576. [CrossRef] [PubMed]
22. Kouzarides, T. Chromatin modifications and their function. *Cell* **2007**, *128*, 693–705. [CrossRef] [PubMed]
23. Wang, Z.; Zang, C.; Cui, K.; Schones, D.E.; Barski, A.; Peng, W.; Zhao, K. Genome-wide mapping of HATs and HDACs reveals distinct functions in active and inactive genes. *Cell* **2009**, *138*, 1019–1031. [CrossRef] [PubMed]
24. Struhl, K. Histone acetylation and transcriptional regulatory mechanisms. *Genes. Dev.* **1998**, *12*, 599–606. [CrossRef]
25. Kuo, M.H.; Allis, C.D. Roles of histone acetyltransferases and deacetylases in gene regulation. *Bioessays* **1998**, *20*, 615–626. [CrossRef]
26. Di Cerbo, V.; Schneider, R. Cancers with wrong HATs: The impact of acetylation. *Brief Funct. Genom.* **2013**, *12*, 231–243. [CrossRef]
27. Shi, M.Q.; Xu, Y.; Fu, X.; Pan, D.S.; Lu, X.P.; Xiao, Y.; Jiang, Y.Z. Advances in targeting histone deacetylase for treatment of solid tumors. *J. Hematol. Oncol.* **2024**, *17*, 37. [CrossRef] [PubMed]
28. Delcuve, G.P.; Khan, D.H.; Davie, J.R. Roles of histone deacetylases in epigenetic regulation: Emerging paradigms from studies with inhibitors. *Clin. Epigenet.* **2012**, *4*, 5. [CrossRef] [PubMed]
29. Varier, R.A.; Timmers, H.T. Histone lysine methylation and demethylation pathways in cancer. *Biochim. Biophys. Acta (BBA)-Rev. Cancer* **2011**, *1815*, 75–89. [CrossRef] [PubMed]
30. Rossetto, D.; Avvakumov, N.; Cote, J. Histone phosphorylation: A chromatin modification involved in diverse nuclear events. *Epigenetics* **2012**, *7*, 1098–1108. [CrossRef] [PubMed]
31. Bartel, D.P. Metazoan MicroRNAs. *Cell* **2018**, *173*, 20–51. [CrossRef]
32. Mohr, A.M.; Mott, J.L. Overview of microRNA biology. *Semin. Liver Dis.* **2015**, *35*, 3–11. [CrossRef] [PubMed]
33. Mao, G.; Xu, Y.; Long, D.; Sun, H.; Li, H.; Xin, R.; Zhang, Z.; Li, Z.; Yang, Z.; Kang, Y. Exosome-transported circRNA_0001236 enhances chondrogenesis and suppress cartilage degradation via the miR-3677-3p/Sox9 axis. *Stem Cell Res. Ther.* **2021**, *12*, 389. [CrossRef] [PubMed]
34. Peng, Y.; Croce, C.M. The role of MicroRNAs in human cancer. *Signal Transduct. Target. Ther.* **2016**, *1*, 15004. [CrossRef]
35. Rhim, J.; Baek, W.; Seo, Y.; Kim, J.H. From Molecular Mechanisms to Therapeutics: Understanding MicroRNA-21 in Cancer. *Cells* **2022**, *11*, 2791. [CrossRef] [PubMed]
36. Statello, L.; Ali, M.M.; Reischl, S.; Mahale, S.; Kosalai, S.T.; Huarte, M.; Kanduri, C. The DNA damage inducible lncRNA SCAT7 regulates genomic integrity and topoisomerase 1 turnover in lung adenocarcinoma. *NAR Cancer* **2021**, *3*, zcab002. [CrossRef] [PubMed]
37. Kopp, F.; Mendell, J.T. Functional Classification and Experimental Dissection of Long Noncoding RNAs. *Cell* **2018**, *172*, 393–407. [CrossRef] [PubMed]
38. Marin-Bejar, O.; Marchese, F.P.; Athie, A.; Sanchez, Y.; Gonzalez, J.; Segura, V.; Huang, L.; Moreno, I.; Navarro, A.; Monzo, M.; et al. Pint lincRNA connects the p53 pathway with epigenetic silencing by the Polycomb repressive complex 2. *Genome Biol.* **2013**, *14*, R104. [CrossRef]
39. Gupta, R.A.; Shah, N.; Wang, K.C.; Kim, J.; Horlings, H.M.; Wong, D.J.; Tsai, M.C.; Hung, T.; Argani, P.; Rinn, J.L.; et al. Long non-coding RNA HOTAIR reprograms chromatin state to promote cancer metastasis. *Nature* **2010**, *464*, 1071–1076. [CrossRef] [PubMed]
40. Anastasiadou, E.; Jacob, L.S.; Slack, F.J. Non-coding RNA networks in cancer. *Nat. Rev. Cancer* **2018**, *18*, 5–18. [CrossRef] [PubMed]
41. Nejadi Orang, F.; Abdoli Shadbad, M. CircRNA and lncRNA-associated competing endogenous RNA networks in medulloblastoma: A scoping review. *Cancer Cell Int.* **2024**, *24*, 248. [CrossRef] [PubMed]
42. Cheng, D.; Wang, J.; Dong, Z.; Li, X. Cancer-related circular RNA: Diverse biological functions. *Cancer Cell Int.* **2021**, *21*, 11. [CrossRef] [PubMed]
43. Nelson, K.M.; Weiss, G.J. MicroRNAs and cancer: Past, present, and potential future. *Mol. Cancer Ther.* **2008**, *7*, 3655–3660. [CrossRef] [PubMed]

44. Wong, C.H.; Lou, U.K.; Fung, F.K.; Tong, J.H.M.; Zhang, C.H.; To, K.F.; Chan, S.L.; Chen, Y. CircRTN4 promotes pancreatic cancer progression through a novel CircRNA-miRNA-lncRNA pathway and stabilizing epithelial-mesenchymal transition protein. *Mol. Cancer* **2022**, *21*, 10. [CrossRef]
45. O'Brien, J.; Hayder, H.; Zayed, Y.; Peng, C. Overview of MicroRNA Biogenesis, Mechanisms of Actions, and Circulation. *Front. Endocrinol.* **2018**, *9*, 402. [CrossRef]
46. Statello, L.; Guo, C.J.; Chen, L.L.; Huarte, M. Gene regulation by long non-coding RNAs and its biological functions. *Nat. Rev. Mol. Cell Biol.* **2021**, *22*, 96–118. [CrossRef]
47. Kumar, S.; Pandey, A.K. Chemistry and biological activities of flavonoids: An overview. *Sci. World J.* **2013**, *2013*, 162750. [CrossRef]
48. Morresi, C.; Cianfruglia, L.; Armeni, T.; Mancini, F.; Tenore, G.C.; D'Urso, E.; Micheletti, A.; Ferretti, G.; Bacchetti, T. Polyphenolic compounds and nutraceutical properties of old and new apple cultivars. *J. Food Biochem.* **2018**, *42*, e12641. [CrossRef]
49. Khan, H.; Belwal, T.; Efferth, T.; Farooqi, A.A.; Sanches-Silva, A.; Vacca, R.A.; Nabavi, S.F.; Khan, F.; Prasad Devkota, H.; Barreca, D.; et al. Targeting epigenetics in cancer: Therapeutic potential of flavonoids. *Crit. Rev. Food Sci. Nutr.* **2021**, *61*, 1616–1639. [CrossRef] [PubMed]
50. Abotaleb, M.; Samuel, S.M.; Varghese, E.; Varghese, S.; Kubatka, P.; Liskova, A.; Busselberg, D. Flavonoids in Cancer and Apoptosis. *Cancers* **2018**, *11*, 28. [CrossRef] [PubMed]
51. Minnelli, C.; Galeazzi, R.; Laudadio, E.; Amici, A.; Rusciano, D.; Armeni, T.; Cantarini, M.; Stipa, P.; Mobbili, G. Monoalkylated Epigallocatechin-3-gallate (C18-EGCG) as Novel Lipophilic EGCG Derivative: Characterization and Antioxidant Evaluation. *Antioxidants* **2020**, *9*, 208. [CrossRef] [PubMed]
52. Kim, J. Circular RNAs: Novel Players in Cancer Mechanisms and Therapeutic Strategies. *Int. J. Mol. Sci.* **2024**, *25*, 10121. [CrossRef]
53. Harper, K.L.; Mottram, T.J.; Whitehouse, A. Insights into the Evolving Roles of Circular RNAs in Cancer. *Cancers* **2021**, *13*, 4180. [CrossRef]
54. Gerhauser, C. Cancer chemoprevention and nutriepigenetics: State of the art and future challenges. *Nat. Prod. Cancer Prev. Ther.* **2013**, *329*, 73–132.
55. Minnelli, C.; Laudadio, E.; Sorci, L.; Sabbatini, G.; Galeazzi, R.; Amici, A.; Semrau, M.S.; Storici, P.; Rinaldi, S.; Stipa, P.; et al. Identification of a novel nitroflavone-based scaffold for designing mutant-selective EGFR tyrosine kinase inhibitors targeting T790M and C797S resistance in advanced NSCLC. *Bioorg. Chem.* **2022**, *129*, 106219. [CrossRef] [PubMed]
56. Mobbili, G.; Romaldi, B.; Sabbatini, G.; Amici, A.; Marcaccio, M.; Galeazzi, R.; Laudadio, E.; Armeni, T.; Minnelli, C. Identification of Flavone Derivative Displaying a 4′-Aminophenoxy Moiety as Potential Selective Anticancer Agent in NSCLC Tumor Cells. *Molecules* **2023**, *28*, 3239. [CrossRef]
57. Somsakeesit, L.O.; Senawong, T.; Senawong, G.; Kumboonma, P.; Samankul, A.; Namwan, N.; Yenjai, C.; Phaosiri, C. Evaluation and molecular docking study of two flavonoids from *Oroxylum indicum* (L.) Kurz and their semi-synthetic derivatives as histone deacetylase inhibitors. *J. Nat. Med.* **2024**, *78*, 236–245. [CrossRef] [PubMed]
58. Pechalrieu, D.; Dauzonne, D.; Arimondo, P.B.; Lopez, M. Synthesis of novel 3-halo-3-nitroflavanones and their activities as DNA methyltransferase inhibitors in cancer cells. *Eur. J. Med. Chem.* **2020**, *186*, 111829. [CrossRef]
59. Middleton, E., Jr.; Kandaswami, C.; Theoharides, T.C. The effects of plant flavonoids on mammalian cells: Implications for inflammation, heart disease, and cancer. *Pharmacol. Rev.* **2000**, *52*, 673–751.
60. Choi, K.C.; Jung, M.G.; Lee, Y.H.; Yoon, J.C.; Kwon, S.H.; Kang, H.B.; Kim, M.J.; Cha, J.H.; Kim, Y.J.; Jun, W.J.; et al. Epigallocatechin-3-gallate, a histone acetyltransferase inhibitor, inhibits EBV-induced B lymphocyte transformation via suppression of RelA acetylation. *Cancer Res.* **2009**, *69*, 583–592. [CrossRef]
61. Fang, M.Z.; Wang, Y.; Ai, N.; Hou, Z.; Sun, Y.; Lu, H.; Welsh, W.; Yang, C.S. Tea polyphenol (-)-epigallocatechin-3-gallate inhibits DNA methyltransferase and reactivates methylation-silenced genes in cancer cell lines. *Cancer Res.* **2003**, *63*, 7563–7570. [PubMed]
62. Kang, Q.; Tong, Y.; Gowd, V.; Wang, M.; Chen, F.; Cheng, K.W. Oral administration of EGCG solution equivalent to daily achievable dosages of regular tea drinkers effectively suppresses miR483-3p induced metastasis of hepatocellular carcinoma cells in mice. *Food Funct.* **2021**, *12*, 3381–3392. [CrossRef] [PubMed]
63. Carotenuto, F.; Albertini, M.C.; Coletti, D.; Vilmercati, A.; Campanella, L.; Darzynkiewicz, Z.; Teodori, L. How Diet Intervention via Modulation of DNA Damage Response through MicroRNAs May Have an Effect on Cancer Prevention and Aging, an in Silico Study. *Int. J. Mol. Sci.* **2016**, *17*, 752. [CrossRef] [PubMed]
64. Rajavelu, A.; Tulyasheva, Z.; Jaiswal, R.; Jeltsch, A.; Kuhnert, N. The inhibition of the mammalian DNA methyltransferase 3a (Dnmt3a) by dietary black tea and coffee polyphenols. *BMC Biochem.* **2011**, *12*, 16. [CrossRef] [PubMed]
65. Berger, A.; Venturelli, S.; Kallnischkies, M.; Bocker, A.; Busch, C.; Weiland, T.; Noor, S.; Leischner, C.; Weiss, T.S.; Lauer, U.M.; et al. Kaempferol, a new nutrition-derived pan-inhibitor of human histone deacetylases. *J. Nutr. Biochem.* **2013**, *24*, 977–985. [CrossRef]

66. Kedhari Sundaram, M.; Hussain, A.; Haque, S.; Raina, R.; Afroze, N. Quercetin modifies 5′CpG promoter methylation and reactivates various tumor suppressor genes by modulating epigenetic marks in human cervical cancer cells. *J. Cell. Biochem.* **2019**, *120*, 18357–18369. [CrossRef] [PubMed]
67. Fang, M.; Chen, D.; Yang, C.S. Dietary polyphenols may affect DNA methylation. *J. Nutr.* **2007**, *137*, 223S–228S. [CrossRef] [PubMed]
68. Pandey, M.; Kaur, P.; Shukla, S.; Abbas, A.; Fu, P.; Gupta, S. Plant flavone apigenin inhibits HDAC and remodels chromatin to induce growth arrest and apoptosis in human prostate cancer cells: In vitro and in vivo study. *Mol. Carcinog.* **2012**, *51*, 952–962. [CrossRef]
69. Krifa, M.; Leloup, L.; Ghedira, K.; Mousli, M.; Chekir-Ghedira, L. Luteolin induces apoptosis in BE colorectal cancer cells by downregulating calpain, UHRF1, and DNMT1 expressions. *Nutr. Cancer* **2014**, *66*, 1220–1227. [CrossRef] [PubMed]
70. Ling, D.; Marshall, G.M.; Liu, P.Y.; Xu, N.; Nelson, C.A.; Iismaa, S.E.; Liu, T. Enhancing the anticancer effect of the histone deacetylase inhibitor by activating transglutaminase. *Eur. J. Cancer* **2012**, *48*, 3278–3287. [CrossRef]
71. Caba, L.; Florea, L.; Gug, C.; Dimitriu, D.C.; Gorduza, E.V. Circular RNA-Is the Circle Perfect? *Biomolecules* **2021**, *11*, 1755. [CrossRef]
72. Fang, M.Z.; Chen, D.; Sun, Y.; Jin, Z.; Christman, J.K.; Yang, C.S. Reversal of hypermethylation and reactivation of p16INK4a, RARbeta, and MGMT genes by genistein and other isoflavones from soy. *Clin. Cancer Res.* **2005**, *11*, 7033–7041. [CrossRef]
73. Vahid, F.; Zand, H.; Nosrat-Mirshekarlou, E.; Najafi, R.; Hekmatdoost, A. The role dietary of bioactive compounds on the regulation of histone acetylases and deacetylases: A review. *Gene* **2015**, *562*, 8–15. [CrossRef]
74. Zhang, Y.; Li, Q.; Chen, H. DNA methylation and histone modifications of Wnt genes by genistein during colon cancer development. *Carcinogenesis* **2013**, *34*, 1756–1763. [CrossRef] [PubMed]
75. Jeong, M.H.; Ko, H.; Jeon, H.; Sung, G.J.; Park, S.Y.; Jun, W.J.; Lee, Y.H.; Lee, J.; Lee, S.W.; Yoon, H.G.; et al. Delphinidin induces apoptosis via cleaved HDAC3-mediated p53 acetylation and oligomerization in prostate cancer cells. *Oncotarget* **2016**, *7*, 56767–56780. [CrossRef] [PubMed]
76. Kuo, H.D.; Wu, R.; Li, S.; Yang, A.Y.; Kong, A.N. Anthocyanin Delphinidin Prevents Neoplastic Transformation of Mouse Skin JB6 P+ Cells: Epigenetic Re-activation of Nrf2-ARE Pathway. *AAPS J.* **2019**, *21*, 83. [CrossRef] [PubMed]
77. Higdon, J.V.; Frei, B. Tea catechins and polyphenols: Health effects, metabolism, and antioxidant functions. *Crit. Rev. Food Sci. Nutr.* **2003**, *43*, 89–143. [CrossRef]
78. Minnelli, C.; Cianfruglia, L.; Laudadio, E.; Mobbili, G.; Galeazzi, R.; Armeni, T. Effect of Epigallocatechin-3-Gallate on EGFR Signaling and Migration in Non-Small Cell Lung Cancer. *Int. J. Mol. Sci.* **2021**, *22*, 11833. [CrossRef] [PubMed]
79. Meeran, S.M.; Patel, S.N.; Chan, T.H.; Tollefsbol, T.O. A novel prodrug of epigallocatechin-3-gallate: Differential epigenetic hTERT repression in human breast cancer cells. *Cancer Prev. Res.* **2011**, *4*, 1243–1254. [CrossRef] [PubMed]
80. Sheng, J.; Shi, W.; Guo, H.; Long, W.; Wang, Y.; Qi, J.; Liu, J.; Xu, Y. The Inhibitory Effect of (-)-Epigallocatechin-3-Gallate on Breast Cancer Progression via Reducing SCUBE2 Methylation and DNMT Activity. *Molecules* **2019**, *24*, 2899. [CrossRef]
81. Bhagwat, S.; Haytowitz, D.B.; Wasswa-Kintu, S.I.; Holden, J.M. USDA Develops a Database for Flavonoids to Assess Dietary Intakes. *Procedia Food Sci.* **2013**, *2*, 81–86. [CrossRef]
82. Hardy, T.M.; Tollefsbol, T.O. Epigenetic diet: Impact on the epigenome and cancer. *Epigenomics* **2011**, *3*, 503–518. [CrossRef]
83. Gilbert, E.R.; Liu, D. Flavonoids influence epigenetic-modifying enzyme activity: Structure-function relationships and the therapeutic potential for cancer. *Curr. Med. Chem.* **2010**, *17*, 1756–1768. [CrossRef] [PubMed]
84. Volate, S.R.; Muga, S.J.; Issa, A.Y.; Nitcheva, D.; Smith, T.; Wargovich, M.J. Epigenetic modulation of the retinoid X receptor alpha by green tea in the azoxymethane-Apc Min/+ mouse model of intestinal cancer. *Mol. Carcinog.* **2009**, *48*, 920–933. [CrossRef] [PubMed]
85. Ahmad, N.; Feyes, D.K.; Nieminen, A.L.; Agarwal, R.; Mukhtar, H. Green tea constituent epigallocatechin-3-gallate and induction of apoptosis and cell cycle arrest in human carcinoma cells. *J. Natl. Cancer Inst.* **1997**, *89*, 1881–1886. [CrossRef] [PubMed]
86. Bhattacharya, R.; Chatterjee, R.; Mandal, A.K.A.; Mukhopadhyay, A.; Basu, S.; Giri, A.K.; Chatterji, U.; Bhattacharjee, P. Theaflavin-Containing Black Tea Extract: A Potential DNA Methyltransferase Inhibitor in Human Colon Cancer Cells and Ehrlich Ascites Carcinoma-Induced Solid Tumors in Mice. *Nutr. Cancer* **2021**, *73*, 2447–2459. [CrossRef]
87. Imran, A.; Butt, M.S.; Xiao, H.; Imran, M.; Rauf, A.; Mubarak, M.S.; Ramadan, M.F. Inhibitory effect of black tea (*Camellia sinensis*) theaflavins and thearubigins against HCT 116 colon cancer cells and HT 460 lung cancer cells. *J. Food Biochem.* **2019**, *43*, e12822. [CrossRef] [PubMed]
88. Maity, S.; Ukil, A.; Karmakar, S.; Datta, N.; Chaudhuri, T.; Vedasiromoni, J.R.; Ganguly, D.K.; Das, P.K. Thearubigin, the major polyphenol of black tea, ameliorates mucosal injury in trinitrobenzene sulfonic acid-induced colitis. *Eur. J. Pharmacol.* **2003**, *470*, 103–112. [CrossRef]

89. Russo, G.L.; Ungaro, P. Chapter 9—Epigenetic Mechanisms of Quercetin and Other Flavonoids in Cancer Therapy and Prevention. *Transl. Epigenet. Epigenet. Cancer Prev. Acad. Press* **2019**, *8*, 187–202.
90. Lu, L.; Wang, Y.; Ou, R.; Feng, Q.; Ji, L.; Zheng, H.; Guo, Y.; Qi, X.; Kong, A.N.; Liu, Z. DACT2 Epigenetic Stimulator Exerts Dual Efficacy for Colorectal Cancer Prevention and Treatment. *Pharmacol. Res.* **2018**, *129*, 318–328. [CrossRef]
91. Zheng, N.G.; Wang, J.L.; Yang, S.L.; Wu, J.L. Aberrant epigenetic alteration in Eca9706 cells modulated by nanoliposomal quercetin combined with butyrate mediated via epigenetic-NF-kappaB signaling. *Asian Pac. J. Cancer Prev.* **2014**, *15*, 4539–4543. [CrossRef]
92. Srinivasan, A.; Thangavel, C.; Liu, Y.; Shoyele, S.; Den, R.B.; Selvakumar, P.; Lakshmikuttyamma, A. Quercetin regulates beta-catenin signaling and reduces the migration of triple negative breast cancer. *Mol. Carcinog.* **2016**, *55*, 743–756. [CrossRef] [PubMed]
93. Struewing, J.P.; Hartge, P.; Wacholder, S.; Baker, S.M.; Berlin, M.; McAdams, M.; Timmerman, M.M.; Brody, L.C.; Tucker, M.A. The risk of cancer associated with specific mutations of BRCA1 and BRCA2 among Ashkenazi Jews. *N. Engl. J. Med.* **1997**, *336*, 1401–1408. [CrossRef]
94. Tao, S.F.; He, H.F.; Chen, Q. Quercetin inhibits proliferation and invasion acts by up-regulating miR-146a in human breast cancer cells. *Mol. Cell. Biochem.* **2015**, *402*, 93–100. [CrossRef]
95. Imran, M.; Saeed, F.; Gilani, S.A.; Shariati, M.A.; Imran, A.; Afzaal, M.; Atif, M.; Tufail, T.; Anjum, F.M. Fisetin: An anticancer perspective. *Food Sci. Nutr.* **2021**, *9*, 3–16. [CrossRef] [PubMed]
96. Pal, H.C.; Pearlman, R.L.; Afaq, F. Fisetin and Its Role in Chronic Diseases. *Adv. Exp. Med. Biol.* **2016**, *928*, 213–244. [PubMed]
97. Dashwood, R.H. Frontiers in polyphenols and cancer prevention. *J. Nutr.* **2007**, *137*, 267S–269S. [CrossRef] [PubMed]
98. Hong, K.S.; Park, J.I.; Kim, M.J.; Kim, H.B.; Lee, J.W.; Dao, T.T.; Oh, W.K.; Kang, C.D.; Kim, S.H. Involvement of SIRT1 in hypoxic down-regulation of c-Myc and beta-catenin and hypoxic preconditioning effect of polyphenols. *Toxicol. Appl. Pharmacol.* **2012**, *259*, 210–218. [CrossRef] [PubMed]
99. Matboli, M.; Ibrahim, D.; Hasanin, A.H.; Hassan, M.K.; Habib, E.K.; Bekhet, M.M.; Afifi, A.M.; Eissa, S. Epigenetic modulation of autophagy genes linked to diabetic nephropathy by administration of isorhamnetin in Type 2 diabetes mellitus rats. *Epigenomics* **2021**, *13*, 187–202. [CrossRef] [PubMed]
100. Barreca, D.; Mandalari, G.; Calderaro, A.; Smeriglio, A.; Trombetta, D.; Felice, M.R.; Gattuso, G. Citrus Flavones: An Update on Sources, Biological Functions, and Health Promoting Properties. *Plants* **2020**, *9*, 288. [CrossRef]
101. Tseng, T.H.; Chien, M.H.; Lin, W.L.; Wen, Y.C.; Chow, J.M.; Chen, C.K.; Kuo, T.C.; Lee, W.J. Inhibition of MDA-MB-231 breast cancer cell proliferation and tumor growth by apigenin through induction of G2/M arrest and histone H3 acetylation-mediated p21(WAF1/CIP1) expression. *Environ. Toxicol.* **2017**, *32*, 434–444. [CrossRef]
102. Wu, H.T.; Lin, J.; Liu, Y.E.; Chen, H.F.; Hsu, K.W.; Lin, S.H.; Peng, K.Y.; Lin, K.J.; Hsieh, C.C.; Chen, D.R. Luteolin suppresses androgen receptor-positive triple-negative breast cancer cell proliferation and metastasis by epigenetic regulation of MMP9 expression via the AKT/mTOR signaling pathway. *Phytomedicine* **2021**, *81*, 153437. [CrossRef] [PubMed]
103. Shoulars, K.; Rodriguez, M.A.; Thompson, T.; Markaverich, B.M. Regulation of cell cycle and RNA transcription genes identified by microarray analysis of PC-3 human prostate cancer cells treated with luteolin. *J. Steroid Biochem. Mol. Biol.* **2010**, *118*, 41–50. [CrossRef] [PubMed]
104. Khan, M.K.; Zill-E-Huma; Dangles, O. A comprehensive review on flavanones, the major citrus polyphenols. *J. Food Compos. Anal.* **2014**, *33*, 85–104. [CrossRef]
105. Wang, S.W.; Sheng, H.; Zheng, F.; Zhang, F. Hesperetin promotes DOT1L degradation and reduces histone H3K79 methylation to inhibit gastric cancer metastasis. *Phytomedicine* **2021**, *84*, 153499. [CrossRef]
106. Zeng, W.; Jin, L.; Zhang, F.; Zhang, C.; Liang, W. Naringenin as a potential immunomodulator in therapeutics. *Pharmacol. Res.* **2018**, *135*, 122–126. [CrossRef]
107. Curti, V.; Di Lorenzo, A.; Rossi, D.; Martino, E.; Capelli, E.; Collina, S.; Daglia, M. Enantioselective Modulatory Effects of Naringenin Enantiomers on the Expression Levels of miR-17-3p Involved in Endogenous Antioxidant Defenses. *Nutrients* **2017**, *9*, 215. [CrossRef] [PubMed]
108. Pandey, M.K.; Gupta, S.C.; Nabavizadeh, A.; Aggarwal, B.B. Regulation of cell signaling pathways by dietary agents for cancer prevention and treatment. *Semin. Cancer Biol.* **2017**, *46*, 158–181. [CrossRef]
109. Panche, A.N.; Diwan, A.D.; Chandra, S.R. Flavonoids: An overview. *J. Nutr. Sci.* **2016**, *5*, e47. [CrossRef] [PubMed]
110. Tham, D.M.; Gardner, C.D.; Haskell, W.L. Clinical review 97: Potential health benefits of dietary phytoestrogens: A review of the clinical, epidemiological, and mechanistic evidence. *J. Clin. Endocrinol. Metab.* **1998**, *83*, 2223–2235.
111. Keinan-Boker, L.; van Der Schouw, Y.T.; Grobbee, D.E.; Peeters, P.H. Dietary phytoestrogens and breast cancer risk. *Am. J. Clin. Nutr.* **2004**, *79*, 282–288. [CrossRef] [PubMed]
112. Xie, Q.; Chen, M.L.; Qin, Y.; Zhang, Q.Y.; Xu, H.X.; Zhou, Y.; Mi, M.T.; Zhu, J.D. Isoflavone consumption and risk of breast cancer: A dose-response meta-analysis of observational studies. *Asia Pac. J. Clin. Nutr.* **2013**, *22*, 118–127. [PubMed]

113. Huser, S.; Guth, S.; Joost, H.G.; Soukup, S.T.; Kohrle, J.; Kreienbrock, L.; Diel, P.; Lachenmeier, D.W.; Eisenbrand, G.; Vollmer, G.; et al. Effects of isoflavones on breast tissue and the thyroid hormone system in humans: A comprehensive safety evaluation. *Arch. Toxicol.* **2018**, *92*, 2703–2748. [CrossRef] [PubMed]
114. Spagnuolo, C.; Russo, G.L.; Orhan, I.E.; Habtemariam, S.; Daglia, M.; Sureda, A.; Nabavi, S.F.; Devi, K.P.; Loizzo, M.R.; Tundis, R.; et al. Genistein and cancer: Current status, challenges, and future directions. *Adv. Nutr.* **2015**, *6*, 408–419. [CrossRef]
115. Adjakly, M.; Bosviel, R.; Rabiau, N.; Boiteux, J.P.; Bignon, Y.J.; Guy, L.; Bernard-Gallon, D. DNA methylation and soy phytoestrogens: Quantitative study in DU-145 and PC-3 human prostate cancer cell lines. *Epigenomics* **2011**, *3*, 795–803. [CrossRef] [PubMed]
116. Zhang, Y.; Chen, H. Genistein, an epigenome modifier during cancer prevention. *Epigenetics* **2011**, *6*, 888–891. [CrossRef] [PubMed]
117. Li, Y.; Meeran, S.M.; Patel, S.N.; Chen, H.; Hardy, T.M.; Tollefsbol, T.O. Epigenetic reactivation of estrogen receptor-alpha (ERalpha) by genistein enhances hormonal therapy sensitivity in ERalpha-negative breast cancer. *Mol. Cancer* **2013**, *12*, 9. [CrossRef] [PubMed]
118. Bosviel, R.; Dumollard, E.; Dechelotte, P.; Bignon, Y.J.; Bernard-Gallon, D. Can soy phytoestrogens decrease DNA methylation in BRCA1 and BRCA2 oncosuppressor genes in breast cancer? *OMICS* **2012**, *16*, 235–244. [CrossRef] [PubMed]
119. Hsieh, P.L.; Liao, Y.W.; Hsieh, C.W.; Chen, P.N.; Yu, C.C. Soy Isoflavone Genistein Impedes Cancer Stemness and Mesenchymal Transition in Head and Neck Cancer through Activating miR-34a/RTCB Axis. *Nutrients* **2020**, *12*, 1924. [CrossRef]
120. Khoo, H.E.; Azlan, A.; Tang, S.T.; Lim, S.M. Anthocyanidins and anthocyanins: Colored pigments as food, pharmaceutical ingredients, and the potential health benefits. *Food Nutr. Res.* **2017**, *61*, 1361779. [CrossRef] [PubMed]
121. Basu, P.; Meza, E.; Bergel, M.; Maier, C. Estrogenic, Antiestrogenic and Antiproliferative Activities of Euphorbia bicolor (Euphorbiaceae) Latex Extracts and Its Phytochemicals. *Nutrients* **2019**, *12*, 59. [CrossRef] [PubMed]
122. Hong, T.; Nakagawa, T.; Pan, W.; Kim, M.Y.; Kraus, W.L.; Ikehara, T.; Yasui, K.; Aihara, H.; Takebe, M.; Muramatsu, M.; et al. Isoflavones stimulate estrogen receptor-mediated core histone acetylation. *Biochem. Biophys. Res. Commun.* **2004**, *317*, 259–264. [CrossRef] [PubMed]
123. Sun, J.; Sun, W.J.; Li, Z.Y.; Li, L.; Wang, Y.; Zhao, Y.; Wang, C.; Yu, L.R.; Li, L.Z.; Zhang, Y.L. Daidzein increases OPG/RANKL ratio and suppresses IL-6 in MG-63 osteoblast cells. *Int. Immunopharmacol.* **2016**, *40*, 32–40. [CrossRef] [PubMed]
124. Shankar, S.; Kumar, D.; Srivastava, R.K. Epigenetic modifications by dietary phytochemicals: Implications for personalized nutrition. *Pharmacol. Ther.* **2013**, *138*, 1–17. [CrossRef] [PubMed]
125. Teiten, M.H.; Eifes, S.; Dicato, M.; Diederich, M. Curcumin-the paradigm of a multi-target natural compound with applications in cancer prevention and treatment. *Toxins* **2010**, *2*, 128–162. [CrossRef] [PubMed]
126. Marcu, M.G.; Jung, Y.J.; Lee, S.; Chung, E.J.; Lee, M.J.; Trepel, J.; Neckers, L. Curcumin is an inhibitor of p300 histone acetylatransferase. *Med. Chem.* **2006**, *2*, 169–174. [PubMed]
127. Sun, M.; Estrov, Z.; Ji, Y.; Coombes, K.R.; Harris, D.H.; Kurzrock, R. Curcumin (diferuloylmethane) alters the expression profiles of microRNAs in human pancreatic cancer cells. *Mol. Cancer Ther.* **2008**, *7*, 464–473. [CrossRef] [PubMed]
128. Yang, J.; Cao, Y.; Sun, J.; Zhang, Y. Curcumin reduces the expression of Bcl-2 by upregulating miR-15a and miR-16 in MCF-7 cells. *Med. Oncol.* **2010**, *27*, 1114–1118. [CrossRef] [PubMed]
129. Mudduluru, G.; George-William, J.N.; Muppala, S.; Asangani, I.A.; Kumarswamy, R.; Nelson, L.D.; Allgayer, H. Curcumin regulates miR-21 expression and inhibits invasion and metastasis in colorectal cancer. *Biosci. Rep.* **2011**, *31*, 185–197. [CrossRef] [PubMed]
130. Adiwidjaja, J.; McLachlan, A.J.; Boddy, A.V. Curcumin as a clinically-promising anti-cancer agent: Pharmacokinetics and drug interactions. *Expert Opin. Drug Metab. Toxicol.* **2017**, *13*, 953–972. [CrossRef]
131. Menezo, Y.; Elder, K.; Clement, A.; Clement, P. Folic Acid, Folinic Acid, 5 Methyl TetraHydroFolate Supplementation for Mutations That Affect Epigenesis through the Folate and One-Carbon Cycles. *Biomolecules* **2022**, *12*, 197. [CrossRef]
132. Banerjee, S.; Kong, D.; Wang, Z.; Bao, B.; Hillman, G.G.; Sarkar, F.H. Attenuation of multi-targeted proliferation-linked signaling by 3,3′-diindolylmethane (DIM): From bench to clinic. *Mutat. Res. Mol. Mech. Mutagen.* **2011**, *728*, 47–66. [CrossRef]
133. Li, Y.; Li, X.; Guo, B. Chemopreventive agent 3,3′-diindolylmethane selectively induces proteasomal degradation of class I histone deacetylases. *Cancer Res.* **2010**, *70*, 646–654. [CrossRef]
134. Cheung, K.L.; Kong, A.N. Molecular targets of dietary phenethyl isothiocyanate and sulforaphane for cancer chemoprevention. *AAPS J.* **2010**, *12*, 87–97. [CrossRef]
135. Wang, L.G.; Beklemisheva, A.; Liu, X.M.; Ferrari, A.C.; Feng, J.; Chiao, J.W. Dual action on promoter demethylation and chromatin by an isothiocyanate restored GSTP1 silenced in prostate cancer. *Mol. Carcinog.* **2007**, *46*, 24–31. [CrossRef]
136. Zhang, T.; Zhang, W.; Hao, M. Phenethyl isothiocyanate reduces breast cancer stem cell-like properties by epigenetic reactivation of CDH1. *Oncol. Rep.* **2021**, *45*, 337–348. [CrossRef]

137. Izzotti, A.; Larghero, P.; Cartiglia, C.; Longobardi, M.; Pfeffer, U.; Steele, V.E.; De Flora, S. Modulation of microRNA expression by budesonide, phenethyl isothiocyanate and cigarette smoke in mouse liver and lung. *Carcinogenesis* **2010**, *31*, 894–901. [CrossRef]
138. Giovannucci, E. Tomatoes, tomato-based products, lycopene, and cancer: Review of the epidemiologic literature. *J. Natl. Cancer Inst.* **1999**, *91*, 317–331. [CrossRef] [PubMed]
139. King-Batoon, A.; Leszczynska, J.M.; Klein, C.B. Modulation of gene methylation by genistein or lycopene in breast cancer cells. *Environ. Mol. Mutagen.* **2008**, *49*, 36–45. [CrossRef] [PubMed]
140. Fazekas, Z.; Gao, D.; Saladi, R.N.; Lu, Y.; Lebwohl, M.; Wei, H. Protective effects of lycopene against ultraviolet B-induced photodamage. *Nutr. Cancer* **2003**, *47*, 181–187. [CrossRef] [PubMed]
141. Srivastava, R.K.; Unterman, T.G.; Shankar, S. FOXO transcription factors and VEGF neutralizing antibody enhance antiangiogenic effects of resveratrol. *Mol. Cell. Biochem.* **2010**, *337*, 201–212. [CrossRef] [PubMed]
142. Pyo, I.S.; Yun, S.; Yoon, Y.E.; Choi, J.W.; Lee, S.J. Mechanisms of Aging and the Preventive Effects of Resveratrol on Age-Related Diseases. *Molecules* **2020**, *25*, 4649. [CrossRef] [PubMed]
143. Izquierdo-Torres, E.; Hernandez-Oliveras, A.; Meneses-Morales, I.; Rodriguez, G.; Fuentes-Garcia, G.; Zarain-Herzberg, A. Resveratrol up-regulates ATP2A3 gene expression in breast cancer cell lines through epigenetic mechanisms. *Int. J. Biochem. Cell Biol.* **2019**, *113*, 37–47. [CrossRef] [PubMed]
144. Tili, E.; Michaille, J.J.; Alder, H.; Volinia, S.; Delmas, D.; Latruffe, N.; Croce, C.M. Resveratrol modulates the levels of microRNAs targeting genes encoding tumor-suppressors and effectors of TGFbeta signaling pathway in SW480 cells. *Biochem. Pharmacol.* **2010**, *80*, 2057–2065. [CrossRef] [PubMed]
145. Clarke, J.D.; Dashwood, R.H.; Ho, E. Multi-targeted prevention of cancer by sulforaphane. *Cancer Lett.* **2008**, *269*, 291–304. [CrossRef] [PubMed]
146. Meeran, S.M.; Patel, S.N.; Tollefsbol, T.O. Sulforaphane causes epigenetic repression of hTERT expression in human breast cancer cell lines. *PLoS ONE* **2010**, *5*, e11457. [CrossRef] [PubMed]
147. Su, X.; Jiang, X.; Meng, L.; Dong, X.; Shen, Y.; Xin, Y. Anticancer Activity of Sulforaphane: The Epigenetic Mechanisms and the Nrf2 Signaling Pathway. *Oxidative Med. Cell. Longev.* **2018**, *2018*, 5438179. [CrossRef]
148. Shankar, S.; Ganapathy, S.; Srivastava, R.K. Sulforaphane enhances the therapeutic potential of TRAIL in prostate cancer orthotopic model through regulation of apoptosis, metastasis, and angiogenesis. *Clin. Cancer Res.* **2008**, *14*, 6855–6866. [CrossRef]
149. Ali Khan, M.; Kedhari Sundaram, M.; Hamza, A.; Quraishi, U.; Gunasekera, D.; Ramesh, L.; Goala, P.; Al Alami, U.; Ansari, M.Z.; Rizvi, T.A.; et al. Sulforaphane Reverses the Expression of Various Tumor Suppressor Genes by Targeting DNMT3B and HDAC1 in Human Cervical Cancer Cells. *Evid. Based Complement Altern. Med.* **2015**, *2015*, 412149. [CrossRef] [PubMed]
150. Hsu, A.; Wong, C.P.; Yu, Z.; Williams, D.E.; Dashwood, R.H.; Ho, E. Promoter de-methylation of cyclin D2 by sulforaphane in prostate cancer cells. *Clin. Epigenet.* **2011**, *3*, 3. [CrossRef]
151. Li, Y.; Buckhaults, P.; Cui, X.; Tollefsbol, T.O. Combinatorial epigenetic mechanisms and efficacy of early breast cancer inhibition by nutritive botanicals. *Epigenomics* **2016**, *8*, 1019–1037. [CrossRef] [PubMed]
152. Li, Y.; Meeran, S.M.; Tollefsbol, T.O. Combinatorial bioactive botanicals re-sensitize tamoxifen treatment in ER-negative breast cancer via epigenetic reactivation of ERalpha expression. *Sci. Rep.* **2017**, *7*, 9345.
153. Paul, B.; Li, Y.; Tollefsbol, T.O. The Effects of Combinatorial Genistein and Sulforaphane in Breast Tumor Inhibition: Role in Epigenetic Regulation. *Int. J. Mol. Sci.* **2018**, *19*, 1754. [CrossRef]
154. Liu, Q.; Loo, W.T.; Sze, S.C.; Tong, Y. Curcumin inhibits cell proliferation of MDA-MB-231 and BT-483 breast cancer cells mediated by down-regulation of NFkappaB, cyclinD and MMP-1 transcription. *Phytomedicine* **2009**, *16*, 916–922. [CrossRef] [PubMed]
155. Cianfruglia, L.; Minnelli, C.; Laudadio, E.; Scire, A.; Armeni, T. Side Effects of Curcumin: Epigenetic and Antiproliferative Implications for Normal Dermal Fibroblast and Breast Cancer Cells. *Antioxidants* **2019**, *8*, 382. [CrossRef]
156. Luan, Y.; Leclerc, D.; Cosin-Tomas, M.; Malysheva, O.V.; Wasek, B.; Bottiglieri, T.; Caudill, M.A.; Rozen, R. Moderate Folic Acid Supplementation in Pregnant Mice Results in Altered Methyl Metabolism and in Sex-Specific Placental Transcription Changes. *Mol. Nutr. Food Res.* **2021**, *65*, e2100197. [CrossRef] [PubMed]
157. Jing-Bo, L.; Ying, Y.; Bing, Y.; Xiang-Bing, M.; Zhi-Qing, H.; Guo-Quan, H.; Hong, C.; Dai-Wen, C. Folic acid supplementation prevents the changes in hepatic promoter methylation status and gene expression in intrauterine growth-retarded piglets during early weaning period. *J. Anim. Physiol. Anim. Nutr.* **2013**, *97*, 878–886. [CrossRef] [PubMed]
158. Duthie, S.J. Epigenetic modifications and human pathologies: Cancer and CVD. *Proc. Nutr. Soc.* **2011**, *70*, 47–56. [CrossRef] [PubMed]
159. Silvestri, S.; Orlando, P.; Armeni, T.; Padella, L.; Bruge, F.; Seddaiu, G.; Littarru, G.P.; Tiano, L. Coenzyme Q10 and alpha-lipoic acid: Antioxidant and pro-oxidant effects in plasma and peripheral blood lymphocytes of supplemented subjects. *J. Clin. Biochem. Nutr.* **2015**, *57*, 21–26. [CrossRef]

160. Li, Y.; VandenBoom, T.G., 2nd; Kong, D.; Wang, Z.; Ali, S.; Philip, P.A.; Sarkar, F.H. Up-regulation of miR-200 and let-7 by natural agents leads to the reversal of epithelial-to-mesenchymal transition in gemcitabine-resistant pancreatic cancer cells. *Cancer Res.* **2009**, *69*, 6704–6712. [CrossRef]
161. Izzotti, A.; Calin, G.A.; Steele, V.E.; Cartiglia, C.; Longobardi, M.; Croce, C.M.; De Flora, S. Chemoprevention of cigarette smoke-induced alterations of MicroRNA expression in rat lungs. *Cancer Prev. Res.* **2010**, *3*, 62–72. [CrossRef] [PubMed]
162. Roy, S.K.; Chen, Q.; Fu, J.; Shankar, S.; Srivastava, R.K. Resveratrol inhibits growth of orthotopic pancreatic tumors through activation of FOXO transcription factors. *PLoS ONE* **2011**, *6*, e25166. [CrossRef] [PubMed]
163. Chen, Q.; Ganapathy, S.; Singh, K.P.; Shankar, S.; Srivastava, R.K. Resveratrol induces growth arrest and apoptosis through activation of FOXO transcription factors in prostate cancer cells. *PLoS ONE* **2010**, *5*, e15288. [CrossRef]
164. Caddeo, C.; Nacher, A.; Vassallo, A.; Armentano, M.F.; Pons, R.; Fernàndez-Busquets, X.; Carbone, C.; Valenti, D.; Fadda, A.M.; Manconi, M. Effect of quercetin and resveratrol co-incorporated in liposomes against inflammatory/oxidative response associated with skin cancer. *Int. J. Pharm.* **2016**, *513*, 153–163. [CrossRef] [PubMed]
165. Li, S.; Yang, Y.; Sargsyan, D.; Wu, R.; Yin, R.; Kuo, H.D.; Yang, I.; Wang, L.; Cheng, D.; Ramirez, C.N.; et al. Epigenome, Transcriptome, and Protection by Sulforaphane at Different Stages of UVB-Induced Skin Carcinogenesis. *Cancer Prev. Res.* **2020**, *13*, 551–562. [CrossRef] [PubMed]
166. Dima, C.; Assadpour, E.; Dima, S.; Jafari, S.M. Bioavailability of nutraceuticals: Role of the food matrix, processing conditions, the gastrointestinal tract, and nanodelivery systems. *Compr. Rev. Food Sci. Food Saf.* **2020**, *19*, 954–994. [CrossRef] [PubMed]
167. Stabrauskiene, J.; Kopustinskiene, D.M.; Lazauskas, R.; Bernatoniene, J. Naringin and Naringenin: Their Mechanisms of Action and the Potential Anticancer Activities. *Biomedicines* **2022**, *10*, 1686. [CrossRef] [PubMed]
168. Wang, X.; Qi, Y.; Zheng, H. Dietary Polyphenol, Gut Microbiota, and Health Benefits. *Antioxidants* **2022**, *11*, 1212. [CrossRef] [PubMed]
169. Molteni, C.; La Motta, C.; Valoppi, F. Improving the Bioaccessibility and Bioavailability of Carotenoids by Means of Nanostructured Delivery Systems: A Comprehensive Review. *Antioxidants* **2022**, *11*, 1931. [CrossRef]
170. Aatif, M. Current Understanding of Polyphenols to Enhance Bioavailability for Better Therapies. *Biomedicines* **2023**, *11*, 2078. [CrossRef]
171. Khan, H.; Ullah, H.; Martorell, M.; Valdes, S.E.; Belwal, T.; Tejada, S.; Sureda, A.; Kamal, M.A. Flavonoids nanoparticles in cancer: Treatment, prevention and clinical prospects. *Semin. Cancer Biol.* **2021**, *69*, 200–211. [CrossRef]
172. Berdasco, M.; Esteller, M. Clinical epigenetics: Seizing opportunities for translation. *Nat. Rev. Genet.* **2019**, *20*, 109–127. [CrossRef]
173. Herman, J.G.; Baylin, S.B. Gene silencing in cancer in association with promoter hypermethylation. *N. Engl. J. Med.* **2003**, *349*, 2042–2054. [CrossRef] [PubMed]
174. Kronfol, M.M.; Dozmorov, M.G.; Huang, R.; Slattum, P.W.; McClay, J.L. The role of epigenomics in personalized medicine. *Expert Rev. Precis. Med. Drug Dev.* **2017**, *2*, 33–45. [CrossRef] [PubMed]
175. Xue, K.; Gu, J.J.; Zhang, Q.; Mavis, C.; Hernandez-Ilizaliturri, F.J.; Czuczman, M.S.; Guo, Y. Vorinostat, a histone deacetylase (HDAC) inhibitor, promotes cell cycle arrest and re-sensitizes rituximab- and chemo-resistant lymphoma cells to chemotherapy agents. *J. Cancer Res. Clin. Oncol.* **2016**, *142*, 379–387. [CrossRef] [PubMed]
176. Baylin, S.B.; Jones, P.A. Epigenetic Determinants of Cancer. *Cold Spring Harb. Perspect. Biol.* **2016**, *8*, a019505. [CrossRef] [PubMed]
177. Mahalmani, V.; Sinha, S.; Prakash, A.; Medhi, B. Translational research: Bridging the gap between preclinical and clinical research. *Indian J. Pharmacol.* **2022**, *54*, 393–396. [CrossRef] [PubMed]
178. el Bahhaj, F.; Dekker, F.J.; Martinet, N.; Bertrand, P. Delivery of epidrugs. *Drug Discov. Today* **2014**, *19*, 1337–1352. [CrossRef]
179. NCT01606124. Randomized Phase II Trial of Polyphenon E vs. Placebo in Patients at High Risk of Recurrent Colonic Neoplasia. 2012. Available online: https://cdek.pharmacy.purdue.edu/trial/NCT01606124/ (accessed on 4 November 2024).
180. NCT00917735. Phase II, Randomized, Double-blind, Placebo-controlled, Study of the Efficacy of Green Tea Extract on Biomarkers of Breast Cancer Risk in High Risk Women with Differing Catechol-O-methyl Transferase (COMT) Genotypes. 2009. Available online: https://cdek.pharmacy.purdue.edu/trial/NCT00917735/ (accessed on 4 November 2024).
181. NCT00666562. A Phase II Randomized, Placebo-Controlled Trial of Polyphenon E to Evaluate Bladder Tissue Levels of EGCG. 2008. Available online: https://cdek.pharmacy.purdue.edu/trial/NCT00666562/ (accessed on 4 November 2024).
182. NCT00573885. Phase II Trial of Polyphenon E in Former Smokers with Abnormal Sputa. 2007. Available online: http://cdek.pharmacy.purdue.edu/trial/NCT00573885/ (accessed on 4 November 2024).
183. NCT00303823. A Phase II Trial of Polyphenon E for Cervical Cancer Prevention. 2006. Available online: https://cdek.pharmacy.purdue.edu/trial/NCT00303823/ (accessed on 4 November 2024).
184. NCT01032031. The Effect of Dietary Bioactive Compounds on Skin Health in Humans In Vivo. 2009. Available online: https://cdek.pharmacy.purdue.edu/trial/NCT01032031/ (accessed on 4 November 2024).

185. NCT02580279. Phase II Study of Topical Epigallocatechin-3-gallate (EGCG) in Patients with Breast Cancer Receiving Adjuvant Radiotherapy. 2015. Available online: https://cdek.pharmacy.purdue.edu/trial/NCT02580279/ (accessed on 4 November 2024).
186. NCT02577393. Epigallocatechin-3-gallate (EGCG) for Esophagus Protection in Patients with Lung Cancer Receiving Radial Radiotherapy. 2015. Available online: https://cdek.pharmacy.purdue.edu/trial/NCT02577393/ (accessed on 4 November 2024).
187. NCT02891538. A Pilot Study to Evaluate the Chemopreventive Effects of Epigallocatechin Gallate (EGCG) in Colorectal Cancer (CRC) Patients with Curative Resections. 2016. Available online: https://cdek.pharmacy.purdue.edu/trial/NCT02891538/ (accessed on 4 November 2024).
188. Pintova, S.; Dharmupari, S.; Moshier, E.; Zubizarreta, N.; Ang, C.; Holcombe, R.F. Genistein combined with FOLFOX or FOLFOX-Bevacizumab for the treatment of metastatic colorectal cancer: Phase I/II pilot study. *Cancer Chemother. Pharmacol.* **2019**, *84*, 591–598. [CrossRef]
189. Ponte, L.G.S.; Pavan, I.C.B.; Mancini, M.C.S.; da Silva, L.G.S.; Morelli, A.P.; Severino, M.B.; Bezerra, R.M.N.; Simabuco, F.M. The Hallmarks of Flavonoids in Cancer. *Molecules* **2021**, *26*, 2029. [CrossRef]
190. NCT03288298. Therapeutic Effect of Luteolin Natural Extract Versus Its Nanoparticles on Tongue Squamous Cell Carcinoma Cell Line: In Vitro Study. 2017. Available online: https://cdek.pharmacy.purdue.edu/trial/NCT03288298/ (accessed on 4 November 2024).

MDPI AG
Grosspeteranlage 5
4052 Basel
Switzerland
Tel.: +41 61 683 77 34

Biomolecules Editorial Office
E-mail: biomolecules@mdpi.com
www.mdpi.com/journal/biomolecules

www.ingramcontent.com/pod-product-compliance
Lightning Source LLC
Chambersburg PA
CBHW061833210926
51850CB00006B/874